Lecture Notes in Mathematics

Edited by A. Dold and B. Eckmann

1370

G. Carlsson R.L. Cohen
H.R. Miller D.C. Ravenel (Eds.)

Algebraic Topology

Proceedings of an International Conference
held in Arcata, California, July 27 – August 2, 1986

Springer-Verlag
Berlin Heidelberg New York London Paris Tokyo

Editors

Gunnar Carlsson
Princeton University, Department of Mathematics
Princeton, NJ 08544, USA

Ralph Cohen
Stanford University, Department of Mathematics
Stanford, CA 94305, USA

Haynes Miller
Massachusetts Institute of Technology, Room 2-237
Cambridge, MA 02139, USA

Douglas Ravenel
University of Rochester, Department of Mathematics
Rochester, NY 14620, USA

Mathematics Subject Classification (1980): 55-06

ISBN 3-540-51118-0 Springer-Verlag Berlin Heidelberg New York
ISBN 0-387-51118-0 Springer-Verlag New York Berlin Heidelberg

Printing and binding: Druckhaus Beltz, Hemsbach/Bergstr.
2146/3140-543210

PREFACE

An international conference on algebraic topology was held July 27 to August 2, 1986, at Humboldt State University in Arcata, California, prior to ICM86 in Berkeley. The conference served in part to celebrate the silver jubilee of the journal Topology, and the sixtieth birthday of Edgar H. Brown. It was supported by a grant from the National Science Foundation and a bequest from Pergammon Press, publishers of Topology.

The Proceedings contain papers submitted by conference participants. All were refereed, and we take this occasion to thank the referees. We would also like to thank the staff of the Jolly Giant Conference Center for its assistance in organizing this conference.

Gunnar Carlsson
Ralph Cohen
Haynes Miller
Doug Ravenel

CONTENTS

1. Ioan M. James, University of Oxford
2. James F. Davis, Indiana University
3. J. M. Boardman, Johns Hopkins University
4. Robert Bruner, Wayne State University
5. Michishige Tesuka, Tokyo Institute of Technology
6. John W. Wood, University of Illinois at Chicago
7. William M. Singer, Fordham University
8. Stanley O. Kochman, York University
9. Ronnie Lee, Yale University
10. Elias Micha, Centro del I.P.N., Mexico
11.
12. Anthony Bak, Universität Bielefeld
13. John C. Harris, University of Washington
14. Michael J. Hopkins, Princeton University
15. Wilberd van der Kallen, State University of Utrecht
16. Masana Harada, Kyoto University
17. Thomas P. Bisson, Canisius College
18. Peter N. Hoffman, University of Waterloo
19. Uwe G. M. Hommel, University of Chicago
20. Larry R. Taylor, University of Notre Dame
21. Alan C. Robinson, University of Warwick
22. Hirotaka Tamanoi, Institute for Advanced Study
23. David Grayson, University of Illinois at Urbana–Champaign
24. Hans J. Munkholm, Odense Universitet
25. Max Karoubi, Université Paris VII
26. James P. Lin, University of California at San Diego
27. Howard Marcum, Ohio State University
28. Eddy Campbell, Queen's University
29. Masaharu Morimoto, Okayama University
30. Paul L. Shick, John Carroll University
31. Masatsugu Nagata, University of Chicago
32. Jack Wagoner, University of California at Berkeley
33. Peter B. Gilkey, University of Oregon
34. Steven A. Mitchell, University of Washington
35. Frank Williams, New Mexico State University
36. Gunnar Carlsson, Princeton University
37. David C. Johnson, University of Kentucky
38. Brayton Gray, University of Illinois at Chicago
39. Carlos Prieto, U.N.A.M., Mexico
40. Thomas Goodwillie, Brown University
41. Keith P. Johnson, Dalhousie University
42. James E. West, Cornell University
43. Bruce Hughes, Vanderbilt University
44. Eldon Dyer, Grad. School and University Center, CUNY
45. Santiago Lopez de Medrano, U.N.A.M., Mexico
46. Kent E. Orr, Brandeis University
47. Ralph L. Cohen, Stanford University
48. Douglas R. Anderson, Syracuse University

49. Ross E. Staffeldt, Pennsylvania State University
50. J. Peter May, University of Chicago
51. Bruce Williams, University of Notre Dame
52. Monica Nicolau, University of California, Berkeley
53. Ethan S. Devinatz, University of Chicago
54. L. Gaunce Lewis, Jr., Syracuse University
55. Edgar H. Brown, Brandeis University
56. Nicholas J. Kuhn, University of Virginia
57. Arunas L. Liulevicius, University of Chicago
58. James A Schafer, University of Maryland
59. Kenshi Ishiguro, University of Chicago
60.
61. Edwin H. Spanier, University of California at Berkeley
62. Charles A. Weibel, Rutgers University
63. Reinhard C. Laubenbacher, New Mexico State University
64. George W. Whitehead, Massachusetts Institute of Technology
65. Nobuaki Yagita, Masashi Institute of Technology
66. J. Frank Adams, University of Cambridge
67. Mark Feshbach, University of Minnesota
68. Jeremy H. C. Gunawardena, University of Cambridge
69. Keith A. Hardie, University of Cape Town
70. Benjamin M. Mann, Clarkson University
71. Richard Miranda
72. Daniel Miranda
73. Clarence W. Wilkerson, Wayne State University
74. Douglas C. Ravenel, University of Washington
75. Robert H. Szczarba, Yale University
76 John H. Ewing, Indiana University
77. A. Jon Berrick, National University of Singapore
78. Donald M. Davis, Lehigh University
79. Mark E. Mahowald, Northwestern University
80. Pierre Vogel, Université de Nantes
81. Don H. Shimamoto, Swarthmore College
82. Dan Burghelea, Ohio State University
83. James E. McClure, University of Kentucky
84. Johan L. Dupont, University of Aarhus
85. Victor P. Snaith, University of Western Ontario
86. Mamoru Mimura, Okayama University
87. Lionel Schwartz, Université Paris–Sud
88. John H. McCleary, Vassar College
89. David J. Pengelley, New Mexico State University
90. John R. Harper, University of Rochester
91. Matthias Kreck, Johannes Gutenberg University
92. F. Thomas Farrell, Columbia University
93. Jean Lannes, Ecole Polytechnique
94. Roderick D. Ball, New Mexico State University
95. Nathan B. Habegger, University of California at San Diego
96. Haynes R. Miller, Massachusetts Institute of Technology

97. David J. Anick, Massachusetts Institute of Technology
98. Frederick R. Cohen, University of Kentucky
99. Alexander E. Koonce, Yale University
100. Kaoru Morisugi, Wakayama University
101. Christopher W. Stark, University of Florida
102. Jean–Claude R. Hausmann, University of Geneva
103. Robert A. Oliver, Aarhus University
104. Andrew Nicas, University of Toronto
105. Samuel Gitler, University of Rochester
106. Alexander Zabrodsky, Hebrew University
107. R. James Milgram, Stanford University
108. Remy Croisille, Ecole Normale Supérieure
109. Francis X. Connolly, University of Notre Dame
110.
111. Wu–Chung Hsiang, Princeton University
112. William G. Dwyer, University of Notre Dame
113. Jeanne Duflot, Colorado State University
114. Toshio Kawashima, Ashikaga Institute of Technology
115. Wen–Hsiung Lin, National Tsing Hua University
116. William Browder, Princeton University,
117. Hitoshi Moriyoshi, University of Tokyo
118. Enric Frau, Massachusetts Institute of Technology
119. Hiroo Shiga, Ryukyu University
120. Graham J. Ellis, University of North Wales
121. Norio Iwase, Kyushu University
122. Marcelo A. Aguilar, U.N.A.M., Mexico
123. Jerry M. Lodder, Stanford University
124. Norihiko Minami, Northwestern University
125. Aleksandar Vucic, University of Washington
126. Alexander Adem, Stanford University
127. Francis Clarke, University College, Swansea
128. Stefan Jackowski, University of Warsaw
129. Tsuneyo Yamanoshita, Masashi Institute of Technology
130. John Scott Maginnis, Stanford University
131. Nigel Ray, Manchester University
132. Stewart R. Priddy, Northwestern University
133. Noriko Minami
134. Tammo tom Dieck, Gottingen
135. Mikio Furuta, University of Tokyo
136. Henry H. Glover, Ohio State University
137. Xueguang Zhou, Nankai University
138. David Benson, University of Oxford
139. Georgia Triantafillou, University of Crete
140. Frank Quinn, Virginia Polytechnic Institute
141. Franklin P. Peterson, Massachusetts Institute of Technology
142. Peter A. Symonds, Ohio State University
143. Ruth M. Charney, Ohio State University
144. Luisa Avez, University of Rochester
145. Joseph A. Neisendorfer, University of Rochester
146. Martin C. Tangora, University of Illinois at Chicago

Topology: past, present and future

I.M. JAMES

Since one of the main purposes of this meeting is to celebrate the Silver Jubilee of our journal Topology, the organizers have invited me, as one who has been an editor of the journal since its inception, to say a few words on the past, present and future of Topology. I hasten to say that the word is to be written with a capital T, not a small t; to discuss the past, present and future of the subject topology would be a mammoth task which might well occupy the whole of the five days we have at our disposal. However if I might have your attention for twenty minutes or so I should like to say something about the journal Topology. In fact to give my remarks an appropriate context I would like to begin with a few words about mathematical journals generally.

Mathematical journals, of course, have much in common with other scientific journals. These originated in the Proceedings of the older Scientific Academies. Pride of place must go to the Philosophical Transactions of the Royal Society of London, which has been appearing regularly since 1665. By comparison the Compte Rendu Hebdomaire of the Academy in Paris (1835) and the Proceedings of the Academy in Washington (1863) are newcomers. In this category one might also include the Proceedings of the Cambridge Philosophical Society (1843).

Journals such as these aim to cover the whole of science, although nowadays they are often issued in parts which concentrate on one area or another. They seem to me to belong to an age when science was more of a unity and it was not unusual for an individual scientist to maintain a serious interest in and indeed to make research contributions to a wide range of different subjects. I would therefore be quite surprised if any more journals of this type were to be founded in future.

The second half of the nineteenth century saw the foundation of learned societies devoted to mathematics alone, rather than science generally. Each of these held regular meetings, at which papers were presented in the form of lectures, and it became usual to publish the proceedings of these meetings, just as the academies did. I believe the earliest example of such a journal is the Proceedings of the London Mathematical Society, which has been published without a break since 1865.

Later these Society journals were expanded to include other papers and gradually that kind of material became the norm so that the formal record of meetings now occupies only a small part of such publications. These days the leading Societies publish several journals of this type, usually including one or more of them in the membership package.

The Learned Society journals cover the whole range of mathematics, on the pure side at any rate. However there are also a number of distinguished independent journals which fulfill much the same role. Of these the oldest, so far as I am aware, are the Journal für die Reine und Ungewandte Mathematik, founded in 1826 and known by the name of Crelle who was its editor for so many years, and the similar Journal de Mathématiques Pures et Appliqués, founded ten years later, which is likewise generally known by the name of its first editor Liouville. Other independent journals of high distinction included the Acta, which is closely associated with the Mittag-Leffler Institute, and the Annals, which draws its Editorial Board largely from members of Princeton University and the Institute for Advanced Study.

It was not until comparatively recently that specialist journals started to appear. An early example is the Fundamenta (1920) which as its name suggests was originally intended to specialize in the foundations of mathematics, particularly mathematical logic and general topology but which has developed into a journal of broad coverage like others I have mentioned earlier. This example, and a few others, dates from the period between the wars. However the idea of having specialist journals did not really catch on until the fifties, when several academic publishing houses began to promote specialist journals in various areas of pure and applied science, including mathematics. Not all of these new journals survived for more than a few issues but among those which did are some of very high repute. In this category I feel confident that Topology can be included and at this stage I would like to say a few words about the way in which it originated.

Shortly before his death in 1960, Henry Whitehead had been engaged in discussions with his old friend Robert Maxwell, the Chairman of Pergamon Press, about the foundation of a new International Journal for Mathematics, to be called Topology. He had got as far as selecting the Editorial Board, the Founder Editors being Michael Atiyah, Raoul Bott, Fritz Hirzebruch, René Thom and myself. After Henry Whitehead's untimely death it fell to me to carry out his intentions and continue negotiations with Pergamon Press about all the things which need to be settled before a new journal can be launched. Robert Maxwell took a very personal interest in the new venture and made a number of valuable suggestions himself - for example at one stage a cover design by Picasso was contemplated and it would be fascinating to have seen what this would have been like.

In fact the first four issues which make up volume one mainly appeared in 1961 - which is why we are celebrating the silver jubilee this year. The call for contributions for the first volume met with an encouraging response, but I must admit that it was quite hard work obtaining material of the right quality for the first few volumes after that. This is often the case with a new journal and it is most important not to compromise standards at this critical stage - if one does so then it can be extremely difficult to recover later. Before long, however, first-class material began to arrive of its own accord and the editors found themselves in the strong position which they have enjoyed ever since.

A perusal of those early volumes leaves an impression of both quality and variety. One finds, for example, René Thom's long article "Topological models in biology", later to form the opening chapter of his well-known book of morphogenesis. On the geometric side of the subject one finds Milnor's paper on Microbundles, the basis for so much subsequent work, and on the algebraic side of the subject the paper in which Brown and Peterson established the existence of the spectrum which bears their name and which may possibly contain the key to some of the more intractable problems of homotopy theory. Right from the outset it was editorial policy to keep the scope of the journal as broad as possible and one finds in these early volumes, just as one does today, articles on algebraic geometry, differential geometry, dynamical systems and many other subjects.

However, in that first decade the theme which seems to stand out in front of everything else is K-theory: so many of the most important papers in the development of that subject appeared in the pages of Topology. One must mention, for example, the classic article by Atiyah, Bott and Shapiro on Clifford modules, with which the Shapiro memorial supplement to volume 3 begins. One must also mention the well-known series of papers on the group J(X) by Adams, which have been the stimulus for so much subsequent work in homotopy theory. In one of these papers Adams formulated the conjecture which bears his name, relating the classification of vector bundles by stable isomorphism to their classification by stable fibre homotopy equivalence of the associated sphere-bundle. And it is particularly gratifying to find that some seven years later Quillen published what I believe to have been the first proof of the Adams conjecture, also in the pages of our journal.

Of the original editors, only Fritz Hirzebruch and I now remain on the Board. The other editors at the present time are Bryan Birch, Simon Donaldson, Blaine Lawson, Larry Siebenmann and Graeme Segal, who has taken over from me as Editor-in-Chief.

The editorial organization has not changed greatly over the years. Normally the individual editors who receive papers arrange refereeing and decide, in the light of reports so obtained, which papers have a fair chance of acceptance. Those which have are then considered further by a committee which meets at Oxford on a monthly basis and takes the final decision. Although all the editors are members of the committee it is usually the Oxford editors who bear the brunt of this work.

At this point I would like to pay tribute to our contributors - no journal could thrive without their support. Some have been loyal to us from the very earliest days, and we are particularly grateful to them for their contributions over the years. However almost every issue, certainly every volume, contains articles by people who have never written for us before. Such papers are always very welcome and we are proud of our record of outstanding "first papers". In fact, looking back over the first 25 years of our journal, I think we can be generally proud of our contribution to the development of some of the most vital areas of mathematical research and I hope we can continue to contribute in the years to come, wherever the future development of the subject may lead.

As with any leading journal, very much depends on the careful and conscientious work of referees, and I would like to take this opportunity to thank the many people who have helped us in this capacity over the years. Authors do not always appreciate referees' suggestions, however tactfully they may be expressed, and yet many an argument has been improved or mistake avoided by this means. In the case of Topology editors do not, as a rule, also act as referees themselves, but generally each of the more promising papers will have been read by one or more of the editors, as well as by the referee, before a final decision is made.

I would also like to express the thanks of the Editors to the staff of Pergamon Press, particularly Mike Church who has done so much for us over the years. The production of a journal is not such an easy matter as some might believe; in the case of Topology practically every difficulty which could conceivably have occurred has occurred. Although occasionally a party of editors has had to set off for Headington in high dudgeon about something or other, on the whole the partnership with the Press works extremely well and it should also be mentioned that Pergamon has contributed with characteristic generosity to the funding of this Silver Jubilee conference.

Among the hundreds of scientific and technical journals published by Pergamon Press, Topology is one of the most successful. It is taken not only by a large number of institutional libraries but also by a large number of individuals who take advantage of the specially attractive subscription rate which is available to them. The Editors try to ensure that every issue is interesting to the readers - to think only of the authors is a recipe for dullness in a journal. I do hope that those present at this meeting who do not already subscribe on an individual basis will consider doing so.

People sometimes ask about the scope of the journal. It is obviously very broad and yet some types of mathematics are unlikely to be found in our pages. The test the editors generally apply is: is this paper likely to interest our readers? There is no hard and fast rule therefore. When a new research area opens up, which passes that simple test, the editors can be relied upon to welcome contributions in that area, even if it means extending the scope of the journal in a new direction. It has never been editorial policy to accept research announcements but articles with an expository content receive favourable consideration. Moreover Topology has published review articles on different subjects from time to time: a fee is payable for these, incidentally, and proposals for further such articles would be welcomed by the editors.

What does the future hold? One can be fairly sure that the picture, for mathematical journals generally, will be very different in the next century from the picture today. Indeed the pace of technical change is so great that it may alter greatly within the next five or ten years. My own view, for what it is worth, is as follows.

According to Mathematical Reviews there are at the present time over 500 mathematical journals of which the contents are regularly reviewed, and at least twice as many more journals which occasionally contain articles of sufficient mathematical interest to justify a review. Some of these journals are well-known to all of us. Others are quite obscure. My guess is that before long there will be many fewer journals of the conventional type. I think we will find the kind of paper which simply places some research on record, and which is intended to be referred to rather than read, will probably go straight into a database. There will surely need to be some form of editorial control, similar to that exercised in the case of a regular journal, and it could be that some databases will earn a reputation for being more discriminating in the material they accept than others. Alternatively it could be that the reviewing journals will undertake an extension of their responsibilities and help to sort out the grain from the chaff. However I feel confident there will remain a fair number of journals of the traditional type, publishing articles which are designed to be read. I very much hope that Topology will be one of these and that there will be another meeting like this, in twenty-five years' time, to celebrate the golden jubilee of our Journal.

THE WORK OF EDGAR H. BROWN, Jr. IN TOPOLOGY

by

George W. Whitehead

It is difficult for me to think of Ed Brown as an "old-timer", for he is some thirteen years my junior in terms of mathematical age. I first met Ed when he came to M. I. T. as a graduate student about 1951. He took my graduate course in topology and soon evinced an interest in working in that subject.

Let me recall something of the state of the subject in the early fifties. It had been known for a long time that the homology groups of a finite complex are finitely generated. Indeed, more is true; they are calculable in the sense that there is a finite algorithm for their determination. With the discovery of the homotopy groups by Hurewicz in the middle thirties, it was natural to ask whether the same was true for them. Now the fundamental group was easily seen to be not merely finitely generated, but even finitely presentable. (Recall that the fundamental group, unlike the higher homotopy groups, may be non-abelian). Thus the problem of its calculability reduces to a group-theoretic problem whose status was then unknown; in later years the latter problem was proved to be unsolvable. It could also be seen that the higher homotopy groups need not be finitely generated, not only as groups, but even as modules over the fundamental group. Indeed, by 1950, about all that was known was that the homotopy groups of a finite complex are countable.

The appearance of Serre's thesis in 1951 effected a revolution in topology by introducing a new and powerful tool--the spectral sequence of a fibration. One consequence of this method was that the homotopy groups of a simply connected finite complex are finitely generated. But the question of their calculability remained open.

The same year marked the discovery in Russia of Postnikov systems. An outline of this theory was published in the form of Doklady notes in Russian, and a full account in English was not available until 1957. Similar results were obtained independently in this country by Zilber, but he never published them. Thus they were only just beginning to be known. However, I had never mentioned them to Ed, and, upon my return to M. I. T. after spending the summer of 1954 in Chicago, I was pleasantly surprised to learn that Ed had not only rediscovered Postnikov systems on his own, but had applied them to obtain a partial solution of the calculability problem. And in a little over a year he had refined his methods to prove the general result: the homotopy groups of a 1-connected finite complex are calculable.

This result, interesting enough in its own right, took on added interest with the years, as a number of plausible sounding problems (for example, the homeomorphism problem for 4-manifolds and the classification of 2-complexes up to homotopy type) were shown by the logicians to be unsolvable.

By this time I was convinced that Ed had the makings of a mathematician, and I looked forward with interest to see what he would come up with next. It should be realized that, at this time, spectral sequences were still new and mysterious tools. If p is a fibre map of a space X upon a space B with fibre F, the Serre spectral sequence is a sequence $\{E^r\}$ of chain complexes, each of which is the homology of the preceding, and which leads to the homology of X; the initial term E^0 is the tensor product of the singular chain complexes S(B) and S(F). Thus the determination of H(X) from B and F involves in principle the calculation of an unlimited number of homology groups. Thus the method, powerful as it is, seemed cumbersome to use, and it was natural to ask whether H(X) could be expressed as the homology of E^0 with respect to some (non-standard) boundary operator. In 1959 Ed showed that this was indeed the case; the difference between the new boundary operator and the old is expressed in terms of certain "twisting cochains". These are homomorphisms $t_q: C_q(B) \dashrightarrow C_{q-1}(W)$ (W is the space of loops of B) satisfying a simple recursion formula. This result made possible new and elegant proofs of many results in fibre space theory.

Eilenberg-Mac Lane spaces had been introduced in the late forties and soon proved to be of the greatest importance in homotopy theory. If G is an abelian group and q a positive integer, the Eilenberg-Mac Lane space K(G,q) is a CW-complex whose homotopy groups all vanish, except for the qth, which is isomorphic with G. One reason for their importance is the result from obstruction theory which shows that the cohomology group $H^q(X;G)$ is isomorphic with the group [X,K(G,q)] of homotopy classes of maps of X into K(G,q). More generally, if Y is any space, the functor [,Y] has many properties resembling those of cohomology functors. In 1962 Ed gave a very simple set of properties of a (contravariant) functor T from a category of CW-complexes to that of based sets which are necessary and sufficient that T be representable; i. e., that T be naturally isomorphic with the functor [,Y] for some space Y. Such a space Y is called a <u>classifying space</u> for the functor T.

The importance of this result cannot be overestimated. Many geometric problems naturally give rise to functors satisfying Brown's axioms. An application of his theorem then reduces the problem in question to a homotopy-theoretic one. And the latter one in turn can often be attacked by studying the cohomology of the resulting classifying space. Cases in point are: (1) vector bundles; (2) microbundles; (3) foliations.

Another application is to the study of cohomology theories satisfying the Eilenberg-Steenrod axioms, except for the dimension axiom. Part of the data for such a theory is a sequence of functors $\{H^n\}$, each of which satisfies Brown's axioms. If Y_n is the classifying space for H^n, the exactness axiom yields a family of maps f_n: SY_n --> Y_{n+1}. Thus the spaces Y_n form a spectrum, and we conclude that every cohomology theory has the form [,Y] for some spectrum Y. Conversely, if Y is a spectrum, then [,Y] defines a cohomology theory.

(Remark: In Ed's proof of this fact, it was necessary to assume that the groups $H^n(S^0)$ are all countable. That this hypothesis is unnecessary was shown shortly afterwards by Adams).

A final application is to homology theories. In 1962 I showed that, if X is a space and Y a spectrum, the spaces $X \wedge Y_n$ form a spectrum whose homotopy groups are the ingredients of a homology theory. Conversely, given a homology theory H, an application of Spanier-Whitehead duality leads to a cohomology theory, and an application of Brown's theorem then yields a spectrum Y. It is then not difficult to show that H is the homology theory arising from the spectrum Y. Thus every homology theory can be obtained in this way. This result was crucial in my proof of Alexander and Poincaré duality for arbitrary theories.

In 1960 Kervaire proved the existence of closed manifolds which do not admit any differential structure. The "Kervaire invariant" of a suitable (4n+2)-manifold M is the Arf invariant of a certain quadratic form with Z_2 coefficients associated with the manifold; it vanishes if the manifold M has a differential structure. And Kervaire exhibits a manifold for which his invariant is 1. In succeeding years, a number of mathematicians proposed generalizations of the Kervaire invariant, none of which was completely satisfactory. Finally, an ingenious idea of Brown led to a complete and satisfying treatment which included the earlier versions as special cases. A curious feature of Ed's treatment is that his invariant takes values in Z_8, rather than in Z_2.

A different aspect of Ed's work has to do with characteristic classes and cobordism theory. If X is a vector bundle over a space B with structural group G, then X corresponds to a map f: B --> BG, where BG is the classifying space. It often happens that the cohomology ring of BG is the polynomial ring in certain classes, called "universal characteristic classes"; their images under f^* are called the characteristic classes of the bundle X. When this is so, there are no relations among the universal classes and therefore no universally valid relations among the characteristic classes. However, the situation is different if attention is confined to the tangent bundles of manifolds of a fixed dimension n. In this case, there are many relations which are forced to hold for algebraic reasons, e. g., Poincaré duality and properties of the Steenrod algebra. In 1964 Ed Brown and Frank Peterson proved that, for the case of the Stiefel-Whitney classes, all relations are algebraic in this sense. In doing so, they inaugurated a long and fruitful collaboration.

Among the characteristic classes of the tangent bundle of a manifold, the characteristic numbers play a special role. These are homogeneous polynomials in the characteristic classes of degree equal to the dimension of the manifold; thus they may be considered as elements of the coefficient ring. Their importance lies in the fact that they are invariants of cobordism. In his fundamental paper on the subject, Thom proved that the characteristic numbers form a complete set of invariants for non-oriented cobordism; this result was later extended by Milnor, Wall and Novikov to oriented cobordism. However, this result fails to be true for the more sophisticated cobordism groups associated with other interesting groups. About 1966 Anderson, Brown and Peterson studied the problem for the groups SU and Spin. They showed that, if one adjoins to the usual characteristic numbers the extraordinary ones defined in a similar fashion, but using real K-theory, then one obtains a complete set of invariants. These results allowed them to give a reasonably complete description of the Spin cobordism ring.

One of the main tools in the above work has been the analysis of the cohomology of the appropriate Thom spectrum M(G) as a module over the Steenrod algebra A. In many cases, $H^*(M(G))$ can be expressed as the direct sum of very simple A-modules. Now the problem of determining which A-modules can be expressed as the cohomology of suitable spectra is one of greatest interest which is very far from being solved. In 1965 Brown and Peterson introduced, for each odd prime p, a spectrum BP whose Z_p cohomology is the quotient of A by the two-sided ideal generated by the Bockstein operator, and proved that the spectrum MU, when localized at p, has the homotopy type of a cluster of iterated suspensions of copies of BP. This gives a strong hold on the spectrum MU and the associated Novikov spectral sequence, and was responsible for a resurgence of interest in that old favorite subject, homotopy groups of spheres, in the early seventies.

Other interesting examples of spectra were found by Ed in collaboration with Sam Gitler in 1971. For each positive integer k, the <u>Brown-Gitler spectrum</u> B(k) has the property that $H^*(B(k);Z_2)$ is isomorphic with the quotient of the mod 2 Steenrod algebra A by the left ideal generated by the elements $c(Sq^i)$ for all $i>k$ (c is the canonical anti-automorphism of A). Part of its importance is due to the fact that certain of the spaces that arise in the solution of the "immersion conjecture" have the same homotopy type as clusters of iterated suspensions of Brown-Gitler spectra. It was the continued joint work of Brown and Peterson over many years that laid the groundwork for its eventual solution by Ralph Cohen in 1981.

As the Brown-Gitler spectra were the subject of a conference at Arcata a year ago, it is hardly necessary to underline its significance; suffice it to say that they played an important role in Mark Mahowald's work on the stable 2-stem.

It is evident that Ed Brown has played a leading part in many of the most interesting developments in our subject. I am proud to have had him as my student.

Homology Representations of Finite Transformation Groups

Alejandro Adem

0. Introduction

Let X be a finite dimensional CW complex with a cellular action of a finite group G. Given prescribed isotropy subgroups (possibly trivial) it is natural to expect restrictions on $H_*(X;R)$ (R a ring) as a graded RG-module.

In this paper we will describe efforts to deal with this problem by systematically applying techniques from group cohomology and modular representation theory. The common strategy is the following: if X is a $G-CW$ complex, then let $C_*(X;R)$ be its cellular chain complex. Then the chain groups are direct sums of permutation modules

$$C_i(X;R) \cong \bigoplus_{\sigma}(\mathbf{Z}G \underset{G_\sigma}{\otimes} \mathbf{Z}_\sigma) \otimes R$$

($\mathbf{Z}_\sigma$ denotes $\mathbf{Z}$ twisted by an orientation character.) Hence C_* may be thought of as a functor from $G-CW$ complexes to "permutation chain complexes". Now given one such chain complex, we can apply algebraic functors or invariants to it, and relate them to properties of $H_*(X;R)$.

We will discuss two distinct approaches within this framework:

(1) The Exponent — equivariant Tate Homology can be used to obtain a numerical invariant for a connected G-chain complex C_* (and hence connected $G-CW$ complexes). This invariant provides restrictions on the torsion in $H^*(G, H_*(C))$ and for a space is determined on the singular set and characterizes free actions.

(2) Growth Rate and Shifted Subgroups — using coefficients in a field, we obtain conditions on the G-cohomological growth rates of $H^*(X)$, for X a finite dimensional connected complex. These can be interpreted in terms of complexity, an invariant from modular representation theory. We also describe a method for extending results about free $(\mathbf{Z}/p)^r - CW$ complexes to arbitrary ones by using the notion of "shifted subgroups". These have important applications to group actions.

The purpose of this note is not only to describe recent developments but also to compare existing results. Most important among them are those due to Browder [4,5], Carlsson [7,8,9], Gottlieb [11] and Heller [12]. There is clearly a common thread and formal similarities; at the end of the paper we carry out a brief comparison of some of these results.

Most of the proofs are omitted, as full details will appear elsewhere [1]. The material presented here is a close version of a lecture presented at Arcata during the conference on Algebraic Topology, in August 1986.

I. The Exponent Approach

We need two definitions

Definition 1.1

For a torsion module M over $\mathbf{Z}$,

$$\exp(M) = \min\{n > 0 \mid nx = 0 \quad \text{for all } x \in M\}$$

Definition 1.2

A complete resolution is an acyclic complex $\mathcal{F}_* = (F_i)_{i \in \mathbf{Z}}$ of projective $\mathbf{Z}G$ modules, together with a map $F_0 \longrightarrow \mathbf{Z}$ such that $\mathcal{F}_+ \longrightarrow \mathbf{Z}$ is a resolution in the usual sense. ■

Now let C_* be a finite dimensional $\mathbf{Z}G$-chain complex; its Tate Homology is defined as

$$\widehat{H}_K(G, C_*) = H_K(\mathcal{F}_* \underset{G}{\otimes} C_*)$$

Assume that C_* is connected, with augmentation

$$C_* \xrightarrow{\epsilon} \mathbf{Z}$$

Then ϵ induces a map

$$\widehat{H}_{-1}(G, C_*) \xrightarrow{\epsilon_*} \widehat{H}_{-1}(G, \mathbf{Z})$$

Definition 1.3

The exponent of C_*, $e_G(C_*)$ is defined as

$$e_G(C_*) = |G|/\exp \operatorname{im} \epsilon_* \ ■$$

The following properties follow directly from this definition.

(1) $e_G(C_*)$ is a positive integer dividing $|G|$.

(2) If $H \subset G$ is a subgroup, then

$$e_H(C_*) \mid e_G(C_*)$$

(3) If $\phi : C_* \longrightarrow D_*$ is a map of connected G-chain complexes, then

$$e_G(D_*) \mid e_G(C_*)$$

$\mathcal{F}_* \underset{G}{\otimes} C_*$ is the total complex associated to a double complex, and hence for C_* finite dimensional, we have two convergent spectral sequences

(A) $E^2_{p,q} = \widehat{H}_p(G, H_q(C)) \Rightarrow \widehat{H}_{p+q}(G, C_*)$

(B) $E^1_{p,q} = \widehat{H}_q(G, C_p) \Rightarrow \widehat{H}_{p+q}(G, C_*)$

Using (A) and (B), we can estimate the exponent for C_*, connected finite-dimensional $\mathbf{Z}G$-chain complex.

Proposition 1.4
(1) If C_* is $\mathbf{Z}G$-acyclic, then

$$e_G(C_*) = |G|$$

(2) $e_G(C_*) \mid \prod_1^\infty \exp H^{i+1}(G, H_i(C))$

(3) If $\phi : C_* \longrightarrow D_*$ is a weak equivalence of connected G-chain complexes, then

$$e_G(C_*) = e_G(D_*) \blacksquare$$

We remark that (2) was due to Browder [5] when (1) holds.

Now if X is a connected, finite dimensional $G-CW$ complex, let

$$e_G(X) = e_G(C_*(X))$$

($C_*(X)$ the cellular chain complex of X).

In this situation, the exponent acquires interesting geometric properties. We list without proof the most important ones.

Properties of $e_G(X)$:
(1) $e_G(X) \mid [G : G_\sigma]$ for all G_σ isotropy subgroups
(2) $e_G(X) = |G| \iff X$ is a free $G-CW$ complex
(3) $e_G(X) \mid \chi(X)$ if X is admissible (i.e., isotropy subgroups fix cells pointwise), where $\chi(X)$ is the Euler characteristic.
(4) If X satisfies Poincare Duality and G preserves the n-dimensional orientation class, then

$$e_G(X) = |G|/\exp \operatorname{im} j^*$$

where $j^* : \widehat{H}^n(G, C^*(X)) \longrightarrow \widehat{H}^\circ(G, H^n(X))$.
(5) In case (4), if $G = (\mathbf{Z}/p)^r$ and X is a manifold, then $e_G(X) =$ co-rank of largest isotropy subgroup.
(6) $e_G(X)$ is determined on the singular set of the action. $\blacksquare$

A couple of remarks: (5) follows from a theorem due to Browder [4]. Some of these properties are similar to those of Gottlieb's trace [11], but their proofs are algebraic and have wider applicability.

We proceed to mention a few examples.
(1) Let M be a Riemann surface, with an orientation preserving action of a p-group

G, with $|G| = p^n$. It can be shown that G has a discrete singular set with cyclic isotropy subgroups (if any).

It is easy to compute the exponent in this case, as it is determined on $C_\circ(M)$, and

$$e_G(M) = p^n / \max\{|G_\sigma|\}$$

Applying 1.4 (2):

$$p^n / \max\{|G_\sigma|\} \mid \exp H^2(G, H_1(M)) \cdot \exp H^3(G, \mathbf{Z})$$

In particular, if $G = (\mathbf{Z}/p)^n$, we have $p^{n-2} \mid \exp H^2(G, H_1(M))$, whether the action is free or not.

This indicates that $H_1(M)$ has an interesting $\mathbf{Z}G$-module structure.

(2) Let $(\mathbf{Z}/p)^r$ act on X, an oriented manifold, trivially in homology. Then

$$\begin{array}{c} \text{co-rank of largest} \\ \text{isotropy subgroup} \end{array} \leq \begin{array}{c} \text{number of non-zero reduced} \\ \text{homology groups of } X \text{ over } \mathbf{Z}_{(p)} \end{array}$$

This theorem was first proved by Browder [4] and applies particularly well to $X = (S^n)^k$.

(3) Let D be the dihedral group of order 8; then it acts on S^3, preserving orientation with its element of order four acting freely. Using the join, we may construct actions on any S^{4n+3} with these properties. Can D act on other S^m in this way? We apply the exponent.

Suppose it does:

$$e_{\mathbf{Z}/4}(S^m) \mid e_D(S^m) \mid \exp H^{m+1}(D, \mathbf{Z}) \Rightarrow 4 \mid \exp H^{m+1}(D, \mathbf{Z}) \Rightarrow m \equiv -1 \bmod 4$$

The answer is no.

II Growth Rate and Shifted Subgroups

Let K be a field and C^* a connected, finite dimensional G-cochain complex. In a manner quite analogous to the proof of 1.4 (2), the following inequality can be verified:

Lemma 2.1

$$\dim \widehat{H}^{k+1}(G, K) \leq \sum_{r=1}^{\infty} \dim \widehat{H}^{k-r}(G, H^r(C)) + \dim_k \widehat{H}^{k+1}(G, C^*) \quad \text{for all } k \in \mathbf{Z}. \blacksquare$$

Here we use Tate Cohomology, defined as

$$\hat{H}^k(G, C^*) = \hat{H}^k(\mathrm{Hom}_G(\mathcal{F}_*, C^*))$$

for $\mathcal{F}_*$ a complete resolution of K over KG.

The term on the extreme right measures how far the cochain complex is from being free. It vanishes for free actions, and in this case a form of this inequality was first proved by Heller [12].

This can of course be applied to $C^*(X; K)$, for X a connected complex. For the free case we mention two interesting applications.

(1) Let $X = S^n \times S^m \quad n \neq m$, then using well-known formulae in group cohomology, (2.1) can be used to show that $(\mathbf{Z}/p)^3$ cannot act freely on X (see [12]).

(2) Let X be any finite complex; we define its free p-rank of symmetry as

$$F_p(X) = \max\{n \mid (\mathbf{Z}/p)^n \text{ acts freely on } X\}$$

Using 2.1, one can show

$$F_p(X) \leq \left(\sqrt{\sum_{i>0} \dim H^i(X)}\right)(\dim X + 1/2) - \dim X + 1$$

This has the virtue of being a global bound; later on we shall see improvements on this.

For a positively graded K-vector space $X_* = \{X_n\}$, we define its growth rate γ as

$$\gamma(X_*) = \min\left\{s \in \mathbf{N} \mid \lim_{n\to\infty} \frac{\dim X_n}{n^s} = 0\right\}$$

A theorem due to Quillen [14] states that

$$\gamma\left(\hat{H}^*(G, C^*(X; K))\right) = \begin{array}{l}\text{largest rank of a } p\text{-elementary}\\ \text{abelian isotropy subgroup}\end{array}$$

where $p = \operatorname{char}(K)$.

There is an analogue of this for modules, which involves the notion of complexity. If M is a finitely generated KG-module, let $P_* \longrightarrow M$ be a minimal projective resolution of M over KG; then the complexity of M is defined as

$$cx_G(M) = \gamma(P_*)$$

This invariant has been shown to have many interesting properties; the one we require is

$$cx_G(M) = \max_{\substack{E\subset G \\ \text{elementary abelian}}} \{\gamma(H^*(E,M))\} \qquad \text{(see [15])}$$

Combining Quillen's result with this algebraic one, and the fact that (2.1) holds for all values of k, we obtain

Theorem 2.2

Let X be a finite dimensional connected $G-CW$ complex such that

$$cx_G(H^i(X,\mathbf{F}_p)) < p\text{-rank of } G \qquad \text{for } i > 0$$

Then there exists an elementary abelian p-subgroup $E \subset G$ of largest rank, such that $X^E \neq \emptyset$. ∎

This says for example that if the modules $H_i(X,\mathbf{F}_p)$ are projective or periodic for $i > 0$ and p-rank $G \geq 2$, then $X^E \neq \emptyset$ for some subgroup $E \cong \mathbf{Z}/p \times \mathbf{Z}/p$ in G.

Next we specialize to actions of elementary abelian p-groups. Let $E = (\mathbf{Z}/p)^n = \langle x_1, \ldots, x_n\rangle$ and K an algebraically closed field of characteristic p. We will define certain subgroups of units in KE which can be applied very fruitfully to group actions.

Let T be an automorphism of $\overset{n}{\underset{1}{\oplus}} K(x_i - 1)$. Then T extends uniquely to an algebra isomorphism $\varphi_T : KE \longrightarrow KE$.

Definition 2.3 (see [6] or [13])

A shifted subgroup S of order p^r in KE is the image of $\langle x_1, \ldots, x_r\rangle$ under φ_T for some $T \in GL_n(K)$. ∎

The importance of shifted subgroups lies in that there is a much wider choice of them than for real subgroups, making restriction arguments much more powerful. The following theorem illustrates this:

Theorem 2.4

Let X be an $E-CW$ complex, where $E \cong (\mathbf{Z}/p)^n$ with isotropy subgroup of largest rank r. Then there exists a shifted subgroup $S \subset KE$ of rank $n-r$, such that $C_*(X,K)|_S$ is free, and S has maximal rank with this property. ∎

The proof of this requires a remarkable theorem due to Kroll [13] which characterizes complexity in terms of shifted subgroups: under the above conditions, if M is a finitely generated KE-module, then $cx_E(M) = rkE - \max\{rkS \mid S \text{ is a shifted subgroup acting freely on } M\}$.

Clearly 2.4 opens the way to generalizing results about free actions to arbitrary ones at one stroke. The following is a list of some of these; let X be a finite dimensional $E-CW$ complex, where $E \cong (\mathbf{Z}/p)^n$ and $r =$ rank of largest isotropy subgroup:

(1) If $X \simeq (S^m)^k$ and the action is trivial in homology, then $n - r \leq k$ (compare with example (2) in I).

(2) If $X \sim S^m \times S^l$, $\quad m \neq l$, then $n - r \leq 2$

(3) If $\overline{H}^*(X,K) \cong_G \begin{cases} M & i = m \\ 0 & \text{otherwise} \end{cases}$ (M a KG-module)

then there exists S, a shifted subgroup of rank $n-r$ in KE, such that

$$\hat{H}^*(S,M) \simeq \hat{H}^{*+m+1}(S,K)$$

(this is a version of the Steenrod Problem)

(4) If $p = 2$, X is finite and $n - r \leq 4$, then

$$2^{n-r} \leq \sum_{i>0} \dim H_i(X;K)$$

(the proof of this in the free case is due to Carlsson [10])

(5) We can estimate the rank of symmetry of a complex with isotropy subgroups of prescribed rank.

To conclude this section we mention that (1) above can be extended to arbitrary $(\mathbf{Z}/2)^n$ actions on $X \simeq (S^m)^k$, provided $m \neq 1,3,7$. As a corollary of this we obtain the first exact value for $F_2((S^m)^k)$:

Corollary 2.5

If $m \neq 1,3,7$

$$F_2\left((S^m)^k\right) = k,$$

necessarily achieved by a homologically trivial action. ∎

III Related Results and Conclusions

Let $G = (\mathbf{Z}/2)^n$, and M a finitely generated KG-module (char $K = 2$). Then we define the Loewy length $\lambda_G(M)$ as

$$\lambda_G(M) = \min\left\{\lambda > 0 \mid I^\lambda M = 0\right\}$$

where $I \subset KG$ is the augmentation ideal. This is well-defined because I is nilpotent, and in fact $\lambda_G(M) \leq n+1$.

Now let C_* be a connected, finite free KG-chain complex. Carlsson [8] proved

Theorem 3.1

Under the above conditions,

$$\sum_{i>0} \lambda_G(H_i(C)) \geq n \blacksquare$$

This result is formally similar to ones mentioned previously but it provides a considerably sharper estimate on the free rank of symmetry of a complex. From the fact that $\lambda_G(M) \leq \dim_K M$, we obtain

$$F_2(X) \leq \sum_{i>0} \dim_K H_i(X;K)$$

Carlsson has in fact conjectured that

$$F_2(X) \leq \log_2(\sum_{i=0} \dim_K H_i(X;K))$$

but this has only been verified for $rk\, G \leq 4$.

It is clear how 3.1 can be painlessly extended to arbitrary finite, connected permutation chain complexes using shifted subgroups. This can be applied to recover example (1) in II.

We have presented an outline of how techniques from group cohomology and representation theory apply successfully to questions about group actions. The methods described here yield a few overlapping results but in general seem to go in different directions. For example, the exponent approach shows that $(\mathbf{Z}/p)^n$ cannot act freely, trivially in homology, on $(S^m)^r$ if $n > r$, but not that $(\mathbf{Z}/p)^3$ cannot act freely on $S^m \times S^l$, $m \neq l$. The approach in II implies the second fact but not the first. Carlsson's result (3.1) provides a sharp estimate on the free 2-rank of symmetry but would not be useful for odd primes (indeed, the Loewy length for $(\mathbf{Z}/p)^n$-modules is bounded by $n(p-1)+1$).

The next stage should be to interpret the algebraic restrictions on the homology of a $G-CW$ complex in terms of more familiar and tractable invariants. However, it first seems necessary to be able to distinguish representations arising from geometric actions on the complex, as the complete family of G-modules (up to isomorphism) is very complicated for almost any group. This has been done only in the simplest of cases (we refer to the work on the Steenrod problem [2], [3], [9]) and a lot remains to be done.

References

[1] A. Adem, "Cohomological Restrictions on Finite Group Actions", to appear in *J. Pure Appl. Algebra.*

[2] A. Assadi, "Homotopy Actions and Cohomology of Finite Groups", *Proc. Poznań Symposium on Transf. Groups* 1985 (LNM 1217 Springer-Verlag).

[3] D. Benson and N. Habegger "Varieties for Modules and a Problem of Steenrod", preprint 1985.

[4] W. Browder, "Actions of Elementary Abelian p-groups", to appear.

[5] W. Browder, "Cohomology and Group Actions", *Inv. Math.* **71** (1983) 599–608.

[6] J. Carlson, "The Varieties and Cohomology Ring of a Module", *J. Algebra* **85** (1983) 104–143.

[7] G. Carlsson, "On the Rank of Abelian groups Acting Freely on $(S^n)^{k}$", *Inv. Math.* **69** (1982).

[8] Carlsson, "On the Homology of Finite Free $(\mathbf{Z}/2)^n$-Complexes", *Inv. Math.* **74** (1983) 139–147.

[9] Carlsson, "A Counterexample to a Conjecture of Steenrod", *Inv. Math.* **64** (1981) 171–174

[10] Carlsson, "Free $(\mathbf{Z}/2)^3$ Actions on Finite Complexes", preprint 1985.

[11] D. Gottlieb, "The Trace of an Action and the Degree of a Map", *T.A.M.S.* **293** (1986) 381–410.

[12] A. Heller, "A Note on Spaces with Operators", *Ill. J. Math.* **3** (1959) 98–100.

[13] O. Kroll, "Complexity and Elementary Abelian p-Groups", *J. Algebra* **88** (1984) 155–172.

[14] D. Quillen, "The Spectrum of an Equivariant Cohomology Ring I", *Ann. Math.* **94** (1971) 549–572.

[15] J. Alperin and L. Evens, "Representations, Resolutions and Quillen's Dimension Theorem", *J. Pure Appl. Algebra* **22** (1981).

Mathematics Department
Stanford University
Stanford, CA 94305

HOMOTOPY EXPONENTS FOR SPACES OF CATEGORY TWO

- David J. Anick -
Department of Mathematics
Massachusetts Institute of Technology
Cambridge, MA 02139

Abstract

Let X be a simply-connected finite CW complex of Lusternik-Schnirelmann category two. There is finite set P of primes, depending upon X, such that the following holds at any $p \notin P$. The total rational homotopy group $\pi_*(X)\otimes\mathbb{Q}$ is finite-dimensional if and only if X has a homotopy exponent at p.

1. Introduction and Summary of Results

We will prove a weak form of John Moore's exponent conjecture, for finite complexes of category two.

Moore's conjecture concerns the existence of "exponents" in the homotopy groups of certain spaces. The space X is said to have exponent dividing p^r at the prime p if and only if for all n, multiplication by p^r annihilates the p-torsion of $\pi_n(X)$. The space X is called elliptic if and only if $\sum_{n \geq 2} \operatorname{rank}(\pi_n(X)) < \infty$, equivalently if $\pi_*(X)\otimes\mathbb{Q}$ is finite-dimensional; otherwise, it is hyperbolic.

Moore's Conjecture. (See [26]). Let X be a simply-connected finite CW complex. Then, for all primes p, X has an exponent at p iff X is elliptic.

Even if Moore's conjecture fails, the elliptic/hyperbolic distinction seems to be an important criterion for the classification of finite complexes. The terminology was introduced in [13], where it was proved for finite hyperbolic spaces that the rational homotopy ranks $\{\dim \pi_n(X)\otimes\mathbb{Q}\}$ increase roughly exponentially with n. Finite elliptic spaces were studied in [16], where it was shown that their rational cohomology rings must satisfy Poincaré duality.

Moore's conjecture has been verified in various special cases [12] [23] [26]. Not surprisingly, it is sometimes easier to verify a weakened form of the conjecture. The weakening consists of relaxing "all p" to "almost all p". To remove any ambiguity about the logic, we restate this weakened form.

Weak Moore conjecture. Let X be a simply-connected finite complex. Then there is a finite set P of primes, depending upon X, such that (a) if X is elliptic then X has an exponent at every $p \notin P$, but (b) if X is hyperbolic then X fails to have an exponent at any $p \notin P$.

Recently McGibbon and Wilkerson proved the weak Moore conjecture for elliptic spaces [21]. That is, they settled part (a) affirmatively. In this paper we will settle part (b) for the class of spaces with category two. This class of spaces, like the class of elliptic spaces, is familiar to rational homotopy theorists for containing many interesting examples while enjoying a fairly tractable theory. In some ways the class is quite general. For instance, Poincaré series of loop spaces for all finite complexes are represented among those for this subclass [6, Theorem 3].

Define a two-cone X to be the cofiber of a map $f:W'\to W$ between two wedges of spheres. The two-cone X is finite if both W' and W are wedges of only finitely many spheres. We will actually prove

Theorem 1.1 (also Theorem 3.7). The weak Moore conjecture holds for finite two-cones.

Theorem 1.1 can be used as follows to show that the weak Moore conjecture holds for any X whose rational localization $X_{\mathbb{Q}}$ has Lusternik-Schnirelmann category two or less. By [14] such an X has the rational homotopy type of a finite two-cone X'. By [8], if X and X' are finite complexes sharing the same rational homtopy type, then X and X' become equivalent upon localizing away from some finite set P" of primes. In particular, for $p \notin P''$, X has an exponent at if and only if X' has an exponent at p. By Theorem 1.1, there is a finite set P' of primes such that X' has an exponent at $p \notin P'$ if and only if X' (equivalently X) is elliptic. Putting $P = P' \cup P''$, we have Moore's conjecture for X at any $p \notin P$. Summarizing, we have shown

Corollary 1.2. The weak Moore conjecture holds for any X such that $\mathrm{cat}(X_{\mathbb{Q}}) \leq 2$.

Because the condition $\mathrm{cat}(X) \leq 2$ may be expressed in terms of the diagonal map $X \to X\times X\times X$ factoring up to homotopy through the fat wedge T^2X [28], we deduce that $\mathrm{cat}(X) \leq 2$ implies $\mathrm{cat}(X_{\mathbb{Q}}) \leq 2$.

Consequently we have

Corollary 1.3. The weak Moore conjecture holds for any X of category ≤ 2.

There is almost no overlap between these results and the results of McGibbon and Wilkerson in the elliptic case. Indeed, the only way X can both be elliptic and have category two or less is for $X_{\mathbb{Q}}$ to coincide with one of the following:

$$(pt),\ (S^m)_{\mathbb{Q}},\ (S^m \times S^n)_{\mathbb{Q}},\ \text{or}\ (S^{2m} \cup_{[\iota,\iota]} e^{4m})_{\mathbb{Q}}.$$

Theorem 1.2 thus asserts that, excluding these four exceptions, an X of rational catagory two or less has no exponent at almost all primes.

We wish to state one more theorem which will be proved as a stepping-stone toward Theorem 1.1. Let R denote a subring of $\mathbb{Q}$; then $R = \mathbb{Z}[P]^{-1}$ for some set of primes. An R-equivalence is a map $f: X \to Y$ which becomes a homotopy equivalence when localized away from P. We will prove

Theorem 1.4 (Combine Corollaries 3.4 and 4.8). Let X be a finite simply-connected two-cone, let $R = \mathbb{Z}[P]^{-1}$, and suppose that $H_*(\Omega X; R)$ is free as an R-module. There is a finite set P' of primes such that, when $R' = \mathbb{Z}[P \cup P']^{-1}$, then $H_*(\Omega X; R')$ is the universal enveloping algebra of $im(h)$, where h denotes the Hurewicz homomorphism

$$h: \pi_*(\Omega X) \otimes R' \to H_*(\Omega X; R').$$

Furthermore, there is a filtration of $im(h)$ whose associated graded Lie algebra fits into a split short exact sequence of bigraded Lie R'-algebras

$$0 \to L' \to gr(im(h)) \to L'' \to 0,$$

where L' is a free Lie R'-algebra and L'' is finitely presented. Lastly, there is an R'-equivalence

$$(\prod_i \Omega S^{2m_i+1}) \times (\prod_j S^{2n_j+1}) \xrightarrow[R']{\simeq} \Omega X, \qquad (1)$$

the left-hand side of (1) denoting a weak infinite product.

2. The Theory of Two-cones

In Section 2 we will recall some of the basic theory concerning loop spaces on two-cones, and we will standardize some notation for working with them.

A two-cone X is the cofiber of a map between two wedges of spheres,

$$\bigvee_{j\in J} S^{d_j+1} \xrightarrow[\alpha=V\alpha_j]{} \bigvee_{i\in I} S^{m_i+1} \xrightarrow{\eta} X,$$

subject to certain constraints. The index sets I and J denote initial segments of the ordered set of positive integers. We assume finite type, i.e., that the sequences $\{m_i\}$ and $\{d_j\}$ are non-decreasing and that $m_i\to\infty$ (resp. $d_j\to\infty$) if I (resp. J) is infinite. The two-cone X is <u>finite</u> if the sets I and J are both finite, in which case we let g and r respectively denote their cardinalities. Since we will only be concerned with simply-connected spaces, we postulate that $m_i\geq 1$ and $d_j\geq 1$. The wedge $\bigvee_{i\in I} S^{m_i+1}$ will always be denoted by W, the attaching maps by $\{\alpha_j\}$, and the inclusion of W into X by η. The adjoint of $\alpha_j:S(S^{d_j})\to W$ is denoted $\hat{\alpha}_j:S^{d_j}\to\Omega W$.

Let R denote a commutative ring with unity. A <u>graded algebra</u> over R is a graded R-module $M = \bigoplus_{n=0}^{\infty} M_n$, together with an associative pairing $M_m\otimes_R M_n \to M_{m+n}$. When each M_n is finitely generated as an R-module we call M <u>locally finite</u>. For $x\in M_n$, we call n the <u>dimension</u> of x and write $|x| = n$. The graded algebra M is <u>connected</u> if there is a unit $1\in M_o$, and if $M_o\approx R$ as R-modules. The graded R-module M is a <u>graded Lie algebra</u> if it has a bilinear bracket which satisfies the Jacobi identities with signs, and it is <u>connected</u> if $M_o=0$.

The loop space homology $H_*(\Omega W;R)$, with Pontrjagin multiplication, is a free associative graded R-algebra with generators $\{a_i\}_{i\in I}$, where $|a_i|=m_i$ [17]. We denote such an algebra as $R\langle a_1,a_2 \ldots\rangle$, and $H_*(\Omega W;R)$ in particular will also be called $F= \bigoplus_{n=0}^{\infty} F_n$. Likewise

$$H_*(\Omega \bigvee_{j\in J} S^{d_j+1};R) = R\langle b_1,b_2,\ldots\rangle,$$

where $|b_j|=d_j$. Let β_j denote the element $(\Omega\alpha)_*(b_j)\in F_{d_j}$. Let B denote the two-sided ideal of F generated by the set $\{\beta_j\}_{j\in J}$.

For any simply-connected space Y and ring R, let h_Y denote the Hurewicz homomorphism

$$h_Y: \pi_*(\Omega Y)\otimes R \to H_*(\Omega Y;R).$$

When $\frac{1}{6}\in R$, $\mathrm{im}(h_Y)$ is a connected graded Lie subalgebra of $H_*(\Omega Y;R)$ [25]. It should be clear that $h_W(\hat{\alpha}_j)=\beta_j$. Consequently we know that each β_j, and hence all of B, must lie in the kernel of

$$(\Omega\eta)_*: F \to H_*(\Omega X;R).$$

The graded algebra we care most about is $H_*(\Omega X;R)$, which we denote by $A=\bigoplus_{n=0}^{\infty} A_n$. When R is a field, the algebra A is fairly well understood [4][19], since it turns out that $B=\ker(\Omega\eta)_*$. Hence the image of $(\Omega\eta)_*$, which we henceforth denote by G, is a graded subalgebra of A and it is isomorphic to the quotient algebra F/B.

When R is a field we may define the Hilbert series of a locally finite graded R-module $M=\bigoplus_{n=0}^{\infty} M_n$ to be the formal power series

$$M(z)=\sum_{n=0}^{\infty} \dim_R(M_n)z^n.$$

Theorem 3.7 of [4] tells us, among other things, that the Hilbert series of A and of G are related by the formula

$$A(z)^{-1} = (1+z)G(z)^{-1}-z[1-\sum_{i\in I} z^{m_i}+\sum_{j\in J} z^{d_j}]. \qquad (2)$$

The main importance of this formula is its implication that $A(z)$ depends only upon $G(z)$. Let us see how this fact can sometimes be used to deduce that A is a free R-module for non-fields R.

Suppose now that R is a subring of $\mathbb{Q}$. Let A denote the Pontrjagin ring of ΩX over R, and let $A^{[p]}$ denote its Pontrjagin ring over the prime field of characteristic p. For p a non-unit in R, the universal coefficient theorem tells us that $A^{[p]}(z)=A^{[o]}(z)$ if and only if A contains no p-torsion. By (2), we have $A^{[p]}(z)=A^{[o]}(z)$ if and only if $G^{[p]}(z)=G^{[o]}(z)$; since G is defined as F/B, this happens if and only if G is a free R-module. We have proved

<u>Lemma 2.1</u>. Fix a subring R of $\mathbb{Q}$. Over R, the Pontrjagin ring A of the loops on a two-cone X is a free R-module if and only if the quotient algebra $G = F/B$ is free as an R-module.

Let us return to the consideration of the Lie algebra $\mathrm{im}(h_Y)$. Let L_W denote the free graded Lie algebra over R, $L_W=\mathbb{L}\langle x_i\rangle_{i\in I}$

where $|x_i|=m_i$. When $\frac{1}{2}\in R$ we have $\mathrm{im}(h_W)\approx L_W$, and the natural inclusion of $\mathrm{im}(h_W)$ into F is the inclusion of L_W into a universal enveloping algebra. Since $\beta_j=h(\hat{\alpha}_j)$ we have $\beta_j\in L_W$, hence the $\{\beta_j\}$ generate a Lie ideal, call it $\langle\beta\rangle$, of L_W. For the two-cone X, define L_X to be $L_W/\langle\beta\rangle$. Note that $(\Omega\eta)_*:L_W \to \mathrm{im}(h_X)$ factors through L_X, giving a natural map

$$L_\eta:\ L_X \to \mathrm{im}(h_X). \tag{3}$$

Put $R=\mathbf{Z}[\frac{1}{6}]$, so that $\pi_*(\Omega Y)\otimes R$, with Samelson product, becomes a graded Lie algebra. Then the Hurewicz map

$$h=h_W:\ \pi_*(\Omega W)\otimes R \to L_W$$

has a right inverse σ defined by $\sigma(x_i)=\hat{\iota}_i$, with commutator brackets being sent to Samelson products [29]. Here ι_i denotes the homotopy class of the inclusion of the ith sphere into the wedge W.

By [22] h and σ are rational isomorphisms, so $\mathrm{im}(\sigma\circ h-1)$ is a torsion group. Put $\hat{\gamma}_j=\sigma h(\hat{\alpha}_j)-\hat{\alpha}_j$. Then $\hat{\alpha}_j=\sigma(\beta_j)-\hat{\gamma}_j$, $h(\hat{\gamma}_j)=0$, $t_j\hat{\gamma}_j=0$, where t_j is the additive order of $\hat{\gamma}_j$ in $\pi_*(\Omega W)\otimes R$. We may view γ_j as the deviation of α_j from being a "pure" R-linear combination of repeated Whitehead products of generators ι_i in $\pi_*(W)\otimes R$. With this notation established, we have

<u>Definition 2.2.</u> The prime p is <u>implicit</u> in the two-cone X if $p=2$ or 3, or if p divides some t_j.

The implicit primes are those for which homotopy p-torsion is implicitly needed in order to construct X. The following properties are immediately apparent.

<u>Lemma 2.3</u>. (a) If each attaching map α_j is a linear combination of Whitehead products of generators ι_i in $\pi_*(W)$, then the only implicit primes are 2 and 3. (b) A finite 2-cone X has only finitely many implicit primes, and an upper bound on their size is $\max(3,\frac{1}{2}(\dim(X)-1))$.

We remark that the set of implicit primes need not be a homotopy invariant, since it depends upon our choice of sphere generators for W. For instance, if $g=2$ with $m_1=2$ and $m_2=9$, then $W=S^3\vee S^{10}$. If γ has order five in $\pi_{10}(S^3)$, we could replace ι_2 by $\iota_2+\iota_1\circ\gamma$ and thereby determine a different splitting σ for h. This change could affect whether or not the prime 5 is implicit in X. Thus the set of implicit primes is actually an invariant of the given CW decomposition for X, and this actually strengthens our results. The set of implicit primes will sometimes serve as the set P of excluded primes for the weak Moore conjecture. Since one is free to choose any two-cone decomposition for X, we need exclude only those primes which are implicit for all choices of wedges-of-spheres cofibration having cofiber X.

We finish this section with a few comments on X's Adams-Hilton model [2], which is a differential graded R-algebra whose homology coincides with $H_*(\Omega X;R)$. As noted in [4] or [20], the Adams-Hilton model for X may be taken to be (C,d), where $C=R\langle x_i,y_j\rangle_{i\in I,j\in J}$, $|x_i|=m_i$ and $|y_j|=d_j+1$, and $d(x_i)=0$ with $d(y_j)=\beta_j$. Because C is a free R-algebra it is automatically the universal enveloping algebra of a free Lie algebra L. We write this as $C=\mathcal{U}L$, where $L=\mathbb{L}\langle x_i,y_j\rangle_{i\in I,j\in J}$ is viewed as an R-submodule of C. Moreover, because each β_j belongs to L, we have $d(L)\subseteq L$, so the differential graded algebra (C,d) is the universal enveloping algebra of the differential graded Lie algebra (L,d). When $R=\mathbb{Q}$ this (L,d) is a valid Quillen model for X [10], but for more general R it has no obvious topological significance. Nevertheless, it will play a central role in helping us to understand $H_*(\Omega X;R)$.

Lastly, we mention that L and C have an additional grading which d respects. This second grading will become a major theme in Section 4. It is obtained by letting each x_i have degree zero and letting each y_j have degree one. We see readily

Lemma 2.4. (a) G is the "degree zero" component of $A=H_*(C,d)$, and $G=\mathcal{U}(L_X)$. (b) If $\frac{1}{2}\in R$, then L_X is the "degree zero" component of $H_*(L,d)$.

By [4, Theorem 3.7] we also have

Lemma 2.5. Suppose R is a field. Then (a) A is generated as an algebra by the "degree zero" and the "degree one" components of A; and (b) the "degree one" component of A has Hilbert series

$$z[G(z)(1-\sum_{i\in I} z^{m_i}+\sum_{j\in J} z^{d_j})-1]G(z).$$

3. A Decomposition Theorem Implies the Weak Moore Conjecture.

In Section 3 we will first recall a very general decomposition lemma. We will then state a theorem, whose proof we postpone, dealing with certain two-cones. Lastly, we will show why the weak Moore conjecture for all finite two-cones is a consequence of the theorem.

Throughout Section 3 we let R denote a subring of $\mathbb{Q}$. Let Y denote any simply-connected CW complex of finite type. For the following discussion, suppose $A=H_*(\Omega Y;R)$ is free as an R-module. Then the R-submodule $\mathrm{im}(h) \subseteq A$ must also be free over R. We may choose elements $\{\lambda_1,\lambda_2 \dots\} \subseteq \pi_*(Y)\otimes R$ such that $\{h(\hat{\lambda}_1),h(\hat{\lambda}_2),\dots\}$ is an R-basis for $\mathrm{im}(h)$. Let n_i denote $|\lambda_i|$. If n_i is odd, let T_i denote ΩS^{n_i} and let $\tilde{\lambda}_i$ denote the loop map $\Omega\lambda_i : T_i \to \Omega Y$. When n_i is even let T_i denote S^{n_i-1} and let $\tilde{\lambda}_i$ denote the adjoint $\tilde{\lambda}_i=\hat{\lambda}_i : T_i \to \Omega Y$. The collection of maps $\{\tilde{\lambda}_i\}$ together define

$$\tilde{\lambda}\colon \prod_i T_i \to \Omega Y. \tag{4}$$

In (4) the left-hand side denotes the weak product when there are infinitely many $\tilde{\lambda}_i$'s, and we must arbitrarily choose an order in which to multiply these maps in ΩY.

If we localized (4) at $\mathbb{Q}$, $\tilde{\lambda}$ would by [22] become a homotopy equivalence, hence $\tilde{\lambda}$ induces a monomorphism of $H_*(\cdot;R)$. If we suppose further that the Pontrjagin ring $A=H_*(\Omega Y;R)$ is generated as an R-algebra by $\mathrm{im}(h)$, then A is filtered by powers of $\mathrm{im}(h)$. If in addition we have $\frac{1}{2}\in R$, then we may use this filtration and [25] to see that λ_* surjects onto A. Thus $\tilde{\lambda}$ induces an isomorphism of $H_*(\cdot;R)$, hence it is an R-equivalence.

The above discussion is summarized in the following "decomposition lemma".

<u>Lemma 3.1</u> (cf. lemma V.3.10 of [9]). Let $R\subseteq\mathbb{Q}$, $\frac{1}{2}\in R$, and let Y be a simply-connected CW complex of finite type. Suppose (a) $H_*(\Omega Y;R)$ is free over R, and (b) $H_*(\Omega Y;R)$ is generated as an R-algebra by $\mathrm{im}(h)$. Then any ordered collection $\{\lambda_1,\lambda_2,\dots\} \subseteq \pi_*(Y)\otimes R$ such that $\{h(\hat{\lambda}_i)\}$ is an R-basis for $\mathrm{im}(h)$ determines an R-equivalence

$$\tilde{\lambda}\colon \prod_i T_i \to \Omega Y \tag{5}$$

in which the restriction of $\tilde{\lambda}$ to the ith factor is $\tilde{\lambda}_i$. Furthermore, Moore's conjecture is true for Y at any prime p which is not a unit in R.

The last conclusion, as to Moore's conjecture, follows because Y is elliptic if and only if the product in (5) has only finitely many factors. By [12] and [15] the latter condition coincides with the existence of a homotopy exponent at odd primes.

An important special case, which also follows from the classical results [17] and [27], is

Lemma 3.2. Let $W = \bigvee_{k\in I} S^{m_k+1}$ be a finite type wedge of spheres. Fix $R \subseteq \mathbb{Q}$ having $\frac{1}{2} \in R$. For any choice $\{\lambda_i\} \subseteq \pi_*(W) \otimes R$ such that $\{h(\hat{\lambda}_i)\}$ is an R-basis for L_W, there exists an R-equivalence

$$\lambda\colon \prod_i T_i \to \Omega W$$

in which the restriction of $\tilde{\lambda}$ to the ith factor T_i is $\tilde{\lambda}_i$.

We will now state a theorem which shows that the hypotheses of the decomposition lemma can be weakened considerably, for two-cones. We postpone the proof until after providing motivation by deducing the weak Moore conjecture as a consequence.

Theorem 3.3. Let X be a two-cone, and let R be a subring of $\mathbb{Q}$ in which all of the implicit primes for X are units. Suppose L_X is a free R-module. Then (a) $H_*(\Omega X;R)$ and im(h) are free R-modules; (b) $H_*(\Omega X;R)$ is generated as an R-algebra by im(h); and (c) the map $L_\eta\colon L_X \to \mathrm{im}(h)$ of (3) is a split injection of Lie algebras.

While Theorem 3.3 has importance in its own right, an immediate consequence of interest is

Corollary 3.4. Under the hypotheses of Theorem 3.3, there is an R-equivalence

$$Q_1 \times Q_2 \to \Omega X,$$

where Q_1 and Q_2 are weak products of T_i's and the factors in Q_1 correspond to an R-basis for L_X. Furthermore, Moore's conjecture is true for X at any prime p which is not a unit in R.

Proof. By 3.3(a) and 3.3(c) we may choose disjoint sets $\{\lambda_i\}$ and $\{\lambda'_j\}$ in $\pi_*(X) \otimes R$ whose adjoints surject under h to R-bases for the free R-modules L_X and $(\mathrm{im}(h)/L_X)$, respectively. Apply lemma 3.1 to $\{\lambda_i\} \cup \{\lambda'_j\}$.

In order to prove the weak Moore conjecture without the hypothesis that L_X be free, we consider the cases $\dim_{\mathbb{Q}}(L_X\otimes\mathbb{Q})<\infty$ and $\dim_{\mathbb{Q}}(L_X\otimes\mathbb{Q})=\infty$ separately.

Lemma 3.5. Let X be a finite two-cone, or more generally suppose that the index set I is finite. If $\dim_{\mathbb{Q}}(L_X\otimes\mathbb{Q})<\infty$, then there exists an integer n_0 such that $L_X\otimes\mathbb{Z}[\frac{1}{n_0}]$ is free over $\mathbb{Z}[\frac{1}{n_0}]$.

Proof. Because I is finite, L_X is a finitely generated connected graded Lie algebra over $\mathbb{Z}$, $L_X = \mathbb{L}\langle x_1,\dots,x_g\rangle/\langle\beta_1,\beta_2,\dots\rangle$. Put $m = \max\{|x_1|,\dots,|x_g|\}$. Because $\dim_{\mathbb{Q}}(L_X\otimes\mathbb{Q})<\infty$, we may put $d=\sup\{j\,|\,(L_X)_j\supseteq\mathbb{Z}\}<\infty$. Let n_0 denote the product of all primes p such that $\mathbb{Z}_p\subseteq\bigoplus_{j=1}^{d+m}(L_X)_j$, and put $R=\mathbb{Z}[\frac{1}{n_0}]$.

By induction on k we will prove that the order of any element $w\in(L_X)_j$, where $d<j\le k$, is a unit in R. This is true for $k=d+m$. Supposing it is true for $k-1$, where $k\ge d+m+1$, let $w\in(L_X)_k$. Because L_X is generated as a Lie algebra by $\{x_1,\dots,x_g\}$, we may write w as

$$w = \sum_{i=1}^{g}[w_i,x_i], \tag{6}$$

where $|w_i|=k-|x_i|\ge d+1$. By our choice of d, we have $t_iw_i=0$ for some $t_i>0$, and by the inductive hypothesis we may assume that $t_i^{-1}\in R$. Putting $t_0 = \prod_{i=1}^{g} t_i$ we obtain from (6) that $t_0w=0$ while $t_0^{-1}\in R$. This completes the inductive step.

Thus $L_X\otimes R$ contains no torsion in dimensions $k\ge d+m$, nor does it contain any in dimensions $\le d+m$. Deduce that $L_X\otimes R$ is free over R.

Proposition 3.6. Let X be a two-cone. Let P be the set of implicit primes for X and put $R=\mathbb{Z}[P]^{-1}$. Suppose $\dim_{\mathbb{Q}}(L_X\otimes\mathbb{Q})=\infty$. Then there is a weak product Q_1 involving infinitely many distinct T_i's which is an R-homotopy retract of ΩX. That is, when the spaces Q_1 and ΩX are localized away from P, there are maps

$$Q_1 \xrightarrow{\tilde{\lambda}} \Omega X \xrightarrow{\rho} Q_1$$

whose composition is homotopic to the identity.

Proof. We cannot apply Theorem 3.3 directly because L_X might contain torsion at infinitely many primes (see [5] or [7] for examples of this phenomenon). The trick is to enlarge X to a space Y for which L_Y is free over R.

Let $L_X = \mathbb{L}\langle x_i\rangle/\langle\beta_j\rangle$ be taken over R, and let D denote the Lie ideal of L_X consisting of all torsion elements in L_X. Since D is a locally finite graded R-module we may write $D=\mathrm{span}\{\tau_k\}_{k\in K}$, where there are only finitely many τ_k's in each dimension. Since L_W surjects onto L_X, we may choose elements $\beta_k'\in L_W$ which surject to τ_k. Let $\hat{\alpha}_k'=\sigma(\beta_k')\in\pi_*(\Omega W)\otimes R$, and let Y be the space obtained by attaching cells to X by the maps $\{\alpha_k'\}$. Since the attaching maps have target W, the space Y is also a two-cone, and there are obvious inclusions

$$W \xrightarrow{\eta} X \xrightarrow{\varphi} Y.$$

The attaching map for the two-cone Y is of course $(\alpha_j)\cdot(\alpha_k')$. Since each $\alpha_k'\in\mathrm{im}(\sigma)$, the new cells introduce no new implicit primes, so all of Y's implicit primes are already invertible in R. Furthermore, $L_Y=L_W/\langle\beta_j,\beta_k'\rangle = L_X/D$ is free over R, so Theorem 3.3 applies to Y. Moreover, $\dim_{\mathbb{Q}}(L_Y\otimes\mathbb{Q})$ is still infinite because the kernel of the surjection $L_X \to L_Y$ is the torsion R-module D.

By Corollary 3.4 choose elements $\lambda_i,\lambda_i'\in\pi_*(Y)\otimes R$ such that $(\tilde{\lambda}_Y\times\tilde{\lambda}_Y'):Q_1\times Q_2 \to \Omega Y$ is an R-equivalence, where the subscript on $\tilde{\lambda}$ emphasizes our choice of target space. Assuming all spaces have been localized away from P, let $(\rho,\rho'):\Omega Y \to Q_1\times Q_2$ denote the homotopy inverse. Now $L_W\to L_Y$ is a surjection between two free R-modules, and $\{h(\hat{\lambda}_i)\}$ surjects to an R-basis for L_Y. We may choose for each λ_i a preimage λ_i^* in $\pi_*(\Omega W)\otimes R$, and we may extend $\{\lambda_i^*\}$ to a set $\{\lambda_i^*\}\cup\{\lambda_i''\}$ whose adjoints surject to an R-basis for L_W. Using Lemma 3.2 we obtain an R-equivalence $(\tilde{\lambda}_W\times\tilde{\lambda}_W''):\ Q_1\times Q_3 \xrightarrow{\simeq} \Omega W$. It should be clear that $\tilde{\lambda}_Y \simeq \Omega(\varphi\eta)\tilde{\lambda}_W$. The composition

$$Q_1 \xrightarrow{(\Omega\eta)\,\circ\,\tilde{\lambda}_W} \Omega X \xrightarrow{\rho\circ(\Omega\varphi)} Q_1$$

is homotopic to $\rho\tilde{\lambda}_Y \simeq \mathrm{id}_{Q_1}$, hence Q_1 is an R-homotopy retract of ΩX. Since the product Q_1 is indexed by an R-basis for the infinite-dimensional R-module L_Y, we are done.

<u>Theorem 3.7</u>. Let X be a finite two-cone. Then the weak Moore conjecture is true for X. That is, there exists a finite set P

of primes (dependent upon X) such that, except perhaps at $p \in P$, X has an exponent at p if and only if X is elliptic.

Proof. Consider the cases $\dim_{\mathbb{Q}}(L_X \otimes \mathbb{Q}) < \infty$ and $\dim_{\mathbb{Q}}(L_X \otimes \mathbb{Q}) = \infty$ separately. When $\dim_{\mathbb{Q}}(L_X \otimes \mathbb{Q}) < \infty$, apply Lemma 3.5 to L_X. Let P consist of all primes which are implicit in X or which divide n_0. By Lemma 3.5 and Corollary 3.4, Moore's conjecture holds away from P.

Suppose instead $\dim_{\mathbb{Q}}(L_X \otimes \mathbb{Q}) = \infty$. Then X is hyperbolic because the rational homotopy $\pi_*(\Omega X) \otimes \mathbb{Q} \approx \mathrm{im}(h) \otimes \mathbb{Q}$, which (by Theorem 3.3 or otherwise) contains $L_X \otimes \mathbb{Q}$, is also infinite-dimensional. Let P consist of the implicit primes only; verifying Moore's conjecture means showing that X has no exponent for primes p outside of P. By Proposition 3.6 there exist arbitrarily high-dimensional spheres whose homotopy groups, localized at $p \notin P$, are summands of $(\pi_*(X))_{(p)}$. By [15] this means that $\mathbb{Z}_{p^r} \subseteq \pi_*(X)$ for any r, i.e., X does not have an exponent at p.

4. Bigraded Differential Lie Algebra

The Pontrjagin algebra for the loops on a two-cone has a natural bigrading. In Section 4 we explore the algebraic consequences of this additional structure. Our main result is Theorem 4.6, which expresses the Pontrjagin ring for certain loop spaces as the enveloping algebra of a certain bigraded Lie algbebra. This will bring us closer to the proof of Theorem 3.3.

In this paper, a bigraded Lie (resp. associative) algebra over a ring R will be viewed as an ordinary graded object which happens to have an additional grading that also respects the pairing. The original grading is called dimension, and the dimension of a bihomogeneous elements x is still denoted $|x|$. The new grading is called degree, and the degree of x is denoted $\deg(x)$. For Lie algebras the two gradings are not interchangeable: the Jacobi identities (with signs) always utilize $|x|$ and not $\deg(x)$. The component of a bigraded object M in dimension j is denoted M_{*j}, the part in degree i is written M_{i*}, and $M_{*j} \cap M_{i*}$ is called M_{ij}. A bigraded Lie (resp. associative) algebra is connected if it is connected with respect to the dimension grading. A differential d on a bigraded Lie (resp. associative) algebra has $|d| = \deg(d) = -1$. "Bigraded connected differential" will henceforth be abbreviated "b.c.d." Bigraded objects generalize ordinary graded objects because any $M=\oplus M_j$ may be made bigraded by defining $\deg(x)=|x|$ for all x.

Two functors which operate on bigraded algebras are $\mathcal{U}$ and $\mathcal{H}$. The universal enveloping functor $\mathcal{U}$ takes bigraded (connected, differential) Lie algebras to bigraded (connected, differential) associative algebras. The homology functor $\mathcal{H}$ sends b.c.d. Lie (resp. associative) algebras to bigraded connected Lie (resp. associative) algebras.

Our first and most difficult lemma measures the failure of $\mathcal{U}$ and $\mathcal{H}$ to commute for b.c.d. Lie algebras over a field. Let (L,d) be a b.c.d. Lie algebra over R. The inclusion $(L,d) \to (\mathcal{U}L,\mathcal{U}d)$ induces a homomorphism of homology $\mathcal{H}L \to \mathcal{H}\mathcal{U}L$, which in turn induces

$$\mu\colon \mathcal{U}\mathcal{H}L \to \mathcal{H}\mathcal{U}L. \tag{7}$$

<u>Lemma 4.1</u>. Suppose R is a field. Put $p=\infty$ if $\mathrm{char}(R)=0$ but let $p=\mathrm{char}(R)$ otherwise. The map μ of (7) is isomorphic in degrees $\leq p-3$ (resp. dimensions $\leq 2p-3$), and it is epimorphic in degree $p-2$ (resp. dimension $2p-2$).

<u>Remark</u>. The term "respectively" is being used in a slightly nonstandard manner in Lemma 4.1. The map μ is to be isomorphic when restricted either to $(\mathcal{U}\mathcal{H}L)_{i*}$ for $i\leq p-3$ or to $(\mathcal{U}\mathcal{H}L)_{j*}$ for $j\leq 2p-3$.

<u>Proof</u>. For a bigraded vector space V over R, let $\mathcal{S}V$ be the symmetric algebra on V if $p=2$, but if $p>2$ let $\mathcal{S}V$ denote the tensor product of the symmetric algebra on the even-dimensional part of V with the exterior algebra on the odd-dimensional part of V. A differential d on V extends to a differential $\mathcal{S}d$ on $\mathcal{S}V$. Note that $\mathcal{S}V$ is actually trigraded, the third grading coming from product length in $\mathcal{S}V$. Let $\mathcal{S}^kV$ denote the span of products of exactly k elements of V, and observe that $(\mathcal{S}d)(\mathcal{S}^kV) \subseteq \mathcal{S}^kV$ while $|\mathcal{S}d|=\deg(\mathcal{S}d)=-1$.

Since we are over a field we may decompose a b.c.d. Lie algebra (L,d) as $L=D'\oplus K\oplus D$, where $D=\mathrm{im}(d)$, $K\oplus D=\ker(d)$, and $d|_{D'}\colon D'\to D$ is an isomorphism, The Poincaré-Birkhoff-Witt theorem gives us an isomorphism $\varphi\colon \mathcal{S}L\to\mathcal{U}L$ as bigraded vector spaces, in which a tensor product $x_{i_1}^{\epsilon_1}\otimes x_{i_2}^{\epsilon_2}\otimes\cdots\otimes x_{i_t}^{\epsilon_t}$ in $\mathcal{S}L$ is sent to the product of the elements $x_{i_1}^{\epsilon_1},\ldots,x_{i_t}^{\epsilon_t}$, multiplied in a particular order, in $\mathcal{U}L$.

It is well-known that

$$F^k\mathcal{U}L=\varphi(\bigoplus_{m\le k}\mathcal{S}^m L)$$

is a filtration on $\mathcal{U}L$, and $(\mathcal{U}d)(F^k\mathcal{U}L)\subseteq F^k\mathcal{U}L$. Furthermore, the associated (tri-)graded module

$$gr(\mathcal{U}L) = \bigoplus_k (F^k\mathcal{U}L/F^{k-1}\mathcal{U}L)$$

is isomorphic with $\bigoplus_k(\mathcal{S}^k L)$, and $gr(\mathcal{U}d)=\mathcal{S}d$. Thus there is a spectral sequence converging to an associated graded module for $\mathcal{H}\mathcal{U}L$, whose E^1 term coincides with the bigraded R-module $\mathcal{H}(\mathcal{S}L,\mathcal{S}d)$.

Let us try to understand $\mathcal{H}(\mathcal{S}L,\mathcal{S}d)$. Computationally,

$$\mathcal{H}\mathcal{S}L = \mathcal{H}\mathcal{S}(K\oplus D'\oplus D) = \mathcal{H}(\mathcal{S}K\otimes\mathcal{S}(D'\oplus D)) = \mathcal{S}K \otimes [\bigotimes_v \mathcal{H}\mathcal{S}(\mathrm{span}(v,dv))], \quad (8)$$

as v runs through a vector space basis for D'. But we are also viewing $\mathcal{H}\mathcal{S}L$ as the E^1 term of a spectral sequence, hence as cycles for $F^k\mathcal{U}L$ modulo $F^{k-1}\mathcal{U}L$. The contribution of $\mathcal{S}^k K$ to this E^1 term comes from $\varphi(\mathcal{S}^k K)$ (where $\mathcal{S}^k K$ is viewed as a submodule of $\mathcal{S}^k L$), and these represent actual cycles in $F^k\mathcal{U}L$, not just cycles modulo $F^{k-1}\mathcal{U}L$. In fact, the $\mathcal{S}K$ in (8) corresponds under φ to the image of μ. This means that all higher differentials in the spectral sequence will vanish on the submodule corresponding to $\mathcal{S}K$ in (8).

Let us compute $\mathcal{H}\mathcal{S}(\mathrm{span}(v,dv))$. Put $V = \mathcal{S}(\mathrm{span}(v,dv))$ for v a basis element of D'. If $|v|$ is odd, then V has

$$\{v(dv)^m,(dv)^m \mid m\ge 0\}$$

as a basis, with $d(v(dv)^m)=(dv)^{m+1}$, so V is acyclic and $\mathcal{H}V$ contributes nothing to the tensor product in (8). When $|v|$ is even, V has

$$\{v^m, v^m dv \mid m\ge 0\}$$

as a basis, where $d(v^m)=mv^{m-1}dv$. The case $\mathrm{char}(R)=0$ can be finished off at once: V is again acyclic, so (8) becomes $E^1=\mathcal{H}\mathcal{S}L=\mathcal{S}K$. Since higher differentials vanish on $\mathcal{S}K$, the spectral sequence degenerates, giving $\mathcal{U}\mathcal{H}L\approx\mathcal{S}K \xrightarrow[\mu\ \varphi]{\approx} \mathcal{H}\mathcal{U}L$.

In the case $p=\mathrm{char}(R)\ne 0$, we obtain

$$\mathcal{H}V=\mathrm{span}\{1, v^{ap}, v^{ap-1}dv \mid a\geq 1\}.$$

Because $dv\neq 0$, however, we must have $\deg(v)\geq 1$, hence $\mathcal{H}V$ vanishes in degrees smaller than $p-1$, and likewise it vanishes in dimensions smaller than $2p-1$.

It follows that all higher differentials in the spectral sequence vanish in degrees $\leq p-2$ (resp. in dimensions $\leq 2p-2$). Contrapositively, non-zero differentials must originate in degrees $\geq p-1$ (resp. in dimensions $\geq 2p-1$), hence they must terminate in degrees $\geq p-2$ (resp. in dimensions $\geq 2p-2$). Thus $(E^1)_{ij}=(E^\infty)_{ij}$ for $i\leq p-3$ and for $j\leq 2p-3$, while $(E^1)_{ij}$ surjects to $(E^\infty)_{ij}$ for $i=p-2$ and for $j=2p-2$. For degrees $\leq p-2$ (resp. dimensions $\leq 2p-2$) we have a commuting diagram

$$\begin{array}{ccccc} \mathcal{S}K & \xrightarrow{\approx} & E^1 & \longrightarrow & E^\infty \\ {\scriptstyle\approx}\downarrow{\scriptstyle\gamma} & & & & \downarrow{\scriptstyle\approx} \\ \mathcal{U}\mathcal{H}L & & \xrightarrow{\quad\mu\quad} & & \mathcal{H}\mathcal{U}L, \end{array}$$

from which the lemma follows.

We remark that much of the above calculation mirrors the work of Cohen-Moore-Neisendorfer in [11, Section 4]. Also, it gives another proof that $\mathcal{U}$ and $\mathcal{H}$ commute over a field of characteristic zero, which was first proved by Quillen in [24]. We will extend Lemma 4.1 to b.c.d. Lie algebras over subrings of $\mathbf{Q}$. We first recall without proof some elementary algebraic facts.

Lemma 4.2. Let R denote a subring of $\mathbf{Q}$, let P denote the set of primes (including zero) which are not units in R, and let $\mathbf{Z}_p$ denote the prime field of characteristic p (with $\mathbf{Z}_0$ denoting $\mathbf{Q}$). Let V and V' denote locally finite bigraded differential R-modules and let M denote a locally finite bigraded connected Lie R-algebra. Then

(a) For $p\in P$, the natural map

$$\tau\colon (\mathcal{U}M)\otimes\mathbf{Z}_p \longrightarrow \mathcal{U}(M\otimes\mathbf{Z}_p)$$

is an isomorphism of bigraded $\mathbf{Z}_p$-algebras.

(b) M is R-free in degrees $\leq m$ (resp. dimensions $\leq m$) if and only if $\mathcal{U}M$ is R-free in degrees $\leq m$ (resp. dimensions $\leq m$).

(c) $\mathcal{H}V$ is R-free in degree (resp. dimension) $m-1$ if and only if, for all $p\in P$, the "universal coefficient homomorphism"

$$\tau\colon (\mathcal{H}V)\otimes\mathbf{Z}_p \longrightarrow \mathcal{H}(V\otimes\mathbf{Z}_p)$$

is isomorphic in degree (resp. dimension) m.

(d) A homomorphism $f\colon V\to V'$ is surjective if and only if, for each $p\in P$, $f_p = f\otimes 1_{\mathbf{Z}_p}: V\otimes\mathbf{Z}_p \to V'\otimes\mathbf{Z}_p$ surjects.

(e) A homomorphism $f: V\to V'$ is isomorphic if and only if, for each $p\in P$, f_p is an isomorphism.

<u>Lemma 4.3</u>. Let R be any subring of $\mathbf{Q}$. Let (L,d) be a locally finite b.c.d. Lie algebra over R. Suppose there is an integer $q\geq 2$ such that $p^{-1}\in R$ for $1\leq p<q$ and also such that $\mathcal{H}\mathcal{U}L$ is R-free in degrees $\leq q-3$ (resp. dimensions $\leq 2q-3$). Then $\mathcal{H}L$ is R-free in degrees $\leq q-3$ (resp. dimensions $\leq 2q-3$), and the natural map

$$\mu: \mathcal{U}\mathcal{H}L \to \mathcal{H}\mathcal{U}L$$

is isomorphic in degrees $\leq q-3$ (resp. dimensions $\leq 2q-3$) and epimorphic in degree $q-2$ (resp. dimension $2q-2$).

<u>Proof</u>. We will prove this for the degree grading only; the proof for the dimension grading is virtually identical.

Retain the notations P, $\mathbf{Z}_p$, τ, and γ from Lemma 4.2. For $p\in P$ let $L^{[p]}$ denote $L\otimes\mathbf{Z}_p$ and consider the following commuting diagram:

$$\begin{array}{ccccccc}
\mathcal{U}\mathcal{H}L & \longrightarrow & (\mathcal{U}\mathcal{H}L)\otimes\mathbf{Z}_p & \xrightarrow{\tau} & \mathcal{U}((\mathcal{H}L)\otimes\mathbf{Z}_p) & \xrightarrow{\mathcal{U}\gamma} & \mathcal{U}\mathcal{H}L^{[p]} \\
\Big\downarrow \mu & & \Big\downarrow \mu\otimes 1_{\mathbf{Z}_p} = \overline{\mu}_p & & & & \Big\downarrow \mu^{[p]} \\
\mathcal{H}\mathcal{U}L & \longrightarrow & (\mathcal{H}\mathcal{U}L)\otimes\mathbf{Z}_p & \xrightarrow{\gamma} & \mathcal{H}((\mathcal{U}L)\otimes\mathbf{Z}_p) & \xrightarrow{\mathcal{H}\tau} & \mathcal{H}\mathcal{U}L^{[p]}
\end{array} \qquad (9)$$

For $0\leq m\leq q-2$ define a homomorphism $f: V\to V'$ between bigraded R-modules to be an <u>m-map</u> if it is isomorphic in degree m when $m\leq q-3$, or if it is epimorphic in degree m when $m=q-2$. We will prove by induction on m, for $0\leq m\leq q-2$, the triple proposition: first, $(\mathcal{H}L)_{m-1,*}$ is R-free; second, $\overline{\mu}_p$ is an m-map; and third, μ is an m-map. The initial step for the induction is provided by the triviality that $(\mathcal{H}L)_{-1,*} = 0$ is R-free.

Assuming that $(\mathcal{H}L)_{m-1,*}$ is R-free, we have by Lemma 4.2(c) that the arrow $\mathcal{U}\gamma$ of (9) is an isomorphism in degree m. Since we hypothesize that $(\mathcal{U}\mathcal{H}L)_{m-1,*}$ is R-free, 4.2(c) also makes the arrow γ of (9) isomorphic in degree m. By 4.2(a) both τ and $\mathcal{H}\tau$ are isomorphisms, and by Lemma 4.1, $\mu^{[p]}$ is an m-map. It follows that $\overline{\mu}_p$ is an m-map for any $p\in P$.

Assuming that $\overline{\mu}_p$ is an m-map for all $p\in P$, we have via 4.2(d) (if $m=q-2$) or by 4.2(e) (if $m\leq q-3$) that μ is an m-map. Lastly, if μ is an m-map for some $m \leq q-3$, then $(\mathcal{U}\mathcal{H}L)_{m*}$ is R-free; by 4.2(b) this makes $(\mathcal{H}L)_{m*}$ R-free. This completes the inductive step, and with it the proof of Lemma 4.3.

We are ready to specialize Lemma 4.3 to the situation described in Theorem 3.3. We will apply Lemma 4.3 to the b.c.d. Lie algebra discussed at the end of Section 2.

Let $R \subseteq \mathbb{Q}$. For a free bigraded R-module M, let $\mathcal{T}M$ denote the bigraded tensor algebra on M, i.e., the R-algebra $\bigoplus_{n=0}^{\infty} (M^{\otimes n})$. Likewise let $\mathcal{L}M$ denote the free bigraded Lie algebra on M. Note that $\mathcal{T}M \approx \mathcal{U}\mathcal{L}M$, and that $\mathcal{L}M$ may be viewed as the Lie algebra of primitives in $\mathcal{T}M$. The objects $\mathcal{T}M$ and $\mathcal{L}M$ are actually trigraded, the third grading being induced by product or bracket length, but in our applications this third grading will always coincide with the degree.

Whenever M is a submodule of an associative R-algebra A, the multiplication in A induces maps $M^{\otimes n} \to A$. Combining these over all n yields a "multiplication map" $\nu\colon \mathcal{T}M \to A$. Likewise, a submodule L' of a Lie algebra L determines $\nu\colon \mathcal{L}L' \to L$.

<u>Proposition 4.4</u>. Let $R \subseteq \mathbb{Q}$ be a subring containing $\frac{1}{6}$. Let (L,d) be a locally finite free b.c.d. Lie R-algebra, generated as a free Lie algebra by elements in degrees zero and one. Put $M=\mathcal{H}(L,d)$ and $A=\mathcal{H}\mathcal{U}(L,d)$, so that $\mu\colon \mathcal{U}M \to A$ is the previously-studied homomorphism. let N_{ij} denote $\mu(M_{ij}) \subseteq A_{ij}$. Suppose that M_{0*} (which corresponds to L_X) is a free R-module. Then A and M_{1*} are R-free, and $\mu|_{M_{1*}}\colon M_{1*} \to N_{1*}$ is an isomorphism. Furthermore, the multiplication map

$$\nu\colon \mathcal{T}N_{1*} \otimes A_{0*} \to A \tag{10}$$

is an isomorphism.

<u>Remark</u>. Because $\frac{1}{6} \in R$, every free b.c.d. Lie R-algebra (L,d) generated in degrees zero and one can be obtained from a two-cone. Using this correspondence and the notation of Section 2, we may write L_X for M_{0*} and G for A_{0*}.

<u>Proof</u>. By Lemma 4.2(b) and Lemma 4.3 we see that M_{0*} being R-free makes $\mathcal{U}M_{0*} \underset{\mu}{\approx} A_{0*}$ R-free. Use Lemma 2.1 to get A to be R-free. As submodules of a free R-module, each N_{ij} is R-free.

Apply Lemma 4.3 with $q=4$ to obtain the isomorphism

$$\mu_{1*}\colon (\mathcal{U}M)_{1*} \approx A_{1*}. \tag{11}$$

It follows that $\mu|_{M_{1*}}$ is one-to-one; since $N_{1*}=\mu(M_{1*})$ is R-free, so is M_{1*}. The Poincaré-Birkhoff-Witt theorem and (11) may be used to show that the homomorphism

$$N_{1*}\otimes A_{0*} \to A_{1*}, \tag{12}$$

induced by the multiplication in A, is an isomorphism.

It remains to show for $t\geq 2$ that the multiplication-induced homomorphisms

$$\nu^t: (N_{1*})^{\otimes t}\otimes A_{0*} \to A_{t*}$$

are isomorphisms. Let P denote the set of primes (including zero) which are not units in R. By Lemma 4.2(d), ν will be surjective if for each $p\in P$

$$\nu_p^t = \nu^t\otimes 1_{\mathbb{Z}_p}: (N_{1*}^{[p]})^{\otimes t} \otimes A_{0*}^{[p]} \to A_{t*}^{[p]}$$

surjects, where $A^{[p]}=A\otimes\mathbb{Z}_p=\mathcal{U}\mathcal{L}(L\otimes\mathbb{Z}_p)$ and $N_{ij}^{[p]}$ denotes the image of $N_{ij} \to A_{ij} \to A_{ij}^{[p]}$.

The surjectivity of (12) may also be expressed as the relation $(N_{1*})(A_{0*})=A_{1*}$. Since $A_{1*}=(A_{0*})(A_{1*})$, we obtain

$$(N_{1*})^2(A_{0*})=(N_{1*})(A_{1*})=(N_{1*})(A_{0*})(A_{1*})=(A_{1*})^2,$$

and by induction on t, $(N_{1*})^t(A_{0*})=(A_{1*})^t$. It follows that $(N_{1*}^{[p]})^t(A_{0*}^{[p]})=(A_{1*}^{[p]})^t$. Combining this with Lemma 2.5(a), we see that $\bar{\nu}_p^t$ surjects, as desired.

It remains to show that ν is one-to-one. Because $\sigma N_{1*}\otimes A_{0*}$ and A are both R-free, it suffices to show that $\bar{\nu}_0=\nu\otimes 1_{\mathbb{Q}}$ is one-to-one. Since $\bar{\nu}_0$ is onto, we need only demonstrate that $\sigma N_{1*}\otimes A_{0*}\otimes\mathbb{Q}$ and $A\otimes\mathbb{Q}$ have the same Hilbert series. We will write $M(z)$ for the dimension Hilbert series $(M\otimes\mathbb{Q})(z)$.

Since L is free and generated in degrees zero and one, we may write $L = \mathbb{L}\langle x_i, y_j\rangle_{i\in I, j\in J}$, where $\deg(x_i)=0$ and $\deg(y_i)=1$. Put $m_i=|x_i|$ and $d_j=|y_j|-1$. By (12) and Lemma 2.5(b) we have

$$N_{1*}(z) = z(q(z)G(z)-1),$$

where $G=A_{0*}$ and

$$q(z) = 1-\sum_{i\in I} z^{m_i}+\sum_{j\in J} z^{d_j}.$$

Then

$$(\sigma N_{1*}\otimes A_{0*})(z) = [1-z(q(z)G(z)-1)]^{-1}G(z),$$

which is easily seen to coincide with (2). This completes the proof.

In the course of proving the main structure theorem for $H_*(\Omega X;R)$ we will encounter the following trivial lemma.

Lemma 4.5. Let $R\subset\mathbb{Q}$. Let $u: L' \to L''$ be a homomorphism of bigraded connected Lie R-algebras, and suppose that L'' embeds via Ψ into a bigraded connected associative R-algebra A. Suppose that L' and A are locally finite and R-free. Suppose further that the composition Ψu is the inclusion of L' into a universal enveloping algebra, and also that $\overline{\Psi}_0=\Psi\otimes 1_{\mathbb{Q}}: L''\otimes\mathbb{Q} \to A\otimes\mathbb{Q}$ is the inclusion of $L''\otimes\mathbb{Q}$ into a universal enveloping algebra. Then u is an isomorphism.

Proof. Since Ψu is one-to-one, u must be one-to-one. Since $A\approx\mathcal{U}L'$, there is a spitting $\mathcal{U}L' \to A$ as R-modules. We may therefore write $L''=L'\oplus(L''/L')$, both summands necessarily being R-free. Because the composition

$$\mathcal{U}L'\otimes\mathbb{Q} \xrightarrow{\mathcal{U}\overline{u}_0} \mathcal{U}L''\otimes\mathbb{Q} \xrightarrow{\overline{\Psi}_0} A\otimes\mathbb{Q}$$

and $\overline{\Psi}_0$ are both isomorphisms, $\mathcal{U}\overline{u}_0$ is isomorphic, forcing $\overline{u}_0$ to be onto. The cokernel L''/L' of u is a torsion R-module, hence $L''/L'=0$.

Theorem 4.6. Under the hypotheses and notation of Proposition 4.4, the inclusion $N=\mathrm{im}(\mu) \to A$ is the inclusion of N into a universal enveloping algebra. Furthermore, there is a split short exact sequence of Lie R-algebras

$$0 \longrightarrow \mathcal{L}(N_{1*}) \longrightarrow N \underset{\longleftarrow}{\longrightarrow} N_{0*} \longrightarrow 0.$$

In particular, N (resp. A) is generated as a Lie (resp. associative) algebra by N_{0*} and N_{1*}.

Proof. Let $[,]_N$ denote the bracket in N. Note that

$$[,]_N: N_{1*}\otimes N_{0*} \to N_{1*}$$

defines a Lie algebra action of N_{0*} on the R-module N_{1*}. We can extend this action via derivations to a Lie algebra action on $\mathcal{L}N_{1*}$. The "crossed product" of N_{0*} and $\mathcal{L}N_{1*}$ under this action is a new bigraded Lie algebra N'. It has these properties: $N'\approx N_{0*}\oplus\mathcal{L}N_{1*}$ as R-modules; N' is generated as a Lie algebra by N_{0*} and N_{1*}; $N'_{0*}\approx N_{0*}$ and $\bigoplus_{n\geq 1} N'_{n*} \approx \mathcal{L}N_{1*}$ as Lie algebras; and there is a split short exact sequence of Lie algebras

$$0 \to \mathcal{L}N_{1*} \to N' \to N_{0*} \to 0.$$

There is also an obvious homomorphism

$$u\colon N' \to N$$

of bigraded Lie algebras, defined such that in degrees zero and one it identifies $N'_{0*} \oplus N'_{1*}$ with $N_{0*} \oplus N_{1*}$. This diagram commutes:

$$\begin{array}{ccc} N'_{1*} \otimes N'_{0*} & \xrightarrow{[\ ,\ ]} & N'_{1*} \\ \downarrow & & \downarrow \\ N_{1*} \otimes N_{0*} & \xrightarrow{[\ ,\ ]_N} & N_{1*}. \end{array}$$

The map u induces a map $\tilde{u}\colon \mathcal{U}N' \to A$, and the Poincaré-Birkhoff-Witt theorem gives us a multiplication-induced isomorphism

$$\Psi\colon \mathcal{T}N_{1*} \otimes A_{0*} = \mathcal{U}\mathcal{L}(N_{1*}) \otimes \mathcal{U}N_{0*} \xrightarrow{\approx} \mathcal{U}N'.$$

Letting ν denote the isomorphism of (10), we see that $\nu = \tilde{u}\Psi$, hence $\tilde{u}$ is isomorphic.

By Lemma 4.3 the composition $M \otimes \mathbb{Q} \xrightarrow{\bar{\mu}_0} N \otimes \mathbb{Q} \longrightarrow A \otimes \mathbb{Q}$ is an inclusion into a universal enveloping algebra. Lemma 4.5, with $L' = N'$ and $L'' = N$, shows u to be an isomorphism. The desired properties of N follow from those of N'.

Remark. Theorem 4.6 was proved for the case $R = \mathbb{Q}$ in [14].

We finish this section by showing how close we are to a proof of Theorem 3.3. In Section 5 we will prove

Claim 4.7. Let X be a two-cone, and suppose that R is a subring of $\mathbb{Q}$ in which the implicit primes are units. If $A = H_*(\Omega X; R)$ is a free R-module, then $N_{1*} \subseteq \mathrm{im}(h) + A_{0*}$.

Corollary 4.8. Under the hypotheses of Theorem 3.3, there is a filtration of $\mathrm{im}(h)$ whose associated bigraded R-module is isomorphic as a Lie algebra with N. Furthermore, the inclusion $\mathrm{im}(h) \to H_*(\Omega X; R)$ is the inclusion of $\mathrm{im}(h)$ into a universal enveloping algebra. In particular, the Pontrjagin ring $H_*(\Omega X; R)$ is generated as an R-algebra by $\mathrm{im}(h)$.

Proof. Put $S = \mathrm{im}(h)$, viewed as a Lie subalgebra of $A = H_*(\Omega X; R)$. We know that S is graded by dimension, and we filter it by degree. Let $\tilde{S}$ denote the associated bigraded Lie algebra, i.e.,

$$\tilde{S}_{mj} = [S_j \cap (\bigoplus_{i=0}^{m} A_{i*})] \,/\, [S_j \cap (\bigoplus_{i=0}^{m-1} A_{i*})].$$

By claim 4.7 we may write each basis element $a \in N_{1*}$ as $a = a_h + a_0$, where $a_h \in S$ and $a_0 \in A_{0*}$. There may be more than one such representation for a given a, but note that

$a = a_h + a_0 = a_h' + a_0'$ implies that $a_h - a_h' \in S \cap A_{0*}$. Thus an R-module homomorphism

$$u_{1*}: N_{1*} \to \tilde{S}_{1*}$$

is well-defined. We also have an inclusion of Lie algebras

$$u_{0*}: N_{0*} = L_X \to \tilde{S}_{0*}$$

Lastly, u_{1*} is a homomorphism of N_{0*}-modules. That is to say, for $a \in N_{1*}$ and $b \in N_{0*}$ we have $u_{1*}([a,b]) = [u_{1*}(a), u_{0*}(b)]$.

These properties indicate that u_{1*} and u_{0*} may be extended to a Lie algebra map $u: N_{**} \to \tilde{S}_{**}$. Furthermore, the inclusion $S \to A$ induces an inclusion of their associated bigraded objects

$$\psi: \tilde{S} \longrightarrow \tilde{A} \xrightarrow{\approx} A.$$

One can easily check that the compositions $\psi \circ u_{0*}$ and $\psi \circ u_{1*}$ are the ordinary inclusions of N_{0*} and of N_{1*} into A, hence $\psi \circ u$ coincides with the inclusion of N. Lemma 4.5 and Theorem 4.6 assure us that u is an isomorphism.

Finally, the filtration on S induces a filtration on $\mathcal{U}S$ such that the diagram

$$\begin{array}{ccccc} \mathcal{U}N & \xrightarrow[\mathcal{U}u]{\approx} & \mathcal{U}\tilde{S} & \xrightarrow{\approx} & (\mathcal{U}S)^{\sim} \\ \approx\downarrow & & \downarrow\psi^{\sim} & & \downarrow \\ A & \xleftarrow{\approx} & \tilde{A} & \xrightarrow{=} & \tilde{A} \end{array}$$

commutes. It follows that $(\mathcal{U}S)^{\sim} \xrightarrow{\approx} \tilde{A}$, whence $(\mathcal{U}S) \xrightarrow{\approx} A$, so A is a universal enveloping algebra for S.

<u>Proof. of Theorem 3.3</u>. Parts (a) and (b) are contained in Results 4.4 and 4.8, respectively. As to part (c), note that N being R-free with $N \approx \tilde{S}$ makes S R-free, with $N \approx S$ as singly-graded R-modules. Since $L_X = N_{0*}$ is an R-module summand of N, it is likewise a summand of S.

<u>Conjecture 4.9</u>. The author conjectures that Claim 4.7 may be strengthened to the statement that $N_{1*} \subseteq \mathrm{im}(h)$. If this is correct, it would follow that $N \approx \mathrm{im}(h)$ as graded Lie algebras. Felix and Thomas [14] have proved this in the special case $R = \mathbb{Q}$.

5. Using the Adams-Hilton Model

In this section we will prove Claim 4.7, thereby completing the proof of the weak Moore conjecture for finite two-cones. Our method requires some technical lemmas concerning the Adams-Hilton construction, but we will be rewarded with a very constructive proof. Throughout this section, R denotes a commutative ring with unity (later we specialize to a subring of $\mathbb{Q}$), and Adams-Hilton models are always taken over R.

We begin by recalling the properties of the Adams-Hilton (henceforth abbreviated to AH) construction. The model for a CW complex X with trivial 1-skeleton is a free associative differential graded R-algebra $(\mathcal{A}(X),d_X)$ having one generator for each non-trivial cell of X. It comes with a quasi-isomorphism of chain algebras $\theta_X:\mathcal{A}(X) \to CU_*(\Omega X)$, where CU_* denotes the cubical singular chain functor with coefficients in R. The model $\varphi_f:(\mathcal{A}(X),d_X) \to (\mathcal{A}(Y),d_Y)$ for a map $f:X \to Y$ is a homomorphism of chain algebras. It comes with an "algebra homotopy" ([see 10, p. 220] for a definition) denoted Φ_f from $CU_*(\Omega f)\circ\theta_X$ to $\theta_Y\circ\varphi_f$. Induction over the skeleta of X is used in defining d_X, θ_X, φ_f, and Φ_f.

A key property concerns the situation where $\{X_i\}$ are subcomplexes of a complex $X = \cup X_i$. As long as the coherency relations

$$d_{X_i}\big|\mathcal{A}(X_i\cap X_j) = d_{X_j}\big|\mathcal{A}(X_i\cap X_j) \text{ and } \theta_{X_i}\big|\mathcal{A}(X_i\cap X_j) = \theta_{X_j}\big|\mathcal{A}(X_i\cap X_j)$$

hold for each pair (X_i, X_j), then the union of the models $\{(\mathcal{A}(X_i), d_{X_i})\}$ defines a valid AH model for X. Suppose further that $f:X \to Y$ is any map. Put $f_i = f|_{X_i}$ and suppose that (φ_i,Φ_i) is a valid model for f_i. As long as the coherency conditions

$$\varphi_i\big|\mathcal{A}(X_i\cap X_j) = \varphi_j\big|\mathcal{A}(X_i\cap X_j) \text{ and } \Phi_i\big|\mathcal{A}(X_i\cap X_j) = \Phi_j\big|\mathcal{A}(X_i\cap X_j)$$

hold for each pair (X_i, X_j), we know that the union of $\{(\varphi_i,\Phi_i)\}$ defines a valid AH model for f.

Where no confusion can result we write d, φ, Φ for d_X, φ_f, Φ_f.

Next we describe AH models for certain elementary CW complexes. When the sphere S^m is viewed as $(pt)\cup e^m$, its AH model is free on one generator $\underline{a}$ of dimension $m - 1$, with $d(a) = 0$. We denote this CW structure for the sphere as S^m_0. When we attach one more cell to obtain the disk $D^{m+1} \approx (pt)\cup e^m \cup e^{m+1}$, its model incorporates a second generator b having $|b| = m$ and $d(b) = a$ [2, p. 315]. We call this CW structure D^{m+1}_0. We presume our chosen AH models to be compatible in the sense that $CU_*(\Omega j)\circ\theta_{S_0^m}(a) = \theta_{D_0^{m+1}}(a)$, where $j:S_0^m \to D_0^{m+1}$ is the inclusion of CW complexes.

A third complex that will arise is an alternate CW structure for the sphere S^{m+1}. For any $s \geq 1$ we may define

$$S_s^{m+1} = (\bigvee_{k=1}^{s} D_0^{m+1}) \cup_{\Sigma\kappa_k} e^{m+1},$$

where the attaching map for the last cell is the sum of the inclusions of the k spheres into $\bigvee_{k=1}^{s} S_0^m$. Its AH model has s generators $a_1,\ldots,a_s$ of dimension $m - 1$ and $s + 1$ generators $b_0, b_1,\ldots,b_s$ of dimension m. The differential satisfies $d(b_i) = a_i$ for $i>0$ and $d(b_0) = a_1 +\ldots+ a_s$. The homeomorphism

$$T:S_0^{m+1} \to S_s^{m+1} \tag{13}$$

has $\varphi_T(a) = b_1+\ldots+ b_s - b_0$. Again, we presume that $\theta_{S_s^{m+1}}$ is compatible with the various inclusions of D_0^{m+1} into S_s^{m+1}.

Our first lemma takes advantage of the flexibility built into the AH model for a map, in particular, the opportunity to make certain choices during the inductive definitions of φ_f and ϕ_f.

<u>Lemma 5.1</u>. Let $r:X \to Y$ denote the inclusion of a subcomplex into a CW complex, fix an AH model for Y, and put $d_X = d_Y|_{A(X)}$, $\theta_X = \theta_Y|_{A(Y)}$. Suppose that r has a right inverse up to homotopy, or more generally that the induced map $(\Omega r)_*:H_*(\Omega X;R) \to H_*(\Omega Y;R)$ surjects. Let $f:S_0^m \to X$ denote any map such that $r\circ f$ is null-homotopic. Let φ_f denote an AH model for f, and put $x = \phi_f(a)\in A(X)$. Then there exists an element $y\in A(Y)$ satisfying $d(y) = x$. Furthermore, for any $y\in A(Y)$ satisfying $d(y) = x$ there is a valid AH model (φ_f', ϕ_f') for an extension of f to $f:D_0^{m+1} \to Y$ such that $\varphi_f'(a) = x$, $\varphi_f'(b) = y$, and $\phi_f'(a)\in CU_*(\Omega X)$.

<u>Proof</u>. Since $r\circ f \simeq 0$, we may extend it over D_0^{m+1} to obtain a map of pairs $f:(D_0^{m+1}, S_0^m) \to (Y,X)$. The AH model φ can be extended over $A(D_0^{m+1})$ without altering $\varphi_f(a)$ or $\phi_f(a)$. So far, we have $\varphi_f(a) = x$, $\phi_f(a)\in CU_*(\Omega X)$, and $d(\varphi_f(b)) = \varphi_f(d(b)) = \varphi_f(a) = x$. This proves the existence of y, since we may take $y = \varphi_f(b)$.

If y' is another element of $A(Y)$ satisfying $d(y') = x$, then $c = y' - y$ is a cycle in $A(Y)$. Since $H_*(A(X),d_X)$ is assumed to surject onto $H_*(\mathcal{A}(Y),d_Y)$, we may write $c = u + d(v)$, where u is a cycle in $\mathcal{A}(X)$ and $v \in \mathcal{A}(Y)$. Put

$$\varphi_f'(a) = x, \quad \varphi_f'(a) = \varphi_f(a) - \theta_X(u),$$
$$\varphi_f'(b) = y', \quad \varphi_f'(b) = \varphi_f(b) - \theta_Y(v).$$

With these alterations the formulas of [2, p. 316] remain true. Thus (φ_f', φ_f') is a valid AH model for $f: D_0^{m+1} \to Y$.

Our next two lemmas will help us understand the AH model for a situation in which a Whitehead product is trivial because one of the factors is trivial. Recall that the Whitehead product is defined using a map $\omega_{pq}: S_0^{p+q-1} \to S_0^p \vee S_0^q$. If the AH generators corresponding to the $(p+q-1)$-, the p-, and the q-spheres are respectively denoted a, a', c', then by Corollary 2.4 of [2] we may take

$$\pm \varphi_{\omega_{pq}}(a) = [a',c'] = a'c' - (-1)^{|a'|\cdot|c'|}c'a'.$$

<u>Lemma 5.2</u>. Denote the generators of the AH model for $D_0^{p+1} \vee S_0^q$ by a',b',c', where c' corresponds to S_0^q and $d(b') = a'$. There is an extension of $\pm\omega_{pq}: S_0^{p+q-1} \to S_0^p \vee S_0^q$ to a map $f: D_0^{p+q} \to D_0^{p+1} \vee S_0^q$ whose AH model may be chosen so as to satisfy

$$\varphi_f(a) = [a',c'], \; \varphi_f(b) = [b',c'], \text{ and } \varphi_f(a) \in CU_*(\Omega(S_0^p \vee S_0^q)).$$

<u>Proof</u>. Apply Lemma 5.1 with $f = \pm\omega_{pq}$, $X = S_0^p \vee S_0^q$, and $Y = D_0^{p+1} \vee S_0^q$.

The next lemma generalizes the result to repeated Whitehead products. We introduce a notation for repeated brackets. For $t \geq 2$, recursively put $[a_0,a_1,\ldots,a_t] = [[a_0,a_1,\ldots,a_{t-1}],a_t]$. We will also use this notation for a repeated Samelson or Whitehead product.

We remark that the AH model for the composition $g \circ f$ may be taken to be $\varphi_{g \circ f} = \varphi_g \circ \varphi_f$, $\varphi_{g \circ f} = CU_*(\Omega g) \circ \varphi_f + \varphi_g \circ \varphi_f$.

<u>Lemma 5.3</u>. Let $W = \bigvee_{i \in I} S_0^{m_i+1}$ be a wedge of spheres, with $\iota_i: S_0^{m_i+1} \to W$ denoting the ith inclusion. Let $\alpha_1: S_0^{d_1+1} \to W$, let $r_1 \in \mathbb{Z}$, and let $\beta_1 = \varphi_{\alpha_1}(a) \in \mathcal{A}(W)$. Let $X' = W \cup_{\alpha_1} e^{d_1+1}$, so that $(\mathcal{A}(X'),d_{X'})$ may be described as $R\langle x_i,y_1\rangle$ with $d(x_i) = 0$ but $d(y_1) = \beta_1$. Let $i_1,\ldots,i_t$ be any list of (not necessarily distinct) indices in I, let $\overline{m}_j$ denote m_{i_j}, and put

$m = d_1 + 1 + \overline{m}_1 + \ldots + \overline{m}_t$. Then there is a map
$g_1: (D_0^{m+1}, S_0^m) \to (X', W)$ such that $g_1|_{S_0^m} = \epsilon_1[r_1\alpha_1, \iota_{i_1}, \ldots, \iota_{i_t}]$,
where $\epsilon_1 = \pm 1$. Furthermore, an AH model (φ_1, φ_1) for g_1 may be chosen satisfying $\varphi_1(a) = r_1[\beta_1, x_{i_1}, \ldots, x_{i_t}]$,
$\varphi_1(b) = r_1[y_1, x_{i_1}, \ldots, x_{i_t}]$, and $\varphi_1(a) \in CU_*(\Omega W)$.

Proof. Proof is by induction on t, the length of the list of indices. When $t = 0$, Theorem 3.2 of [2] tells us that the AH model $(\varphi_{\alpha_1}, \varphi_{\alpha_1})$ may be extended over $A(D_0^{d_1+2})$ such that the generators $\underline{a}$ and b are sent to β_1 and to y_1. Composing this with the degree r_1 map from $D_0^{d_1+2}$ to itself gives the map g_1 for which $\varphi_1(a) = r_1\beta_1$ and $\varphi_1(b) = r_1 y_1$.

For the inductive step, suppose the result to be true for the list $i_1, \ldots, i_{t-1}$. Putting $m' = d_1 + 1 + \overline{m}_1 + \ldots + \overline{m}_{t-1}$, the inductive hypothesis gives us a map

$$g': (D_0^{m'+1}, S_0^{m'}) \to (X', W)$$

satisfying $\varphi_{g'}(a') = r_1[\beta_1, x_{i_1}, \ldots, x_{i_t}]$, $\varphi_{g'}(b') = r_1[y_1, x_{i_1}, \ldots, x_{i_t}]$, and $\varphi_{g'}(a') \in CU_*(\Omega W)$. We obtain the map

$$g'' = g' \vee \iota_{\overline{m}_t}: (D_0^{m'+1} \vee S_0^{\overline{m}_t+1}, S_0^{m'} \vee S_0^{\overline{m}_t+1}) \to (X', W)$$

in which we have $\varphi_{g''}(c') = x_{i_t}$, c' being the generator corresponding to $S_0^{\overline{m}_t+1}$. Let f denote the map of Lemma 5.2, where $p = m'$ and $q = \overline{m}_t + 1$, and set $g_1 = f \circ g''$. Then g_1 has the desired properties.

Up to now our choice of ground ring R has been irrelevant, but in the remaining results R plays a key role. In fact, it is best henceforth to assume that all spaces are localized away from a set P of primes, that $R = \mathbb{Z}[P]^{-1}$, and that CW complexes are built out of local spheres and local disks. The Adams-Hilton construction works equally well for local CW complexes, giving a differential graded R-algebra with one generator for each positive-dimensional local cell. We will henceforth interpret S_0^m, D_0^{m+1}, X, etc. as referring to these R-local objects.

Recall from Section 4 the notations L, M, N, A, and W associated with a two-cone X. Let $u \in L_{1*}$; then u is an R-linear combination of repeated brackets of the generators of $L = \mathbb{L}\langle x_i, y_j \rangle$. Because $\deg(u) = 1$, each repeated bracket in this linear combination must have exactly one factor of the form y_j, the remainder being x_i's. Using the Jacobi identities we may therefore write

$$u = \sum_{k=1}^{s} r_k u_k, \tag{14}$$

where $r_k \in R$ and each u_k has the form

$$u_k = [y_j, x_{i_1}, \ldots, x_{i_t}]. \tag{15}$$

<u>Proposition 5.4</u>. Let X be a two-cone and suppose that all the implicit primes are units in R. Let $u \in L_{1m}$ be any cycle, and write $u = \sum_{k=1}^{s} r_k u_k$ as in (14) and (15). Then there exists a map

$$g: S_s^{m+1} \to X$$

which has an AH model φ_g satisfying $\varphi_g(b_k) = r_k u_k$ for $1 \le k \le s$, and $\varphi_g(b_0) \in \mathcal{A}(W)$.

<u>Proof</u>. The requirement in Lemma 5.3 that $r_1 \in \mathbb{Z}$ may be relaxed to $r_1 \in R$ now that $R \subseteq \mathbb{Q}$ and everything is R-local. We may therefore use Lemma 5.3 to construct a map.

$$g' = g_1 \vee g_2 \vee \ldots \vee g_s : X_1 = \bigvee_{k=1}^{s} D_0^{m+1} \to X$$

satisfying $\varphi_{g'}(b_k) = r_k u_k$. Further properties are that

$$g_k \big| S_0^m = \epsilon_k [r_k \alpha_j, \iota_{i_1}, \ldots, \iota_{i_t}], \ \epsilon_k = \pm 1,$$

and that $\varphi_{g'}(a_k) \in CU_*(\Omega W)$. The restriction of g' to $X_0 = \bigvee_{k=1}^{s} S_0^m$ may be viewed as a map $g_0: X_0 \to W$, and $(\varphi_{g'}, \varphi_{g'})$ extends a valid AH model $(\varphi_{g_0}, \varphi_{g_0})$ for g_0.

We observe next that g_0 can also be extended to a map

$$g'': X_2 = (\bigvee_{k=1}^{s} S_0^m) \cup_{\Sigma \kappa_k} e^{m+1} \to W.$$

It will follow at once that there is an AH model for g'' which extends $(\varphi_{g_0}, \varphi_{g_0})$ and has $\varphi_{g''}(b_0) \in \mathcal{A}(W)$, b_0 denoting the m-dimensional generator of $\mathcal{A}(X_2)$. To prove the existence of g'', it suffices to show that $g_{0\#}(\sum_{k=1}^{s} \kappa_k)$ vanishes in $\pi_m(W)$. Let $[f_k] = [r_k \alpha_j, \iota_{i_1}, \ldots, \iota_{i_t}] \in \pi_m(W)$, and let $\epsilon_k = \pm 1$ be chosen so that the adjoint $[\hat{f}_k]$ of the repeated Whitehead product $[f_k]$ equals the repeated Samelson product

$\epsilon_k[r_k\hat{\alpha}_j,\hat{\iota}_{i_1},\ldots,\hat{\iota}_{i_t}]\in\pi_{m-1}(\Omega W)$. Then $g_{0\#}(\kappa_k) = \epsilon_k[f_k]$, so

$$g_{0\#}(\sum_{k=1}^{s}\kappa_k) = \sum_{k=1}^{s}\epsilon_k r_k[\alpha_j,\iota_{i_1},\ldots,\iota_{i_t}]. \tag{16}$$

We need only to verify that the right-hand side of (16), which we henceforth denote by $[f']$, is zero in $\pi_m(W)$.

Because u is a cycle in L_{1m} we have the relation

$$0 = d(u) = \sum_{k=1}^{s} r_k[\beta_j,x_{i_1},\ldots,x_{i_t}] \tag{17}$$

in L_W. Applying the Lie algebra map σ of Section 2 to (17) gives

$$0 = \sum_{k=1}^{s} r_k[\sigma(\beta_j),\sigma(x_{i_1}),\ldots,\sigma(x_{i_t})] = \sum_{k=1}^{s} r_k[\hat{\alpha}_j+\hat{\gamma}_j,\hat{\iota}_{i_1},\ldots,\hat{\iota}_{i_t}]$$

in $\pi_{m-1}(\Omega W)$. Hence

$$\begin{aligned} 0 &= \sum_{k=1}^{s}\epsilon_k r_k[\alpha_j+\gamma_j,\iota_{i_1},\ldots,\iota_{i_t}] \\ &= [f'] + \sum_{k=1}^{s}\epsilon_k r_k[\gamma_j,\iota_{i_1},\ldots,\iota_{i_t}] \end{aligned}$$

in $\pi_m(W)$. Because all the implicit primes are units in R, we have that each $\gamma_j = 0$, hence $[f'] = 0$ as needed. (We remark that this is the only place in the entire paper where we actually use the hypothesis that the implicit primes have been inverted.)

Now g' and g'' are compatible extensions of g_0, so together they define a map $g\colon X_1\cup X_2 = S_s^{m+1} \to X$. Furthermore,

$$\varphi_{g'}\big|_{A(X_0)} = \varphi_{g_0} = \varphi_{g''}\big|_{A(X_0)} \text{ and } \varphi_{g'}\big|_{A(X_0)} = \varphi_0 = \varphi_{g''}\big|_{A(X_0)},$$

so $\varphi_{g'}$ and $\varphi_{g''}$ together define a valid AH model for g. This model φ_g clearly enjoys the requisite properties.

<u>Proof. of Claim 4.7</u>. Let $\bar{u}\in N_{1*} = \mu(M_{1*})$, and let $u\in L_{1*}$ be a cycle representing $\bar{u}$. Compose the maps $\bar{f}$ of (13) and g of Proposition 5.4 to obtain a map

$$g\bar{f}\colon S_0^{m+1} \to X$$

in which $\varphi_{g\bar{f}}(a) = (\sum_{k=1}^{s} r_k u_k) - \varphi_g(b_0) = u - \varphi_g(b_0)$.

Put $w = \varphi_{g\bar{f}}(a)$ and $v = \varphi_g(b_0)$, each of which is a cycle in $A(X)$. It is clear that $\bar{v}\in A_{0*}$ (because $v\in A(W)$) and that $\bar{w}\in\mathrm{im}(h_X)$ and that $\bar{u} = \bar{v}+\bar{w}$ in $A = H_*(\Omega X;R)$.

References

1. J.F. Adams, "On the Groups J(X)-IV", Topology Vol. 5(1966), pp. 21-71.
2. J.F. Adams and P.J. Hilton, "On the Chain Algebra of a Loop Space", Comm. Math. Helv. 30(1955), pp. 305-330.
3. D. Anderson, "Localizing CW Complexes", Ill. J. Math. 16(1972), pp. 519-525.
4. D. Anick, "A Counterexample to a Conjecture of Serre", Annals of Math. 115(1982), pp 1-33.
5. D. Anick, "A Loop Space whose Homology Has Torsion of All Orders", Pac. J. Math. Vol. 123 No 2(1986), pp. 257-262.
6. D. Anick and T. Gulliksen, "Rational Dependence among Hilbert and Poincaré Series", J. Pure and Appl. Alg. Vol. 38(1985), pp. 135-157.
7. L. Avramov, "Torsion in Loop Space Homology", Topology Vol. 25 No. 2(1986), pp. 155-157.
8. H.J. Baues, "Rationale Homotopietypen", Manuscr. Math. Vol. 20 Fasc. 2(1977), pp. 119-131.
9. H.J. Baues, Commutator Calculus and Groups of Homotopy Classes, London Math. Soc. Lecture Note Series 50, Cambridge Univ. Press, 1981.
10. H.J. Baues and J.-M. Lemaire, "Minimal Models in Homotopy Theory", Math. Ann. 225(1977), pp. 219-242.
11. F.R. Cohen, J.C. Moore, and J.A. Neisendorfer, "Torsion in Homotopy Groups", Annals of Math. 109(1979), pp. 121-168.
12. F.R. Cohen, J.C. Moore, and J.A. Neisendorfer, "The Double Suspension and Exponents of the Homotopy Groups of Spheres", Annals of Math. 110(1979), pp. 549-565.
13. Y. Felix and S. Halperin, "Rational L.-S. Category and its Applications", Trans. A.M.S. Vol. 273 No. 1(1982), pp. 1-37.
14. Felix and J.-C. Thomas, "Sur la Structure des Espaces de L.S. Catégorie Deux", Ill. J. Math., Vol. 30 No. 4(1986), pp. 574-593.
15. B. Gray, "On the Sphere of Origin of Infinite Families in the Homotopy Groups of Spheres", Topology Vol. 8(1969), pp. 219-232.
16. S. Halperin, "Finiteness in the Minimal Models of Sullivan", Trans. A.M.S. Vol. 230(1977), pp. 173-199.
17. P. J. Hilton, "On the Homotopy Groups of the Union of

Spheres", J. London Math. Soc. Vol.30(1955), pp. 154-172.

18. I.M. James, "Reduced Product Spaces, Annals of Math. 62(1955), pp. 170-197.

19. J.-M. Lemaire, Algèbres Connexes et Homologie des Espaces de Lacets, Lecture Notes in Math. 422, Springer-Verlag, 1974.

20. J.-M. Lemaire, "Anneaux locaux et espaces de lacets à séries de Poincaré irrationnelles", Lecture Notes in Math. 901, Springer-Verlag, pp. 149-156.

21. C.A. McGibbon and C.W. Wilkerson, "Loop Spaces of Finite Complexes at Large Primes", Proc. A.M.S. Vol. 96 No. 4(1986), pp. 698-702.

22. J. Milnor and J.C. Moore, "On the Structure of Hopf Algebras", Annals of Math. 81(1965), pp. 211-264.

23. J.A. Neisendorfer and P. Selick, "Some Examples of Spaces with or without Exponents", Canad. Math. Soc. Conf. Proceedings Vol. 2, Part 1(1982), pp. 343-357.

24. D.G. Quillen, "Rational Homotopy Theory", Annals of Math. 90(1969), pp. 205-295.

25. H. Samelson, "A Connection between the Whitehead and the Pontrjagin Product", Amer. J. Math. 75(1953), pp. 744-752.

26. P. Selick, "On Conjectures of Moore and Serre in the Case of Torsion-Free Suspensions", Math. Proc. Camb. Phil. Soc. 94(1983), pp. 53-60.

27. J.-P. Serre, "Groupes d'homotopie et classes de groupes abeliens", Annals of Math. 58(1953), pp. 258-294.

28. G.W. Whitehead, "The Homology Suspension", Colloque de Topologie Algébrique, tenu à Louvain, juin 1956, Masson, Paris, 1957.

29. G.W. Whitehead, Elements of Homotopy Theory, Springer-Verlag, New York, 1978.

On the complex bordism of classifying spaces

MARTIN BENDERSKY
DONALD M. DAVIS

In this short note we make several observations regarding the following conjecture of Gilkey. ([G])

CONJECTURE 1. *If G is any spherical space-form group, there is an isomorphism of graded abelian groups*

$$MU_*(BG) \approx ku_*(BG) \otimes \mathbf{Z}[x_{2i} : i \geq 2],$$

where x_{2i} is a generator of grading $2i$ in a polynomial algebra.

Here $MU_*(\ \ )$ is complex bordism, and $ku_*(\ \ )$ is connective complex K-theory. It is well known ([CF1]) that $MU_i(BG)$ is isomorphic to the group of bordism classes of i-dimensional stably almost complex manifolds with free G-action.

In an earlier version of this paper, we gave a rather computational proof of

THEOREM 2. *Conjecture 1 is true if G is a cyclic group.*

After submitting the paper, we discovered that the bulk of our work had been done much earlier by Flynn. ([F]) Indeed, a comparison of his result for the graded abelian group structure of $MU_*(B\mathbf{Z}/p^r)$ with that of [H] for $ku_*(B\mathbf{Z}/p^r)$ establishes Theorem 2.

Although a nice generator-and-relation description of these groups is well-known (see (3) below), the explicit abelian group structure is rather complicated. Indeed,

$$ku_*(B\mathbf{Z}/p^r) \approx \bigoplus_{s=0}^{r-1} \bigoplus_{j=p^s}^{p^{s+1}-1} \bigoplus_{m=0}^{\infty} \Sigma^{2j+2m-1}\mathbf{Z}/p^{r-s+[m/(p-1)p^s]},$$

where $[\ \ ]$ denotes the greatest integer function, and $\Sigma^n G$ is the graded abelian group whose only nonzero group is G in grading n. A noncomputational proof of Theorem 2, due to Tony Bahri, can be given as follows.

Let $L^n = L^n(p^r)$ denote the lens space which is the $(2n+1)$-skeleton of $B\mathbf{Z}/p^r$. By Poincare duality,

$$MU_{2n-1}(B\mathbf{Z}/p^r) \approx MU_{2n-1}(L^n) \approx MU^2(L^n)$$
$$\text{and} \quad ku_{2n-1}(B\mathbf{Z}/p^r) \approx ku^2(L^n) \approx \widetilde{K}(L^n).$$

Let η denote the canonical line bundle over L^n, and $z = c_1(\eta)$ its first Conner-Floyd class. By the Gysin sequence, there is an isomorphism in even dimensions

(3) $$MU^{ev}(L^n) \approx MU^*[z]/([p^r](z), z^{n+1}),$$

The second author was supported by a National Science Foundation research grant

where $[p^r](z) = c_1(\eta^{p^r})$, and the coefficient ring $MU^* \approx \mathbf{Z}[x_{2i} : i \geq 1]$. Note that x_{2i} has degree $2i$ when thought of as a homology class, and degree $-2i$ when thought of as a cohomology class.

Let S denote the set of finite sets of integers ≥ 2 (including the empty set). If $I = \{i_j\} \in S$, let $|I| = \sum i_j$ and $x_I = \prod x_{2i_j} \in MU^*$. To prove Theorem 2, we will show the following homomorphism is bijective.

$$\phi_n : \bigoplus_{I \in S} \widetilde{K}(L^{n-|I|}) \to MU^2(L^n)$$

$$\phi_n(\theta, I) = x_I c_1(\theta) z^{|I|}.$$

Here (θ, I) denotes $\theta \in \widetilde{K}(L^{n-|I|})$.

We will prove ϕ_n is surjective, by induction on n. Then an easy counting argument, using the Atiyah-Hirzebruch spectral sequence, shows the groups have the same order, and so ϕ_n is bijective.

In the commutative diagram

$$\begin{array}{ccc} \bigoplus_I \widetilde{K}(L^{n-|I|}) & \xrightarrow{\phi_n} & MU^2(L^n) \\ \Big\downarrow \bigoplus i^*_{n-|I|} & & \Big\downarrow i^*_n \\ \bigoplus_I \widetilde{K}(L^{n-1-|I|}) & \xrightarrow{\phi_{n-1}} & MU^2(L^{n-1}) \end{array}$$

the epimorphisms i^*_j are induced by the inclusion maps. We show $\ker(i^*_n) \subset \operatorname{im}(\phi_n)$, from which surjectivity of ϕ_n follows by diagram chasing. By (3), $\ker(i^*_n)$ is spanned by elements of the form $x_1^{n-|I|-1} x_I z^n = x_I z^{|I|} \cdot x_1^{n-|I|-1} z^{n-|I|}$. If I is nonempty, then by induction $x_1^{n-|I|-1} z^{n-|I|}$ can be written as $\sum_J a_J x_J c_1(\theta_{n-|I|-|J|}) z^{|J|}$ with $a_J \in \mathbf{Z}$, and so

$$x_I z^{|I|} \cdot x_1^{n-|I|-1} z^{n-|I|} = \sum_J \phi_n(a_J \theta_{n-|I|-|J|}, I \cup J).$$

This leaves $x_1^{n-1} z^n$, which is in $\operatorname{im}(\phi_n)$ since $x_1^{n-1} z^n + c_1(\eta^n)$ is in the kernel of the Conner-Floyd map ([**CF2**]) $\mu : MU^2(L^n) \to \widetilde{K}(L^n)$, and this kernel is spanned by elements divisible by x_i for $i > 1$, which have already been shown to be in $\operatorname{im}(\phi_n)$. ∎

To prove Conjecture 1, it is equivalent to prove it after localizing at an arbitrary prime p. Hence using Quillen's splitting of $MU_{(p)}$ as a wedge of Brown-Peterson spectra BP ([**Q**]) and Adams' ([**A**]) splitting of $bu_{(p)}$ as a wedge of copies of a spectrum often called $BP\langle 1\rangle$ or ℓ (or G in [**J**]), it is equivalent to prove

$$BP_*(BG) \approx BP\langle 1\rangle_*(BG) \otimes \mathbf{Z}_{(p)}[v_i : i \geq 2],$$

where v_i has degree $2(p^i - 1)$. This can be studied via the spectral sequence

$$BP\langle 1\rangle_*(X) \otimes \mathbf{Z}_{(p)}[v_i : i \geq 2] \Longrightarrow BP_*(X) \tag{4}$$

introduced in [**J**].

We recall the following result from [**CE**] and [**L**].

THEOREM 5. *The following are equivalent for a finite group G.*

(i) *Every abelian subgroup of G is cyclic;*
(ii) *Every Sylow subgroup of G is cyclic or generalized quaternionic;*
(iii) *G has periodic cohomology;*
(iv) $H_n(BG;\mathbf{Z}) = 0$ *for all even* $n > 0$;
(v) *The Atiyah-Hirzebruch spectral sequence* $H^*(BG;\mathbf{Z}) \Rightarrow K^*(BG)$ *collapses;*
(vi) $MU_*(BG) \to H_*(BG;\mathbf{Z})$ *is surjective;*
(vii) $\text{hom}\,\dim_{MU_*}(MU_*BG) \leq 1$.

For all such groups, the spectral sequence (4) will collapse for dimensional reasons. The question is whether all extensions are trivial in the spectral sequence.

The group $G = \mathbf{Z}_2 \times \mathbf{Z}_2$ does not satisfy the conditions of Theorem 5, and it does not satisfy Conjecture 1. However, the spectral sequence (4) with $X = B(\mathbf{Z}_2 \times \mathbf{Z}_2)$ does collapse. It is nontrivial extensions in the spectral sequence (4) which cause its failure to satisfy Conjecture 1, and differentials in the spectral sequence of Theorem 5(v) which cause its failure to satisfy the conditions of Theorem 5.

Noting that $\Sigma B(\mathbf{Z}_2 \times \mathbf{Z}_2) \simeq \Sigma(B\mathbf{Z}_2 \wedge B\mathbf{Z}_2) \vee \Sigma B\mathbf{Z}_2 \vee \Sigma B\mathbf{Z}_2$, we sketch below a portion of the spectral sequence of (4) for $X = B\mathbf{Z}_2 \wedge B\mathbf{Z}_2$, indicating in position (i,j) the group $ku_i(X) \otimes P_j$, where P_j denotes the grading j part of the polynomial algebra $\mathbf{Z}_{(2)}[v_i : i \geq 2]$. Dots are $\mathbf{Z}_2$'s and a number k indicates $\mathbf{Z}/2^k$. Diagonal lines indicate nontrivial extensions in passing from $\oplus E^{\infty}_{i,n-i}$ to $BP_n(X)$. That these are present is a consequence of [D] or [JW1].

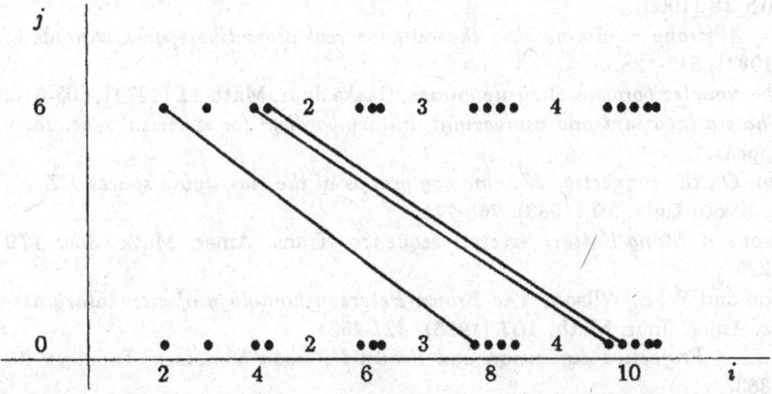

Thus, for example, $BP_8(B\mathbf{Z}_2 \wedge B\mathbf{Z}_2) \approx \mathbf{Z}_2 \oplus \mathbf{Z}_2 \oplus \mathbf{Z}_2 \oplus \mathbf{Z}_4$, while $ku_8(B\mathbf{Z}_2 \wedge B\mathbf{Z}_2) \approx \mathbf{Z}_2 \oplus \mathbf{Z}_2 \oplus \mathbf{Z}_2 \oplus \mathbf{Z}_2$.

Work in progress with Bahri and Gilkey suggests that Conjecture 1 is true for all groups satisfying the conditions of Theorem 5. This includes all space-form groups and others ([W]). The idea of a proof is to verify it for cyclic groups (Theorem 2) and generalized quaternionic groups, and then use a transfer argument to pass from the validity for Sylow subgroups to validity for G.

Condition (vii) of Theorem 5 suggests that Conjecture 1 should perhaps be extended to the class of spaces X for which $\operatorname{hom}\dim_{MU_*} MU_*(X) \leq 1$. We see no way to attack such a conjecture at the moment. Such a conjecture would be the case $n = 1$ of a broader conjecture:

There is an isomorphism of graded abelian groups

$$BP_*(X) \approx BP\langle n\rangle_*(X) \otimes \mathbf{Z}_{(p)}[v_i : i > n]$$

$$\text{if } \operatorname{hom}\dim_{BP_*} BP_*(X) \leq n.$$

Here $BP\langle n\rangle$ are the spectra studied in [**JW2**]. Their characterization that

$$\operatorname{hom}\dim_{BP_*}(BP_*X) \leq n \text{ iff } v_n \text{ acts injectively on } BP\langle n\rangle_*(X)$$

seems somewhat close to our conjecture, since in our example $X = B\mathbf{Z}_2 \wedge B\mathbf{Z}_2$ the nontrivial extension on the element in (8,0) in the spectral sequence could have been subsumed by a nontrivial v_1 action on the element in (6,0).

References

A. J. F. Adams, *Lectures on generalized cohomology*, Springer Verlag Lecture Notes in Math **99** (1969), 1-138.

CE. H. Cartan and S. Eilenberg, "Homological algebra," Princeton Univ. Press, 1956.

CF1. P. E. Conner and E. E. Floyd, *Periodic maps which preserve a complex structure*, Bull Amer. Math. Soc. **70** (1964), 574-579.

CF2. ———, *The relation of cobordism to K-theories*, Springer Verlag Lecture Notes in Math **28** (1966).

D. D. M. Davis, *A strong nonimmersion theorem for real projective spaces*, Annals of Math **120** (1984), 517-528.

F. T. Flynn, *The complex bordism of cyclic groups*, Osaka Jour. Math **11** (1974), 503-516.

G. P. Gilkey, *The eta invariant and equivariant unitary bordism for spherical space form groups*, to appear.

H. S. Hashimoto, *On the connective K-homology groups of the classifying spaces $B\mathbf{Z}/p^r$*, Publ. RIMS, Kyoto Univ. **19** (1983), 765-771.

J. D. C. Johnson, *A Stong-Hattori spectral sequence*, Trans. Amer. Math. Soc. **179** (1973), 211-225.

JW1. D. C. Johnson and W. S. Wilson, *The Brown-Peterson homology of elementary abelian p-groups*, Amer. Jour. Math. **107** (1985), 427-453.

JW2. ———, *Projective dimension and Brown-Peterson homology*, Topology **12** (1973), 327-353.

L. P. S. Landweber, *Complex bordism of classifying spaces*, Proc. Amer. Math. Soc. **27** (1971), 175-179.

Q. D. G. Quillen, *On the formal groups laws of unoriented and oriented cobordism*, Bull. Amer. Math. Soc. **75** (1969), 1293-1298.

W. J. A. Wolf, "Spaces of constant curvature," McGraw-Hill, 1967.

Rider College, Lawrenceville, NJ 08648

Lehigh University, Bethlehem, Pa. 18015

ON EQUIVARIANT MAPS AND IMMERSIONS OF REAL PROJECTIVE SPACES

by

A. J. Berrick and Duane Randall

National University of Singapore
Kent Ridge 0511, Singapore

Pontificia Universidade Catolica
Rio de Janeiro, Brazil

Let $V_k(\mathbf{R}^m)$ denote the Stiefel manifold of orthonormal k-frames in Euclidean m-space. The antipodal involution σ on $V_k(\mathbf{R}^m)$ given by $\sigma(v_1,\ldots,v_k) = (-v_1,\ldots,-v_k)$ generalizes the antipodal map on spheres. A classical construction [3] establishes a bijection between the set of maps $f\colon S^n \to V_k(\mathbf{R}^m)$ equivariant with respect to σ and the set of sections in the bundle of k-frames associated to $m\xi_n$ where ξ_n denotes the Hopf line bundle over n-dimensional real projective space P^n. In particular, P^n immerses differentiably in $\mathbf{R}^{n+k}$ for $k > 0$ if and only if there exists a map $f\colon S^n \to V_{n+1}(\mathbf{R}^{n+k+1})$ equivariant with respect to σ. The homotopy class of an equivariant map $f\colon S^n \to V_k(\mathbf{R}^m)$ yields an obstruction in a certain subquotient $\pi_{n,(m,k)}$ of $\pi_n(V_k(\mathbf{R}^m))$ to the existence of an equivariant map $g\colon S^{n+1} \to V_k(\mathbf{R}^m)$ which coincides with f on S^{n-1}. The first author determined $\pi_{n,(m,k)}$ explicitly in [1]. In this note we interpret $\pi_{n,(m,k)}$ in terms of indices of vector bundle monomorphisms with finite singularities and use it to derive some immersions of projective spaces in Euclidean space. We also consider the representability of the stable homotopy of spheres by immersed projective spaces.

The singularity method developed extensively by Koschorke relates the indices of vector bundle monomorphisms with finite singularities over manifolds to the normal bordism obstruction to the existence of a vector bundle *monomorphism*. (See Chapter 13 of [7].) Propositions 1.2 and 1.5 of this paper present not merely alternative proofs for results in [1], but also make important identifications with equivariant maps for applications of the singularity method to the immersion problem for projective spaces.

1. Equivariant Maps and Indices

Let $\mathcal{R}^k$ denote the trivial k-dimensional vector bundle. We shall first show how to make an identification between (i) indices of vector bundle monomorphisms $\mathcal{R}^k \to m\xi_n$ with finite singularities and (ii) homotopy classes containing equivariant maps $S^{n-1} \to V_k(\mathbf{R}^m)$.

Let $W_k(m\xi_n)$ denote the bundle of nonsingular k-frames associated to $m\xi_n$ with fiber $W_k(\mathbf{R}^m)$, the Stiefel manifold of nonsingular k-frames in $\mathbf{R}^m$. Let x_0 denote the image of $\pm e_{n+1}$ under the double covering $S^n \to P^n$. Given an equivariant map $f\colon S^{n-1} \to V_k(\mathbf{R}^m)$, we shall define a section s_f of the bundle of nonsingular k-frames associated to $m\xi_n$ over $P^n - \{x_0\}$ such that the index of

1980 Mathematics Subject Classification: Primary 57R42, 55R25.
The second author was supported by $FINEP$ and CNP_q of Brazil.

s_f, ind s_f, in $\pi_{n-1}(V_k(\mathbf{R}^m))$ is $[f]$. We first define $g\colon S^n - \{\pm e_{n+1}\} \to W_k(\mathbf{R}^m)$ as follows. Given (x,t) in $S^n \subset \mathbf{R}^n \times \mathbf{R}^1$, set

$$g(x,t) = \begin{cases} (1-t)f\left(\frac{x}{\|x\|}\right) & \text{for } 0 \le t < 1 \\ (1+t)f\left(\frac{x}{\|x\|}\right) & \text{for } -1 < t \le 0 \end{cases} \tag{1.1}$$

Clearly g extends f on S^{n-1} and $g(-x,-t) = -g(x,t)$. We define the section

$$s_f : P^n - \{x_0\} \to S^n \times_{\mathbf{Z}/2} W_k(\mathbf{R}^m)$$

by $s_f([x,t]) = [(x,t), g(x,t)] = [(-x,-t), g(-x,-t)]$. Now ind s_f in $\pi_{n-1}(V_k(\mathbf{R}^m))$ is just the local obstruction at x_0, which is clearly $[f]$. (We identify the homotopy of $V_k(\mathbf{R}^m)$ and $W_k(\mathbf{R}^m)$ via the inclusion and an orthonormalization map $\Gamma\colon W_k(\mathbf{R}^m) \to V_k(\mathbf{R}^m)$.)

Suppose that α in $\pi_{n-1}(V_k(\mathbf{R}^m))$ occurs as the index for some bundle monomorphism $\mathcal{R}^k \to m\xi_n$ with finite singularities. We may choose a section with a unique singularity

$$s\colon P^n - \{x_0\} \to S^n \times_{\mathbf{Z}/2} W_k(\mathbf{R}^m)$$

whose index is α. Now s determines an equivariant map $g\colon S^n - \{\pm e_{n+1}\} \to W_k(\mathbf{R}^m)$ by $s([x,t]) = [(x,t), g(x,t)]$. Further, the index of s is $[\Gamma \circ g]$ with orientations properly chosen. Thus we have proved the following.

PROPOSITION 1.2. *The elements of $\pi_{n-1}(V_k(\mathbf{R}^m))$ which occur as indices of vector bundle monomorphisms $\mathcal{R}^k \to m\xi_n$ with finite singularities are precisely those elements represented by equivariant maps.*

We shall now derive Lemma 6.5 of [1]. Given an equivariant map $f\colon S^{n-1} \to V_k(\mathbf{R}^m)$, the map g in (1.1) yields a section

$$s_g\colon S^n - \{\pm e_{n+1}\} \to S^n \times W_k(\mathbf{R}^m)$$

of the bundle $W_k(\mathbf{R}^m)$ on S^n. Clearly the local obstruction at e_{n+1} is $[f]$. Now the local obstruction at $-e_{n+1}$ can be expressed in terms of the index at e_{n+1} by $[\sigma \circ g \circ A] = (-1)^{n+1}\sigma_*[f]$ where A is the antipodal map on S^n. So ind $s_g = [f] + (-1)^{n+1}\sigma_*[f]$. But ind $s_g = 0$ by uniqueness of indices on S^n for the trivial bundle $\mathcal{R}^m$. So we conclude that $[f]$ belongs to

$$\operatorname{Ker}(1-\sigma_*) \text{ for } n \text{ even and } \operatorname{Ker}(1+\sigma_*) \text{ for } n \text{ odd.} \tag{1.3}$$

We recall from [6] that the canonical automorphism μ of the Stiefel manifolds comprises the involutions which change the sign of any fixed row of an $(m \times k)$-matrix. Now $\sigma = \mu^m$ and $\mu_* = 1 - \psi$ on $\pi_{n-1}(V_k(\mathbf{R}^m))$ where ψ denotes the composite morphism $i_* \circ \Sigma \circ \partial$

$$\pi_{n-1}\, V_k(\mathbf{R}^m) \xrightarrow{\partial} \pi_{n-2}(S^{m-k-1}) \xrightarrow{\Sigma} \pi_{n-1}(S^{m-k}) \xrightarrow{i_*} \pi_{n-1}\, V_k(\mathbf{R}^m)\,.$$

Here ∂ denotes the boundary operator for the fibration $V_{k+1}(\mathbf{R}^m) \to V_k(\mathbf{R}^m)$, Σ denotes the suspension morphism, and i denotes the fiber inclusion in the fibration $S^{m-k} \subset V_k(\mathbf{R}^m) \to V_{k-1}(\mathbf{R}^m)$.

LEMMA 1.4. *The indeterminacy subgroup in $\pi_{n-1}(V_k(\mathbf{R}^m))$ for indices associated to vector bundle monomorphisms $\mathcal{R}^k \to m\xi_n$ with finite singularities is given by*

n even		*n odd*
$\mathrm{Im}(2-\psi)$	*mk odd*	$\mathrm{Im}\,\psi$
$2\pi_{n-1}(V_k(\mathbf{R}^m))$	*mk even*	0

Proof. As an obstruction cohomology class, the index of a bundle monomorphism $\mathcal{R}^k \to m\xi_n$ with finite singularities belongs to $H^n(P^n;\ \pi_{n-1}(V_k(\mathbf{R}^m))_{w_1})$. The local coefficient system $\pi_{n-1}(V_k(\mathbf{R}^m))_{w_1}$ is twisted if and only if $w_1(m\xi_n)$ is nonzero, that is, if and only if m is odd. Every orientation-reversing loop in the base of a nonorientable m-plane bundle acts by μ_* on $\pi_*(V_k(\mathbf{R}^m))$ in the associated bundle of k-frames, since the action of the generator of $\pi_1(BO(m))$ on the homotopy of the fiber in the fibration $V_k(\mathbf{R}^m) \subset BO(m-k) \to BO(m)$ is induced through matrix multiplication on the left by a non-rotation by [11, p.306]. Let α generate $\pi_1(P^n)$. Thus $\alpha_*(\beta) = \mu_*(\beta)$ for all β in $\pi_{n-1}(V_k(\mathbf{R}^m))$ for m odd while $\alpha_*(\beta) = \beta$ for m even. Let $\mathbf{Z}_{w_1(P^n)}$ denote the local coefficient system of integers twisted by $w_1(P^n)$. That is, $\alpha_*(1) = \mu_*(1) = -1$ for n even while $\alpha_*(1) = 1$ for n odd. By Poincaré duality and evaluation [11, p.275] we obtain the isomorphisms

$$\begin{aligned} H^n\left(P^n;\ \pi_{n-1}(V_k(\mathbf{R}^m))_{w_1}\right) &\cong H_0\left(P^n;\ \pi_{n-1}(V_k(\mathbf{R}^m))_{w_1} \otimes \mathbf{Z}_{w_1(P^n)}\right) \\ &\cong \pi_{n-1}(V_k(\mathbf{R}^m)) \otimes \mathbf{Z}/G. \end{aligned}$$

The indeterminacy subgroup G for indices is the subgroup generated by $\beta \otimes 1 - \alpha_*(\beta \otimes 1)$ for all β in $\pi_{n-1}(V_k(\mathbf{R}^m))$. The action of α on $\beta \otimes 1$ is given by

$\mu_*(\beta) \otimes -1$	for m odd, n even	$\beta \otimes -1$	for m even, n even
$\mu_*(\beta) \otimes\ 1$	for m odd, n odd	$\beta \otimes\ 1$	for m even, n odd

Thus the indeterminacy subgroup G is given by $2\pi_{n-1}(V_k(\mathbf{R}^m))$ for m and n even, and is trivial for m even and n odd. For m odd and k even, the morphism $1+\mu_*$ is multiplication by 2 while $1-\mu_*$ is trivial. For mk odd, $1+\mu_* = 2-\psi$ and $1-\mu_* = \psi$. The result follows.

The proof of the following proposition follows immediately from Proposition 1.2, Lemma 1.4 and (1.3). We emphasize the identification of the classical obstruction group for this problem with the given subquotients of homotopy groups of Stiefel manifolds in our proof of this result, which is essentially Lemma 6.5 of [1].

PROPOSITION 1.5 (Berrick [1]). *Let $s: \mathcal{R}^k \to m\xi_n$ denote any vector bundle monomorphism with finite singularities on P^n. Evaluation of the associated classical cohomology obstruction on the fundamental homology class of P^n (twisted integers for n even) yields an element* Ind s *in the vector space $\pi_{n-1,(m,k)}$ over $\mathbf{Z}/2$ defined below.*

	$n-1$ *odd*		$n-1$ *even*
	Ker $\psi/\mathrm{Im}(2-\psi)$	*mk odd*	$\mathrm{Ker}(2-\psi)/\mathrm{Im}\psi$
	$\pi_{n-1}(V_k(\mathbf{R}^m))/2$	*mk even*	$\mathrm{Ker}(\cdot 2: \pi_{n-1}(V_k(\mathbf{R}^m)) \supset)$.

Now Ind $s = 0$ in $\pi_{n-1,(m,k)}$ if and only if there exists a bundle *monomorphism* $\mathcal{R}^k \to m\xi_n$ which agrees with s on P^{n-2}. Equivalently, given any equivariant map $f: S^{n-1} \to V_k(\mathbf{R}^m)$, there exists an equivariant map $g: S^n \to V_k(\mathbf{R}^m)$ which agrees with f on S^{n-2} if and only if [f] projects trivially into $\pi_{n-1,(m,k)}$.

2. Generalised J-Morphism and Representability

In this section we consider briefly the respresentability of elements of the stable homotopy of spheres through immersions of projective spaces in Euclidean space. We recall from [2] that α in π_k^s is represented by an immersion $P^n \subseteq \mathbf{R}^{2n-k}$ if $\alpha = J([f])$ where $f: S^n \to V_{n+1}(\mathbf{R}^{2n-k+1})$ denotes the Smale invariant of the immersion and $J: \pi_n(V_{n+1}(\mathbf{R}^{2n-k+1})) \to \pi_{2n+1}(S^{2n-k+1})$ denotes the generalized J-morphism. The following lemma is an immediate generalization of Proposition 1.1. of [10] and so the proof is omitted. The condition $a+b < 2c$, which seems necessary in (1.1) of [10] also, insures that a self-map of degree -1 on S^c will induce multiplication by -1 on $\pi_{a+b-1}(S^c)$.

LEMMA 2.1. *Let $H(f): S^{a+b-1} \to S^c$ denote the Hopf construction on a nonsingular biskew map $f: \mathbf{R}^a \times \mathbf{R}^b \to \mathbf{R}^c$. If $a+b < 2c$ and either $a+c$ or $b+c$ is odd, then $[H(f)]$ has order at most 2.*

The following result gives us the opportunity to correct a parity error in Theorem 4(b)(ii) of [2], and at the same time clarify the argument used there.

THEOREM 2.2. *Suppose α in π_k^s is represented by some immersion $P^n \subseteq \mathbf{R}^{2n-k}$. If either $n+k$ is odd or else the immersion is self-adjoint with nk odd, then $2\alpha = 0$.*

Proof. By (1.5) the Smale invariant of the immersion has order ≤ 2 in $\pi_n(V_{n+1}(\mathbf{R}^{2n-k+1}))$ for n even and k odd. The result for n odd and k even follows by Lemma 2.1, since the Smale invariant yields a nonsingular biskew map $\mathbf{R}^{n+1} \times \mathbf{R}^{n+1} \to \mathbf{R}^{2n-k+1}$.

Next suppose nk odd and the immersion self-adjoint. Now the Smale invariant $f: S^n \to V_{n+1}(\mathbf{R}^{2n-k+1})$ produces a biskew map $\hat{f}: S^n \times S^n \to S^{2n-k}$ such that $\alpha = J([f])$ is represented by the Hopf construction $H(\hat{f})$. Consider the following diagram.

$$\begin{array}{ccc} S^n * S^n & \xrightarrow{\overline{T}} & S^n * S^n \\ \downarrow H(\hat{f}) & & \downarrow H(\hat{f}) \\ S^{2n-k+1} & \xrightarrow{r_1} & S^{2n-k+1} \end{array}$$

For x and y in S^n and $\theta \in [0, \pi/2]$, $\overline{T}(x\cos\theta,\ y\sin\theta) = (y\sin\theta,\ x\cos\theta)$ while $H(\hat{f})(x\cos\theta, y\sin\theta) = (\cos 2\theta,\ \hat{f}(x,y)\sin 2\theta)$ and $r_1(z_1, z_2, \cdots, z_{2n-k+2}) = (-z_1, z_2, \cdots, z_{2n-k+2})$. The involution T interchanging factors on $S^n \times S^n$ is homotopic to $\hat{f} \circ T$ by the hypothesis of self-adjointness. This makes the above diagram homotopy commutative. The result is immediate, since $\overline{T}$ and r_1 have degrees $(-1)^{n+1}$ and -1 respectively.

3. Immersing Projective Spaces

THEOREM 3.1. *Suppose that P^n immerses in $\mathbf{R}^{2n-12}$ for $n \equiv 4(8)$. Then $P^{n+1} \subseteq \mathbf{R}^{2n-12}$. In addition, P^{n+2} immerses in $\mathbf{R}^{2n-11}$ for $n \equiv 12(16)$.*

Proof. By hypothesis there exists an equivariant map $f: S^n \to V_{n+1}(\mathbf{R}^{2n-11})$ with $n \equiv 4(8)$. We remark that $n \geq 60$ from known results in low dimensions. Since the 2-primary components of $\pi_n(V_{n+1}(\mathbf{R}^{2n-11}))$ and $\pi_n(V_{n+2}(\mathbf{R}^{2n-11}))$ are trivial by [9], there exists an immersion $P^{n+1} \subseteq \mathbf{R}^{2n-12}$ by [1, Prop. 6.6].

Suppose now that $n \equiv 12(16)$. Since $\pi_{n+1}(V_{n+2}(\mathbf{R}^{2n-11})) = 0$ by [9], the Smale invariant of any immersion $P^{n+1} \subseteq \mathbf{R}^{2n-12}$ extends to an equivariant map $S^{n+2} \to V_{n+2}(\mathbf{R}^{2n-11})$. We recall from [5, Theorem 1.2] that an equivariant map $S^i \to V_k(\mathbf{R}^m)$ produces an equivariant map $S^{k-1} \to V_{i+1}(\mathbf{R}^m)$ provided $k+2 \leq 2(m-i)$. Thus we obtain an equivariant map $g: S^{n+1} \to V_{n+3}(\mathbf{R}^{2n-11})$. Finally, the equivariant map $j \circ g$ where $j: V_{n+3}(\mathbf{R}^{2n-11}) \hookrightarrow V_{n+3}(\mathbf{R}^{2n-10})$ denotes the natural inclusion extends to an equivariant map $S^{n+2} \to V_{n+3}(\mathbf{R}^{2n-10})$ because $\pi_{n+1}(V_{n+3}(\mathbf{R}^{2n-10})) = 0$ by [9].

Let $\alpha(n)$ and $\nu(n)$ denote the number of 1s in the dyadic expansion of n and the highest power of 2 dividing n respectively. Now Davis proved in [4] that P^n immerses in $\mathbf{R}^{2n-12}$ for $n \equiv 12(16)$ with $\alpha(n) \geq 5$ and $\nu(n+4) < 7$. Thus we obtain the following.

COROLLARY 3.2. *$P^n \subseteq \mathbf{R}^{2n-14}$ for $n \equiv 13(16)$ with $\alpha(n) \geq 6$ and $\nu(n+3) < 7$. $P^n \subseteq \mathbf{R}^{2n-15}$ for $n \equiv 14(16)$ with $\alpha(n) \geq 6$ and $\nu(n+2) < 7$.*

The above results improve those of [1] by 2 and 1 dimensions respectively. For example, $P^{189} \subseteq \mathbf{R}^{364}$ by (3.2) whereas $P^{189} \subseteq \mathbf{R}^{366}$ by [1] and $P^{189} \not\subseteq \mathbf{R}^{359}$ by [4]. Further $P^{190} \subseteq \mathbf{R}^{365}$ by (3.2) while $P^{190} \subseteq \mathbf{R}^{366}$ by [1] and $P^{190} \not\subseteq \mathbf{R}^{360}$ from the table of [4].

REFERENCES

[1] A. J. Berrick, The Smale invariants of an immersed projective space, *Math. Proc. Camb. Phil. Soc.* **86** (1979), 401-411.

[2] A. J. Berrick, Consequences of the Kahn-Priddy Theorem in homotopy and geometry, *Mathematika* **28** (1981), 72-78.

[3] P. E. Conner and E. E. Floyd, Fixed point free involutions and equivariant maps, *Bull. Amer. Math. Soc.* **66** (1960), 416-441.

[4] D. M. Davis, Some new immersions and nonimmersions of real projective spaces, *Proceedings of Homotopy Theory Conference,* Contemp. Math. 19, Amer. Math. Soc. (1983), 51-64.

[5] A. Haefliger and M. W. Hirsch, Immersions in the stable range, *Ann. of Math.* **75** (1962), 231-241.

[6] I. M. James, *The Topology of Stiefel Manifolds*, London Math. Soc. Lecture Note Series no. 24, Cambridge Univ. Press, 1976.

[7] U. Koschorke, *Vector Fields and Other Vector Morphisms - a Singularity Approach,* Springer-Verlag Lecture Notes no. 847, 1981.

[8] K. Y. Lam, Nonsingular bilinear maps and stable homotopy classes of spheres, *Math. Proc. Camb. Phil. Soc.* **82** (1977), 419-425.

[9] M. E. Mahowald, *The Metastable Homotopy of S^n*, Mem. Amer. Math. Soc. **72** (1967).

[10] L. Smith, Nonsingular bilinear forms, generalized J homomorphisms, and the homotopy of spheres I, *Indiana Univ. Math. J.* **27** (1978), 697-737.

[11] G. W. Whitehead, *Elements of Homotopy Theory,* Springer-Verlag, 1978.

COGROUPS WHICH ARE NOT SUSPENSIONS

By

Israel Berstein and John R. Harper[1]

In memory of

Tudor Ganea (1922-1971)

In one of his last papers, Ganea makes a fundamental study of associativity, in the sense $(1 \vee \sigma) \circ \sigma \simeq (\sigma \vee 1) \circ \sigma$ for comultiplications σ on a space X, [G]. A co-H-space X is called a cogroup if it supports an associative comultiplication σ and an inversion $\eta : X \to X$, namely a map such that both composites $\nabla \circ (1 \vee \eta) \circ \sigma$ and $\nabla \circ (\eta \vee 1) \circ \sigma$ are null-homotopic. It is well known that any simply connected co-H-space has left and right inversions, which agree up to homotopy, if σ is associative. A map $f : (X,\sigma) \to (Y,\psi)$ of co-H-spaces is primitive provided $(f \vee f) \circ \sigma \simeq \psi \circ f$.

Ganea addresses the relation between cogroups and suspensions with the desuspension problem; given a cogroup (X,σ), when is there a space W and a primitive homotopy equivalence $e : (\Sigma W,\tau) \to (X,\sigma)$, where τ is the standard suspension comultiplication. Ganea proves that any (n-1)-connected cogroup of dimension $\leq 4n-5$ has the primitive homotopy type of a suspension. He also proves the existence of cogroup structures on $S^3 \vee S^{15}$ which are not of the primitive homotopy type of any suspension. However, left open was the question whether there are cogroups where the underlying space fails to have the homotopy type of any suspension. The purpose of this paper is to provide examples answering this question.

---

[1]Both authors received support from NSF grants.

Our examples have three cells and are described as follows. Recall that ${}_p\pi_{2p+2}S^5 \cong Z/pZ$ with generator α_1 and ${}_p\pi_{6p-3}S^5 \cong Z/pZ$ for $p \geq 5$ and $Z/9Z$ for $p = 3$. We denote a generator by γ. Fix an odd prime p and form the pushout

$$\begin{array}{ccc} S^5 & \longrightarrow & S^5 \cup_\gamma e^{6p-2} \\ \downarrow & & \downarrow \\ S^5 \cup_{\alpha_1} e^{2p+3} & \longrightarrow & Y(\gamma) . \end{array}$$

<u>Theorem A</u>. <u>For each</u> $p \geq 3$, $Y(\gamma)$ <u>is a cogroup</u>. <u>If</u> $p \equiv 1 \bmod 3$ <u>then</u> $Y(\gamma)$ <u>is not homotopy equivalent to any suspension</u>.

The need for three cells is present, because none of the complexes $S^5 \cup e^{6p-2}$ support cogroup structures. The case for $p = 3$ is the theorem, proved by Barratt and Chan [BC] that $S^5 \cup_{\beta_1} e^{16}$ admits no cogroup structure. We augment their result with

<u>Theorem B</u>. <u>For each</u> $p \geq 5$, $S^5 \cup_\gamma e^{6p-2}$ <u>admits no cogroup structure</u>.

Ganea was concerned whether the bound in his desuspension theorem is sharp. The relevant case of Theorem A is for $p = 3$, where $\gamma = \beta_1$. We are not able to settle this question but do obtain some information.

<u>Proposition C</u>. <u>There is a complex</u> $W = HP^2 \cup e^{15}$ <u>such that</u> $Y(\beta_1)$ <u>and</u> ΣW <u>have the same homotopy type, but there is no primitive homotopy equivalence of</u> $(\Sigma W, \tau)$ <u>with</u> $(Y(\beta_1), \sigma)$, <u>where</u> σ <u>is the cogroup structure of Theorem A</u>.

The result would show that Ganea's bound is sharp except there may be other ways to desuspend $Y(\beta_1)$.

Theorem A is far from the most general result we can prove. However, by confining attention to this case, the argument is substantially shorter.

One can ask whether every 2-cell complex which admits some cogroup structure has the homotopy type of a suspension. We answer this question with another example. From [T, p.139] we obtain the information that the 3-primary part of $\pi_{34}S^5$ is cyclic of order 3. A generator ϵ suspends to a $\alpha_1\beta_2$ in the stable range (ibid p. 141). From the same source, we obtain the fact that $_3\pi_{32}S^3 = 0$. Thus the complex

$$Z = S^5 \cup_\epsilon e^{35}$$

does not have the homotopy type of any suspension. Now the third James-Hopf invariant of ϵ lies in $_3\pi_{34}S^{13} = 0$ (this group is already stable). Thus Z is a co-H-space, (see 1.10).

<u>Theorem D</u>. The space Z admits the structure of a cogroup, and is not homotopy equivalent to any suspension.

The results of this paper answer the questions raised in [BC].

It is a pleasure to acknowledge useful remarks and suggestions from M.C. Barratt, J. Neisendorfer and M. Steinberger. We are especially grateful to Michael Barratt for detecting a serious error in an earlier version of this work, and for sharing, in extensive correspondence, his insights into the subtleties of non-associative comultiplications.

§1. Co-H-structures on mapping cones.

Our proofs are based on some direct constructions, involving attaching maps. Spaces have non-degenerate basepoints. All maps and homotopies preserve basepoints and cones, suspensions, etc. are reduced. The 5-tuple (X, τ, L, M, η) will denote a suspension co-H-structure, where X is the suspension of some space,

$\tau : X \to X \vee X$ is the suspension comultiplication, η is the usual inversion and the homotopies L, M are from $(1 \vee \tau) \circ \tau$, $j \circ \tau$ to $(\tau \vee 1) \circ \tau$, Δ respectively

$$\begin{array}{ccc} X & \xrightarrow{\tau} & X \vee X \\ & \searrow^{\Delta} & \downarrow j \\ & & X \times X . \end{array}$$

Some of our constructions involve piecing together strings of homotopies. We denote this process as follows: Let $H_i : X \times I \to Y$, $1 \leq i \leq n$ satisfy $H_i(x,1) = H_{i+1}(x,0)$. Then $\{H_1, \ldots, H_n\} : X \times I \to Y$ is defined by

$$\{H_1, \ldots, H_n\} (x,t) = H_{k+1}(x, nt-k) , \quad k \leq nt \leq k+1 .$$

We shall use the same notation for maps from ΣX or $X \times S^1$, when S^1 is regarded as the unit interval with its endpoints identified. Occasionally, $H(x,t) = (f(x),t) = \Sigma f$ for some map f . The symbol H' denotes the homotopy obtained from H by reversing the I-variable,

$$H'(x,t) = H(x, 1-t) .$$

We also need to classify homotopies. Let $f_0, f_1 : X \to Y$ and $H_i : X \times I \to Y$, $i = 1,2$ be homotopies from f_0 to f_1 , that is $H_1 = H_2$ on $X \times \dot{I}$. Then, because our maps preserve basepoints, we can regard $\{H_1, H_2'\}$ as defining a map

$$\{H_1, H_2'\} : (X, *) \to (Y^{S^1}, c)$$

where c is the constant map sending everything to the basepoint.

Lemma 1.0. Homotopy classes rel $X \times \dot{I}$ of homotopies from f_0 to f_1 are in one-to-one correspondence with $[X, \Omega Y]$ provided Y is path connected and X is a suspension.

Proof. Consider the fibration $\Omega Y \xrightarrow{i} Y^{S^1} \xrightarrow{e_0} Y$. Then e_0 has a section so i induces a monomorphism of generalized homotopy groups. Since $e_0\{H_1, H_2'\} = f_0$, the lemma follows from $e_0^{-1}[f_0] \cong [X, \Omega Y]$.

Let $f : (X_1, \tau_1) \to (X_2, \tau_2)$ be a primitive map with mapping cone C. It follows from general pinciples ([HMR$_1$]4.1) that C receives the structure of a co-H-space. But we want to relate co-H-structures on C to homotopies from $(f \vee f) \circ \tau_1$ to $\tau_2 \circ f$. Then $\sigma : C \to C \vee C$ defined by

$$(1) \qquad \sigma(x_1, t) = \begin{Bmatrix} (\tau_1(x), 2t) & 0 \leq 2t \leq 1 \\ H(x_1, 2t-1) & 1 \leq 2t \leq 2 \end{Bmatrix}, \quad \sigma(x_2) = \tau_2'(x_2)$$

(we could write $\{\Sigma\tau_1, H\}$) will define a co-H-structure provided that $j_2 \circ H$ is homotopic rel $X \times \overset{\circ}{I}$ to $\{(f \times f) \circ M_1, M_2' \circ f\}$. We call a homotopy H with this additional property primitive.

Lemma 1.1. If $f : (X_1, \tau_1) \to (X_2, \tau_2)$ is a primitive map, then $(f \vee f) \circ \tau_1$ is connected to $\tau_2 \circ f$ by a primitive homotopy.

We defer the proof until the end of this section.

The obstruction to associativity of σ defined by (1) is a function $d : \Sigma X_1 \to Z = C \vee C \vee C$. A specific representative is

$$(2) \qquad d = \{\Sigma((1 \vee \sigma) \circ \sigma), L_2 \circ f, \Sigma((\sigma \vee 1) \circ \sigma)'\}.$$

We denote the action of the group $[\Sigma X_1, Z]$ on $[C, Z]$ by $\oplus$. Then

$$(3) \qquad (1 \vee \sigma) \circ \sigma \simeq (\sigma \vee 1) \circ \sigma \oplus d.$$

Let $q : C \to \Sigma X_1$ denote the pinch map, and assume C is simply connected. From σ, $[C,Z]$ receives the structure of an algebraic loop, denoted by $+$, [HMR_1]. Then

$$(\sigma \vee 1) \circ \sigma \oplus d \simeq (\sigma \vee 1) \circ \sigma + dg .$$

Hence σ is associative if and only if the homotopy class of d lies in $\operatorname{im} \Sigma f^*$. If $e : \Sigma X_1 \to Z$ is another map satisfying (3), then

$$(\sigma \vee 1) \circ \sigma \oplus (d + \eta e) \simeq (\sigma \vee 1) \circ \sigma .$$

Hence, up to homotopy, d is characterized modulo $\operatorname{im} \Sigma f^*$ by (3).

To simplify the analysis of f, we introduce a construction which has the effect of compressing the essential part of the image of d into $Y = X_2 \vee X_2 \vee X_2$. Define $1^* : C \to C$ by

$$1^*(x_1,t) = \begin{cases} (x_1,2t) & 0 \le 2t \le 1 \\ f(x_1) & 1 \le 2t \le 2 \end{cases}, \quad 1^*(x_2) = x_2 .$$

Then $1^* \simeq 1 \text{ rel } X_2$ and σ is associative if and only if $(1^* \vee \sigma) \circ \sigma \simeq (\sigma \vee 1^*) \circ \sigma$. The difference element d^* is defined by (2) with 1 replaced by 1^*. Next we express $(1^* \vee \sigma) \circ \sigma$ in terms of the data defining σ. The result is

(4) $(1^*V\sigma) \circ \sigma \simeq \{\Sigma((1V\tau_1) \circ \tau_1), (fVH) \circ \tau_1, (1V\tau_2) \circ H\} \text{ rel } X_1 \times \dot{I}$ on CX_1 and

$$(1^*V\sigma) \circ \sigma(x_2) = (1V\tau_2) \circ \tau_2(x_2), \text{ and similarly for } (\sigma V1^*) \circ \sigma .$$

The expression for d^* is obtained from these formulae using (2).

Closely related to d^* is a function known as the A-deviation of f,H and written .

$$A(f,H) : (X_1 \times S^1, * \times S^1) \to (Y,*) .$$

It is defined by

$$(5) \quad A(f,H) = \{(fVH) \circ \tau_1, (1V\tau_2) \circ H, L_2 \circ f, (\tau_2V1) \circ H', (H'Vf) \circ \tau_1, (fVfVf) \circ L_1'\} .$$

The dependence of $A(f,H)$ on the homotopy H is as follows. Given primitive homotopies $H_i : X_1 \times I \to X_2 \vee X_2$, $i = 1,2$, we have $\omega = Ad\{H_1, H_2'\} : X_1 \times S^1 \to X_2 \vee X_2$. We define a linear operator

$$\delta : [X_1, X_2 \vee X_2] \to [X_1, X_2 \vee X_2 \vee X_2]$$

by

$$\delta(f) = E_2 \circ f + (1 \vee \tau_2) \circ f - (\tau_2 \vee 1) \circ f - E_1 \circ f$$

where $E_i : X_2 \vee X_2 \to X_2 \vee X_2 \vee X_2$ for $i = 1,2$ are the inclusions in the front and back ends, and addition is with respect to X_1. Linearity of δ is routine, provided S_1 is a double suspension. Direct calculation yields

$$(6) \qquad A(f,H_1) - A(f,H_2) \simeq \delta\omega.$$

Since H is primitive, the inclusions into cartesian products annihilate the homotopy classes of $A(f,H)$ and d^*.

To compare d^* and $A(f,H)$, consider the diagram

$$\begin{array}{ccc} [\Sigma X_1, Y] & \longrightarrow & [\Sigma X_1, Z] \\ \downarrow & & \downarrow \\ [X_1 \times S^1, Y] & \longrightarrow & [X_1 \times S^1, Z] \end{array}$$

where the vertical maps are injective.

<u>Lemma 1.2</u>. <u>The images of</u> $A(f,H)$ <u>and</u> d^* <u>are equal in</u> $[X_1 \times S^1, Z]$.

<u>Proof</u>. We refer to Figure 1 and leave the details, using (2), (4), (5) to the reader.

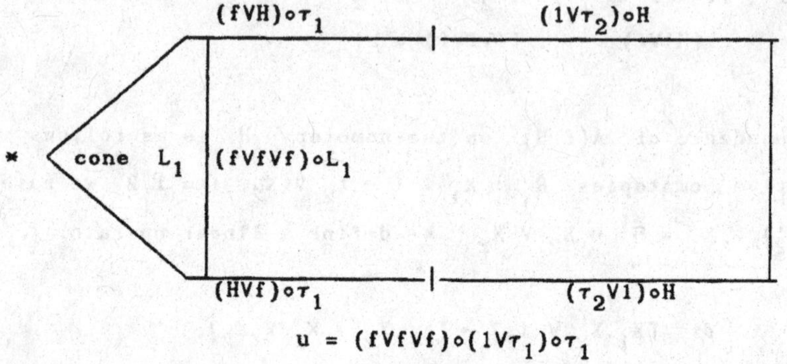

Figure 1

Next, we endeavor to pull $A(f,H)$ back into $[\Sigma X_1, Y]$. Consider the "constant" function $u : X_1 \times S^1 \to Y$ given by

$$u(x_1, t) = (f \vee f \vee f) \circ (1 \vee \tau_1) \circ \tau_1(x_1) .$$

Since $u(x_1, 0) = A(f,H)(x_1, 0)$, the difference $A(f,H)-u$ (obtained by adding in the first variable) is the image of a map

$$A^{\dagger} : \Sigma X_1 \to Y$$

which is unique up to homotopy. Note also that the composition

$$X_1 \times S^1 \xrightarrow{u} Y \to Z$$

is null-homotopic. Hence we have

Lemma 1.3. The image of $A^{\dagger}$ equals d^{*} in $[\Sigma X_1, Z]$.

We shall also need some simple composition properties of $A(f,H)$. Let $g : (X_2, \tau_2) \to (X_3, \tau_3)$ be the suspension of a basepoint preserving map. Then $(g \vee g) \circ \tau_2 = \tau_3 \circ g$ and $(g \vee g \vee g) \circ L_2 = L_3 \circ g$. Hence $(g \vee g) \circ H$ is a homotopy from $(gf \vee gf) \circ \tau_1$ to $\tau_3 \circ gf$. Similar considerations apply if $g : (X_3, \tau_3) \to (X_2, \tau_2)$, and we have

Lemma 1.4. $A(gf, (g \vee g) \circ H) = (g \vee g \vee g) \circ A(f,H)$

$A(fg, H \circ g) = A(f,H) \circ (g \times 1)$.

and similarly for the corresponding $A^{\dagger}$, with equality replaced by $\simeq$.

The proof of Theorem A is based on 1.3 and 1.4. For Theorem B we need further information relating co-H-structures on mapping cones with primitive attaching maps.

A comultiplication on a mapping cone need not be induced by a primitive attaching map. An extreme example is provided by the composition of the pinch map with the Hopf map in

$$CP^2 \to S^4 \to S^3 .$$

Using homology, one sees that the mapping cone has the homotopy type of $(S^3 \vee S^3) \cup e^5$. By the classification of (n-1)-connected (n+2)-dimensional complexes, this space has the homotopy type of $S^3 \vee \Sigma CP^2$. However, there are relationships if C is the mapping cone of $f : S^n \to X$ and $\dim X < n$. Suppose C is endowed with a comultiplication σ. We can assume that σ is cellular, with restriction $\sigma' : X \to X \vee X$. Then σ' is a comultiplication because there is a cellular homotopy from $j \circ \sigma$ to Δ. In the next three lemmas this data is understood. We shall also assume that X is simply connected.

<u>Lemma 1.5</u>. $f : (S^n, \tau) \to (X, \sigma')$ <u>is primitive</u>.

<u>Proof</u>. The n-skeleton of the fibre of $X \vee X \to C \vee C$ has the homotopy type of $S^n \vee S^n$ and the lemma follows by general principles.

<u>Lemma 1.6</u>. σ <u>is homotopic to a comultiplication induced by a primitive homotopy from</u> $(f \vee f) \circ \tau$ to $\sigma' \circ f$.

<u>Proof</u>. Using 1.5 and 1.1, we have a primitive homotopy from $(f \vee f) \circ \tau$ to $\sigma' \circ f$ inducing some comultiplication ψ on C, extending σ'. Let $d \in \pi_{n+1}(C \vee C)$ be the difference between σ and ψ. In $[C, C \vee C]$ we have $\psi + dq = \sigma$. Since the image of dq in $C \times C$ is 0 and $\pi_{n+1}(C \vee C)$ maps surjectively to $\pi_{n+1}(C \times C)$, we can choose d so that its image in $\pi_{n+1}(C \times C)$ is 0 . Thus d pulls back to $\pi_{n+1}(\Omega C * \Omega C)$. By a standard construction, the difference element for a pair of comultiplications on C induced by a pair of primitive homotopies can be identified with an element of $\pi_{n+1}(\Omega X * \Omega X)$. So to prove the lemma, it is enough to pull d back into this group. Since the pair (C,X) is n-connected and X is 1-connected, the pair $(\Omega C * \Omega C, \Omega X * \Omega X)$ is (n+1)-connected and the lemma follows.

We now can assume that σ is induced by a primitive homotopy H and turn to the corresponding $A^{\dagger} : S^{n+1} \to X \vee X \vee X$. In the next lemma, we localize at an odd prime, and set $X = S^r$.

<u>Lemma 1.7</u>. <u>Suppose</u> $f : S^n \to S^r$ <u>with</u> $r \geq 3$. <u>Then</u> σ <u>is associative if and only if</u> $A^{\dagger} \sim *$.

<u>Proof</u>. If $A^{\dagger} \sim *$, then σ is associative by 1.3. For the converse we examine the diagram

$$\pi_{n+1}(Y) \xrightarrow{i} \pi_{n+1}(Z) \xleftarrow{\Sigma f^*} \pi_{r+1}(Y) .$$

Then $\ker i$ is the image of $\pi_{n+2}(Z,Y) \cong \pi_{n+2}(S^{n+1} \vee S^{n+1} \vee S^{n+1})$ by the Blakers-Massey Theorem and the latter group is 0. Similarly $\Sigma f^* = 0$ and the lemma follows.

We turn to the proof of 1.1. First we prove a lemma from fibre space theory. We remind the reader that all maps preserve basepoints.

<u>Lemma 1.8</u>. <u>Let</u> X <u>be a suspension and</u> $F \xrightarrow{i} E \xrightarrow{p} B$ <u>a fibration with</u> i <u>inducing a monomorphism on generalized homotopy groups</u>. <u>Suppose</u> $f_0, f_1 : X \to E$ <u>are homotopic and</u> $H : X \times I \to B$ <u>is a homotopy such that</u> $H(x,i) = pf_i(x)$, $i = 0,1$. <u>Then there is a</u> $G : X \times I \to E$ <u>from</u> f_0 <u>to</u> f_1 <u>such that</u> $pG = H$.

<u>Proof</u>. Let $\bar{H} : X \times I \to E$ lift H and start at f_0. Define $h_1 = \bar{H}(\ ,1)$. Let $\theta : X \times I \to X$ be a homotopy from $1 + \eta$ to $*$. Then there is a homotopy $\bar{\theta} : X \times I \to E$ starting at $f_1 + h_1\eta$ and satisfying $p\bar{\theta} = pf_1\theta$. Define $\varphi = \bar{\theta}(\ ,1)$, $\varphi : X \to F$. By the hypothesis on i and $f_0 \simeq f_1$, we have $\varphi \sim *$ and we denote a null-homotopy from $*$ to φ by K. We now have a basepoint preserving map

$$G_1 : X \times I \times \{0\} \cup X \times \dot{I} \times I \to E$$

obtained by piecing $K + \bar{H}$ on $X \times I \times \{0\}$ together with $\{f_0 \circ \theta' + f_0,\ f_0 \circ L,\ f_0 + f_0 \circ \psi\}$ and $\{\bar{\theta}' + h_1,\ \nabla_3 \circ (f_1 \vee h_1\eta \vee h_1) \circ L,\ f_1 + h_1 \circ \psi\}$ on $X \times \{0\} \times I$ and $X \times \{1\} \times I$ respectively and an extension of pG_1 to $G_2 : X \times I \times I \to B$ given by $\{H \circ \theta' + H,\ H \circ L,\ H + H \circ \psi\}$ where $L, \psi : X \times I \to X$ are homotopies from $(1 + \eta) + 1$, $\eta + 1$ to $1 + (\eta + 1)$, $*$ respectively, and $\nabla_3 : B \vee B \vee B \to B$. We obtain G by the homotopy lifting property.

The proof of 1.1. is obtained by first replacing j_2 by a fibration

$$\begin{array}{ccc} E & \overset{\lambda_0}{\rightleftarrows} & X_2 \vee X_2 \\ \downarrow & \diagup_{j_2} & \\ X_2 \times X_2 & & \end{array}$$

and apply 1.8 to the data $f_0 = \lambda_0 \circ (f \vee f) \circ \tau$, $f_1 = \lambda_0 \circ \tau_2 \circ f$ and $H_1 : X_1 \times I \to X_2 \times X_2$.

$$H_1 = \{(f \times f) \circ M_1, M_2' \circ f, j_2 \circ H', j_2 \circ H\}$$

where H is a homotopy from $(f \vee f) \circ \tau_1$ to $\tau_2 \circ f$.

Next we develop some elementary material enabling us to decide when the attaching map of a two-cell complex is primitive.

A theorem proved in [BH] asserts that the two-cell complex $C = S^n \cup_\alpha e^{r+1}$ is a co-H-space, if and only if $\alpha \in \pi_r S^n$ is primitive. Here, we combine this result with the fact that after localization at an odd prime p, for n odd, S^n is an H-space. Let λ be an integer and λ_i a self-map of degree λ on the i-sphere, $i = r,n$. Write σ_i for the standard cogroup structure on the i-sphere. We have $\lambda_n \circ \alpha = \alpha \circ \lambda_r$, and hence

(7) $$\lambda(\sigma_n \circ \alpha) = \sigma_n \circ (\alpha \circ \lambda_r) = \sigma_n \circ (\lambda_n \circ \alpha)$$

where it is understood that spaces are p-local and we are following the practice of identifying maps with their homotopy classes. Subtracting the first term from the third in (7), and expanding by means of the Hilton-Milnor theorem [W] we obtain

(8) $$0 = \sum_{j=3}^{\infty} (\lambda^{|w_j|} - \lambda) w_j(\iota_1, \iota_2) \circ h_{j-3}(\alpha).$$

where $|w_j|$ denotes the weight of the basic Whitehead product w_j. Thus, each summand is 0. Here

$$h_{j-3} : \pi_r S^n \to \pi_r S^{w(n-1)+1} \quad \text{where} \quad w = |w_j|.$$

<u>Lemma 1.9</u>. <u>The non-zero obstructions to primitivity of</u> α <u>correspond to terms of the form</u> $w_j(\iota_1, \iota_2) \circ h_{j-3}(\alpha)$ <u>where</u> $|w_j| \equiv 1$ <u>mod</u> $(p-1)$ <u>and</u> <u>stem</u>$(h_{j-3}(\alpha)) > 0$.

<u>Proof</u>. Pick $\lambda > 0$ such that its mod p reduction is a primitive $(p-1)$-st root of unity. Then $\lambda^k \equiv \lambda^i$ mod p, if and only if $k \equiv i$ mod$(p-1)$. If $h_{j-3}(\alpha)$ lies in the 0-stem, then $\lambda^{|w_j|} - \lambda \neq 0$, so $h_{j-3}(\alpha) = 0$, and the latter condition in 1.9 is established. If $|w_j| \equiv 1$ mod$(p-1)$, then there is an integer $i \neq 1$, $0 \leq i \leq p-1$, such that $\lambda^{|w_j|} = \lambda^i + cp$ and $\lambda^i - \lambda$ is a unit mod p. From (8) it follows that

$$(9) \qquad 0 = (\lambda^i - \lambda) w_j(\iota_1, \iota_2) \circ h_{j-3}(\alpha).$$

The positive stems are p-groups, so (9) yields $h_{j-3}(\alpha) = 0$.

Now the p-th James-Hopf invariant factors through the space of loops on a bouquet,

$$H_p : \Omega S^{2n+1} \to \Omega(S^{2n+1} \vee S^{2n+1}) \to \Omega S^{2np+1}$$

with fibre $J_{p-1} S^{2n}$. Let $\alpha \in \Pi_q S^{2n+1}$ and α^* denote its adjoint.

<u>Proposition 1.10</u>. α <u>is primitive if and only if</u> $H_p(\alpha^*) = 0$.

<u>Proof</u>. One direction is immediate from 1.9. If $H_p(\alpha^*) = 0$, then α^* factors through $J_{p-1}S^{2n}$. Hence the remaining $h_{j-3}(\alpha) = 0$ for dimensional reasons.

<u>§2. Proofs of the theorems.</u>

Spaces are p-local. The passage from local to global results is based on general results in [HMR_2]. We let the context distinguish between maps and their homotopy classes.

<u>Proof that</u> $Y(\gamma)$ <u>is a cogroup</u>.

The element $\gamma \in \pi_{6p-3} S^5$ is represented by the Toda bracket $\langle \alpha_1(5), \alpha_1(2p+2), \alpha_1(4p-1) \rangle$. We extract the following diagram from this construction

$$\begin{array}{ccccc} S^{4p-1} & \longrightarrow & S^{2p+2} & \xrightarrow{\alpha} & S^5 \\ & & \downarrow & & \\ S^{6p-2} & \xrightarrow{g} & B & & a \end{array}$$

where $B = S^{2p+2} \cup_{\alpha_1} e^{4p}$ and all the spaces and maps, except a, are suspensions. Then the composition $a \circ g$ represents γ. Let $q : B \to S^{4p}$ be the pinch map. Then $q \circ g = \alpha_1(4p)$.

<u>Lemma 2.1</u>. <u>Any choice of</u> a <u>is primitive</u>. <u>The corresponding</u> $A^{\dagger} : \Sigma B \to S^5 \vee S^5 \vee S^5$ <u>factors</u> $A^{\dagger} \simeq \tilde{A} \circ \Sigma q$ <u>where</u> $\tilde{A} \in \pi_{4p+1}(S^5 \vee S^5 \vee S^5)$.

<u>Proof</u>. Consider the exact sequence induced by $S^{2p+2} \to B \xrightarrow{q} S^{4p}$

$$[B,Y] \xleftarrow{q^*} \pi_{4p}Y \xleftarrow{\alpha_1^*} \pi_{2p+3}Y \longleftarrow [\Sigma B,Y] \longleftarrow \pi_{4p+1}(Y)$$

where Y is either a 2 or 3 fold bouquet of S^5. Use of the Hilton-Milnor theorem yields that α_1^* is an ismorphism, and both parts follow.

Hence $\tilde{A} = \Sigma k_j w_j$, where w_j are basic Whitehead products in ι_5 of weight p and k_j are integers.

Our model for $Y(\gamma)$ is $S^5 \cup e^{2p+3} \cup e^{6p-2}$ with the top cell attached by the composition $k \circ a \circ g$ where $k : S^5 \to S^5 \cup_{\alpha_1} e^{2p+3}$ is the inclusion. We have

$$A^{\dagger} \circ \Sigma g \simeq \tilde{A} \circ \alpha_1(4p+1) .$$

Since $\alpha_1(4p+1)$ is a (4p-2)-fold suspension, the Barcus-Barratt theorem [BB] implies that any $w_j \circ \alpha_1(4p+1)$ is equal up to sign to a similar Whitehead product, where one of the factors is $\alpha_1(5)$ and the others are ι_5. Since $k \circ \alpha_1(5) = 0$, we have

$$(k \vee k \vee k) \circ w_j \circ \alpha_1(4p+1) = 0 .$$

It now follows from 1.3 and 1.4 that $Y(\gamma)$ is a cogroup.

<u>Proof that</u> $Y(\gamma)$ <u>is not equivalent to any suspension</u>, if $p \equiv 1$ mod 3.

The first step is to show that the attaching map $k \circ \gamma$ of the top cell is essential. The complex $S^5 \cup e^{2p+3}$ is a sub complex of the total space E of an S^5-fibration over S^{2p+3}. In the homotopy sequence

$$\pi_{6p-2}\, S^{2p+3} \xrightarrow{\partial} \pi_{6p-3}\, S^5 \longrightarrow \pi_{6p-3}\, E$$

the lefthand group is generated by $\alpha_2(2p+3)$ and

$$\partial\, \alpha_2(2p+3) = \alpha_1(5) \circ \alpha_2(2p+2) .$$

This element is 0 if $p \geq 5$ and $\pm 3\beta_1$ for $p = 3$. Thus γ is mapped non-zero to E and hence, non-zero to $S^5 \cup e^{2p+3}$.

By connectivity and homology considerations, we can assume that a putative desuspension (W,e) of $Y(\gamma)$ has cell structure

$$W = S^4 \cup e^{2p+2} \cup e^{6p-3}$$

and $e : \Sigma W \to Y$ is cellular. Then the adjoint, Ade_* is non-zero in the following diagram

$$\begin{array}{ccc} \pi_{6p-4}(S^4 \cup e^{2p+2}) & \xrightarrow{Ade_*} & \pi_{6p-4}(\Omega(S^5 \cup e^{2p+3})) \\ \varphi \uparrow & & \uparrow \\ \pi_{6p-4}(S^4) & \xrightarrow{\pm E} & \pi_{6p-4}(\Omega S^5) \end{array} .$$

Now $\mathrm{im}\, E = \mathrm{im}\, E^2 = 0$ for $p \geq 5$ by [T]. Thus (W,e) is ruled out by

<u>Lemma 2.2</u>. <u>For</u> $p \equiv 1 \bmod 3$, φ <u>is surjective</u>.

<u>Proof</u>. We use the results of [Gr] or [C] on the relative James construction. If X is obtained by attaching an $(r+1)$-cell to a simply connected complex X', then a cell decomposition for the fibre of the pinch map $X \to S^{r+1}$ has cells in dimensions $ir + k$ for each k-cell of X', $k > 0$ and $i \geq 0$. Starting with $X = S^4 \cup e^{2p+2}$, we take the fibre of the pinch map, restrict to the 6p-4 skeleton, pinch off a top cell and repeat this process. By induction, we obtain a factorization of φ (we abbreviate $\pi_{6p-4}(X)$ by (X))

$$\begin{array}{ccccccc} (S^4) & \to \cdots \to & (F) & \to \cdots & (S^4 \cup e^{2p+2}) \\ & & \downarrow & & \downarrow \\ & & (S^\delta) & & (S^{2p+2}) \end{array}$$

where

$$\delta = 3x + (2p+1)\, y+1 \leq 6p-4$$

for positive integers x, y. Since all stems are $\leq 4p-6$, the only non-zero stems for the groups in the lower line are 0 and $2p-3$. Substituting $y = 1$ or 2 and solving for p in

$$6p-4 - \delta = \{0, 2p-3\}$$

yields either that $3|p$ or $p \equiv 2 \bmod 3$. Thus the lemma follows and the proof of Theorem A is complete.

<u>Proof of Theorem B</u>.

If $S^5 \cup_\gamma e^{6p-2}$ is a cogroup, then by 1.6 and 1.7, the comultiplication is induced by a primitive homotopy for $\gamma = a \circ g$, with zero A-deviation. We next observe that such a homotopy factors as $H \circ g$ where H is a primitive homotopy for a. In general the primitive homotopies for $f : X_1 \to X_2$ are classified by $[X_1, \Omega X_2 * \Omega X_2]$. For the case $X_1 = S^{6p-3}, B$ and $X_2 = S^5$, we have

$$\begin{array}{ccc} \tilde{\pi}_{6p-2}(S^5 \vee S^5) & & \\ {\scriptstyle g^*}\uparrow & \nwarrow {\scriptstyle \alpha_1^*} & \\ [B, \Omega S^5 * \Omega S^5] & \xleftarrow[q^*]{} & \tilde{\pi}_{4p+1}(S^5 \vee S^5) \end{array} .$$

where we have identified $\pi_*(\Omega S^5 * \Omega S^5)$ with

$$\mathrm{Ker}\{\pi_*(S^5 \vee S^5) \to \pi_*(S^5 \times S^5)\} \equiv \tilde{\pi}_*(S^5 \vee S^5) .$$

By the Hilton-Milnor theorem, α_1^* is an isomorphism and g^* is surjective.

We obtain a contradiction by showing that $\tilde{A} \circ \alpha_1(4p+1) \neq 0$, where $\tilde{A}$ is derived from H as in 2.1. This, in turn, follows provided one of the coefficients k_j in $\tilde{A} = \Sigma k_j w_j$ is not divisible by p. To see this, consider

$$k : S^5 \to P^6(p) = S^5 \cup_p e^6 .$$

We show $(k \vee k \vee k) \circ \tilde{A} \neq 0$ by proving

<u>Lemma 2.3</u>. <u>The complex</u> $Y = P^6(p) \cup_{k \circ a} CB$ <u>is not a cogroup for</u> $p \geq 5$.

<u>Proof</u>. If Y is a cogroup, then by [B_2], $S^{-1}H_*(Y)$ has the structure of a coalgebra such that there is an isomorphism of Hopf algebras

$$H_*(\Omega Y) \cong T(S^{-1}H_*(Y)) .$$

We now argue as in [B_1]. Since $S^{-1}H_*(Y)$ is concentrated in dimensions 4, 5, 2p+2, 4p and $p \geq 5$ the submodule of primitives of $S^{-1}H_*Y$ equals all of $S^{-1}H_*Y$, thus $H_*(\Omega Y)$ is primitively generated. But $H^*(\Omega Y)$ has a non-zero p-th power because

$$\mathcal{P}^2 : H^5(Y) \to H^{4p+1}(Y)$$

is non-zero and $\Sigma\Omega Y \to Y$ induces a monomorphism in cohomology because Y is a co-H-space.

<u>Proof of Proposition C</u>.

In this argument, spaces are localized at 3. The attaching map of the top cell is $k \circ \beta_1$, where $\beta_1 \in \langle \alpha_1(5), \alpha_1(8), \alpha_1(11) \rangle$ and is represented by a composition

$$b : S^{15} \to S^8 \cup e^{12} \to S^5 ,$$

with different choices given by

$$\alpha_1(5) \circ \pi_{15}\, S^8 + \pi_{12}\, S^5 \circ \alpha_1(12)\ .$$

The summands are generated by $\alpha_1(5) \circ \alpha_2(8)$ and $\alpha_2(5) \circ \alpha_1(12)$, both of which are double suspensions from $\pi_{13}\, S^3 \cong Z/3Z$. Hence these generators are equal up to sign. We can factor $k \circ b = h \circ \alpha_1(12)$

$$\begin{array}{ccccc} S^{15} & \longrightarrow & S^8 \cup e^{12} & \longrightarrow & S^5 \\ & {\scriptstyle \alpha_1(12)} \searrow & \downarrow & & \downarrow {\scriptstyle k} \\ & & S^{12} & \xrightarrow{\ h\ } & S^5 \cup e^9\ . \end{array}$$

Hence, it is enough to desuspend h.

<u>Lemma 2.4</u>. $\pi_{12}(S^5 \cup e^9) \cong Z/9Z$ <u>and there is a short exact sequence</u>

$$0 \to \pi_{12}\, S^5 \to \pi_{12}(S^5 \cup e^9) \to \pi_{12}\, S^9 \to 0\ .$$

<u>Proof</u>. Let E be the S^5-fibration over S^9 classified by α_1. The pair $(E, S^5 \cup e^9)$ is 13-connected, so the exact sequence follows from the homotopy sequence for E. The asserted extension is proved in [MT] Th. 3.2 or is an easy calculation with the unstable Adams spectral sequence.

To use the lemma, we observe that the composition

$$S^{12} \xrightarrow{\ h\ } S^5 \cup e^9 \to S^9$$

is $\alpha_1(9)$, by construction of the Toda bracket. Thus to desuspend h, it is enough to find a map $S^{11} \to S^4 \cup e^8$ satisfying

(a) $\Sigma(S^4 \cup e^8) \simeq S^5 \cup_{\alpha_1} e^9$

(b) $S^{11} \to S^4 \cup e^8 \to S^8$ is $\pm\, \alpha_1(8)$.

<u>Lemma 2.5</u>. <u>The standard fibre map</u> $p : S^{11} \to HP^2$ <u>satisfies</u> (a) <u>and</u> (b).

<u>Proof</u>. Since $\mathcal{P}^1$ is non-trivial in $H^*(HP^2)$, $\Sigma HP^2 \simeq S^5 \cup e^9$ with non-trivial attaching map and part (a) follows because $\pi_8 S^5$ is generated by $\alpha_1(5)$. To prove (b), consider the diagram of cofibrations, where the vertical maps are induced by pinching the bottom 4-cell to a point,

$$\begin{array}{ccccc} S^{11} & \xrightarrow{p} & HP^2 & \longrightarrow & HP^3 \\ & & \downarrow & & \downarrow \\ & & S^8 & \longrightarrow & S^8 \cup e^{12} . \end{array}$$

Since $\mathcal{P}^1 : H^8(HP^3) \to H^{12}(HP^3)$ is non-zero, the 12-cell is attached essentially. So (b) is satisfied because $\pi_{11} S^8$ is generated by $\alpha_1(8)$.

Thus $Y(\beta_1) \simeq \Sigma W$, where $W = HP^2 \cup e^{15}$ with the top cell attached by $h' \circ \alpha_1(11)$, $\Sigma h' = h$.

To prove that $(Y(\beta_1),\sigma)$ and $(\Sigma W,\tau)$ do not have the same primitive homotopy type, we observe that this is not even the case rationally. Working over the rationals, we have

$$W \simeq S^4 \cup e^8 \cup e^{15}$$

with the 8-cell attached by the Whitehead product $[\iota_4,\iota_4]$. Then

$$\Sigma W \simeq S^5 \vee S^9 \vee S^{16}$$

but the inclusion of S^9 into ΣW is not primitive. On the other hand, the cogroup structure on $Y(\beta_1)$ is constructed to extend the double suspension comultiplication on $S^5 \cup e^9$. Thus after localization, Y has the primitive homotopy type of a bouquet of spheres. A check of dimensions yields that ΣW has no self-equivalences which would make the inclusion of S^9 primitive, concluding the proof.

Remark. There is choice of b so that $k \circ b = \Sigma(p \circ \alpha_1(11))$.

<u>Lemma 2.8</u>. <u>There is a comultiplication</u> σ <u>on</u> Z <u>such that</u> $2I$ <u>is primitive</u>.

<u>Proof</u>. According to Ganea, the statement is equivalent to the existence of a coretraction

$$\gamma : Z \to \Sigma\Omega Z$$

such that

$$\Sigma\Omega(2I)\circ\gamma = \gamma\circ 2I.$$

Note that the evaluation map $p : \Sigma\Omega Z \to Z$ induces isomorphisms in homology of dimensions 5, 35. Since composition with p equalizes these maps, we may write

$$\Sigma\Omega(2I)\circ\gamma = \gamma\circ 2I \oplus d$$

for a unique d in $\pi_{35}(\Sigma\Omega S^5)$. If γ is varied, say $\gamma' = \gamma+\alpha$ for α in $\pi_{35}(\Sigma\Omega S^5)$ then

$$\Sigma\Omega(2I)\circ\gamma' = \gamma\circ 2I \oplus d'$$

and routine calculation, using 2.7, yields

$$d' = d + \Sigma\Omega(2I)\circ\alpha - \alpha\circ 2I.$$

Writing

$$\pi_{35}\Sigma\Omega S^5 = \underset{\omega_j}{\oplus} \pi_{35}S^{4j+1}$$

where $\omega_j : S^{4j+1} \to \Sigma\Omega S^5$ is a Whitehead product in generators from the James decomposition, then we have

$$\alpha = \sum_j \omega_j\beta_j.$$

In terms of the James decomposition, each ω_j has the form

$$\omega_j = \omega_j(\iota_{n_1}, \iota_{n_2}, \ldots, \iota_{n_k}) n_i \equiv 1 \bmod 4$$

where the n_i denote dimension. Applying $\Sigma\Omega(2I)$ to each generator ι_{4m+1} multiplies it by 2^m. Hence

$$\Sigma\Omega(2I)\circ\omega_j = 2^j\omega_j.$$

Hence

$$d' = d + \sum_j (2^j-2)\omega_j\circ\beta_j.$$

Thus α can be chosen to cancel the components of d corresponding to spheres of dimension $8j+1$. Having made this choice, the remaining obstructions to primitivity of $2I$ correspond to elements of $\pi_{35}S^{8j+5}$ for $j = 1,2,3$. But these groups (already stable) are zero. We remind the reader that, below 33, the non-zero stable stems are 10,13,20,26,29,30 and all stems congruent to 3 mod 4.

<u>Lemma 2.9</u>. Any comultiplication satisfying 2.8 determines a cogroup structure on Z.

<u>Proof</u>. Let γ be a coretraction from 2.8. According to Ganea, it is enough to show

$$S\circ\gamma = (\gamma \vee \gamma)\circ\sigma$$

where S is the suspension comultiplication on $\Sigma\Omega Z$ and σ is the comultiplication corresponding to γ. After composition with the evaluation map, these maps are equal, so we can apply 2.6 to write

$$(\gamma \vee \gamma)\circ\sigma = S\circ\gamma \oplus d$$

for d in $\pi_{35}(\Sigma\Omega S^5 \vee \Sigma\Omega S^5)$. Routine calculation using 2.7 and 2.8 yields

$$(\Sigma\Omega(2I) \vee \Sigma\Omega(2I))\circ d = d\circ 2I.$$

As before

$$d = \sum_j \omega_j \circ \beta_j$$

with ω_j in $\pi_{4j+1}(\Sigma\Omega Z \vee \Sigma\Omega Z)$, β_j in $\pi_{35}S^{4j+1}$. Arguing as in the proof of 2.8, we obtain

$$0 = \sum_j (2^j-2)\omega_j \circ \beta_j .$$

Thus if j is even, $\beta_j = 0$ and if j is odd, β_j lies in the same groups as in the proof of 2.8. Thus d = 0 and the proof is complete.

REFERENCES

[BB] W.D. Barcus and M.G. Barratt, On the homotopy classification of extensions of a fixed map, Trans. Amer. Math. Soc. 88 (1958) 57-74.

[BC] M.G. Barratt and P.H. Chan, A note on a conjecture of Ganea, J. London Math. Soc. 20 (1979) 544-548.

[B_1] I. Berstein, A note on spaces with non-associative comultiplications, Proc. Cambridge Phil. Soc. 60 (1964) 353-354.

[B_2] ___________, On co-groups in the category of graded algebras, Trans. Amer. Math. Soc. 115 (1965) 257-269.

[BH] I. Berstein and P. J. Hilton, On suspensions and comultiplications, Topology, 2 (1963), 73-82.

[C] G. Cooke, Thickenings of CW complexes of the form $S^m \cup e^n$. Trans. Amer. Math. Soc. 247 (1979) 177-209.

[G] T. Ganea, Cogroups and suspensions, Invent. Math. 9 (1970) 185-197.

[Gr] B. Gray, On homotopy groups of mapping cones, Proc. London Math. Soc. 26 (1973) 497-520.

[HMR_1] P.J. Hilton, G. Mislin, J. Roitberg, On co-H-spaces, Comment. Math. Helvetice 53 (1978) 1-14.

[HMR_2] ___________, Localization of Nilpotent Groups and Spaces, North Holland (1975).

[MT] M. Mimura and H. Toda, Cohomology Operations and the homotopy of compact Lie groups, I, Topology 9 (1970) 317-336.

[T] H. Toda, On iterated suspensions I, J. Math. Kyoto Univ. 5 (1965) 87-142.

[W] G. W. Whitehead, Elements of Homtopy Theory, Springer- Verlag (1978).

Cornell University, Ithaca, NY 14853
University of Rochester, Rochester, NY 14627

Instantons and Homotopy

Charles P. Boyer*
Benjamin M. Mann†
Department of Mathematics and Computer Science
Clarkson University
Potsdam, New York 13676

Over the last ten years examination of the Yang-Mills equations in mathematical physics has led to exciting new problems and results in analysis, geometry and topology. The solutions to these equations are certain distinguished connections, called instantons by the physicists, on principal bundles over smooth four manifolds. Donaldson's celebrated work [11,13,14] demonstrates that in many cases the moduli space of these instantons determines the homeomorphism or even the diffeomorphism type of the underlining smooth manifold. These moduli spaces themselves have been extensively studied in recent years by many people, including Atiyah [1,2], Atiyah, Hitchin and Singer [5], Atiyah, Drinfeld, Hitchin and Manin [4], Atiyah and Bott [3], Atiyah and Ward [7], Donaldson [11,12,13,14], Taubes [23,24,25,26] and Uhlenbeck [28,29], using various techniques from algebraic geometry, complex manifold theory, global analysis and twistor theory. This rich influx of ideas into topology from other areas of mathematics has inspired advances based on more classical topological techniques, as exemplified by the work of Fintushel and Stern [15,16] and it is natural to see how homotopy theoretic techniques may be used to study instantons.

One remarkable aspect of the mathematical development of the Yang-Mills theory has been the on-going program of Cliff Taubes [23,24,25,26], in which he has used powerful techniques from global analysis and the theory of partial differential equations to obtain deep results on the homotopy type of these moduli spaces of instantons. Taubes' techniques are far removed from the standard tools of a homotopy theorist but his results strongly suggest that homotopy theory has non-trivial things to say about instantons. Briefly put, one may go from the space of instantons to the space of all connections by forgetting the analytic structure. The forgetful functor here takes one from a finite dimensional smooth manifold to an infinite dimensional CW complex. Moreover, the former provides a good homotopy approximation to the latter. Motivated by the foundational paper of Atiyah and Jones [6] and using a result of Taubes [26], we showed [9] that the disjoint union of these moduli spaces over the four sphere (where the union is taken over all positive instanton numbers k) behaves homologically like a

*Partially supported by NSF grant DMS-8508950.
†Partially supported by NSF grant DMS-8701539.

four-fold iterated loop space (more precisely like a C_4 little cubes operad space in the sense of May [19]) with associated iterated loop space operations. We then used these operations to obtain new information on the homology of the moduli spaces.

In this paper we concentrate on the Yang-Mills theory associated to the four sphere. However, Taubes' analysis holds for the Yang-Mills theory associated to an arbitrary compact, connected, closed, orientable Riemannian four manifold and much of the structure (and hence the corresponding calculations) developed in [9] for S^4 transport to the general case, obviously with some major modifications.

In section one we briefly recall the differential geometric formulation of the Yang-Mills gauge theory associated to a principal G bundle P_k over S^4, where G is a compact, connected, simple Lie group. We then define our main objects of study, $\mathcal{M}_k$, the moduli space of instantons with instanton number k and $i_k : \mathcal{M}_k \longrightarrow \mathcal{C}_k$, the natural inclusion of $\mathcal{M}_k$ into $\mathcal{C}_k$, the moduli space of all connections on P_k. We conclude the section with a construction of 't Hooft which associates elements of certain configuration spaces to instantons.

Section two briefly reviews what is known about the topology of $\mathcal{M}_k$, highlighting the work of Atiyah and Jones and Taubes. We summarize the foundational results of Atiyah and Jones and state the topological questions that arise naturally from their work, including the Atiyah-Jones conjecture on the low dimensional homotopy type of $\mathcal{M}_k$. Next we state Taubes' strong version of the Atiyah-Jones surjection theorem and highlight two of his analytic constructions which have strong homotopy theoretic content. We then show precisely how the disjoint union of the $\mathcal{M}_k$, over $k > 0$, is surrounded by four-fold loop spaces. This, in turn, suggests that the moduli spaces $\mathcal{M}_k$ have a rich homological structure. Finally, we observe that the existence of such a C_4 structure on $\coprod \mathcal{M}_k$ is not immediate on the analytic level.

Sections three and four briefly review the main constructions and results of [9] and represent a summary of the talk given by one of us at the Arcata conference. In the last section we relate a special case of a recent result of Taubes [26], to the results of the previous sections and we conclude the paper with three new corollaries and a conjecture.

We would like to thank Jim Stasheff for very helpful discussions relating our instanton constructions to operads and to thank Cliff Taubes for communicating some of his recent results to us and for valuable discussions which have improved our understanding of these moduli spaces.

1 Yang-Mills Instantons

We begin with a very brief review of the differential geometric formulation of the Yang-Mills theory over the four sphere. Good references on the foundational material in this area include [1], [4], [5], [6], [7], and [20].

Let G be a compact, connected, simple Lie group (we are interested mainly in the compact, simple classical groups) and let $\pi : P \longrightarrow S^4$ be a principal G bundle over the four sphere. Recalling that such bundles are indexed by the integers, we write P_k for the bundle classified by the map $S^4 \longrightarrow BG$ of degree k (recall $\pi_3(G) = Z$). There are two natural spaces associated to P_k, the space of all connections, $\mathcal{A}_k$, and the gauge group, $\mathcal{G}(P_k)$. $\mathcal{A}_k$ is well-known to be an affine space and the gauge group is defined as follows:

Definition 1.1 $\mathcal{G}(P_k) = \{f | f : P_k \longrightarrow P_k$ *is a bundle automorphism which covers the identity map on* $S^4\}$.

Definition 1.2 $\mathcal{G}^b(P_k)$, *the based gauge group of* P_k, *is the normal subgroup of* $\mathcal{G}(P_k)$ *given by all* $f \in \mathcal{G}(P_k)$ *such that* f *is the identity map on the fibre over a distinguished base point.*

If we think of $\omega \in \mathcal{A}_k$ as a pseudotensorial 1-form then $\mathcal{G}(P_k)$, and thus $\mathcal{G}^b(P_k)$, acts on $\mathcal{A}_k$ via the pullback; that is

$$\omega \longmapsto f^*(\omega) = ad_{f^{-1}}\omega + f^{-1}df \tag{1.3}$$

Here we have identified $\mathcal{G}(P_k)$ with ad-equivariant maps $f : P_k \longrightarrow G$. While the action of all of $\mathcal{G}(P_k)$ on $\mathcal{A}_k$ is not free, this action restricted to the base gauge group $\mathcal{G}^b(P_k)$ is free. This observation and a local slice analysis [6], [22], show

Proposition 1.4 $\mathcal{A}_k \longrightarrow \mathcal{A}_k/_{\mathcal{G}^b(P_k)} = \mathcal{C}_k$ *is a principal* $\mathcal{G}^b(P_k)$ *bundle where* $\mathcal{C}_k$ *may be identified with* $B\mathcal{G}^b(P_k)$, *the classifying space of* $\mathcal{G}^b(P_k)$. *Furthermore,* $\mathcal{C}_k$ *is homotopy equivalent to* $\Omega^3_k G \simeq \Omega^4_k BG$.

The fact that $\mathcal{C}_k$ may be identified, up to homotopy, with a four fold loop space plays a key role in what follows.

Given a representation $\rho : G \longrightarrow AutV$ of G on a vector space V we may form the associated vector bundle $E_k = P_k \times_G V$. Of course, connections in P_k give rise to connections in E_k and vice versa. Now let $\omega \in \mathcal{A}_k$. Its curvature $F^\omega = D^\omega\omega$ is a section of the vector bundle $(P_k \times_G g) \otimes \Lambda^2(S^4)$ where g is the Lie algebra of G and $\rho : G \longrightarrow Aut(g)$ is the adjoint representation. There is a natural bilinear form on $(P_k \times_G g) \otimes \Lambda^2(S^4)$ given by the Hodge inner product on $\Lambda(S^4)$ (with respect to the standard metric on S^4) and the Killing form on g. The corresponding norm gives the Yang-Mills functional on $\mathcal{A}_k$:

$$\mathcal{YM}(\omega) = \int_{S^4} \| F^\omega \|^2 \tag{1.5}$$

Furthermore, F^ω orthogonally splits into self-dual, F^ω_+, and anti-self-dual, F^ω_-, components (with respect to the Hodge decomposition) and we may rewrite the integral as

$$\mathcal{YM}(\omega) = \int_{S^4} \| F^\omega_+ \|^2 + \| F^\omega_- \|^2 \tag{1.6}$$

On the other hand, Chern-Weil theory yields

$$p_1(g) = \frac{1}{4\pi^2}\int_{S^4} \| F^\omega_+ \|^2 - \| F^\omega_- \|^2 \tag{1.7}$$

where $p_1(g)$ is the first Pontrjagin number of the adjoint bundle $P_k \times_G g$. Therefore, self-dual (for $k > 0$) and anti-self-dual (for $k < 0$) connections give absolute minima of $\mathcal{YM}$. These self-dual and anti-self-dual connections are called instantons and anti-instantons, respectively. As any orientation reversing diffeomorphism will pull back P_k to P_{-k} and pull back instantons to anti-instantons (or anti-instantons to instantons depending on the sign of k) it suffices to restrict our attention to $k > 0$ and instantons. We do so for the remainder of the paper.

Let $I_k \subset \mathcal{A}_k$ denote the subspace of instanton (self-dual) connections in $\mathcal{A}_k$. Further, we let $\hat{\mathcal{A}}_k \subset \mathcal{A}_k$ denote the subspace of all irreducible connections on P_k and set $\hat{I}_k = I_k \cap \mathcal{A}_k$. For $G = SU(2) = Sp(1)$ and $k > 0$ it follows from the fact that there are no harmonic 2-forms on S^4 that $\mathcal{A}_k = \hat{\mathcal{A}}_k$ and thus $I_k = \hat{I}_k$. However, these equalities are definitiely false for all other compact simple Lie groups.

Most importantly, a direct calculation shows that the Yang-Mills functional, $\mathcal{YM}$, is invariant under the action of the gauge group $\mathcal{G}(P_k)$ on $\mathcal{A}_k$ given in 1.3. Thus, we obtain the following moduli spaces:

Definition 1.8 $\mathcal{M}_k = I_k/_{\mathcal{G}^b(P_k)}$ *is the based moduli space of all instantons.*

Definition 1.9 $\mathcal{M}_k' = I_k/_{\mathcal{G}(P_k)}$ *is the moduli space of all instantons.*

Definition 1.10 $\hat{\mathcal{M}}_k = \hat{I}_k/_{\mathcal{G}^b(P_k)}$ *is the based moduli space of all irreducible instantons.*

Definition 1.11 $\hat{\mathcal{M}}_k' = \hat{I}_k/_{\mathcal{G}(P_k)}$ *is the moduli space of all irreducible instantons.*

A theorem of Atiyah, Hitchin and Singer [5] shows that $\hat{\mathcal{M}}_k'$ is either empty or is a smooth manifold of dimension $p_1(g)$ - $\dim(g)$. Furthermore, the factor group $\mathcal{G}(P_k)/_{\mathcal{G}^b(P_k)}$ is naturally identified with G/Z and by 1.3 we have a principal bundle

$$G/Z \longrightarrow \hat{\mathcal{M}}_k \longrightarrow \hat{\mathcal{M}}_k' \tag{1.12}$$

where $\hat{\mathcal{M}}_k$ is a smooth manifold of dimension $p_1(g)$ and Z is the center of G. We also have the obvious commutative diagram

$$\begin{array}{ccc} \hat{\mathcal{C}}_k & \longrightarrow & \mathcal{C}_k \\ \uparrow & & \uparrow \\ \hat{\mathcal{M}}_k & \longrightarrow & \mathcal{M}_k \\ \downarrow & & \downarrow \\ \hat{\mathcal{M}}_k' & \longrightarrow & \mathcal{M}_k' \end{array} \tag{1.13}$$

For any compact simple G, $p_1(g)$ can be computed in terms of the degree k and the rank of G, see [5]. In fact, $p_1(g) = a(g)k$ where the proportionality constant depends only on g. For example, if $G = Sp(n)$ then $a(g) = 4(n+1)$ so $p_1(g) = 4(n+1)k$, and thus dim $\hat{\mathcal{M}}_k = 4(n+1)k$, and dim $\hat{\mathcal{M}}'_k = 4(n+1)k - n(2n+1)$. If there exist irreducible self-dual connections on $Sp(n) \longrightarrow P_k \longrightarrow S^4$ then $k \geq n$ [5]. Of special interest is the case $G = Sp(1)$ for then $\hat{\mathcal{M}}_k = \mathcal{M}_k$ and $\hat{\mathcal{M}}'_k = \mathcal{M}'_k$ are non-empty smooth manifolds of dimension $8k$ and $8k-3$ respectively for all $k > 0$.

We conclude this section with a construction of a subclass of $Sp(1)$ instantons due to 't Hooft, [27]. Recall that the set of k-distinct points in R^4 is an open submanifold of R^4 which admits an obvious free action of the symmetric group Σ_k. The quotient space $C_k(R^4)$ is precisely the configuration space of k unordered points in R^4 which has played such a key role in the homology theory of iterated loop spaces. Viewing R^4 as the quaternionic plane H^1 and S^4 as the quaternionic projective plane $HP(1)$ we may construct an instanton as follows: Choose k distinct points $(a_1, a_2, \ldots, a_k)$ in H^1 and define a map $f : HP(1) \longrightarrow HP(k)$ by the formula $f(x) = (1, (x-a_1)^{-1}, \ldots, (x-a_k)^{-1})$. Fixing a standard connection ω on the Hopf fibration $S^3 \longrightarrow S^{4k+3} \longrightarrow HP(k)$ we obtain a connection $f^*(\omega)$ on $P_k \longrightarrow S^4 = HP(1)$ which can be shown to be self-dual. This assignment gives rise to the first inclusion j_k in

$$C_k(R^4) \xrightarrow{j_k} \mathcal{M}_k \xrightarrow{i_k} C_k \tag{1.14}$$

Atiyah and Jones [6] first noticed 1.14 and in the next section we review their work relating the topology of $\mathcal{M}_k$ to the topology of the better known spaces $C_k(R^4)$ and $C_k \simeq \Omega^3_k S^3 \simeq \Omega^4_k BS^3$.

2 On the Topology of Instantons

In this section we review some results of Atiyah and Jones [6] and Taubes [26] and describe the basic idea behind the main program of [9].

We begin with $G = Sp(1)$ and start with the inclusions $C_k \xrightarrow{j_k} \mathcal{M}_k \xrightarrow{i_k} C_k$ described at the end of section one. Segal [21] and Boardman and Vogt [8] have shown there are maps $S_k : C_k(R^n) \longrightarrow \Omega^n_k S^n$ which, for all k and n, are homology (although not homotopy) equivalences through a range (that depend on k and n). Furthermore, Atiyah and Jones showed that the following diagram commutes for all k:

$$\begin{array}{ccccc}
C_k(R^3) & & \xrightarrow{S_k} & & \Omega^3_k S^3 \\
& \searrow & & \nearrow & \\
\Sigma \downarrow & & \mathcal{M}_k \xrightarrow{i_k} C_k & & \downarrow \Sigma \\
& \nearrow{\scriptstyle j_k} & & \searrow & \\
C_k(R^4) & & & & \Omega^4_k S^4
\end{array} \tag{2.1}$$

where Σ represent the standard suspension maps. Using this commutative diagram and the Segal, Boardman-Vogt result, Atiyah and Jones obtained the following:

Theorem 2.2 *[6]. Let $G = Sp(1)$. Then $(i_k)_q : H_q(\mathcal{M}_k) \longrightarrow H_q(C_k)$ is a surjection for $q << k$.*

Atiyah and Jones then pose the following questions:

1. Is $(i_k)_q$ actually a homology isomorphism through a range?
2. Can the range of the surjection (isomorphism) $q = q(k)$ be explicitly determined as a function of k?
3. Is $(i_k)_q$ a surjection on homotopy groups through a range?
4. Is i_k a homotopy equivalence through a range?
5. Are similar results true if $Sp(1)$ is replaced by more general compact simple G? Notice that immediately there arises a new complication as $\hat{\mathcal{M}}_k$ is properly contained in $\mathcal{M}_k$ for all $G \neq Sp(1)$. In fact, $\hat{\mathcal{M}}_k$ may be empty while at the same time $\mathcal{M}_k$ may be quite large.

Question 4, which is now commonly known as the Atiyah-Jones conjecture, remains open at this time. However, substantial progress has been made toward answering these questions. The following theorem of Taubes [26] represents a major step forward in understanding the low dimensional homotopy and homology of $\mathcal{M}_k$.

Theorem 2.3 *(Taubes [26]). Let G be any compact, connected, simple Lie group. Then $(i_k)_q : \pi_q(\mathcal{M}_k) \longrightarrow \pi_q(C_k)$ and $(i_k)_q : H_q(\mathcal{M}_k) \longrightarrow H_q(C_k)$ are surjections for $q << k$.*

Although it is impossible in any brief general exposition to do justice to Taubes' work in this area, we would like to highlight two aspects which we feel are especially valuable for homotopy theorists to know, namely Taubes' patching and the Taubes' tubular neighborhood theorem. We defer discussion of patching to section five and turn our attention to the following theorem.

Theorem 2.4 *(Taubes, [24]). Let $P_k \longrightarrow S^4$ be a principal G-bundle with $k \geq 0$. Then there exists $\epsilon > 0$ and a strong deformation retract C_k^ϵ onto $\mathcal{M}_k$.*

Here $C_k^\epsilon = \mathcal{YM}^{-1}([0,\epsilon))$ where we have normalized $\mathcal{YM} : C_k \longrightarrow R$ as in [24] so that $\mathcal{YM}^{-1}(0) = \mathcal{M}_k$.

We wish to stress that this theorem, in its full generality [26], is fundamental in the homotopy theortic study of instantons because it shows that the property of being an "almost" instanton is a precise homotopy notion. That is, the theorem allows one to make many natural homotopy constructions on $\mathcal{M}$ which do not hold on the analytic level. While Theorems 2.3 and 2.4 are very powerful existence theorems they do not give explicit computational information. In [9] we were able, for $G = Sp(1)$, to determine an explicit bound for the surjection dimension $q(k)$ in terms of k, to show there are non-zero classes in $H_{6k-3}(\mathcal{M}_k)$ for infinitely many k, and, in general, to characterize large families of non-trivial homology classes in $H_q(\mathcal{M}_k)$ in terms of Dyer-Lashof operations (see 4.11, 4.13, 4.14, 4.20, 4.22, 4.23 and 5.3 in particular). The remainder of this section is devoted to explaining the motivation behind our program.

As the homology of $C_k(R^n)$ and $\Omega^n_k S^n$ is well-known, it is possible to use diagram 2.1 to construct non-zero classes in the image $(i_k)_q : H_q(\mathcal{M}_k) \longrightarrow H_q(C_k)$. It is easy to see that if one works with homology with $Z/2$ coefficients then this method yields non-zero classes for $q \leq 2^{r+1} - 2$ when $k = 2^r$. However, it is not possible to directly use this method to obtain interesting classes in $H_q(\mathcal{M}_k)$ when $q > k/4$.

For reasons that will quickly become apparent it is natural to take the union, over all positive values of k, of diagram 2.1 to obtain the following commutative diagram:

$$\begin{array}{ccccccc} C(R^3) & & & \xrightarrow{\iota \circ S} & & & \Omega^3 S^3 \\ & \searrow & & & & \nearrow & \\ \Sigma\downarrow & & \mathcal{M} & \xrightarrow{i} & C & & \downarrow\Sigma \\ & \nearrow^{j} & & & & \searrow & \\ C(R^4) & & & & & & \Omega^4 S^4 \end{array} \qquad (2.5)$$

where $C(R^n) = \coprod_{k>0} C_k(R^n)$, $\mathcal{M} = \coprod_{k>0} \mathcal{M}_k$, $C = \coprod_{k>0} C_k$, and ι is the natural inclusion of the positively indexed path components into the total iterated loop space. The inclusion $\iota \circ S$ is well known to be a group completion [8,21] and thus we find that $\coprod \mathcal{M}_k = \mathcal{M}$ is completely surrounded by iterated loop spaces (or spaces that group complete to iterated loop spaces). The homology theory of such spaces is well-known to be very rich and it is clear that if one could enrich $\mathcal{M}$ with an "iterated loop space structure" such that diagram 2.5 respects the additional structure (at least up to homotopy) then it would be possible to learn much more about $H_*(\mathcal{M})$ and $H_*(\mathcal{M}_k)$.

If we replace $G = Sp(1)$ by any compact, connected, simple Lie group G we obtain the composition of inclusions

$$\coprod C_k(R^4) \longrightarrow \mathcal{M} \longrightarrow C \longrightarrow \Omega^4 BG \qquad (2.6)$$

where every space but $\mathcal{M}$ is known to be a C_4 space and the total composite map is known to be a C_4 map. Here, C_4 is May's little cubes operad space (when $n = 4$), [19]. In the next section we will summarize results of [9] which show that, up to homotopy, $\mathcal{M}$ behaves like a C_4 space and that, up to homotopy, the inclusions in 2.6 behave like C_4 maps of spaces with compatible C_4 structures.

Notice, however, that any naive attempt to impose a compatible C_4 structure on $\mathcal{M}$ by directly minimizing the little cubes action on $\Omega^4 BG$ fails. This is evident even at the loop sum level in trying to construct a map $\mathcal{M}_k \times \mathcal{M}_l \xrightarrow{*} \mathcal{M}_{k+l}$ that commutes with the map $\Omega^4_k BG \times \Omega^4_l BG \xrightarrow{*} \Omega^4_{k+l} BG$ given by $\vartheta_2(c, f, g) = f * g$ for any fixed $c \in C_4(2)$. Essentially, the problem arises because $\vartheta_2(c, f, g)$ is constant on a rather large set while for an appropriate choice of gauge any $\omega \in \mathcal{M}_{k+l}$ can be seen to be analytic. Thus, some care must be taken in constructing a compatible C_4 structure on $\mathcal{M}$.

3 Operad Structures on Instantons

The next two sections are based on a talk presented by one of us at the Arcata conference. Proofs and further details may be found in [9]. We begin with a key technical result, namely the existence of a "loop sum" map

$$* : \mathcal{M}_k \times \mathcal{M}_l \longrightarrow \mathcal{M}_{k+l} \tag{3.1}$$

that is homotopy compatible with the loop sum map on $\Omega^4 BSp(n)$.

Theorem 3.2 *There is a loop sum map*

$$* : \mathcal{M}_k \times \mathcal{M}_l \longmapsto \mathcal{M}_{k+l} \tag{3.3}$$

such that the following diagram commutes up to homotopy

$$\begin{array}{ccc} \Omega^4_k BSp(n) \times \Omega^4_l BSp(n) & \xrightarrow{*} & \Omega^4_{k+l} BSp(n) \\ \uparrow & & \uparrow \\ \mathcal{M}_k \times \mathcal{M}_l & \xrightarrow{*} & \mathcal{M}_{k+l} \\ \uparrow & & \uparrow \\ C_k(R^4) \times C_l(R^4) & \xrightarrow{*} & C_{k+l}(R^4) \end{array} \tag{3.4}$$

where the vertical arrows are given by the the compositions of the natural inclusions i_* and any homotopy equivalence of $C_j \longrightarrow \Omega^4_j BSp(n)$, and $*$ is the standard loop sum map on $\Omega^4 BSp(n)$.

In order to construct the map $*$ we construct a map ϕ from $\mathcal{M}_k \times \mathcal{M}_l$ to the Taubes tubular neighborhood $\mathcal{C}^\epsilon_{k+l}$ of $\mathcal{M}_{k+l}$ (recall theorem 2.4). Then, using the strong deformation retraction $\mathcal{T}_{k+l} : \mathcal{C}^\epsilon_{k+l} \longrightarrow \mathcal{M}_{k+l}$ given in 2.4, we obtain the following commutative diagram

$$\begin{array}{ccc} & & \mathcal{C}^\epsilon_{k+l} \\ & \overset{\phi}{\nearrow} & \downarrow \mathcal{T}_{k+l} \\ \mathcal{M}_k \times \mathcal{M}_l & \xrightarrow{*} & \mathcal{M}_{k+l} \end{array} \tag{3.5}$$

The construction of our map ϕ depends on a deep result of Atiyah, Drinfeld, Hitchin and Manin [4] which gives a linear algebraic description of all instantons on S^4.

We next extend the loop sum maps $* : \mathcal{M}_k \times \mathcal{M}_l \longrightarrow \mathcal{M}_{k+l}$ defined to homotopy C_4 operad maps

$$\vartheta : C_4(p) \times_{\Sigma_p} (\mathcal{M}_k)^p \longrightarrow \mathcal{M}_{pk} \tag{3.6}$$

with the following essential property:

Theorem 3.7 *The following diagram commutes up to homotopy:*

$$\begin{array}{ccc} C_4(p)\times_{\Sigma_p}(\Omega^4_k BSp(n))^p & \xrightarrow{\vartheta} & \Omega^4_{pk}BSp(n) \\ \uparrow & & \uparrow \\ C_4(p)\times_{\Sigma_p}(\mathcal{M}_k)^p & \xrightarrow{\vartheta} & \mathcal{M}_{pk} \\ \uparrow & & \uparrow \\ C_4(p)\times_{\Sigma_p} C_k(R^4) & \xrightarrow{\vartheta} & C_{pk}(R^4) \end{array} \tag{3.8}$$

We should point out that the constructions of $*$ and ϑ are not well-defined if we mod out by the full gauge group, rather than the based gauge group, and thus our construction of the $*$ and ϑ maps does not descend to the $\mathcal{M}'$ level.

4 Homology Computations

Let $G = Sp(1)$. Then $\mathcal{M}_1$ is homeomorphic to $SO(3)\times B^5$, [5]. Furthermore, the natural inclusion $i_1 : \mathcal{M}_1 \longrightarrow C_1$ is equivalent, in homotopy, to the classical J-homomorphism $J: SO(3) \longrightarrow \Omega^3_1 S^3$, [9].

Theorems 3.2 and 3.7 show that $(\mathcal{M}\ ,\ \vartheta)$ is homotopy compatible with the usual C_4 operad structures on $\Omega^4 BSp(n)$. We assume that the reader is familiar with the definitions and properties of the standard homology operations on C_4 spaces (see, for example, Cohen's concise yet encyclopedic treatment [10, section 1, pages 213-219]), and recall the following classical facts:

Theorem 4.1 *a.* $H_*(SO(3), Z/2) \simeq E(x_1, x_2)$, *an exterior algebra over Z/2 where* $|x_1| = 1$, $|x_2| = 2$ *and the Pontrjagin product in* $(H_*(SO(3), Z/2)$ *is induced by the Lie group multiplication on* $SO(3)$.

b. $H_*(SO(3), Z/p) \simeq E(x_3)$, *an exterior algebra over Z/p on a single generator of dimension 3. Here p is an odd prime.*

Theorem 4.2 $H_*(\Omega^{m+1}S^{m+1}, Z/2) \cong Z/2([1], Q_I(1))$, *a polynomial algebra over Z/2, under the loop sum Pontrjagin product, on generators [1] and* $Q_I(1) = Q_{i_1}Q_{i_2}\ldots Q_{i_k}(1)$ *where* $I = (i_1, \ldots, i_k)$ *satisfies* $0 < i_1 \le i_2 \le \ldots \le i_k \le m$.

Theorem 4.3 *Let p be an odd prime. As algebras, under the loop sum Pontrjagin product:*

a. $H_*(\Omega^4 BS^3, Z/p) \cong H_*(\Omega^3 S^3, Z/p) \cong \bigwedge([1], Q^I(1))$ *where I is admissible,* $\epsilon(I) + b(I) > 0, s_k \le 1$.

b. $H_*(\Omega^4 S^4, Z/p) \cong \bigwedge([1], Q^I(1), Q^J(\lambda_3(1,1)))$ *where I, J admissible,* $\epsilon(I) + b(I) > 0, e(J) \ge 3, s_k(I) \le 1, s_k(J) \le 3$.

Theorem 4.4 *Let $J : SO(3) \longrightarrow \Omega_1^3 S^3$ be the classical J-homomorphism. Then $J_* : H_*(SO(3), Z/p) \longrightarrow H_*(\Omega_1^3 S^3, Z/p)$ is given by the following formulae:*

A. If $p = 2$:

$$J_*(x_1) = Q_1(1) * [-1]$$

$$J_*(x_2) = Q_2(1) * [-1]$$

$$J_*(x_1 x_2) = Q_2(1) * Q_1(1) * [-3] + Q_1 Q_1(1) * [-3] + (Q_1(1))^3 * [-5]$$

B. If $p = 3$

$$J_*(x_3) = Q_3(1) * [-2] = -\beta Q^I(1) * [-2]$$

C. If $p \geq 5$

$$J_* \equiv 0$$

Thus, we easily obtain the following two corollaries.

Corollary 4.5 *The natural inclusion $i : \mathcal{M}_1 \longrightarrow C_1$ induces a monomorphism in $Z/2$ and $Z/3$ homology and is trivial in Z/p homology for $p \geq 5$. Furthermore, if we write x_3 for $x_1 x_2 \in H_3(SO(3), Z/2)$, and z_q for the q-dimensional generator of $H_*(\mathcal{M}_1)$, then $i_*(z_q) = J(x_q)$ for all q and primes p.*

Corollary 4.6 *The natural inclusion $i : \mathcal{M}_1 \longrightarrow C_1$ induces an isomorphism for $q = 1$ and an injection for $q = 2$ in homology whereas $i_* : H_3(\mathcal{M}_1, Z) \longrightarrow H_3(C_1, Z)$ is isomorphic to the natural epimorphism of Z onto a $Z/12$ summand. Hence $ker(i_1)_3$ is isomorphic to Z.*

The operad maps of the previous section can be used to construct many new non-trivial classes in $H_*(\mathcal{M}_k, Z/2)$. Theorems 3.2 and 3.7 imply that the following diagrams commute:

$$\begin{array}{ccc} H_s(\Omega_k^4 BS^3, Z/2) \otimes H_t(\Omega_l^4 BS^3, Z/2) & \xrightarrow{*} & H_{s+t}(\Omega_{k+l}^4 BS^3, Z/2) \\ i_{k*} \otimes i_{l*} \uparrow & & \uparrow i_{k+l*} \\ H_s(\mathcal{M}_k, Z/2) \otimes H_t(\mathcal{M}_l, Z/2) & \xrightarrow{*} & H_{s+t}(\mathcal{M}_{k+l}, Z/2) \end{array} \tag{4.7}$$

and, for $i \leq 3$,

$$\begin{array}{ccc} H_s(\Omega_k^4 BS^3, Z/2) & \xrightarrow{Q_i} & H_{2s+i}(\Omega_{2k}^4 BS^3, Z/2) \\ i_{k*} \uparrow & & \uparrow i_{pk*} \\ H_s(\mathcal{M}_k, Z/2) & \xrightarrow{Q_i} & H_{2s+i}(\mathcal{M}_{2k}, Z/2) \end{array} \tag{4.8}$$

We may use our structure maps $\vartheta_j : C_4(j) \times \mathcal{M}_k^j \longrightarrow \mathcal{M}_{jk}$ to define classes in $H_*(\mathcal{M})$. These diagrams imply that the images of our classes in $H_*(\Omega^4 BS^3)$ are fully compatible with the usual C_4 homology operations in $\Omega^4 BS^3$. Thus, by abuse of notation we will use the standard notation for the operations to designate our homology classes in $H_*(\mathcal{M})$.

Proposition 4.9 *$z_i * [1] = Q_i(1) \in H_i(\mathcal{M}_2, Z/2)$ for $i = 1,2,3$.*

Corollary 4.10 *$Q_3(1) = Q_1Q_1(1) * [-2] + Q_2(1) * Q_1(1) * [-2] + Q_1(1)^3 * [-4]$ in $H_3(\Omega_2^4 BS^3, Z/2) \simeq H_3(\Omega_2^3 S^3, Z/2)$.*

Adopting the conventions $z_0 = [1]$ and $Q_\phi(z_i) = z_i$ for all i, we have

Theorem 4.11 *$H_*(\mathcal{M}_k, Z/2)$ contains elements of the form*

$$z = z(I_1, \ldots, I_n, j_1, \ldots, j_n) = Q_{I_1}(z_{j_1}) * \ldots * Q_{I_n}(z_{j_n}) \tag{4.12}$$

for all sequences $(I_1, \ldots, I_n, j_1, \ldots, j_n)$ such that $\sum_{m=1}^n 2^{l(I_m)} \leq k$. (Here each $I_m = (i_1, \ldots, i_{l(I_m)})$ is an admissible sequence $0 \leq i_1 \leq \ldots \leq i_{l(I_m)} \leq 3$ and $0 \leq j_a \leq 3$ for all $1 \leq a \leq n$). Furthermore the image of z in $H_*(C_k, Z/2)$ is given by replacing z_1, z_2 and z_3 in 4.12 by $Q_1(1) * [-1]$, $Q_2(1) * [-1]$, and $(Q_1Q_1(1) * [-3] + Q_2(1) * Q_1(1) * [-3] + (Q_1(1))^3 * [-5])$ respectively.

Corollary 4.13 *Let $k = 2^j$. Then $Q_3Q_3 \ldots Q_3(z_3) \in H_{6k-3}(\mathcal{M}_k, Z/2)$ has non-zero image in $H_{6k-3}(C_k, Z/2)$.*

Corollary 4.14 *With $Z/2$ coefficients i_* is a surjection for $q \leq k$.*

It is natural to ask if the bound $q = q(k) \leq k$ through which i_* is a surjection can be improved. The answer is no (at least using the methods developed in [9]). For example, $(Q_1(1))^{k+1} * [-k-2] \in H_{k+1}(C_k, Z/2)$ cannot be shown to be an image (i_*) for $k \leq 4$ using theorem 4.11 (and we conjecture the restriction $k \leq 4$ can be dropped). Of course, since $\mathcal{M}_k$ is finite dimensional and C_k has homology in every dimension surjectivity must fail sooner or later. One of the reasons it apparently fails sooner $(q = k+1)$ rather than later $(q > k+1)$ is the decomposability of Q_3 in $\Omega^3 S^3$ which also forces candidates for ker i_* (for $q > k$), as follows.

Proposition 4.15 *$Q_1(z_1) + z_2 * z_1 + z_3 * [1]$ and $Q_2(z_1) + z_3 * z_1$ are in the kernel of $i_* : H_*(\mathcal{M}_2, Z/2) \longrightarrow H_*(C_2, Z/2)$.*

$H_*(\mathcal{M}_k, Z/p)$ for p an odd prime can be analyzed using the same techniques. Again, the following diagrams commute:

$$\begin{array}{ccc} H_r(\Omega_k^4 BS^3, Z/p) \otimes H_t(\Omega_l^4 BS^3, Z/p) & \xrightarrow{*} & H_{r+t}(\Omega_{k+l}^4 BS^3, Z/p) \\ i_{k*} \otimes i_{l*} \uparrow & & \uparrow i_{k+l*} \\ H_r(\mathcal{M}_k, Z/p) \otimes H_t(\mathcal{M}_l, Z/p) & \xrightarrow{*} & H_{r+t}(\mathcal{M}_{k+l}, Z/p) \end{array} \tag{4.16}$$

$$\begin{array}{ccc} H_r(\Omega^4_k BS^3, Z/p) & \xrightarrow{Q^s} & H_{2s(p-1)+r}(\Omega^4_{pk} BS^3, Z/p) \\ i_{k*}\uparrow & & \uparrow i_{pk*} \\ H_r(\mathcal{M}_k, Z/p) & \xrightarrow{Q^s} & H_{2s(p-1)+r}(\mathcal{M}_{pk}, Z/p) \end{array} \tag{4.17}$$

and

$$\begin{array}{ccc} H_r(\Omega^4_k BS^3, Z/p) \otimes H_t(\Omega^4_l BS^3, Z/p) & \xrightarrow{\lambda_3} & H_{r+t+3}(\Omega^4_{k+l} BS^3, Z/p) \\ i_{k*} \otimes i_{l*}\uparrow & & \uparrow i_{k+l*} \\ H_r(\mathcal{M}_k, Z/p) \otimes H_t(\mathcal{M}_l, Z/p) & \xrightarrow{\lambda_3} & H_{r+t+3}(\mathcal{M}_{k+l}, Z/p) \end{array} \tag{4.18}$$

where the last diagram corresponds to the Browder operation λ_3 which exists in all C_4 spaces.

Proposition 4.19 $z_3 * [2] = -\beta Q^1(1) \in H_3(\mathcal{M}_3, Z/3)$.

Since the J homomorphism, $J : SO(3) \longrightarrow \Omega^3_1 S^3$, is trivial in Z/p homology if $p \geq 5$ the analog of 4.12 splits into two cases and we have

Theorem 4.20 *$H_*(\mathcal{M}_k, Z/p)$ contains elements of the form*

$$z = z(I_1, \ldots, I_n, j_1, \ldots, j_n) = Q^{I_1}(z_{j_1}) * \cdots * Q^{I_n}(z_{j_n}) \tag{4.21}$$

for all sequences $(I_1, \ldots, I_n, j_1, \ldots, j_n)$ such that $\sum_{m=1}^n p^{l(I_m)} \leq k$. (Here each I_m is an admissible sequence with $s_l = 1, \epsilon(I) + b(I) > 0$ and $j_a = 0$ or 3 for all $1 \leq a \leq n$). Furthermore, the image of z in $H_*(C_k, Z/p)$ is given as follows:
A. If $p = 3$ replace z_0 by $[1]$ and z_3 by $-\beta Q^1(1) * [-2]$.
B. If $p \geq 5$ then the image is zero if any $j_a = 3$. Otherwise, replace each z_0 by $[1]$.

Corollary 4.22 *With Z/p coefficients i_* is a surjection for $q \leq k$.*

For many choices of k and p, q can be chosen to be greater than k in 4.22. For example, if $p \leq k < 2p$ then i_* is a surjection for $q < 4p - 5$.

Corollary 4.23 *Let $k = 3^j$. Then $Q^{3^j} Q^{3^{j-1}} \ldots Q^9 Q^3(z_3) \in H_{6k-3}(\mathcal{M}_k, Z/3)$ has non-zero image in $H_{6k-3}(C_k, Z/3)$.*

Corollary 4.24 *For $p \geq 5$ $i_* : H_*(\mathcal{M}_k, Z/p) \longrightarrow H_*(C_k, Z/p)$ has a non-trivial kernel.*

5 Taubes' Patching and Loop Sums

Let (X^4, m) be a smooth, compact, connected, closed four manifold with Riemannian metric m. The Yang-Mills theory naturally extends to principal G bundles over X^4 where G is any simple compact Lie group. Of course, the notion of self-duality depends on the Hodge inner product on $\Lambda^2(X^4)$ and thus on the choice of metric m (up to conformal equivalence). In this general situation it is not at all clear which principal G bundles over which Riemannian manifolds (X^4, m) carry a self-dual connection and, if so, which of these moduli spaces of instantons are manifolds. It is the fundamental work of Taubes [23,24,25,26] that shows that for sufficiently high instanton number k (depending on the group G) these moduli spaces of instantons are *always* non-empty. As mentioned in the introduction, much beautiful and deep mathematics has been generated by studying the geometry of these *non-empty* spaces.

Taubes' method for constructing instantons on general four manifolds, which he called patching, is to transport known instantons on S^4 to discs in (X^4, m) via a smooth gluing procedure that a topologist would recognize as a "conformally close connected sum". Many treatments of patching are given in the literature; for example, one can consult [23], [11], [17], or [18]. Perhaps the most general version and the one of most interest to homotopy theorists is the Taubes' patching on conformal connected sums given by Donaldson [14]. In all cases one starts with two manifolds and two self-dual connections on those manifolds. In the patching procedure the two connections are used to build a connection over the connected sum of the two manifolds; however, during the initial gluing the self-duality condition is destroyed. Delicate analysis of the relevant partial differential equations is then used to show that this "almost instanton" may be deformed to an actual instanton. As mentioned in section two, Taubes' tubular neighborhood theorem [24,26] (2.4 is the version over S^4) explains precisely how the deformation may be accomplished. The operad constructions on $\mathcal{M}$ if [9] are easily seen to be global versions of Taubes patching using the Atiyah-Drinfeld-Hitchin-Manin characterization of instantons for input.

In [26] Taubes studies the stable behavior of the relative homotopy and homology of the pairs $(C_k, \mathcal{M}_k)$ for all (X^4, m), see [26, Theorem 2]. Applying his second main result to the case when $G = Sp(1)$ and $(X^4, m) = (S^4, m_0)$ where m_0 is the standard conformally flat metric yields:

Theorem 5.1 *Taubes, [26]. Let* $*[l] : (C_k, \mathcal{M}_k) \longrightarrow (C_{k+l}, \mathcal{M}_{k+l})$ *denote the map of pairs induced from the maps in diagram 3.4 by fixing a point in* $\mathcal{M}_l$. *Then for every* $x \in H_*(C_k, \mathcal{M}_k)$ *there is an* l_0 *depending on* x *and* k *such that for all* $l > l_0$, $*[l](x) = x * [l] = 0 \in H_*(C_{k+l}, \mathcal{M}_{k+l})$.

Example 5.2 *Using proposition 4.9 one can directly check theorem 5.1 on the classes given in proposition 4.15 and obtain*

$$(Q_1(z_1) + z_2 * z_1 + z_3 * [1]) * [4]$$

and

$$(Q_2(z_1) + z_3 * z_1) * [4]$$

are actually zero in $H_*(\mathcal{M}_6, Z/2)$.

Combining theorem 5.1 with the results of sections three and four yields the following corollaries:

Corollary 5.3 *For every* $x \in H_*(\mathcal{M}_k, Z/p)$ *there is an* l_0 *depending on* x *and* k *such that for all* $l > l_0$, $*[l](x) = x * [l] \in H_*(\mathcal{M}_{k+l}, Z/p)$ *is completely expressible in terms of Dyer-Lashof operations on* $[1]$.

Corollary 5.4 *For every non-zero* $x \in H_*(\mathcal{M}_k, Q)$ *there is an* l_0 *depending on* x *and* k *such that for all* $l > l_0$, $*[l](x) = x * [l] = 0 \in H_*(\mathcal{M}_{k+l}, Q)$.

Corollary 5.5 $\mathcal{M}_\infty = lim_{k \to \infty} \mathcal{M}_k$ *is stably rationally trivial.*

Finally, observe that every non-zero class $x \in H_*(\mathcal{M}_k, Q)$ generates a non-trivial unstable summand of the kernel of $i_* : H_*(\mathcal{M}_k, Z) \longrightarrow H_*(\mathcal{C}_k, Z)$. Corollary 4.6 exhibits this phenomenon when $k = 1$ and $* = 3$. We have a conjecture with Sam Gitler that this phenomenon occurs for every k with $* = 4k - 1$.

References

[1] M.F. Atiyah, *The Geometry of Yang-Mills Fields*, Scuola Normale Sup., Pisa, 1979.

[2] M.F. Atiyah, *Instantons in Two and Four Dimensions*, Comm. Math. Phys., 93 (1984), 437-451.

[3] M.F. Atiyah and R. Bott, *The Yang-Mills Equations over Riemann Surfaces*, Philos. Trans. Roy. Soc. London, Ser.A, 308 (1982), 523-615.

[4] M.F. Atiyah, V.G. Drinfeld, N.J. Hitchin and Y.I. Manin, *Construction of Instantons*, Phys. Lett. A, 65 (1978), 185-187.

[5] M.F. Atiyah, N.J. Hitchin and I. Singer, *Self-duality in Four Dimensional Riemannian Geometry*, Proc. Roy. Soc. London, Ser. A, 362 (1978), 425-461.

[6] M.F. Atiyah and J.D. Jones, *Topological Aspects of Yang-Mills Theory*, Comm. Math. Phys., 61 (1978), 97-118.

[7] M.F. Atiyah and R.S. Ward, *Instantons and Algebraic Geometry*, Comm. Math. Phys., 55 (1977), no. 2, 117-124.

[8] J.M. Boardman and R.M. Vogt, *Homotopy Everything H-spaces*, Bul. A.M.S., 79 (1973), 1236-1241.

[9] C.P. Boyer and B.M. Mann, *Homology Operations on Instantons*, J. Diff. Geo. 28 (1988), 423-465.

[10] F.R. Cohen, T.J. Lada and J.P. May, *The Homology of Iterated Loop Spaces*, L.N.M. 533 (1976), Springer-Verlag.

[11] S.K. Donaldson, *An Application of Gauge Theory to Four Dimensional Topology*, J. Diff. Geo., 18 (1983), 279-315.

[12] S.K. Donaldson, *Anti-Self-Dual Yang-Mills Connections over Complex Algebraic Surfaces and Stable Vector Bundles*, Proc. London Math. Soc., 50 (1) (1985), 1-26.

[13] S.K. Donaldson, *La Topologie Differentielle des Surfaces Complexes*, C.R. Acad. Sc. Paris Ser. I Math., 301 (1985), 317-320.

[14] S.K. Donaldson, *Connections, Cohomology and the Intersection Forms of 4-Manifolds*, J. Diff. Geo., 24 (1986), 275-341.

[15] R. Fintushel and R. Stern, *$SO(3)$-Connections and the Topology of Four-Manifolds*, J. Diff. Geo., 20 (2), (1984), 523-539.

[16] R. Fintushel and R. Stern, *Pseudofree Orbifolds*, Ann. Math., 122 (1985), 335-364.

[17] D. Freed and K.K. Uhlenbeck, *Instantons and Four-Manifolds*, Springer-Verlag, New York, 1984.

[18] H. B. Lawson, *The Theory of Gauge Fields in Four Dimensions*, A.M.S., CBMS 58 (1985).

[19] J.P. May, *The Geometry of Iterated Loop Spaces*, L.N.M. 271 (1972), Springer-Verlag.

[20] R.L. Mills and C.N. Yang, *Conservation of Isotopic Spin and Isotopic Gauge Invariance*, Phys. Rev., 96 (1954), 191.

[21] G. Segal, *Configuration Spaces and Iterated Loop-Spaces*, Invent. Math., 21 (1973), 213-221.

[22] I.M. Singer, *Some Remarks on the Gribov Ambiguity*, Comm. Math. Phys., 60 (1978), 7-12.

[23] C.H. Taubes, *Self-Dual Connections on Non-Self-Dual 4-Manifolds*, J. Diff. Geo., 17 (1982), 139-170.

[24] C.H. Taubes, *Path-Connected Yang-Mills Moduli Spaces*, J. Diff. Geo., 19 (1984), 337-392.

[25] C.H. Taubes, *Self-Dual Connections on 4-Manifolds with Indefinite Intersection Matrix*, J. Diff. Geo., 19 (1984), 517-560.

[26] C.H. Taubes, *The Stable Topology of Self-Dual Moduli Spaces*, preprint (1986), Harvard University.

[27] G.'t Hooft, *On the Phase Transition Toward Permanent Quark Confinement.*

[28] K.K. Uhlenbeck, *Connections with L^p-Bounds on Curvatures*, Comm. Math. Phys., 83 (1982), 31-42.

[29] K.K. Uhlenbeck, *Removable Singularities in Yang-Mills Fields*, Comm. Math. Phys., 83 (1982), 11-30.

ON REAL HOMOTOPY THEORY

EDGAR H. BROWN, JR.[1] AND ROBERT H. SZCZARBA

INTRODUCTION.

One of the most important tools in the study of foliations is Haefliger's classifying space $B\Gamma_q$. (See [5].) The importance of this space lies in the fact that it "classifies" codimension q foliations of smooth manifolds (see [5],[14]). Furthermore, a notion of continuous cohomology can be defined for $B\Gamma_q$ and all known characteristic classes of foliations come from these continuous cohomology classes. On the other hand, $B\Gamma_q$ is an enormous space; given q, there are integers $\{j\}$ and surjective homomorphisms of $\pi_j B\Gamma_q$ onto the real number R. (See [13] and [6].)

Our goal is to develop a context in which to study spaces like $B\Gamma_q$. The appropriate setting seems to be the category of simplicial spaces with continuous cohomology as the homology functor. Roughly speaking, we establish an equivalence between the category of simplicial spaces localized at R and the category of topological commutative differential graded algebras over R. (Compare [1], [9], [12].) The topics considered include continuous cohomology, continuous de Rham cohomology, the Kan extension condition, homotopy relations, real homotopy type and its relation to differential graded algebras.

In this note, we give our basic definitions and state our main results. The proofs will appear in a subsequent paper [3].

1. STATEMENT OF RESULTS.

For a category $\mathcal{C}$, $\Delta\mathcal{C}$ will denote the category of simplicial $\mathcal{C}$'s; $\mathcal{S}$ denotes the category of sets and hence $\Delta\mathcal{S}$ is the category of simplicial sets: $\mathcal{T}$ denotes the category of compactly generated, Hausdorff topological spaces [11]. If A, B are objects in $\mathcal{C}$, (A, B) denotes the set of morphisms from A to B. For $X, Y \in \Delta\mathcal{T}$, we make (X, Y) an object in $\mathcal{T}$ by viewing it as a subspace of $\Pi(X_q, Y_q)$ where the product topology and the function space topology on (X_q, Y_q) are in the sense of products and function spaces in $\mathcal{T}$.

---

[1]Partially supported by the National Science Foundation.

If $X \in \Delta\mathcal{T}$ and G is a topological abelian group, we define the *continuous cohomology groups of X with coefficients in G* by

$$H^q(X;G) = H_q(C^*(X;G), \delta)$$

where $C^q(X;G)$ is the set of all continuous $u : X_q \to G$ satisfying $us_i = 0, 0 \leq i < q$, and

$$(\delta u)(x) = \sum_{i=0}^{q+1} (-1)^i u(\partial_i x).$$

Here ∂_i and s_i are the face and degeneracy mappings of X. If G is a topological ring, we define cup products in $C^*(X;G)$ in the usual way. Our main concern is with $G = R$, the real numbers; we define

$$H^*(X) = H^*(X;R).$$

This is a graded commutative algebra over R.

Note that if X is a simplicial space with the discrete topology (that is, a simplicial set), then $H^*(X)$ is the ordinary real cohomology ring of X.

Remark. Many authors have defined notions of continuous cohomology in various contexts. (See Mostow [8], pp. vi, 9-10, for a discussion of some of these.) In particular, Mostow considers continuous cohomology theories on a category whose objects are continuous mappings $i : Y' \to Y$ of topological spaces. His theories satisfy four axioms and, in some cases, are characterized by these axioms. The relation between Mostow's continuous cohomology and ours can be described as follows. Let X be a simplicial space aid, let X^δ denote X in the discrete topology. If T^* is any continuous cohomology theory in the sense of Mostow and if each X_q is a paracompact, then,

$$T^*(\|X^\delta\| \to \|X\|) = H^*(X)$$

where $\| \ \|$ is the fat (or unnormalized) geometric realization functor (see [**10**] or [8], p. 68) and $\|X^\delta\| \to \|X\|$ is induced by the natural mapping $X^\delta \to X$. (See Corollary 7.5 of [8].)

Let $\Delta[q]$ be the simplicial standard q-simplex, that is, all tuples $(i_0, i_1, \ldots, 1_p)$ where $0 \leq i_0 \leq i_1 \leq \cdots \leq i_p \leq q$. Let $e_i : \Delta[q-1] \to \Delta[q]$ and $d_i : \Delta[q-1] \to \Delta[q]$ be the usual face inclusions and degeneracy projections. We form a simplicial Eilenberg-MacLane space, $K(G,n) \in \Delta T$ in the usual way, namely

$$K(G,n)_q = Z^n(\Delta[q]; G)$$

where Z^n denotes normalized cocycles topologized via the topology on G, $\partial_i = e_i^*$ and $s_i = d_i^*$. Note that $K(G,n)$ is actually a simplicial topological abelian group.

We now define a continuous version of the Kan condition. Let X be a simplicial space and $I = \{i_1, \ldots, i_\ell\}$, $0 \le i_1 < i_2 < \cdots < i_\ell \le q$, $1 \le \ell \le q$. Let $X(q,I)$ be the subspace of the ℓ-fold produce $(X_{q-1})^\ell$ of ℓ-tuples $(x_{i_1}, \ldots, x_{i_\ell})$ such that $\partial_i x_j = \partial_{j-1} x_i$ for $i, j \in I$, $i < j$, and let

$$p_{q,I} : X_q \to X(q,I)$$

be given by $p_{q,I}(x) = (\partial_{i_1} x, \ldots, \partial_{i_\ell} x)$.

Definition 1.1. A simplicial space X is *Kan* if the mappings $p_{q,I}$ have continuous sections $\lambda_{q,I}$ which satisfy the following:

(i) If $k, k+1 \notin I$ and $J = \{j_1, \ldots, j_\ell\}$ is defined by

$$\begin{aligned} j_m &= i_m \quad \text{for} \, i_m < k, \\ &= i_m - 1 \quad \text{for} \, i_m > k, \end{aligned}$$

then

$$s_k \lambda_{q,I}(x_{i_1}, \ldots, x_{i_\ell}) = \lambda_{q+1,J}(s_{k-1} x_{i_1}, \ldots, s_{k-1} x_{i_s}, s_k x_{i_{s+1}}, \ldots, s_k x_{i_\ell})$$

where $i_s < k < i_{s+1}$.

(ii) If k or $k+1 \in I$ and $x \in X_{q-1}$, then

$$\lambda_{q,I}(\partial_{i_1} s_k x, \ldots, \partial_{i_\ell} s_k x) = s_k x.$$

Conditions (i) and (ii) assert that $\lambda_{q+1}(x_{i_1}, \ldots, x_{i_\ell})$ is degenerate if the x_{i_j} are such as to make this possible in which case the λ's commute with the degeneracy operators.

Remark. It follows from Lemma 6.8 of [**7**] that our continuous version of the Kan condition reduces to the usual version for simplicial sets. The definition was chosen so that if X is Kan, then the simplicial function space $\mathcal{F}(Y, X)$ is Kan for any simplicial space Y; see Theorem 1.17.

If $Y \in \Delta\mathcal{T}$ is Kan, it can be shown that homotopy is an equivalence relation among maps from X to Y for any $X \in \Delta\mathcal{T}$. (A *homotopy* in $\Delta\mathcal{T}$ is a map $F : X \times \Delta[1] \to Y$.) If $[X, Y]$ denotes the set of homotopy classes of maps from X to Y, the usual arguments yield:

Theorem 1.2. *The simplicial space $K(G, n)$ is Kan and there is a natural isomorphism*

$$H^n(X; G) \simeq [X, K(G, n)]$$

where the group structure on $[X, K(G, n)]$ is defined using the group structure on $K(G, n)$.

The following result makes possible our analysis of real homotopy types.

Theorem 1.3. *According as n is even or odd, $H^*(K(R, n); R)$ is a polynomial or an exterior algebra on one generator in dimension n.*

Our original proof of this result outlined in [**2**], made use of a version of the Van Est Theorem for simplicial Lie groups. Our present proof uses a continuous cohomology analogue of the Serre spectral sequence of a twisted cartesian product of simplicial spaces and the Van Est Theorem for $n = 1$.

We next develop a de Rham approach to continuous cohomology. Let $\mathcal{A}$ denote the category of differential graded, topological algebras with unit over R which satisfy the following conditions: if $A \in \mathcal{A}$, $A = \Sigma A^p$, then $A^p = 0, p < 0$, $d : A^p \to A^{p+1}$, each A^p is a locally convex Hausdorff topological vector space over R, and multiplication $A^p \times A^q \to A^{p+q}$ is continuous. We denote by $\mathcal{AC}$ the full subcategory of $\mathcal{A}$ consisting of those algebras which are commutative in the graded sense:

$$ab = (-1)^{pq} ba$$

where $a \in A^p$, $b \in A^q$.

Following the techniques developed in [1], a central object for our study is the algebra of differential forms on the standard simplex:

$$\Delta^q = \{(t_0, t_1, \ldots, t_q) \in R^{q+1} \mid \Sigma t_i = 1, t_i \geq 0\}$$

Let ∇_q^p denote the C^∞ differential p forms on Δ^q with the C^∞ topology. The face inclusions and degeneracy projections, $e_i : \nabla^{q-1} \to \Delta^q$, $d_i : \Delta^{q+1} \to \Delta^q$ induce face and degeneracy maps $\partial_i : \nabla_q^p \to \nabla_{q-1}^p$ and $s_i : \nabla_q^p \to \nabla_{q+1}^p$. Then, for fixed p,

$$\nabla^p = \{\nabla_q^p, \partial_i, s_i\}$$

is in $\Delta\mathcal{T}$ and, for fixed q,

$$\nabla_q = \sum_p \nabla_q^p$$

is in $\mathcal{AC}$ (with differential the exterior differential). Combining these two structures, we obtain

$$\nabla = \{\nabla_q, \partial_i, s_i\}$$

in $\Delta\mathcal{AC}$. (Note that $\nabla_q^p \in \mathcal{T}$ because it is metrizable [11].)

If $X \in \Delta\mathcal{T}$, let $\langle X, \nabla^p\rangle$ be the set of simplicial mappings, $\langle X, \nabla^p\rangle$ topologized as a subset of the cartesian product $\Pi(X_q, \nabla_q^p)$ where (X_q, ∇_q^p) has the compact open topology. The linear structure on ∇^p makes $\langle X, \nabla^p\rangle$ onto a locally convex Hausdorff space. (We have not passed to compactly generated topologies on $\langle X_q, \nabla_q^p\rangle$ because this might destroy the local convexity.) Let $\mathcal{A}(X) \in \mathcal{AC}$ be defined by

$$\mathcal{A}^p(X) = \langle X, \nabla^p\rangle$$

An element $\omega \in \mathcal{A}^p(X)$ assigns to each q-simplex $x \in X_q$ a differential p-form on Δ_q and these forms fit together along faces, $e_i^*\omega(x) = \omega(\partial_i x)$, as well as respecting degeneracy. We think of $\mathcal{A}(X) \in \mathcal{AC}$ as the algebra of differential forms on X.

We define the *continuous de Rham cohomology of* X by

$$H^q_{dR}(X) = H_q(\mathcal{A}(X)).$$

Let $\psi : \mathcal{A}(X) \to C^*(X)$ be given by

$$\psi(\omega)(x) = \int_{\Delta^p} \omega(x)$$

for $\omega \in \mathcal{A}^p(X)$, $x \in X_p$. (Compare [**16**].)

Theorem 1.4. *The map ψ induces a ring isomorphism*

$$\Psi : H^*_{dR}(X) \to H^*(X).$$

We next describe a simplicial space $\Delta(A)$ for an algebra A in $\mathcal{AC}$ which has the property that, if A is free, R-nilpotent (see definition below) and of finite type, then

$$H_*(A) \approx H_*(\mathcal{A}(\Delta(A))).$$

If $A, B \in \mathcal{AC}$, we form $A_p \otimes B_q$ algebraically in the usual way with the projective topology on $A_p \otimes B_q$ ([**15**]; the projective topology in the strongest topology making $A_p \otimes B_q$ into a locally convex topological vector space such that $A_p \times B_q \to A_p \otimes B_q$ is continuous.)

If $A, B \in \mathcal{A}$ we topologize (A, B) so that it is in $\mathcal{T}$ as follows: For a topological space U let $k(U)$ be U with its compactly generated topology. For U and V in $\mathcal{T}$ their $\mathcal{T}$ product is k of their cartesian product and (U, V) is topologized by k of the compact open topology. Then, working in $\mathcal{T}$, (A, B) is topologized as a subspace of $\Pi(A_p, B_p)$.

In general, for $\Gamma \in \Delta\mathcal{A}$, $X \in \Delta\mathcal{T}$ and $A \in \mathcal{A}$, we form $\Gamma \otimes A \in \Delta\mathcal{A}$, $\langle X, \Gamma \rangle \in \mathcal{A}$ and $(A, \Gamma) \in \Delta\mathcal{T}$ by

$$(\Gamma \otimes A)_q = \Gamma_q \otimes A$$

$$\langle X, \Gamma \rangle^p = \langle X, \Gamma^p \rangle$$

$$(X, \Gamma)_q = (A, \Gamma_q)$$

For $A \in \mathcal{AC}$ we define its *real simplicial form* by

$$\Delta(A) = (A, \nabla).$$

For example, if $A = R[x_n]$ is the polynomial or exterior algebra on a generator x in dimension n with $dx = 0$, then

$$\Delta(R[x])_q = \{\omega \in \nabla_q^n \mid d\omega = 0\}$$

which is a differential form model of $K(R,n)$. If $A = R[x,y]$, $dx = 0$, and $dy = x^k$, then

$$\Delta(A)_q = \{(u,v) \in \nabla_q^n \times \nabla_q^{kn-1} \mid du = 0, dv = u^k\}$$

is a two stage Postnikov system.

Note that $\Delta : \mathcal{AC} \to \Delta\mathcal{T}$ and $\mathcal{A} : \Delta\mathcal{T} \to \mathcal{AC}$ are contravariant functors.

We next define the notion of an algebra $A \in \mathcal{AC}$ being FNF, that is free, nilpotent and of finite type. If $A \in \mathcal{AC}$, $a \in A^{n+1}$ and $da = 0$, we define $A[x] = A_a[x] \in \mathcal{AC}$ to be $A \otimes R[x]$ as a graded topological algebra, $\dim x = n$, and we define the differential by

$$d(b \otimes x) = db \otimes x + (1)^p (ba \otimes 1)$$

for $b \in A^p$.

Definition 1.5. An algebra $A \in \mathcal{AC}$ is *FNF (free, nilpotent and of finite type)* if it is isomorphic to $\cup A_n$ where $A_0 = R$, and $A_n = A_{n-1}[x]$. We also require that given m, there is an n so that $A^m = A_n^m$. We say A is *minimal* if, in addition, da is decomposable for all $a \in A$. An algebra $B \in \mathcal{AC}$ is *R-nilpotent and of finite type* if there is an FNF algebra A and $f : A \to B$ inducing an isomorphism on homology.

Theorem 1.6. *If A is R-nilpotent and of finite type, there is a minimal algebra M and $f : M \to A$ such that $f_* : H_*(M) \approx H_*(A)$. If $f' : M' \to A$ is another such pair, there is an isomorphism $h : M \to M'$ such that $f'h$ and f are homotopic (see below for the definition of homotopy in $\mathcal{A}$). If $A \in \mathcal{AC}$ is of finite type, $H_1(A) \approx 0$, then A is R-nilpotent.*

If $A \in \mathcal{AC}$, we have a map $i : A \to \mathcal{A}(\Delta(A))$ defined by $i(a)(u) = u(a)$.

Lemma 1.7. *If for each p, A^p has compactly generated topology (for example, if it is metrizable) then $i : A \to \mathcal{A}(\Delta(A))$ is continuous.*

Using the Serre spectral sequence for continuous cohomology, we prove

Proposition 1.8. *If A it is FNF and each A^p has compactly generated topology, then*

$$i_* : H_q(A) \to H_q(\mathcal{A}(\Delta(A)))$$

is an isomorphism for all $q \geq 0$.

For any simplicial space X, define $j : X \to \Delta\mathcal{A}(X)$ by the formula

$$j(x)(w) = w(x) \tag{1.9}$$

for $x \in X$, $w \in \mathcal{A}(X) = (X, \nabla)$.

Lemma 1.10. *The mapping $j : X \to \Delta\mathcal{A}(X)$ defined above is continuous.*

Theorem 1.11. *If $\mathcal{A}(X)$ is R-nilpotent and of finite type, there is a minimal algebra M and $f : X \to \Delta(M)$ such that*

$$f^* : H^*(\Delta(M)) \to H^*(X)$$

is an isomorphism. Furthermore, if M' and f' are another such pair there is an isomorphism $h : M \to M'$ such that $f\Delta(h)$ and f' are homotopic.

Since the proof is short we include it here. Suppose X is R-nilpotent and of finite type. By Theorem 1.6 there is a minimal algebra M and a map $g : M \to \mathcal{A}(X)$ such that g induces an isomorphism on H_*. Let $f : X \to \Delta(M)$ be the composite:

$$X \xrightarrow{j} \Delta(\mathcal{A}(X)) \xrightarrow{\Delta(g)} \Delta(M).$$

By Theorem 1.4 it is sufficient to show that $\mathcal{A}(f) : \mathcal{A}(\Delta(M)) \to \mathcal{A}(X)$ induces an isomorphism on H_*. Consider the commutative diagram

$$\begin{array}{ccccc} \mathcal{A}\Delta(M) & \xrightarrow{\mathcal{A}\Delta(g)} & \mathcal{A}\Delta(\mathcal{A}(x)) & \xrightarrow{\mathcal{A}(jx)} & \mathcal{A}(X) \\ {\scriptstyle i_M}\uparrow & & \uparrow{\scriptstyle i_{\mathcal{A}(X)}} & & \\ M & \xrightarrow{g} & \mathcal{A}(X) & & \end{array}$$

One easily checks that $\mathcal{A}(jx)i_{\mathcal{A}(X)}$ is the identity. By Proposition 1.8, i_M induces an isomorphism in continuous cohomology as does g by construction. Hence $\mathcal{A}(jx\Delta(g)) = \mathcal{A}(f)$ induces an isomorphism in continuous cohomology. The last part of Theorem 1.11 follows from Theorem 1.6 and the fact that Δ preserves homotopy (see Theorem 1.20 below).

The relation between the homotopy types of X and $\Delta(M)$ in $\Delta\mathcal{T}$ and the relation between their associated simplicial sets (forgetting the topology) is as present unclear to the authors. For example, it is probably true that $|\Delta(M)|$ is contractible. We next give very stringent hypotheses on X which will ensure that X and $\Delta(M)$ are related as one would most optimistically expect (see Theorem 1.13).

Suppose $X \in \Delta\mathcal{T}$ is Kan, $x_0 \in X$ is a base point and $\pi_n(X, x_0)$ denotes the usual homotopy groups of X, as a simplicial set, with the topology coming from the topology on X_n; $\pi_n(X, x_0)$ is a quotient of $\{x \in X_n \mid \partial_i x = x_0, 0 \leq i \leq n\}$ with the quotient topology. Let

$$\pi^n(X, x_0) = \mathrm{Hom}_{\mathrm{cont}}(\pi_n(X, x_0), R).$$

Suppose π is a topological abelian group, $X \in \Delta\mathcal{T}$, $u \in C^{n+1}(X, \pi)$, and $\delta u = 0$. Let $p : E(u) \to X$ be the fibration

$$E(u)_q = \{(x, v) \in X_q \times C^n(\Delta[q]; \pi) \mid t_x^\star u = \delta v\},$$

where $p(x, v) = x$, and $t_x : \Delta[q] \to X$ is the unique map such that $t_x(0, \ldots, q) = x$.

Definition 1.12. A simplicial space X has a *simple Postnikov system* if it has the homotopy type of X' in $\Delta\mathcal{T}$ where $X' = \varprojlim X_n$, $X_n = E(u_n)$, $u_n \in C^{n+1}(X_{n-1}; \pi_n)$, π_n is a topological abelian group and $X_0 = pt$. If in addition, each π_n is locally Euclidean we say that X is of *finite type*. Note that $\pi_n(X') \approx \pi_n$ as topological groups.

Theorem 1.13. *If X has a simple Postnikov system and is of finite type, and if $f : X \to \Delta(M)$ is as in Theorem 1.11, then*

$$f^* : \pi^i(\Delta(M)) \approx \pi^i(X).$$

Furthermore,

$$\pi^i(\Delta(M)) \approx (M/\hat{M})^i$$

where $\hat{M} \subset M$ is the ideal of decomposable elements.

Remark. If $X \in \Delta\mathcal{S}$ is simple and of finite type, then, viewed as an object in $\Delta\mathcal{T}$ with the discrete topology, it has a simple Postnikov system and is of finite type.

We next investigate how maps behave under the functor Δ. Suppose $A, B \in \mathcal{AC}$. We define a function space $\mathcal{F}(A,B) \in \Delta\mathcal{T}$ by

$$\mathcal{F}(A,B) = (A, \nabla \otimes B).$$

Thus a q-simplex in $\mathcal{F}(A,B)$ is a DGA map $A \to \nabla_q \otimes B$. Note

$$\Delta(A) = \mathcal{F}(A,R).$$

As usual, we define $\mathcal{F}(X,Y) \in \Delta\mathcal{T}$, for $X, Y \in \Delta\mathcal{T}$ by

$$\mathcal{F}(X,Y)_q = (\Delta[q] \times X, Y).$$

Let composition

$$\mathcal{O} : \mathcal{F}(A,kB) \times \mathcal{F}(B,C) \to \mathcal{F}(A,C) \tag{1.14}$$

be defined by

$$(A, \nabla_q \otimes B) \times (B, \nabla_q \otimes C) \xrightarrow{id \times \beta} (A, \nabla_q \otimes B) \times (\nabla_q \otimes B, \nabla_q \otimes C) \xrightarrow{\mathcal{O}_0} (A, \nabla_q \otimes C)$$

where $\mathcal{O}_0$ is the usual composition and $\beta(u)(\omega \otimes b) = (\omega \otimes b)(u(b))$.

Lemma 1.15. *The map $\mathcal{O}$ is continuous.*

The proof of this result uses the fact that we are working with compactly generated spaces.

Taking $C = R$ in (1.14) and taking the adjoint of $\mathcal{O}$, we obtain a continuous map

$$\Delta : \mathcal{F}(A, B) \to \mathcal{F}(\Delta(B), \Delta(A))$$

which extends the map $\Delta : (A, B) \to (\Delta(B), \Delta(A))$.

Let $\gamma : \mathcal{F}(A, \mathcal{A}(X)) \to \mathcal{F}(X, \Delta(A))$ be the adjoint of

$$\mathcal{F}(A, \mathcal{A}(X)) \times X \xrightarrow{1 \times j} \mathcal{F}(A, \mathcal{A}(X)) \times \mathcal{F}(\mathcal{A}(X), R) \xrightarrow{\mathcal{O}} \mathcal{F}(A, R).$$

One of the main results is that Δ and γ are weak equivalences when A and B are FNF and X is connected.

Definition 1.16. If X, Y in $\Delta\mathcal{T}$ are Kan, then $f : X \to Y$ is a *weak equivalence* if for all base points $x_0 \subset X$ and all $n \geq 0$, $f_\star : \pi_n(X, x_0) \to \pi_n(Y, f(x_0))$ is a group isomorphism for $n > 0$ and a set isomorphism when $n = 0$. ($f_\star$ is automatically continuous but need not be a homeomorphism.)

Remark. We note that a weak equivalence $f : X \to Y$ need not induce isomorphisms on continuous cohomology. For example, the "identity" mapping from $K(R^\delta, n)$ onto $K(R, n)$ is certainly a weak equivalence but is not a continuous cohomology isomorphism. However, if X and Y have simple Postnikov systems and $f : X \to Y$ is a weak equivalence with the property that each $f_q : \pi_q(X, x_0) \to \pi_q(Y, f(x_0))$ is a homeomorphism, then f can be shown to induce isomorphisms on continuous cohomology.

We have the following two results:

Theorem 1.17. *If $X, Y \in \Delta\mathcal{T}$ and Y is Kan, then $\mathcal{F}(X, Y)$ is Kan.*

Theorem 1.18. *If $A, B \in \mathcal{AC}$ and A is FNF, then $\mathcal{F}(A, B)$ and $\Delta(A)$ are Kan.*

The usual arguments and Theorem 1.17 show that homotopy for mapping $f : X \to Y$ is an equivalence relation when Y is Kan and $[X, Y] = \pi_0(\mathcal{F}(X, Y))$. For $A, B \in \mathcal{AC}$, define a homotopy to be a map $F : A \to \nabla_1 \otimes B$. Then Theorem 1.18 yields the fact that homotopy is an equivalence relation in $\mathcal{A}$ and $[A, B] = \pi_0(\mathcal{F}(A, B))$ for A FNF. Since Δ maps function spaces to function spaces we have

Lemma 1.19. *Δ induces a map $[A, B] \to [\Delta(B), \Delta(A)]$ when A is FNF.*

Theorem 1.20. *If A and $B \in \mathcal{AC}$ are FNF then*

$$\Delta : \mathcal{F}(A, B) \to \mathcal{F}(\Delta(B), \Delta(A))$$

is a weak equivalence; in particular, Δ induces a bijection $[A, B] \simeq [\Delta(B), \Delta(A)]$. If, in addition, $X \in \Delta\mathcal{T}$ is O-connected ($\pi_0(X) = pt$), then

$$\gamma : \mathcal{F}(A, \mathcal{A}(X)) \to \mathcal{F}(X, \Delta(A))$$

is a weak equivalence. Finally, if $A \in \mathcal{AC}$ is FNF and $B, C, f : B \to C$ are in $\mathcal{AC}$ with $f_* : H_*(B) \approx H_*(C)$, then

$$f_\sharp : \mathcal{F}(A, B) \to \mathcal{F}(A, C)$$

is a weak equivalence.

We can recast our results into a categorical form as in Quillen [9] as follows: let $\mathcal{AC}_0$ be the full subcategory of $\mathcal{AC}$ consisting of objects which are FNF. Let $\Delta\mathcal{T}_0$ be the full subcategory of $\Delta\mathcal{T}$ consisting of those X such that $\mathcal{A}(X)$ is FNF.

Definition 1.21. A mapping $h : A \to B$ in $\mathcal{AC}_0$ is an *R-equivalence* if $h_* : H_*(A) \approx H_*(B)$. A mapping $f : X \to Y$ in $\Delta\mathcal{T}_0$ is an *R-equivalence* if $f^* : H^*(Y) \approx H^*(X)$.

Let $\Delta\mathcal{T}_{0R}$ and $\mathcal{AC}_{0R}$ be the categories $\Delta\mathcal{T}_0$ and $\mathcal{AC}_0$ localized with respect to their R-equivalences ([9]). We define a functor $\Delta_h : \mathcal{AC}_{0R} \to \Delta\mathcal{T}_{0R}$ as follows: For each $A \in \mathcal{AC}_0$ choose $f_A : M_A \to A$ where M is free and f_A is an R-equivalence. Let $\Delta_h(A) = \Delta(M_A)$. By Theorem 1.20, if $g : B \to C$ is a map, there is a unique map $\bar{g} : M_B \to M_C$ in $\mathcal{A}_{0R}$ such that $f_C\bar{g} = gf_B$. Let $\Delta_h(g) = \bar{g}$. Then Theorems 1.11 and 1.20 immediately give:

Theorem 1.22. $\Delta_h : \mathcal{AC}_{0R} \to \Delta\mathcal{T}_{0R}$ *is an equivalence of categories in the sense that* Δ_h *is an isomorphism on morphism sets and each object of* $\Delta\mathcal{T}_{0R}$ *is isomorphic to* $\Delta_h(M)$ *for some* M.

Remark. A casual reading of Theorem 1.22 might suggest that if $A, B \in \mathcal{AC}_0$ and $f : A \to B$ induces an isomorphism on $H_\star$, then $\Delta(f) : \Delta(B) \to \Delta(A)$ induces an isomorphism on $H^\star$. If A and B are not free this may not be true. It is true that if f is a homotopy equivalence i.e. there is a map $g : B \to A$ such that fg and gf are homotopic to the identity, then $\Delta(f)$ is a homotopy equivalence because Δ extends to $\mathcal{F}(A, B) \to \mathcal{F}(A(B), \Delta(A))$.

Let $\alpha : \Delta\mathcal{S} \to \Delta\mathcal{T}$ be the functor which assigns to each simplicial set X, the simplicial space X with the discrete topology. One may localize with respect to the rationals to form $\Delta\mathcal{S}_{0Q}$, exactly as above and α induces a functor

$$\bar{\alpha} : \Delta\mathcal{S}_{0Q} \to \Delta\mathcal{T}_{0R}.$$

Our next result shows that this functor is neither injective nor surjective.

Theorem 1.23. *There are simply connected simplicial sets* X_1 *and* X_2 *each of finite type such that* X_1 *and* X_2 *are not isomorphic in* $\Delta\mathcal{S}_{0Q}$ *but* $\bar{\alpha}(X_1)$ *and* $\bar{\alpha}(X_2)$ *are isomorphic. In addition, there is a FNF* $A \in \mathcal{AC}$ *such that* $\Delta(A)$ *is not isomorphic to anything in the image of* $\bar{\alpha}$.

We conclude with a formulation of a main result of [4] in our context.

If B is a commutative graded algebra of finite type, we view it as being in $\mathcal{AC}$ with zero differential. Let $M(B)$ be a minimal algebra and $\gamma : M(B) \to B$ a map such that γ induces an isomorphism $H_\star(M(B)) \to H_\star(B) = B$.

Definition 1.24. A simplicial space $X \in \Delta\mathcal{T}$ is *R-formal* if there is a map $g : X \to \Delta(M(H^\star(X)))$ inducing an isomorphism in cohomology.

Theorem 1.25. *If* N *is a compact nilpotent Kähler manifold, then the total singular complex* $\Delta(N)$ *in the discrete topology is R-formal. Hence there is a minimal algebra*

$M(H^\star(N)) \in \mathcal{A}C$ and a map $g : \Delta(N) \to \Delta(M(H^\star(N)))$ inducing isomorphisms on $H^\star$ and $\pi^\star$.

References

1. A.K. Bousfield and V.K.A.M. Gugenheim, *On PL de Rham theory and rational homotopy type*, Memoirs of A.M.S., Providence **179** (1976).
2. E.H. Brown and R.H. Szczarba, *Continuous cohomology*, Contemporary Mathematics **58, Part II** (1987), 15–19.
3. E.H. Brown and R.H. Szczarba, *Real homotopy theory and continuous cohomology*, to appear in Trans. Amer. Math. Soc..
4. P. Deligne, P. Griffiths, J. Morgan, and D. Sullivan, *Real homotopy theory of Kähler manifolds*, Invent. Math. **29** (1975), 245–274.
5. A. Haefliger, *Feuilletages sur les variétés ouvertes*, Topology 9 **No. 2** (1970), 183–194.
6. S. Hurder, *On the classifying spaces of smooth foliations*, Ill. J. Math. **29** (1985), 108–133.
7. J.P. May, "Simplicial Objects in Algebraic Topology," The University of Chicago Press, Chicago, Ill, 1982.
8. M.A. Mostow, *Continuous cohomology of spaces with two topologies*, Memoirs of A.M.S. **175** (1976).
9. D.G. Quillen, *Rational homotopy theory*, Ann. of Math. **90** (1969), 205–295.
10. G. Segal, *Categories and cohomology theories*, Topology **13** (1974), 293–312.
11. N.E. Steenrod, *A convenient category of topological spaces*, Michigan Math. Jour. **14** (1967), 133–152.
12. D. Sullivan, *Infinitesimal computations in topology*, Publ. Math. I.H.E.S. **47** (1977), 269–331.
13. W.P. Thurston, *Noncobordant foliations of S^3*, Bull. Amer. Math. Soc. **78** (1972), 511–514.
14. W.P. Thurston, *The theory of foliations of codimension greater than one*, Comment. Math. Helv. **49** (1974), 214–231.
15. F. Treves, "Topological Vector Spaces, Distributions, and Kernels," Academic Press, New York, 1967.
16. J.L. Dupont, *Simplicial de Rham cohomology and characteristic classes of flat bundles*, Topology **15** (1976), 233-245.

SOME REMARKS ON THE SPACE IM J

by

F.R. Cohen and F.P. Peterson

Section 1, Preface.

In this note we give some results about the homology of the general linear groups and orthogonal groups which follow directly from work of Quillen [3] and Fiedorowicz and Priddy [1]. As an immediate consequence, one gets very simple proofs that the p-local space Im J is a retract of many natural spaces. Examples include $(BM)^+$ where M is the stable automorphism group of the free group or its outer-automorphism group. Other examples include Tornehave's splitting of SG as Im J × Coker J [4] or the Harris-Segal results on the K-theory of number rings [2].

The proofs here just use the homology of the cyclic group as a coalgebra over the Steenrod algebra. The idea of the proof is very close to that given in Tornehave's thesis with a slight twist given by Lemma 2.1. The methods here ought to apply to the other "image of J" spaces [1], but we make no systematic attempt to carry this out here.

Section 2, Results.

Let p and q be primes such that $q^i - 1 \neq 0 \bmod p$ if $0<i<p-1$ and $q^{p-1} - 1 = pv$ for $(p,v) = 1$. Write $GL_n(R)$ and $O_n(R)$ for the general linear group and orthogonal groups over R; GL(R) and O(R) denote the natural colimits. If F_q is the prime field of characteristic q, then Quillen gives $H_*(GL(F_q);F_p)$ as a Hopf algebra over the Steenrod algebra [3]; Fiedorowicz and Priddy give $H_*(O(F_3);F_2)$ as a Hopf algebra over the Steenrod algebra [1]. The following is a rather long exercise in morphisms of coalgebras over the Steenrod algebra, and is proved in section 3.

Lemma 2.1.

(1) If p is an odd prime and $f:X \to Y$ is any map of spaces which non-zero on $H_{2p-3}(\ ;F_p)$, then f_* is a mod-p homology isomorphism where X and Y are spaces whose mod-p homology is isomorphic to that of $BGL(F_q)$ as a coalgebra over the Steenrod algebra.

(2) If $f:X \to Y$ is any map of spaces which induces a mod-2 homology isomorphism in dimensions 1, 2, and 3, then f_* is a mod-2 homology isomorphism where X and Y are spaces whose mod-2 homology

is isomorphic to that of $BO(F_3)$ as a coalgebra over the Steenrod algebra.

In what follows we shall check that $BGL(F_q)^+$ is a retract of many spaces. One useful feature of these retractions is the fact that $BGL(F_q)^+$ is homotopy equivalent to $J_{(p)}$ where this last space denotes the localization at p, $p>2$, of the fibre of $\psi^q - 1: BU \to BU$. If $p = 2$, let $JO(3)$ denote the fibre of $\psi^3 - 1: BO \to BSO$. This fact is proven in [1,3], and we state it as Lemma 2.2.

<u>Lemma 2.2</u>. If $p>2$, there is a map $BGL(F_q) \to J_{(p)}$ which induces a p-local equivalence $BGL(F_q)^+ \to J_{(p)}$. There is a map $BO(F_3) \to JO(3)$ which induces a 2-local homotopy equivalence $BO(F_3)^+ \to JO(3)$.

Let M be a topological space. We say that <u>M is J-retractable at p</u> if there are maps

$$\lambda: BG \to M \quad \text{and} \quad \rho: M \to B\pi^+$$

where (1) $G = \Sigma_\infty$ is either the natural colimit of the symmetric groups Σ_n if $p>2$ or $O(F_3)$ if $p = 2$ and (2) π is either $GL(F_q)$ if $p>2$ or $O(F_3)$ if $p = 2$, such that

(i) if $p>2$, $\rho\cdot\lambda$ is non-zero on $H_{2p-3}(\ ;F_p)$ or

(ii) if $p = 2$, $\rho\cdot\lambda$ is an isomorphism on $H_j(\ ;F_2)$ for $j \leq 3$.

The proof of the following corollary is given in section 3.

<u>Corollary 2.3</u>. If M admits the plus construction abelianizing $\pi_1(M)$ and M is J-retractable at p, then $J_{(p)}$ is a retract of M^+. Furthermore if $M^+ \to J_{(p)}$ is an H-map in <u>any</u> H-structure, then M^+ is homotopy equivalent to $J_{(p)} \times$ another space. The retraction is natural for maps $f: M \to M'$ giving homotopy commutative diagrams

$$\begin{array}{ccc} M & \xrightarrow{f} & M' \\ & {\scriptstyle\rho}\searrow \quad \swarrow{\scriptstyle\rho'} & \\ & B\pi^+ & \end{array} \qquad \text{and} \qquad \begin{array}{ccc} & B\Sigma_\infty & \\ & {\scriptstyle\lambda}\swarrow \quad \searrow{\scriptstyle\lambda'} & \\ M & \xrightarrow{f} & M'. \end{array}$$

<u>Remark 2.4</u>. By inspecting the proof of 2.3, we see that a sufficient condition for M to be J-retractable at p is that the composite $\rho\cdot\lambda\cdot Bi$, for $i: Z/pZ \to \Sigma_\infty$ with $i(1) = (1,\ldots,p)$, induces a non-zero map on $H_{2p-3}(\ ;F_p)$ for p an odd prime. A similar statement follows for $p = 2$ with Z/pZ replaced by the dihedral group of order 8.

We list some examples satisfying 2.3. It follows that $\pi_* J_{(p)}$ is a summand of the homotopy groups in the following examples.

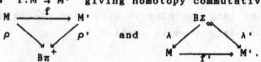

Example 2.5. Let Aut(n) denote the automorphism group of the free group on n letters and Aut the colimit of Aut(n) under the natural inclusions. Let Out denote the outer automorphism group of Aut. Define

$$\lambda: B\Sigma_\infty \to BAut$$

to be given by permutations of basis vectors. Abelianization induces a map $i: Aut \to GL(Z)$ which factors through Out. Finally, reduction mod-q gives $GL(Z) \to GL(F_q)$ and $\rho: Aut \to GL(F_q)$ is the composite of these last two maps. Thus if $p>2$, $BAut^+$ is homotopy equivalent to $J_{(p)} \times$ another space and $H_*(Out,F_p)$ has $H_*(GL(F_q);F_p)$ as a summand.

Example 2.6. Consider the natural inclusion $\lambda: \Sigma_\infty \to GL(F_q)$ and the map $\theta: GL_n(F_q) \to \Sigma_{q^n}$ defined by Quillen [3]: $GL_n(F_q)$ acts on the left of the vector space $(F_q)^n$ considered as a finite set. Thus one gets a map $\theta: BGL(F_q) \to B\Sigma_\infty^+$ as in section 3 here. We check that Σ_∞ is J-retractable at p and so one gets Tornehave's theorem that SG is homotopy equivalent to $J_{(p)} \times$ Coker J [4].

Example 2.7. Let R be a ring which admits a ring homomorphism

$$\overline{\rho}: R \to F_q.$$

If $p>2$, it follows that the natural inclusion $\Sigma_\infty \subset GL(R)$ followed by the reduction map $\rho = GL(\overline{\rho}): GL(R) \to GL(F_q)$ implies that if $p>2$, GL(R) is J-retractable at p. Thus $BGL(R)^+ \simeq J_{(p)} \times$ another space. Thus one gets the odd-primary results in [2]. Modifications are required when $p=2$.

Example 2.8. Let X be a pointed CW complex and $\vee_n X$ the n-fold wedge. Let $Homeo_n(X)$ denote the based homeomorphisms of $\vee_n X$, $Aut_n(X)$ the based homotopy equivalences of $\vee_n X$, and $GL_n(\overline{H}_*X)$ the group of graded linear isomorphisms of $\overline{H}_*(\vee_n X;R)$. Let $Homeo_\infty(X)$, $Aut_\infty(X)$, $GL_\infty(\overline{H}_*X)$ denote the natural colimits. Permutation of coordinates induces an inclusion $\lambda: \Sigma_\infty \to Homeo_\infty(X)$. If $\overline{H}_j(X_j;F_q) \cong F_q$ for some j, then $GL_\infty(\overline{H}_*X)$ projects to $GL(F_q)$. Thus there is a composite of homomorphisms (*) $Homeo_\infty(X) \longrightarrow Aut_\infty(X) \longrightarrow \pi_0 Aut_\infty(X) \xrightarrow{\rho} GL(F_q)$. For example if $X = S^1$, then $\pi_0 Aut_\infty(X)$ is example 2.4. Thus if $p>2$, then homomorphisms of monoids in (*) give that all classifying spaces of monoids here are J-retractable at $p>2$.

Un-example 2.9. One might wonder whether $B\Gamma^+$ has $J_{(p)}$ as a retract where Γ is the stable mapping class group. It is easy to see that $J_{(p)}$ is a stable retract of $B\Gamma$ through finite sketeta. (The proof of this will appear elsewhere.) However a direct calculation with Euler characteristics gives that a homomorphism $\Sigma_n \to \Gamma^0_{g,0}$ is zero on $H_{2p-3}(\ ;F_p)$ if $n \geq p^2$.

Section 3, Proofs.

We first prove lemma 2.1. The proof naturally breaks up into two cases, the first of which is where p is an odd prime.

By [3], $H_*(GL(F_q);F_p) \cong S[H_*(\Sigma_p;F_p)]$ as a Hopf algebra over the Steenrod algebra, where $S[\quad]$ is the symmetric algebra functor. There is basis for $H_*(\Sigma_p;F_p)$ given by

(1) e_{jq} and e_{jq-1}, $q = 2p - 2$, and $|e_i| = i$,

(2) $\Delta(e_{jq}) = \Sigma\, e_{iq} \otimes e_{(j-i)q}$

$\Delta(e_{jq-1}) = \Sigma\, e_{iq} \otimes e_{(j-1)q-1} + \Sigma\, e_{iq-1} \otimes e_{(j-i)q}$,

(3) $\beta(e_{jq}) = (e_{jq-1})$ and

(4) $P^i_*(e_{(k+i)q}) = \begin{bmatrix} k(p-1) \\ i \end{bmatrix} e_{kq}$,

$P^i_*(e_{(k+i)q-1}) = \begin{bmatrix} k(p-1)-1 \\ i \end{bmatrix} e_{kq-1}$

Thus the primitives in $H_*(GL(F_q);F_q)$ are given by the Newton polynomials in the e_i. The even degree primitives are x_k of degree kq where

$$x_1 = e_q \quad \text{and}$$
$$x_k = ke_{kq} - \sum_{i=1}^{k-1} e_{iq}x_{k-1}$$

The odd degree primitives are given by b_k of degree $kq - 1$ where

$$b_1 = e_{q-1} \quad \text{and}$$
$$b_k = e_{kq-1} - \sum_{i=1}^{k-1} e_{iq}b_{k-i}.$$

Notice that there is at most one primitive in any fixed degree. Furthermore, the even degree primitives are given by $x_1^{p^n}$ and $(x_j)^{p^n}$, $j>1$, $j \not\equiv 0 \bmod p$.

Notice that if $j \not\equiv 1 \pmod p$, then $P^1_*x_j \neq 0$ and so $P^{p^n}_*(x_j)^{p^n} \neq 0$. If $j \equiv 1 (\mathrm{mod}\ p)$, $j>0$, then $P^2_*x_j \neq 0$ and so $P^{2p^n}_*(x_j)^{p^n} \neq 0$. If k

$\not\equiv 0 \pmod p$, then $P^1_* b_k \neq 0$ and $P^p_*(b_{np+1}) \neq 0$.

Let $\alpha: GL(F_q) \to GL(F_q)$ be any homomorphism with $\alpha_*(e_q) = e_q$. (There are two different q's here!) Thus $\alpha_*(x_i) \neq 0$ for $i<p$ by commutation with P^1_*. Assume that

$$\alpha_*(x_i) \neq 0,\ i \leq N,\ i \not\equiv 0 \pmod p.$$

If $N + 1 \not\equiv 0,1 \pmod p$, then $\alpha_*(x_{N+1}) \neq 0$ by commutation with P^1_* and induction. If $N + 1 \equiv 1 \pmod p$, then $P^2_* x_{N+1} \neq 0$. Hence

$$\alpha_*(x_i) \neq 0 \quad \text{for} \quad i \not\equiv 0 \pmod p.$$

We next check that $\alpha_*(x_i^{p^n}) \neq 0$ if $i \not\equiv 0 \pmod p$ by induction on n. The above paragraph is the case $n = 0$. Also notice that by the above paragraph, it suffices to check that $\alpha_*(x_1^{p^n}) \neq 0$ and that our inductive hypothesis gives $\alpha_*(x_1^{p^{n-1}}) \neq 0$. By commutation with the coproduct one has

$$\alpha_*(x_1^{p^n-1}\cdot x_2) = e(x_1^{p^n-1}\cdot x_2) + \text{"others"}$$

for $e \neq 0$. Notice that "others" is a linear combination of monomials not involving $x_1^{p^n-1}\cdot x_2$. Apply P^1_* to this equation to get $\alpha_*(x_1^{p^n}) \neq 0$.

By commutation with the Bockstein, $\alpha_*(e_{q-1}) \neq 0$. Assume $\alpha_*(b_i) \neq 0$, $i \leq N$. If $N + 1 \not\equiv 0 \pmod p$, then $\beta x_{N+1} \neq 0$ and $\alpha_*(b_{N+1}) \neq 0$ by commutation with β (and the above two paragraphs). If $N + 1 \equiv 0 \pmod p$), then $\alpha_*(b_{N+2}) \neq 0$ by applying β to $\alpha_*(x_{N+2})$. Thus $\alpha_*(b_{N+1}) \neq 0$ by applying P^1_* to $\alpha_*(b_{N+2}) \neq 0$.

Next assume that $p = 2$.

First recall $H_*(O(F_3);F_2)$. By [1,p 17] $H_*(O(F_3);F_2) \cong F_2[v_1,v_2,\ldots] \otimes \Lambda[u_1,u_2,\ldots]$ as an algebra. Furthermore

(1) the degree of v_i and u_i is i,

(2) $\Delta(v_k) = \Sigma v_i \otimes v_{k-i}$, $v_0 = 1$,

$\Delta(u_k) = \Sigma u_i \otimes u_{k-i}$, $u_0 = 1$, and

(3) $Sq^i_*(v_{k+i}) = \binom{k}{i} v_k$, and

$Sq^i_*(u_{k+i}) = \binom{k}{i} u_k$.

The primitives are given by Newton polynomials

$$x_1 = v_1$$

$$x_k = kv_k + \sum_{i=1}^{k-1} v_i x_{k-i},$$

$$b_1 = u_1$$

$$b_{2k+1} = u_{2k+1} + \sum_{i=1}^{2k} u_i b_{2k+1-i}.$$

The odd degree primitives are x_{2k+1} and b_{2k+1} while there is precisely one even degree primitive $x_{2k} = x_k^2$.

Let $\alpha\colon O(F_3) \to O(F_3)$ be a group homomorphism which induces an isomorphism up to and including dimension 3. The primitives through dimension 3 are given by

(i) v_1 and u_1 in degree 1,

(ii) $x_2 = v_1^2$ in degree 2, and

(iii) x_3 and b_3 in degree 3.

As α_* is an isomorphism in degree 2, $\alpha_*(x_2) = x_2$. Since $x_3 = v_3 + v_2v_1 + v_1^3$, and $b_3 = u_3 + u_1u_2$, we have $Sq_*^1(x_3) = x_2$ and $Sq_*^1(b_3) = 0$. Thus by naturality $\alpha_*(x_3) = x_3$ + others. Furthermore, $\alpha_*(b_3) \neq 0$ by hypothesis. Hence $\alpha_*(b_3)$ is a multiple of b_3 by commutation with Sq_*^1 and so $\alpha_*(b_3) = b_3$. Next notice that

$$Sq_*^2(x_{4k+1}) = x_{4k-1},$$
$$Sq_*^4(x_{4k+1}) = x_{4k-3},$$
$$Sq_*^2(b_{4k+1}) = b_{4k-1}, \text{ and}$$
$$Sq_*^4(b_{4k+1}) = b_{4k-3}.$$

Inductively assume that $\alpha_*(x_{4k+3}) = x_{4k+3}$ + others for $k \leq N - 1$. But $\alpha_*(x_{4k+5}) = Ax_{4k+5} + Bb_{4k+5}$. Apply Sq_*^2 to get $\alpha_*(x_{4k+3}) = Ax_{4k+3} + Bb_{4k+3}$. Thus $A = 1$. Applying Sq_*^4 to $\alpha_*(x_{4k+9}) = Cx_{4k+9} + Db_{4k+9}$ gives $C = 1$. Applying Sq_*^2 to $\alpha_*(x_{4k+9}) = x_{4k+9} + Db_{4k+9}$ gives $\alpha_*(x_{4k+7}) = x_{4k+7}$ + others. Hence

$$\alpha_*(x_{2j+1}) = x_{2j+1} + \text{others}.$$

Similarly assume that $\alpha_*(b_{4k+3}) = b_{4k+3}$. Apply Sq_*^2 to $\alpha_*(b_{4k+5}) = Ex_{4k+5} + Fb_{4k+5}$ to get $E = 0$ and $F = 1$. Apply Sq_*^4 to $\alpha_*(b_{4k+9}) = Gx_{4k+9} + Hb_{4k+9}$ to get $G = 0$ and $H = 1$. Apply Sq_*^2 to $\alpha_*(b_{4k+9}) = b_{4k+9}$ to get $\alpha_*(b_{4k+7}) = b_{4k+7}$. Hence

$$\alpha_*(b_{2j+1}) = b_{2j+1},\ j > 0.$$

Finally, we check that $\alpha_*(x_{2k}) = x_{2k}$. Assume that $x_{2k} = x_j^{2^n}$ where j is odd. Notice that $\alpha_*(x_1^2) = x_1^2$ by hypothesis. Assume inductively that $\alpha_*(x_1^{2^i}) = x_1^{2^i}$ for $i \leq n$. We claim that the inductive hypothesis implies that $\alpha_*(x_{2j+1}^{2^i}) = x_{2j+1}^{2^i}$ for all j with i strictly less than n and, in turn, that this last conclusion implies that $\alpha_*(x_1^{2^{n+1}}) = x_1^{2^{n+1}}$. Thus the inductive step follows from the claim. The argument in the previous paragraph with Sq_*^2 and Sq_*^4 gives $\alpha_*(x_{2j+1}^{2^{n-1}}) = x_{2j+1}^{2^{n-1}}$ provided we know $\alpha_*(x_3^{2^{n-1}}) = x_3^{2^{n-1}}$ for $n>2$. But $Sq_*^{2^{n-1}}(x_3^{2^{n-1}}) = x_1^{2^n}$ and so $\alpha_*(x_3^{2^{n-1}}) = x_3^{2^{n-1}}$ for $n>2$ by naturality. Thus it suffices to check that $\alpha_*(x_1^{2^{n+1}}) = x_1^{2^{n+1}}$. Here observe that $\alpha_*(x_5^{2^{n-1}}) = x_5^{2^{n-1}}$ and $Sq_*^{2^{n-1}}(x_5^{2^{n-1}}) = x_1^{2^{n+1}}$ as $Sq_*^1 x_5 = x_1^4$. The lemma follows.

We now prove Corollary 2.3. Let $G = GL(F_q)$ if $p>2$ or $O(F_3)$ if $p=2$. By definition and Lemma 2.1, it suffices to give a map $\theta: BGL(F_q) \to B\Sigma_\infty^+$ such that $\alpha = \rho\cdot\lambda\cdot\theta$ satisfies the hypotheses of Lemma 2.1. The map θ is defined as follows in case p is an odd prime. By inspection of the definitions in example 2.6, there is a commutative diagram of groups and homomorphisms

$$\begin{array}{ccc} GL_n(F_q) & \longrightarrow & \Sigma_{q^n} \\ \sigma\big\downarrow & & \big\downarrow\Delta^q \\ GL_{n+1}(F_q) & \longrightarrow & \Sigma_{q^{n+1}} \end{array}$$

where σ is stabilization and Δ^q is the qth diagonal followed by the natural inclusion. [We include a check of this last point at the request of the referee. Define $\sigma: GL_n(F_q) \to GL_{n+1}(F_q)$ and $\theta: GL_n(F_q) \to \Sigma_{q^n}$ by $\sigma(\alpha) = \left[\begin{array}{c|c} \alpha & 0 \\ \hline 0 & 1 \end{array}\right]$ and $\sigma(\alpha)(\sum_1^n x_i e_i) = \alpha(\sum_1^n x_i e_i)$. Thus $(\theta\cdot\sigma)(\alpha)(\sum_1^{n+1} x_i e_i) = \alpha(\sum_1^n x_i e_i) + x_{n+1}e_{n+1}$. Regard $(\Sigma_{q^n})^q$ as a subgroup of $\Sigma_{q^{n+1}}$ by fixing $x_{n+1}e_{n+1}$ and sending $\sum_1^n x_i e_i$ to a linear combination $\sum_1^n x_i' e_i$. The above diagram commutes by inspection.] Thus one obtains a homotopy commutative diagram

$$\begin{array}{ccc} BGL_n(F_q) & \longrightarrow & B\Sigma_\infty^+ \\ {\scriptstyle B\sigma}\downarrow & & \downarrow{\scriptstyle q} \\ BGL_{n+1}(F_q) & \longrightarrow & B\Sigma_\infty^+ \end{array}$$

where q denotes the qth power map on the H-space $B\Sigma_\infty^+$. Thus one gets a map from $BGL(F_q)$ to $B\Sigma_\infty^+[1/q]$. As $B\Sigma_\infty^+$ is a bouquet of p-local spaces, we obtain the map θ. It suffices to check the homological properties given in the next paragraph.

Let $p>2$. Define $f:Z/pZ \to GL_p(F_q)$ to be given by the subgroup generated by

$$A = \begin{bmatrix} 0 & 0 & \cdots & 0 & 1 \\ 1 & 0 & & 0 & 0 \\ 0 & 1 & & & \\ \vdots & 0 & & \vdots & \vdots \\ & \vdots & & 0 & \\ 0 & 0 & & 1 & 0 \end{bmatrix}.$$

Notice that $H_{2p-3}(GL(F_q);F_p)$ and $H_{2p-3}(\Sigma_\infty;F_p)$ are both F_p. Consider $\theta \cdot Bf$ and notice that this factors as $BZ/pZ \to BGL_p(F_q) \xrightarrow{\theta} B\Sigma_{q^p} \to B\Sigma_\infty^+$. The matrix A acts on an element $\sum_1^p x_i e_i$ by $A(\Sigma x_i e_i) = \Sigma x_i e_{i+1}$ where $e_{p+1} = e_1$, and $e_1, \ldots, e_p$ is an F_q-basis for F_q^p. Thus the elements fixed by A are $x(\Sigma e_i)$, $x \in F_q$. Furthermore Z/pZ acts freely on the set of points $\Sigma x_i e_i$, $x_i \neq x_j$ for some $i \neq j$. Thus the representation $\theta_p \cdot f$ is conjugate to $\frac{1}{p}(q^p - q)$ copies of the regular representation plus trivial representations. Hence if $\frac{1}{p}(q^p - q)$ is prime to p, $(\theta \cdot Bf)_*$ is non-zero on $H_{2p-3}(\ ;F_p)$. But $\frac{1}{p}(q^p - q)$ is indeed prime to p by hypothesis.

Finally, we give a proof for the examples. Let $p>2$. Notice that all monoids N in the examples fit in to a commutative diagram

$$\begin{array}{ccc} \Sigma_\infty & \longrightarrow & N \\ & {\scriptstyle \rho}\searrow & \downarrow \\ & & GL(F_q) \end{array}$$

where ρ is given by permutation matrices. The proof follows from that given in Corollary 2.3.

References

1. Z. Fiedorowicz and S. Priddy, Homology of Classical Groups Over Finite Fields and Their Associated Infinite Loop Spaces, Springer-Verlag, L.N.M. v. 674, Springer-Verlag, Heidelberg, 1978.

2. B. Harris and G. Segal, K_i groups of rings of algebraic integers, Annals of Math., 101(1975), 20-33.

3. D. Quillen, On the cohomology and K-theory of the general linear groups over a finite field, Annals of Math., 96(1972), 552-586.

4. J. Tornehave, The spitting of spherical fibration theory at odd primes, prepint.

5. J.P. May, E_∞ Ring spaces and E_∞ Ring Spectra, Springer-Verlag, L.N.M. v. 577, Springer-Verlag, Heidelberg, 1977.

F.R. Cohen
University of Kentucky
Lexington, KY 40506

F.P. Peterson
Massachusetts Institute of Technology
Cambridge, MA 02139

A new spectrum related to 7-connected cobordism

DONALD M. DAVIS
MARK MAHOWALD

Abstract. A new 2-local spectrum Y is constructed so that H^*Y is a cyclic A-module which in degrees ≤ 23 is the quotient of the Steenrod algebra by the left ideal generated by Sq^1, Sq^2, and Sq^4. In order to show that in this range Y splits off $MO\langle 8\rangle$, the groups $\pi_i(MO\langle 8\rangle)$ are calculated for $i \leq 23$. This includes a novel Adams differential. When $i \geq 16$, these are the cobordism groups of 7-connected manifolds.

A sketch of the applicability of Y to obstruction theory is given.

1. INTRODUCTION

Let $BO8(= BO\langle 8\rangle)$ denote the classifying space for vector bundles trivial on the 7-skeleton and $MO8(= MO\langle 8\rangle)$ the localization at 2 of the associated Thom spectrum. Then $\pi_*(MO8)$ is the 2-primary component of the cobordism ring of 7-connected manifolds in dimensions > 15 (see[**26**,p.16]).

Let A_n be the subalgebra of the mod 2 Steenrod algebra A generated by $\{Sq^i : i \leq 2^n\}$, and $A//A_n$ the quotient by the left ideal generated by A_n. There is a splitting of A-modules $H^*(MO8) \approx A//A_2 \oplus N$ for some N ([**1**]); however there is no spectrum whose cohomology is $A//A_2$ ([**10**]). All cohomology groups have $\mathbf{Z}_2$-coefficients unless indicated otherwise. In this paper we construct a new spectrum which through degree 23 agrees with $A//A_2$ and splits off $MO8$.

THEOREM 1.1. *There is a Thom spectrum Y satisfying*

i) $H^i(Y) \approx \begin{cases} \mathbf{Z}_2 & \text{if } i \equiv 0,4,6,7 \pmod 8, i \geq 8, \text{ or } i = 0 \\ 0 & \text{otherwise} \end{cases}$

with $Sq^i I_0 \neq 0$ *if* $i \equiv 0,4,6,7 \pmod 8, i \geq 8$.

ii) *There is an epimorphism* $A//A_2 \to H^*Y$ *which is an isomorphism in degree* ≤ 23.

iii) *There is a natural map* $Y \xrightarrow{u} MO8$ *which splits through degree 23, i.e. there is a map* $v : MO8^{(23)} \to Y$ *so that* $v \circ u|Y^{(23)}$ *is homotopic to the inclusion* $Y^{(23)} \to Y$.

We will expand upon these properties in Section 2. Y is a spectrum quite similar to those considered in [**9**],[**19**], and [**18**].

While proving (iii) above, we also calculate $\pi_* MO8$ through degree 23, extending work of [**12**]. Of course, the Ext groups are easily calculated in this range (using [**12**] and [**11**]); it is differentials in the Adams spectral sequence (ASS) which require care. In fact, the proof of the differential in the 20-stem is the most novel feature of the paper.

Both authors were supported by National Science Foundation research grants

THEOREM 1.2.

i) *In the ASS converging to $\pi_* MO8$, all possible d_2-differentials (not ruled out by h_0- or h_1-naturality) on stems ≤ 24 are nonzero. The only other nonzero differentials in this range are d_3 on elements in the 17- and 18-stems, and a differential hitting the element of filtration 9 in the 23-stem.*

ii) *$\pi_i MO8$, and hence the 2-component of the cobordism group of i-dimensional 7-connected manifolds, is*

$i = 16$	17	18	19	20	21	22	23
$\mathbf{Z}\oplus\mathbf{Z}$	$\mathbf{Z}_2\oplus\mathbf{Z}_2\oplus\mathbf{Z}_2$	$\mathbf{Z}_2\oplus\mathbf{Z}_2$	$\mathbf{Z}_2$	$\mathbf{Z}\oplus\mathbf{Z}\oplus\mathbf{Z}_8$	$\mathbf{Z}_2$	$\mathbf{Z}_2\oplus\mathbf{Z}_2$	0

Groups for $i < 16$ are not listed because they were computed in [**12**]. The entire spectral sequence in $t - s < 24$ will be depicted in Section 2.

The spectrum Y can be used in obstruction theory. In [**8**],[**2**], and [**3**], MO8 was utilized (indirectly) in obstruction theory to obtain immersions and nonimmersions of real projective spaces. A major limitation of this method was that $H^* MO8$ becomes larger than desired (larger than $A//A_2$) beginning in degree 16. The splitting map in 1.1 (iii) allows us to utilize Y through 23 dimensions, and by 1.1 (ii) it will agree with $A//A_2$ in this range. In Section 3 we sketch an argument which implies

THEOREM 1.3. *If $n \equiv 6(16)$ and $\alpha(n) \geq 5$, then $\mathbf{R}P^n \subseteq \mathbf{R}^{2n-14}$.*

Here $\alpha(n)$ denotes the number of 1's in the binary expansion of n. This improves upon previous best immersions ([**8**]) by 2 dimensions if $\alpha(n) = 5$ or 6 and is in dimension 4 (resp. 6) greater than that of the best known nonimmersions ([**4**]) if $\alpha(n) = 5$ (resp. 6). Bruner has produced computer-generated tables of minimal A_2-resolutions of $H^* P_N$. These lend some hope to finding a general pattern for these immersions and nonimmersions; however, proof of such a general result does not seem near at hand because of irregular obstructions.

2. CONSTRUCTION OF Y.

Let $Z = BO8^{(15)}/BO8^{(8)} = S^{12} \cup_\eta e^{14} \cup_2 e^{15}$, where η is the Hopf map in $\pi_{n+1}(S^n)$. Note that $BO8^{(8)} = S^8$, and consider the cofiber sequence

$$S^8 \to BO8^{(15)} \to Z \xrightarrow{f} S^9 \xrightarrow{g'} (\Sigma BO8)^{(16)}.$$

Let g denote the composite of g' followed by

$$(\Sigma BO8)^{(16)} \to \Sigma BO8 = \Sigma\Omega B^2O8 \to B^2O8.$$

If Q denotes the fibre of f, there is a commutative diagram of fibrations

$$\begin{array}{ccccc} Q & \longrightarrow & Z & \xrightarrow{f} & S^9 \\ \downarrow{\scriptstyle \beta} & & \downarrow & & \downarrow{\scriptstyle g} \\ BO8 & \longrightarrow & * & \longrightarrow & B^2O8 \end{array}$$

Let Y denote the Thom spectrum of β. The following result contains parts (i) and (ii) of 1.1.

THEOREM 2.1. *Let h be the nontrivial map from Y into the Eilenberg- MacLane spectrum $H\mathbf{Z}_2$.*

i) *$H^*(h)$ is injective with image*

$$\langle \zeta_1^{8i}\cdot\zeta_2^{4t}\cdot\zeta_3^{2s}\cdot\zeta_4^{r} : r+s+t\le 1, i\ge 0\rangle.$$

Here ζ_i is the conjugate of the Milnor element ξ_i in $H_{2^i-1}(H\mathbf{Z}_2)\subset A_*$.

ii) *Dually, $h^*(\chi\,\mathrm{Sq}(R))\ne 0$ if $r_1\equiv 0 \pmod 8$, $r_2=4t, r_3=2s, t+s+r\le 1$ and $r_i=0$ for $i\ge 5$.*

iii) *A basis for $\mathrm{im}(h^*)$ is $\{h^*\,\mathrm{Sq}^i : i\equiv 0,4,6,7 \pmod 8), i\ge 8\}$.*

iv) *There is a short exact sequence of A-modules in degree less than 32*

$$0\to H^*(\Sigma^{24}\bar{B}(2))\to A//A_2 \xrightarrow{\Phi} H^*Y\to 0,$$

where $\bar{B}(2)$ is the integral Brown-Gitler spectrum of [20], [6], or [15], satisfying $H^*\bar{B}(2)\approx\langle \mathrm{Sq}^i : 0\le i\le 7, i\ne 1\rangle$.

PROOF: The Serre spectral sequence of $\Omega S^9\to Q\to Z$ shows that $\dim(H^iQ)$ and hence also $\dim(H^iY)$ is 1 if $i\equiv 0,4,6,7 \pmod 8$ and 0 otherwise. There is a map of principal fibrations

$$\begin{array}{ccccc} \Omega S^9 & \longrightarrow & Q & \longrightarrow & Z \\ \downarrow{\scriptstyle \alpha} & & \downarrow{\scriptstyle \beta} & & \downarrow \\ BO8 & \xrightarrow{=} & BO8 & \longrightarrow & * \end{array}$$

and hence compatible actions

$$\begin{array}{ccc} \Omega S^9\times Q & \xrightarrow{m} & Q \\ \downarrow{\scriptstyle \alpha\times\beta} & & \downarrow{\scriptstyle \beta} \\ BO8\times BO8 & \longrightarrow & BO8. \end{array}$$

Let X_9 denote the Thom complex of α as in [19] or [9]. The map m induces a pairing of Thom complexes so that

$$\begin{array}{ccc} X_9\wedge Y & \longrightarrow & Y \\ {\scriptstyle T\alpha\wedge T\beta}\downarrow & & \downarrow{\scriptstyle T\beta} \\ HZ_2\wedge HZ_2 & \longrightarrow & HZ_2 \end{array}$$

commutes. $(T\beta)^*$ sends $Sq^8, Sq^4Sq^8, Sq^2Sq^4Sq^8$, and $Sq^3Sq^4Sq^8$ nontrivially. Dual to this is the stated $\mathrm{im}((T\beta)_*) \subset H_i(HZ_2)$, $i \le 15$. Since $\mathrm{im}((T\alpha)_*) = \langle \xi_1^{8i} \rangle$, the pairing establishes that the entire $\mathrm{im}((T\beta)_*)$ is as claimed.

Part (ii) is immediate, (iii) follows since $\langle \zeta_1^{8i}\zeta_3, Sq^{8i+7}\rangle \neq 0$, and (iv) follows since $\mathrm{coker}(\phi^*) = \langle \zeta_2^8, \zeta_2^4\zeta_3^2, \zeta_2^4\zeta_4, \zeta_3^4, \zeta_3^2\zeta_4, \zeta_4^2, \zeta_5\rangle$ in this range. ∎

THEOREM 2.2. *The following chart gives all differentials in the ASS for $\pi_* MO8$ which terminate in $t-s \le 23$. There are no nontrivial $\cdot 2$ extensions in this range, but there is a nontrivial $\cdot\eta$ extension indicated by a dotted line.*

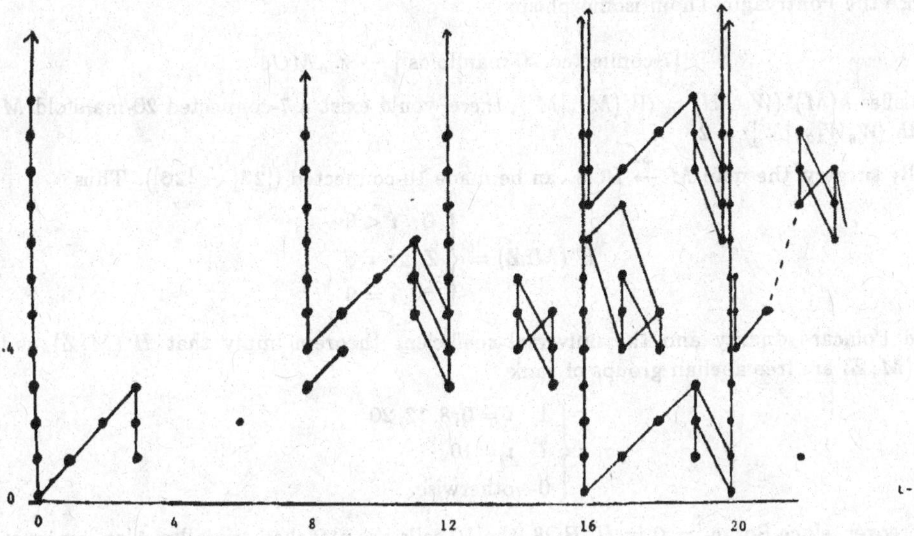

Remark As usual (e.g. [24], [17]), this ASS chart depicting $\mathrm{Ext}_A^{s,t}(H^*MO8, \mathbf{Z}_2)$ has dots in position $(t-s, s)$ representing Z_2's, vertical lines multiplication by h_0, / lines multiplication by h_1, and \ lines differentials in the ASS. Nontrivial extensions referred to in 2.2 are relations in $\pi_*(\ \ )$ not seen by h_0 or h_1 in Ext.

PROOF: Giambalvo ([12],[13]), showed that in degree ≤ 32, there is a splitting of A-modules

$$H^*MO8 \approx A//A_2 \oplus \Sigma^{16} A//A_2 \oplus \Sigma^{20} A/A(Sq^1, Sq^5, Sq^6),$$

and that the above table correctly lists differentials terminating in stems ≤ 17. The d_2's in high filtration on stems 20 and 23 follows from d_2's 8 stems earlier since they are products by the class in (8,4) using the multiplication of $MO8$. The exotic η-extensions on the 21-stem is present in $\pi_*(S^0)$ ([22, 2.1.1]), in the image from which both of these elements lie. The element in (23,9) must be hit by a differential since it is $\eta^3\bar{\kappa} = 4\nu\bar{\kappa}$ in a group of order at most 2. The differential is probably a d_3.

The extensions in the 17- and 18-stems are trivial since $2\eta = 0$, in 22 since $2\nu^2 = 0$, and in 20 since $8\bar{\kappa} = 0$ in $\pi_{20}S^0$. Finally, it remains to verify the d_2-differential from (20,1). To do this, we show that if the element in (20,1) survived the ASS to give a nontrivial element of $\pi_{20}MO8$, then the corresponding cobordism class would contain a manifold whose attaching maps contradict Toda's analysis ([**25**]) of $\pi_{19}S^9$.

For $i = 8$, 12, and 20, $H^i(BO8;\mathbf{Z})$ is $\mathbf{Z}$, $\mathbf{Z}$, and $\mathbf{Z}\oplus\mathbf{Z}$, respectively, with generators W_8, W_{12}, W_{20}, and W_8W_{12}, whose mod 2 reductions are the Stiefel-Whitney classes. The tower in $\mathrm{Ext}_A^{*,*+20}(H^*MO8,\mathbf{Z}_2)$ arises from $w_8w_{12}U$, and hence if the element in (20,1) survives the ASS, the corresponding map $S^{20}\xrightarrow{f}MO8$ satisfies $f^*(W_8W_{12}U) = 2\cdot$ gen. Since the Pontryagin-Thom isomorphism

$$[\text{7-connected 20-manifolds}\,]\xrightarrow{h}\pi_{20}MO8$$

satisfies $h(M)^*(W\cup U) = \langle W(M),[M]\rangle$, there would exist a 7-connected 20-manifold M with $\langle W_8W_{12},[M]\rangle = 2$.

By surgery, the map $M\xrightarrow{\phi}BO8$ can be made 10-connected ([**23**] or [**26**]). Thus

$$H^i(M;\mathbf{Z}) = \begin{cases} 0 & i<8\\ \mathbf{Z} & i=8\\ 0 & i=9\end{cases}$$

and Poincare duality and the universal coefficient theorem imply that $H^i(M;\mathbf{Z})$ and $H_i(M;\mathbf{Z})$ are free abelian groups of rank

$$\begin{cases} 1 & i=0,8,12,20\\ ? & i=10\\ 0 & \text{otherwise.}\end{cases}$$

Moreover, since $\mathrm{Sq}^2\,w_8 = 0$ in H^*BO8, the 10-cells are attached trivially. Also, we must have $\phi^*(W_8) = g_8$ and $\phi(W_{12}) = 2g_{12}$, where g_i generates $H^i(M;\mathbf{Z})$, since $\phi^*(W_8W_{12}) = 2g_8g_{12}$ and the mod 2 reduction $\phi^*(w_{12}) = \mathrm{Sq}^4\,\phi^*(w_8)$ cannot be nonzero if $\phi^*(w_8)$ is zero.

The restriction of ϕ to the 12-skeleton is a map

$$(S^8\vee\bigvee S^{10})\cup_\alpha e^{12}\xrightarrow{\phi}S^8\cup_\nu e^{12}$$

of degree 1 on the 8-cell and 2 on the 12-cell. This implies that the attaching map α is $\pm 2\nu$ (+ perhaps η). To see this, we note that if ϕ sends a 10-cell nontrivially (by η^2) and the 12-cell is attached to this 10-cell (by η), this would just account for a degree 4 in $H_{12}(\quad)$, because $\eta^3 = 4\nu$.

We may collapse the 10-cells of M, obtaining a 3-cell complex $S^8\cup_{2\nu}e^{12}\cup_\beta e^{20}$ in which Sq^8 is nonzero [because this is equivalent to $0\neq\chi(\mathrm{Sq}^8)U = (\mathrm{Sq}^8+\mathrm{dec})U = w_8U$, and $w_8\neq 0$]. Consideration of the Puppe sequence

$$\begin{array}{ccccc} & & S^{19} & & \\ & \beta\swarrow & & \searrow\sigma & \\ S^8\cup_{2\nu}e^{12} & \xrightarrow{p} & S^{12} & \xrightarrow{2\nu} & S^9\end{array}$$

with $p \circ \beta$ equal to the Hopf map σ (since $\mathrm{Sq}^8 \neq 0$) implies that $2\nu\sigma = 0$ in $\pi_{19}S^9$. But [25, 7.20] shows that $2\nu\sigma \neq 0$ in $\pi_{19}S^9$. Indeed, the appropriate part of the EHP sequence has

$$\begin{array}{ccccc} \pi_{20}(S^{17}) & \longrightarrow & \pi_{18}(S^8) & \longrightarrow & \pi_{19}(S^9) \\ \mathbf{Z}_8 & & \mathbf{Z}_8 \oplus \mathbf{Z}_8 \oplus \mathbf{Z}_2 & & \mathbf{Z}_8 \oplus \mathbf{Z}_2 \\ & & \sigma\nu \quad \nu\sigma \quad \eta\mu & & \sigma\nu \quad \eta\mu \\ \nu & \longmapsto & 2\sigma\nu + \nu\sigma & & \\ & & 2\nu\sigma & \longmapsto & 4\sigma\nu \neq 0. \end{array}$$

Now that the analysis of $\pi_* MO8$ has been completed, the proof of 1.1 (iii) is completed quite easily. Let $W = S^{16} \cup_{2\nu} e^{20} \cup_\eta e^{22} \cup_2 e^{23} \cup_\nu e^{24}$. We will show there is a map $W \xrightarrow{w} MO8$ of degree 1 on the bottom cell. Then it is easy to check that $Y \vee W \xrightarrow{u \vee w} MO8$ induces an isomorphism in $H^i(\ ;\mathbf{Z})$ for $i \leq 23$ and an epimorphism for $i = 24$. Thus the inclusion of $MO8^{(23)}$ factors as

$$MO8^{(23)} \xrightarrow{\tilde{v}} Y \vee W \xrightarrow{u \vee w} MO8.$$

Our desired map v is $\tilde{v}$ followed by the pinching into Y. The composite $Y^{(23)} \to Y$ induces an epimorphism in $H^*(\ ;\mathbf{Z}_2)$ since H^*Y is a cyclic A-module.

We now construct w. The ASS calculations just completed shows that there is a map $f_1 : W_1 = S^{16} \cup_{2\nu} e^{20} \to MO8$ of degree 1 on the bottom cell. This map f_1 can be varied by a multiple of $W_1 \xrightarrow{p} S^{20} \xrightarrow{\tilde{\kappa}} MO8$, where $\tilde{\kappa}$ is the filtration 4 generator of $\pi_{20}MO8$. Let $M_{21} = S^{21} \cup_2 e^{22}$, and $a : M_{21} \to W_1$ the attaching map in W. We will show that f_1 can be chosen so that the composite $f_1 \circ a$ is null. This f_1 extends over all but the top cell of W, and since $\pi_{23}MO8 = 0$, the extension over the top cell follows for free.

$[M_{21}, W_1] \approx \mathbf{Z}_4 \oplus \mathbf{Z}_2 \approx [M_{21}, MO8]$, where in both cases the generator of the $\mathbf{Z}_4$ is an extension of the nonzero element γ of $\pi_{21}(\ )$, and 2·(this generator) is $\gamma \circ \eta \circ p$, where $p : M_{21} \to S^{22}$ is the pinch map. The $\mathbf{Z}_2$'s are the filtration 2 elements of $\pi_{22}(\ )$ preceded by the pinch. The attaching map a in $[M_{21}, W_1]$ can be chosen to be any element whose first component in $\mathbf{Z}_4 \oplus \mathbf{Z}_2$ is a generator. Any f_1 induces a homomorphism $[M_{21}, W_1] \to [M_{21}, MO8]$ sending $g_2 \to G_2$, but $f_{1*}(g_1)$ may be changed by any multiple of G_1, since the composite $W_1 \xrightarrow{p} S^{20} \xrightarrow{\tilde{\kappa}} MO8$ induces an isomorphism in $\pi_{21}(\ )$. Thus if in one choice of f_1 and a, $f_{1*}(a)$ is nonzero in the first component, f_1 can be rechosen to make it 0, and if it is nonzero in the second component, a can be rechosen to make it 0.

3. APPLICATION TO IMMERSIONS

In [5] it was shown that in the stable range there is an orientation $BO8/BO_N 8 \to \Sigma P_N \wedge MO8$, where $P_N = RP^\infty / RP^{N-1}$. If we are studying a lifting question

$$\begin{array}{ccc} & & BO_N 8 \\ & & \downarrow \\ X & \longrightarrow & BO8 \end{array}$$

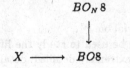

with $N + 16 < \dim < N + 25$, it should simplify calculations to follow by the map $MO^{(23)} \to Y$, and use Y in obstruction theory. The theory set up in [21] allows us to form a modified Postnikov tower (MPT) with respect to Y, i.e. a diagram through dimension $< \min(N + 24, 2N - 1)$

$$
\begin{array}{ccccc}
BO_N 8 & \longrightarrow & \vdots & & \\
 & & \downarrow & & \\
 & & E_3 & & \\
 & & \downarrow & & \\
\Sigma^{-1} P_N \wedge Y \wedge \overline{Y} & \longrightarrow & E_2 & \xrightarrow{c_2} & \Sigma^{-1} P_N \wedge Y \wedge \overline{Y} \wedge \overline{Y} \\
 & & \downarrow & & \\
P_N \wedge Y & \longrightarrow & E_1 & \xrightarrow{c_1} & P_N \wedge Y \wedge \overline{Y} \\
 & & \downarrow & & \\
 & & BO8 & \xrightarrow{c_0} & \Sigma P_N \wedge Y
\end{array}
\tag{3.1}
$$

in which $\overline{Y} = \text{cof}(S^0 \to Y)$, $\longrightarrow \downarrow_{\longrightarrow}$ are fiber sequences, and $BO_N 8 \to E_i$ is $(N + 7i)$-connected.

This applies to proving immersions by using Hirsch's theorem ([16]), which involves the geometric dimension (gd) of stable vector bundles. $\text{gd}(\theta)$ is the smallest N such that $X \xrightarrow{\theta} BO$ lifts to BO_N. If ξ_n denotes the Hopf bundle over RP^n, then Hirsch's theorem applied to the normal bundle minus ξ_n implies

$$P^n \subseteq R^{n+N+1} \quad \text{if } \text{gd}((-n-2)\xi_n) \leq N$$

Thus the immersion of 1.3 will follow from showing that the lifting ℓ in (3.2) exists when $N = 16A - 9$, $n = 16A + 6$, $\alpha(A) \geq 3$, and θ classifies $(-16A - 8)\xi$.

$$
\begin{array}{ccc}
 & & BO_N \\
 & \ell \nearrow & \downarrow \\
P^n & \xrightarrow[\theta]{} & BO8
\end{array}
\tag{3.2}
$$

We may use $BO8$ since 8ξ is trivial on P^7.

The method of [3] and [8] can be used to study the lifting of the map θ of (3.2) in the diagram (3.1). The argument is extremely detailed; details are available from the authors upon request.

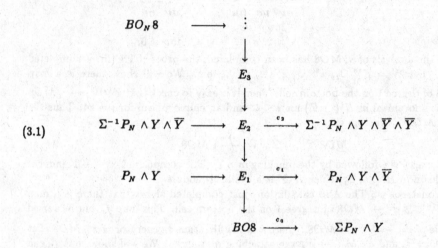

REFERENCES

1. A. P. Bahri and M. Mahowald, *A direct summand in $H^*MO\langle 8\rangle$*, Proc. Amer. Math. Soc. **78** (1980), 295-298.
2. D. M. Davis, *Connective coverings of BO and immersions of projective spaces*, Pac. Jour. Math. **76** (1978), 33-42.
3. ———, *Some new immersions and nonimmersions of real projective spaces*, Contemp. Math. **19** (1983), 33- 42.
4. ———, *A strong nonimmersion theorem for real projective spaces*, Annals of Math **120** (1984), 517-528.
5. D. M. Davis, S. Gitler, W. Iberkleid, and M. Mahowald, *The orientability of vector bundles with respect to certain spectra*, Bol. Soc. Mat. Mex **268** (1981), 49-55.
6. D. M. Davis, S. Gitler, and M. Mahowald, *The stable geometric dimension of vector bundles over real projective spaces*, Trans. Amer. Math. Soc. **268** (1981), 39-62.
7. D. M. Davis and M. Mahowald, *The geometric dimension of some vector bundles over projective spaces*, Trans. Amer. Math. Soc. **205** (1975), 295-316.
8. ———, *The immersion conjecture for $RP^{8\ell+7}$ is false*, Trans. Amer. Math. Soc. **236** (1978), 361-383.
9. ———, *v_1- and v_2-periodicity in stable homotopy theory*, Amer. Jour. Math. **103** (1978), 615-659.
10. ———, *The nonrealizability of the quotient $A//A_2$ of the Steenrod algebra*, Amer. Jour. Math. **104** (1982), 1211-1216.
11. ———, *Ext over the subalgebra A_2 of the Steenrod algebra for stunted real projective spaces*, Current Trends in Algebraic Topology, Conference, Proc. of Canadian Math. Soc. **2** (1982), 297-343.
12. V. Giambalvo, *On $\langle 8\rangle$ cobordism*, Ill. Jour. Math. **15** (1971), 533-541.
13. ———, *Correction to [**12**]*, Ill. Jour. Math. **16** (1972), p. 704.
14. S. Gitler and M. Mahowald, *The geometric dimension of real stable vector bundles*, Bol. Soc. Mat. Mex. **11** (1966), 85-107.
15. P. Goerss, J. Jones and M. Mahowald, *Some generalized Brown-Gitler spectra*, Trans Amer. Math. Soc. **294** (1986), 113-132.
16. M. W. Hirsch, *Immersion of manifolds*, Trans. Amer. Math. Soc. **93** (1959), 242-276.
17. M. Mahowald, *The metastable homtopy of S^n*, Mem. Amer. Math. Soc. **72** (1967).
18. ———, *A new infinite family in ${}_2\pi^*$*, Topology **16** (1977), 249-256.
19. ———, *Ring spectra which are Thom complexes*, Duke Math. Jour. **46** (1979), 549-559.
20. ———, *bo-resolutions*, Pac. Jour. Math. **92** (1981), 365-383.
21. M. Mahowald and R. Rigdon, *Obstruction theory with coefficients in a spectrum*, Trans. Amer. Math. Soc. **204** (1975), 365-385.
22. M. Mahowald and M. Tangora, *Some differentials in the Adams spectral sequence*, Topology **6** (1967), 349-369.
23. J. Milnor, *A procedure for killing homotopy groups of differentiable manifolds*, Proc. Symp. Pure Math. **3** (1961), 39-55.
24. M. Tangora, *On the cohomology of the Steenrod algebra*, Math. Zeit **116** (1970), 18-64.

25. H. Toda, "Composition methods in the homotopy groups of spheres," Ann. of Math. Studies vol 45, Princeton Univ. Press, 1962.
26. C. T. C. Wall, "Surgery on compact manifolds," Academic Press, 1970.

1980 *Mathematics subject classifications*: 57R42,57R90,55P42.

Lehigh University, Bethlehem, PA 18015
Northwestern University, Evanston, IL 60201

ASPHERICAL MANIFOLDS WITHOUT SMOOTH OR PL STRUCTURE

Michael W. DAVIS and Jean-Claude HAUSMANN

ABSTRACT. One construct closed aspherical PL-manifolds which are not homotopy equivalent to closed smooth manifolds. Examples of closed aspherical TOP-manifolds which are not homeomorphic to closed PL-manifolds are also given.

A space X is **aspherical** if it is homotopy equivalent to a CW-complex and if its universal covering is contractible (in other words : X is homotopy equivalent to the Eilenberg-McLane space $K(\pi_1(X),1)$). Closed aspherical manifolds form an interesting class of aspherical spaces. Classical examples come from Lie group theory and differential geometry and are smooth manifolds. A new kind of construction of closed aspherical manifolds appeared in [D1], giving rise to closed aspherical smooth manifolds with universal coverings not homeomorphic to $\mathbf{R}^n$. Using these techniques, we prove in this note the following results :

Theorem 1 For each $n \geqslant 13$, there exists an aspherical closed PL-manifold M of dimension n which does not have the homotopy type of a closed smooth manifold.

We prove Theorem 1 by showing that the Spivak bundle of M admits no linear reduction.

Theorem 2 For each $n \geqslant 8$, there exists an aspherical closed topological manifold M of dimension n such that M is not homeomorphic to a closed PL-manifold.

1980 Mathematics Subject Classification : 55D20, 57C99, 57A99, 20E40.

We do not know whether M has the homotopy type of a closed PL-manifold (See Remark 4).

Recall that a group G is a **Poincaré duality group** if there exists a ZG-module structure Z on the abelian group of integers and a class $e \in H_n(G;Z)$ so that the cap product homomorphism $-\cap e : H^i(G;B) \longrightarrow H_{n-i}(G;B \otimes_Z Z)$ is an isomorphism for all G-modules B (definition of Bieri and Eckmann [BE]). The fundamental groups of the manifolds M of our theorem are the first examples of Poincaré duality groups G such that K(G,1) is not homotopy equivalent to a closed smooth manifold. Recall that a strong version of the Novikov conjecture says that for a Poincaré duality group G, the space K(G,1) should be homotopy equivalent to a closed topological manifold (see Remark 2).

We are grateful to R. Schultz and S. Cappell for valuable help in improving the dimensional restriction in Theorem 1 and in adapting our original proof of Theorem 1 in order to obtain Theorem 2.
The proof of Theorem 1 is given is Section 2, that of Theorem 2 in Section 3, while Section 1 is devoted to recalling some facts about the construction of [D1] (and also [D2]).

1. CONSTRUCTING ASPHERICAL MANIFOLDS WITHOUT CAT-STRUCTURES USING REFLECTION GROUPS

This section is a development of [D1, Remark 15.9]. The notations and the terminology are from [D1], except that, following W. Thurston's terminology, we use the word "mirror" instead of "panel".

Let Q be (finite) CW-complex, P a subcomplex of Q and L a triangulation of P. Replacing L by its barycentric subdivision if necesary, we can assume that L is "determined by its 1-skeleton". (This means that for any set T of vertices in L, if any two distinct elements of T bound an edge, then T spans a simplex of L.) Under the assumption that L is determined by its 1-skeleton, there is a Coxeter system (Γ,V) with L as associated complex (see [D1, Section 11]). In particular, V is the set of vertices of L. For instance, one could take Γ generated by V with the relations

$$v^2 = 1 \text{ for all } v \in V$$

$$(vw)^2 = 1 \text{ if } v \text{ and } w \text{ bound an edge in } L.$$

The "canonical mirror structure on L" is then defined as follows. For each $v \in V$, let P_v denote the closed star of v in the barycentric subdivision L' of L; each P_v is called a **mirror**. Form the complex $X = Q \times \Gamma/\sim$, where $(x,g)\sim(y,h)$ if $x = y$ and $h^{-1}g$ is in the subgroup of Γ generated by those v such that $x \in P_v$. We recall the following facts from [D1] :

(1.1.a) $X = \bigcup X_i$, with $X_1 = Q$ and X_i is the union of X_{i-1} with with a translated of Q (we write $X_i = X_{i-1} \cup Q$ for simplicity). The intersection $X_{i-1} \cap Q$ is a union of mirrors and can be identified with the closed star in L' of a certain simplex in L [D1, Section 8]. In particular, $X_{i-1} \cap Q$ is contractible.

(1.1.b) The group Γ operates properly on X as a reflection group with finite isotropy groups, and $X/\Gamma = Q$.

(1.1.c) As a Coxeter group, Γ contains a torsion-free subgroup Γ' of finite index. Then $X \to X/\Gamma' = M$ is a covering projection.

This has the following immediate consequences :

(1.2.a) If P is a polyhedral homology m-manifold, then each $X_{i-1} \cap Q$ is a compact contractible polyhedral homology m-manifold with boundary. (Recall that a polyhedral homology m-manifold is a m-dimensional simplicial complex such that the link of any k-simplex has the homology of S^{m-k-1}).

(1.2.b) If P is a topological manifold, then each $X_{i-1} \cap Q$ is a compact contractible manifold with boundary.

(1.2.c) If P is a PL-manifold and L is a PL-triangulation of P, then each $X_{i-1} \cap Q$ is a PL m-cell.

From now on, we suppose that (Q,P) is a Poincaré pair of formal dimension n and that the simplicial complex L (=P) is a polyhedral homology manifold. In this special case, (1.1) and (1.2) above have the following consequences :

(1.3.a) Each X_i is a Poincaré space, X is an (infinite) Poincaré space and M is a closed Poincaré space.

(1.3.b) If Q is a topological n-manifold with boundary $\partial Q = P$, then each X_i is a manifold with boundary, X is a manifold and M is a closed topological manifold.

(1.3.c) If Q is a triangulated manifold and L is a triangulation of ∂Q, then the barycentric subdivision of the triangulation of Q extends to a triangulation of M. In other words, if Q has a TRI-structure in the sense of [GS], then the closed manifold M inherits a TRI-structure.

(1.3.d) Similarly, if Q is a PL-manifold and L is a PL-triangulation of ∂Q, then X is a PL-manifold, Γ acts through PL-automorphisms, and hence M is a PL-manifold.

(1.3.e) If Q is a smooth manifold and L is a smooth triangulation of ∂Q, then Q can be given the structure of a smooth orbifold (see [D1, Section 17]), and hence M is a smooth manifold.

Essentially, (1.3) says that if Q is a CAT-manifold, then so is M (where CAT = DIFF, PL, TRI or TOP). Moreover, M contains Q as a codimension zero Poincaré space or submanifold.

Finally, we deduce the following facts :

(1.4) If Q is aspherical, then M is aspherical (since $X_i \cap Q$ is contractible, X is aspherical and $X \to M$ is a covering projection).

(1.5) If the map $\nu_Q : Q \to BG$ classifying the Spivak bundle of Q does not lift through BCAT (CAT = DIFF, PL, or TRI), then neither does $\nu_M : M \to BG$ (since $\nu_M|Q = \nu_Q$) and then M does not have the homotopy type of a CAT-manifold.

(1.6) Suppose that Q is a topological manifold. If the map $\tau_Q : Q \to BTOP$ classifying the stable tangent micro-bundle of Q does not lift through BCAT (CAT = DIFF, PL, or TRI), then neither does $\nu_M : M \to BTOP$ and M is not homeomorphic to a CAT-manifold.

Remark : The argument of [D2, Proposition 1.4] shows that there is a Γ-equivariant embedding $f : X \hookrightarrow Q \times \mathbf{R}^N$ with trivial normal bundle, where Γ acts on $\mathbf{R}^N$ as a linear reflection group. Moreover, the composition of f with the projection on the first factor is the orbit map $\pi : X \longrightarrow Q$. It follows that $\nu_M : M \longrightarrow BCAT$ factors as $M \xrightarrow{\pi} Q \xrightarrow{\nu_Q} BCAT$. This gives a sort of a converse statement to (1.4)-(1.6).

2. PROOF OF THEOREM 1

Observe that, if M is a closed aspherical PL-manifold with Spivak bundle $\nu_M : M \longrightarrow BG$ admitting no linear reduction, so is $M \times S^1$. Therefore, in order to prove Theorem 1, it is enough, using (1.5) and (1.6), to construct a compact aspherical PL-manifold Q of dimension 13 such that $\nu_Q : Q \longrightarrow BG$ has no linear reduction.

Our technique to construct such manifolds Q is the following. Let a be an element of $\pi_k(BPL)$ so that its image α in $\pi_k(BG)$ is not in the image of the natural homomorphism $\pi_k(BO) \longrightarrow \pi_k(BG)$. One has $BPL = B\widetilde{PL} = \lim(B\widetilde{PL}_i)$, where $B\widetilde{PL}_i$ is the classyfing space for PL-block bundles of rank i [RS1]. Let $a_r : S^k \longrightarrow B\widetilde{PL}_r$ represent a. Let T^k be the torus of dimension k. Take a degree one map $T^k \longrightarrow S^k$ and compose it with a_r to get $b_r : T^k \longrightarrow B\widetilde{PL}_r$, or with α to get $\beta : T^k \longrightarrow BG$. By the classification of "abstract regular PL-neighbourhoods" over a PL-manifold [RS1, Corollary 4.7], there is a compact PL-manifold Q of dimension $k + r$ containing T^k as a codimension r-submanifold such that :

1) Q collapses onto T^k. Therefore, Q is aspherical.

2) The map b_r classifies the normal block-bundle of T^k into Q.

As T^k is parallelizable, it follows from 2) that the composition $Q \longrightarrow T^k$ with β classifies the inverse of the Spivak bundle ν_Q of Q. Suppose that ν_Q admits a lifting through BO. The spaces BO and BG are known to be infinite loop spaces, so one can write $BO = \Omega(\Omega^{-1}BO)$ and $BG = \Omega(\Omega^{-1}BG)$. If ν_Q lifts through BO, the adjoint map $ad(\nu_Q) : \Sigma T^k \longrightarrow \Omega^{-1}BG$ would lift through $\Omega^{-1}BO$. But ΣT^k is homotopy equivalent to $S^{k+1} \vee A$, where A is a wedge of spheres of dimension k, and $ad(\nu_Q)|S^{k+1}$ is $ad(-\alpha)$. This contradicts the fact that α does not lift through BO.

We now give an example of an element a with the above properties. The group $\pi_9(BG) = \pi_8(G) = \pi_8^s$ is isomorphic to $Z_2 \oplus Z_2$. Observe that an element of $\pi_i(BG)$ lifts to $\pi_i(BG_{k+1})$ if and only if the correrspondíng element of π_{i-1}^s lifts to $\pi_{i+k-1}(S^k)$. Therefore, the generators of $\pi_9(BG)$ are $\bar{\nu}$ coming from $\pi_9(BG_7)$ and ε coming from $\pi_9(BG_4)$ [To, Theorem 7.1]. The homomorphism $Z_2 = \pi_9(BO) \to \pi_9(BG)$ can be identified with the J-homomorphism $J : \pi_8(SO) \to \pi_8^s$. The group $\pi_7(SO)$ is infinite cyclic, generated by w, and $J(w) = \sigma$, where σ is represented by the Hopf map $S^{15} \to S^8$. Let η be the non-zero element of π_1^s. One has $J(w \circ \eta) = J(w) \circ \eta = \sigma \circ \eta = \eta \circ \sigma = \bar{\nu} + \varepsilon$ (for the last equality, see [To, Theorem 14.1]).
Take $\alpha = \varepsilon$. The homomorphism $\pi_9(BPL) \to \pi_9(BG)$ is onto since $\pi_8(G/PL) = Z$. Therefore, there exists an element $a \in \pi_9(\widetilde{BPL})$ having image α. The following diagram

$$\begin{array}{ccc} \widetilde{BPL}_r & \longrightarrow & \widetilde{BPL} \\ \downarrow & & \downarrow \\ BG_r & \longrightarrow & BG \end{array}$$

is a pull-back diagram for $r \geq 3$ [RS2, Theorem 1.10]. Therefore, a is the image of an element $a_4 \in \pi_9(\widetilde{BPL}_4)$. Thus, the above construction of Q can be performed with dimQ = 13, which proves Theorem 1.

3. PROOF OF THEOREM 2

If M is a closed aspherical TRI-manifold so that the topological stable tangent microbundle $\tau_M : M \to BTOP$ does not lift through BPL, then $M \times S^1$ has the same property. Therefore, using (1.6), it is enough to construct a compact aspherical TRI-manifold Q of dimension 7 so that τ_Q does not lift through BPL.

Consider the diagram

$$\begin{array}{ccccc} \pi_4(BTRI) & \longrightarrow & \pi_3(TRI/PL) & \longrightarrow & \pi_3(PL) = 0 \\ \downarrow & & \downarrow & & \\ \pi_4(BTOP) & \longrightarrow & \pi_3(TOP/PL) & & \end{array}$$

The homomorphism $\Omega_H^3 \cong \pi_3(TRI/PL) \longrightarrow \pi_3(TOP/PL) = Z_2$ can be identified with the Rohlin invariant [GS, Section 6] and is therefore surjective. On the other hand, the homomorphism $\pi_4(BPL) \longrightarrow \pi_4(BG)$ is also surjective, since $\pi_3(G/PL) = 0$. Therefore, there exists a map $\bar{a} : S^4 \longrightarrow BTRI$ inducing a non-zero class in $\pi_3(TOP/PL)$ and the zero class in $\pi_4(BG)$. Denote by a the composition of $\bar{a}$ with the map $BTRI \longrightarrow BTOP$. The diagram

$$\begin{array}{ccc} BTOP_r & \longrightarrow & BTOP \\ \downarrow & & \downarrow \\ BG_r & \longrightarrow & BG \end{array}$$

is a pull-back diagram for $r \geqslant 3$ [RS3, Corollary 2.5]. Therefore, one can find $a_3 : S^4 \longrightarrow BTOP_3$ giving a when composed with $BTOP_3 \longrightarrow BTOP$. Take a degree one map $T^4 \longrightarrow S^4$ and compose it with a_3 to get $b_3 : T^4 \longrightarrow BTOP_3$. Using the classification of topological abstract regular neighbourhoods [RS3, Theorem 3.2], one shows, as in the proof of Theorem 1, that there is a compact topological manifold Q of dimension 7 containing T^4 as a codimension 3 submanifold and satisfying :

1) The inclusion $T^4 \subset Q$ is a homotopy equivalence.

2) $b : Q \simeq T^4 \longrightarrow BTOP$ is homotopic to τ_Q.

As in the proof of Theorem 1, one shows that τ_Q admits no lifting through BPL. But, as τ_Q lifts through BTRI, Q admits a TRI-structure [GS, Theorems 1 and 1.5]. We have thus constructed a manifold Q with the required properties.

REMARKS :

1) We do not know whether 13 and 7 are the smallest dimensions for which Theorems 1 and 2 are respectively true.

2) It is tempting to use the above method to construct an aspherical Poincaré complex which is not homotopy equivalent to a closed topological manifold. This would contradict a (folklore) strong version of the Novikov conjecture. The problem would be to find a fundamental chamber Q which is a Poincaré complex, so that the Spivak bundle ν_Q admits no TOP-reduction, but with $P = \partial Q$ homotopy equivalent to a closed

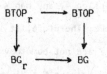

polyedral homology manifold.

3) Other examples of aspherical manifolds for Theorems 1 and 2 are obtainable as follows : in the proofs, replace the degree one map $T^k \rightarrow S^k$ by a map $f : K \rightarrow S^k$ inducing an isomorphism on integral homology, where K is a finite aspherical polyhedron of dimension k (K and f exist by [Ma]). The manifold Q will then be a thickening of K with $\tau_Q = a \circ f$, which exists in the stable range.

4) By obstruction theory, if K is a complex of dimension 4, any map $K \rightarrow BG$ which lifts through BTOP admits a lifting through BPL. Therefore, it is not possible to assert that the manifolds M of Theorem 2 are not homotopy equivalent to closed PL-manifolds. But if a homotopy equivalence $f : M' \rightarrow M$ existed with M' a closed PL-manifold, then f would yeld a homotopy equivalence between aspherical closed manifolds which is not homotopic to a homeomorphism. This would be a negative answer to a question of A. Borel.

REFERENCES

[BE] BIERI R.-ECKMANN B. Groups with homological duality generalizing Poincaré duality. Inv. Math. 20 (1973) 103-124.

[D1] DAVIS M.W. Groups generated by reflections and aspherical manifolds not covered by Euclidean space. Ann. of Math. 117 (1983) 293-324.

[D2] DAVIS M.W. Some aspherical manifolds. Duke Math. J. 55 (1987) 105-140

[GS] GALEWSKI D.-STERN R. Classification of simplicial triangulations of topological manifolds. Annals of Math. 111 (1980), 1-34.

[Ma] MAUNDER C.R.F. A short proof of a theorem of Kan-Thurston. Bull. London Math. Soc. 13 (1981) 325-327.

RS1] ROURKE C.-SANDERSON B. Block bundles I. Ann. of Math. 87 (1968) 1-28.

[RS2] ROURKE C.-SANDERSON B. Block bundles III. Ann. of Math. 87 (1968) 431-483..

[RS3] ROURKE C.-SANDERSON B. On topological neighbourhoods. Compositio Math. 22 (1970) 387-424.

[To] TODA H. Composition methods in homotopy groups of spheres. Ann. of Math. Studies 49, Princeton Univ. Press 1962.

Ohio State University, Colombus, Ohio. University of Geneva, Switzerland.

Homology with simplicial coefficients

W. G. DWYER AND D. M. KAN

§1. INTRODUCTION

1.1 Summary. This paper deals with *homology of simplicial sets over a fixed simplicial set*, say L, and in particular with the *simplicial coefficient systems* for such homology.

A simplicial coefficient system (over L) consists of a collection of abelian groups and homomorphisms between them indexed by the simplices of L and the simplicial operators between them, i.e., it is an abelian group object in the category of simplicial sets over L. If all the homomorphisms between the abelian groups are actually isomorphisms, then such a simplicial coefficient system is essentially a *local coefficient system* in the usual sense.

It turns out that, given a simplicial coefficient system A and a weak (homotopy) equivalence $K \to K'$ of simplicial sets over L, the induced map on homology $H_*(K; A) \to H_*(K'; A)$ need *not* be an isomorphism unless either *the structure map $A \to L$ is a fibration of simplicial sets* or *both of the structure maps $K \to L$ and $K' \to L$ are so.* This suggests calling a map $A \to A'$ between simplicial coefficient systems a *weak equivalence* whenever it induces an isomorphism $H_*(K; A) \cong H_*(K; A')$ for every simplicial set K over L for which the structure map $K \to L$ is a fibration, and asking whether, given any simplicial coefficient system A, there exists a weak equivalence $A \to A'$ such that the structure map of A' is a fibration (and every weak equivalence $K \to K'$ of simplicial sets over L thus induces an isomorphism $H_*(K; A') \cong H_*(K'; A')$). We give a positive answer to this question by showing that *the category* **ab**/L *of simplicial coefficient systems over L admits a closed simplicial model category structure in the sense of Quillen* in which the weak equivalences are as above and in which the fibrant objects are exactly those simplicial coefficient systems for which the structure map is a fibration.

In the remainder of the paper we compare the model catcgories **ab**/L and **ab**/L' for weakly equivalent L and L' and we observe that, for connected L, the weak equivalence classes of the simplicial coefficient systems over L are in a natural $1-1$ correspondence with the weak equivalence classes of the simplicial modules over the loop group GL of L (and hence with the weak equivalence classes of non-negatively graded differential modules over the chains on GL).

1.2 Organization of the paper. After fixing some notation and terminology (in §2), we define (in §3) the *homology* $H_*(K; A)$ of a simplicial set K over L with simplicial coefficients A, and obtain some of its basic properties. *Weak equivalences* betweeen simplicial coefficient systems then are introduced in §4, where we also give a positive answer to the question which was raised in 1.1. In §5, we establish the *closed simplicial model category structure* on **ab**/L. The proof is more difficult than one would expect and requires a *Bousfield cardinality argument.* The remaining two sections are devoted to the results which were mentioned at the end of 1.1.

1.3 Application. The arguments which establish the closed simplicial model category structure on **ab**/L will be used in [**3**,§6] to obtain closed simplicial model category structures *on the category of abelian group objects* over a fixed *simplicial diagram of simplicial sets* and *on the category of abelian group objects* over a fixed *small simplicial category*. An understanding of these structures is necessary for our study of Hochschild-Mitchell cohomology [**3**].

§2. NOTATION, TERMINOLOGY, ETC.

We will use among others the following notation, terminology and results:

This research was in part supported by the National Science Foundation.

2.1 Simplicial sets. As usual ([5] and [2, Ch. VIII]) $\mathbf{S}$ will denote the category of *simplicial sets*; for every integer $n \geq 0$, $\Delta[n] \in \mathbf{S}$ will be the *standard n-simplex* (i.e., the simplicial set freely generated by a single n-simplex i_n), $\dot{\Delta}[n] \subset \Delta[n]$ will be its subcomplex spanned by the faces of i_n and, for every pair of integers (k,n) with $0 \leq k \leq n$, $V[n,k] \subset \Delta[n]$ will be the subcomplex spanned by the faces $d_j i_n$ with $j \neq k$. If $L \in \mathbf{S}$ and $x \in L$ is an n-simplex, then $\Delta[x] : \Delta[n] \to L \in \mathbf{S}$ will denote the unique map which sends i_n to x and $\dot{\Delta}[x] : \dot{\Delta}[x] \to L \in \mathbf{S}$ and $V[x,k] : V[n,k] \to L \in \mathbf{S}$ will be the restrictions of $\Delta[x]$ to $\dot{\Delta}[n]$ and $V[n,k]$, respectively.

2.2 The over category $\mathbf{S}/L$. For $L \in \mathbf{S}$, we write $\mathbf{S}/L$ for its *over category* (which has as objects the maps $K \to L \in \mathbf{S}$). An object $(K \to L) \in \mathbf{S}/L$ will often be denoted by K alone, without its *structure map* $K \to L$. To avoid confusion we therefore use $\times_L$ for the *product* in $\mathbf{S}/L$.

2.3 A model category structure for $\mathbf{S}/L$. The category $\mathbf{S}/L$ admits a closed simplicial model category structure [6, Ch.II] in which the simplicial structure is the obvious one and in which the fibrations, the cofibrations and the weak equivalences are induced by those of $\mathbf{S}$ [2, Ch.VIII]. Thus (2.2) an object $K \in \mathbf{S}/L$ is fibrant iff its structure map $K \to L \in \mathbf{S}$ is a fibration (in $\mathbf{S}$).

2.4 Abelian group objects in $\mathbf{S}/L$. An abelian group object in $\mathbf{S}/L$ consists of an object $(f : K \to L) \in \mathbf{S}/L$ together with a *multiplication* map $m : K \times_L K \to K$, a *unit* map $u : L \to K$ and an *inverse* map $i : K \to K$ in $\mathbf{S}/L$ satisfying the usual abelian group axioms. These abelian group objects in $\mathbf{S}/L$ form an abelian category which we denote by $\mathbf{ab}/L$.

2.5 A pair of adjoint functors $\mathbf{S}/L \leftrightarrow \mathbf{ab}/L$. The forgetful functor $U : \mathbf{ab}/L \to \mathbf{S}/L$ has as left adjoint the functor $Z_L : \mathbf{S}/L \to \mathbf{ab}/L$ which sends an object $K \in \mathbf{S}/L$ (2.2) to the object $Z_L K$, consisting of the disjoint union of the free abelian groups on the inverse images (in K) of the simplices of L.

Using this pair of adjoint functors, one can assign to each object $A \in \mathbf{ab}/L$ its *simplicial resolution* $(Z_L U)^{*+1} A$ which is the simplicial object over $\mathbf{ab}/L$ which, in dimension n, consists of $(Z_L U)^{n+1} A$ and which has the property that

$$\pi_0 (Z_L U)^{*+1} A \cong A \text{ and } \pi_i (Z_L U)^{*+1} A = 0 (i > 0).$$

2.6 Homotopy categories. If $\mathbf{C}$ is a closed model category, then [6, Ch. 1, §1] ho($\mathbf{C}$) will denote its homotopy category, i.e., the category obtained by formally inverting all weak equivalences.

§3. Homology with simplicial coefficients

In this section, we define homology with simplicial coefficients and prove some of its basic properties. First some

3.1 Preliminaries. Let (2.1) $L \in \mathbf{S}$ and (2.4) $A, A' \in \mathbf{ab}/L$. Then one can form the *tensor product* $A \otimes A' \in \mathbf{ab}/L$ which assigns to every simplex of L the tensor product of its inverse images in A and A', and note that, for $K, K' \in \mathbf{S}/L$, *there is a natural isomorphism* (2.5) $Z_L K \otimes Z_L K' \cong Z_L(K \times_L K')$.

Another useful construction assigns to an object $A \in \mathbf{ab}/L$ the simplicial abelian group $\bigoplus_L A$ which, in dimension n, consists of the direct sum of the inverse images (in A) of the n-simplices of L. For $K \in \mathbf{S}/L$, *the simplicial abelian group* $\bigoplus_L Z_L K$ *is just the free simplicial abelian group on the simplices of* K.

Now we can define

3.2 Homology with simplicial coefficients. Given $K \in \mathbf{S}/L$ and $A \in \mathbf{ab}/L$, *the homology* $H_*(K;A)$ *of* K *with simplicial coefficients* A will be

$$H_*(K;A) = \pi_* \bigoplus_L (Z_L K \otimes A)$$

and this definition readily implies:

3.3 PROPOSITION. *If $K, K' \in \mathbf{S}/L$, then $H_*(K; Z_L K')$ is just the ordinary integral homology of $K \times_L K'$.*

3.4 PROPOSITION. *Let $A \in \mathbf{ab}/L$ and $K \in \mathbf{S}/L$ and let $K_1, K_2 \subset K$ be subcomplexes. Then there is a natural long exact (Mayer-Vietoris) sequence*

$$\to H_n(K_1 \cap K_2; A) \to H_n(K_1; A) \oplus H_n(K_2; A) \to H_n(K_1 \cup K_2; A) \to H_{n-1}(K_1 \cap K_2; A) \to$$

3.5 PROPOSITION. *Let $K \in \mathbf{S}/L$ and let $0 \to A'' \to A \to A' \to 0$ be a short exact sequence in $\mathbf{ab}/L$. Then there is a natural long exact sequence*

$$\cdots \to H_n(K; A'') \to H_n(K; A) \to H_n(K; A') \to H_{n-1}(K; A'') \to \cdots$$

Less obvious is

3.6 PROPOSITION. *Let $A \in \mathbf{ab}/L$ and let $g : K \to K' \in \mathbf{S}/L$ be a weak equivalence (2.3). Then g induces an isomorphism $H_*(K; A) \cong H_*(K'; A)$ whenever one of the following conditions is satisfied:*

(1) *the structure maps $K \to L$ and $K' \to L$ are both fibrations (in $\mathbf{S}$) or,*
(2) *the structure map $A \to L$ is a fibration (in $\mathbf{S}$).*

PROOF: If $A = Z_L K''$ for some $K'' \in \mathbf{S}/L$, then part (1) is an easy consequence of 3.3 and the general case now follows readily from (2.5) and the existence of simplicial resolutions.

Part (2) is proved in a similar manner using the following lemma.

3.7 LEMMA. *If the structure map of $K \in \mathbf{S}/L$ is a fibration (in $\mathbf{S}$), then so is the structure map of $Z_L K$.*

PROOF: Given a pair of integers (k, n) with $0 \leq k < n$ (resp. $0 < k \leq n$), an n-simplex $x \in L$ and an $(n-1)$-simplex $y_k \in Z_L K$ over $d_k x$ such that $d_i y_k = 0$ for $i < k$ (resp. $k < i$), a careful calculation (which uses the fact that the structure map $K \to L \in \mathbf{S}$ is a fibration) yields an n-simplex $y \in Z_L K$ over x such that $d_k y = y_k$ and $d_i y = 0$ for $i < k$ (resp. $k < i$). The rest of the proof now is straightforward.

§4. WEAK EQUIVALENCES BETWEEN SIMPLICIAL COEFFICIENT SYSTEMS

Next we discuss the notion of weak equivalence between simplicial coefficient systems which was mentioned in 1.1 and give a positive answer (4.6) to the question which was raised there. We start with the definition of

4.1 Weak equivalences between simplicial coefficient systems. A map $A \to A' \in \mathbf{ab}/L$ will be called a *weak equivalence* if, for every fibrant (2.3) object $K \in \mathbf{S}/L$, it induces an isomorphism $H_*(K; A) \cong H_*K; A')$.

Using 3.4 and 3.6, one then readily shows

4.2 PROPOSITION. *Let $A, A' \in \mathbf{ab}/L$ be such that (2.5) $UA, UA' \in \mathbf{S}/L$ are fibrant. Then a map $A \to A' \in \mathbf{ab}/L$ is a weak equivalence iff the underlying map $UA \to UA' \in \mathbf{S}/L$ is a weak equivalence (2.3).*

One also has

4.3 PROPOSITION. *A map $A \to A' \in \mathbf{ab}/L$ is a weak equivalence if the underlying map $UA \to UA' \in \mathbf{S}/L$ is a weak equivalence as well as a fibration.*

PROOF: The map $A \to A'$ fits into a short exact sequence $0 \to A'' \to A \to A' \to 0$ such that the underlying map of the obvious map $A'' \to L \in \mathbf{ab}/L$ (recall that L is the zero object of $\mathbf{ab}/L$) is a weak equivalence as well as a fibration. By 4.2 the map $A'' \to L \in \mathbf{ab}/L$ is a weak equivalence and the desired result now readily follows from 3.5.

Furthermore 3.3 implies

4.4 PROPOSITION. *If a map $K \to K' \in \mathbf{S}/L$ is a weak equivalence, then so is the induced map (2.3) $Z_L K \to Z_L K' \in \mathbf{ab}/L$.*

Applying this to the maps $V[x,k] \to \Delta[x] \in \mathbf{S}/L$ (2.1), one can construct as follows

4.5 The extension functor $E : \mathbf{ab}/L \to \mathbf{ab}/L$. For $A \in \mathbf{ab}/L$, let $EA \in \mathbf{ab}/L$ be determined by the push out diagram

$$\begin{array}{ccc} \coprod Z_L V[x,k] & \longrightarrow & \coprod Z_L \Delta[x] \\ \downarrow & & \downarrow \\ A & \longrightarrow & EA \end{array}$$

in which the sums are taken over all 4-tuples (k,n,x,g), where k and n are integers such that $0 \le k \le n$, x is an n-simplex of L and g is a map $g : Z_L V[x,k] \to A \in \mathbf{ab}/L$. Then 3.5 implies that the map $A \to EA \in \mathbf{ab}/L$ *is a weak equivalence* and hence *so is the resulting map*

$$A \to E^\infty A = \varinjlim E^n A \in \mathbf{ab}/L\,.$$

This last statement immediately provides a positive answer to the question which was raised in 1.1 as one has, almost by definition:

4.6 PROPOSITION. *For every object $A \in \mathbf{ab}/L$, the structure map of $E^\infty A$ is a fibration of simplicial sets, i.e. $UE^\infty A \in \mathbf{S}/L$ is fibrant.*

We end with observing that the above results also readily imply the following *characterization of weak equivalences in* $\mathbf{ab}/L$.

4.7 PROPOSITION. *A map $A \to A' \in \mathbf{ab}/L$ is a weak equivalence iff the induced map $UE^\infty A \to UE^\infty A' \in \mathbf{S}/L$ is a weak equivalence.*

4.8 PROPOSITION. *Let $P \to L \in \mathbf{S}$ be a path fibration (i.e., a fibration such that (i) the induced map $\pi_0 P \to \pi_0 L$ is an isomorphism and (ii) each component of P is contractible). Then a map $A \to A' \in \mathbf{ab}/L$ is a weak equivalence iff the induced map $H_*(P;A) \to H_*(P;A')$ is an isomorphism.*

§5. A MODEL CATEGORY STRUCTURE FOR $\mathbf{ab}/L$

The preceding results suggest

5.1 PROPOSITION. *The category $\mathbf{ab}/L$ admits a closed simplicial model category structure [**6**, Ch. II] in which the simplicial structure is the obvious one, the weak equivalences are as in 4.1 and a map $X \to Y$ is a trivial fibration (i.e., a fibration as well as a weak equivalence) whenever the underlying map $UX \to UY \in \mathbf{S}/L$ is so (2.3).*

This, of course, implies the following rather formal

5.2 Definition of cofibrations and fibrations in $\mathbf{ab}/L$.

(1) The *cofibrations* in $\mathbf{ab}/L$ are the maps which have the left lifting property [**6**, Ch. I, §5] with respect to the maps $X \to Y$ for which the underlying map $UX \to UY \in \mathbf{S}/L$ is a trivial fibration.

(2) The *fibrations* in $\mathbf{ab}/L$ are the maps which have the right lifting property [**6**, Ch. I,§5] with respect to the trivial cofibrations (i.e., the cofibrations which are weak equivalences).

A more useful description of the cofibrant objects and the cofibrations is

5.3 PROPOSITION.

(1) *An object $A \in \mathbf{ab}/L$ is cofibrant iff it is free (i.e. iff the inverse image in A of each simplex in L is a free abelian group).*

(2) *A map $A \to B \in \mathbf{ab}/L$ is a cofibration iff it is relatively free (i.e., it fits into a short exact sequence $0 \to A \to B \to C \to 0$ in which C is free).*

Proof: A map in **ab**$/L$ clearly has a trivial fibration in **S**$/L$ as underlying map iff it has the right lifting property with respect to all inclusions (2.1) $Z_L\dot{\Delta}[x] \to Z_L\Delta[x] \in \mathbf{ab}/L$. In view of 5.2(i) and the small object argument of [6, Ch. II, §3], this implies that *the cofibrations in* **ab**$/L$ *are the retracts of the maps* $A \to B \in \mathbf{ab}/L$ *which admit (possibly transfinite) factorizatons*

$$A = A_1 \to \cdots \to A_s \to A_{s+1} \to \cdots \to \varinjlim^s A_s = B$$

in which each map $A_s \to A_{s+1}$ *is obtained by pushing out an inclusion* $Z_L\dot{\Delta}[x] \to Z_L\Delta[x]$ *and in which, for every limit ordinal* t *involved,* $A_t = \varinjlim^{s<t} A_s$. The desired result now follows readily.

For fibrations one can, in general, do no better than 5.2(ii). However, for fibrant objects and fibrations between them, one has:

5.4 **Proposition.**

(1) *An object* $Y \in \mathbf{ab}/L$ *is fibrant iff the underlying object* $UY \in \mathbf{S}/L$ *is fibrant (i.e., the structure map* $Y \to L \in \mathbf{S}$ *is a fibration).*
(2) *Let* $X, Y \in \mathbf{ab}/L$ *be fibrant. Then a map* $X \to Y \in \mathbf{ab}/L$ *is a fibration iff the underlying map* $UX \to UY \in \mathbf{S}/L$ *is a fibration.*

5.5 **Corollary.** *Let* $X, Y \in \mathbf{ab}/L$ *be fibrant. Then (4.2) a map* $X \to Y \in \mathbf{ab}/L$ *is a weak equivalence iff the underlying map* $UX \to UY \in \mathbf{S}/L$ *is a weak equivalence.*

Proof: In view of 4.4 and 5.2(ii) a fibration $X \to Y \in \mathbf{ab}/L$ has the right lifting property with respect to the maps $Z_L V[x,k] \to Z_L\Delta[x]$ and hence its underlying map $UX \to UY \in \mathbf{S}/L$ is also a fibration. It thus remains to show that a map $X \to Y \in \mathbf{ab}/L$, for which the underlying map $UX \to UY \in \mathbf{S}/L$ is a fibration between fibrant objects, has the right lifting property with respect to all trivial cofibrations in **ab**$/L$. Because UX and UY are fibrant (in **S**$/L$), a commutative diagram in **ab**$/L$

$$\begin{array}{ccc} A & \longrightarrow & X \\ \downarrow & & \downarrow \\ B & \longrightarrow & Y \end{array}$$

in which the map $A \to B$ is a trivial cofibration, admits a factorization (4.4)

$$\begin{array}{ccccc} A & \longrightarrow & E^\infty A & \longrightarrow & X \\ \downarrow & & \downarrow & & \downarrow \\ B & \longrightarrow & E^\infty B & \longrightarrow & Y \end{array}$$

in which the map $E^\infty A \to E^\infty B$ is a trivial cofibration. Moreover (4.5) $UE^\infty A$ and $UE^\infty B$ are fibrant objects of **S**$/L$ and hence (4.2) the underlying map $UE^\infty A \to UE^\infty B \in \mathbf{S}/L$ is a weak equivalence. Using this fact it is now not difficult to obtain a lifting $E^\infty B \to X$ which, composed with the map $B \to E^\infty B$, yields the desired lifting $B \to X$.

We end with observing that the pair of adjoint functors $Z_L : \mathbf{S}/L \leftrightarrow \mathbf{ab}/L : U$ behaves as expected, i.e., (4.4 and 5.3-5).

5.6 **Proposition.** *The functor* $Z_L : \mathbf{S}/L \to \mathbf{ab}/L$ *preserves cofibrations and weak equivalences and the functor* $U : \mathbf{ab}/L \to \mathbf{S}/L$ *preserves fibrations and weak equivalences between fibrant objects.*

It thus remains to give a

5.7 Proof of 5.1. One has to verify axioms CM1-5 of [**2**, Ch. VIII, §2]. This is straightforward, except for axioms CM4(ii) and CM5(ii).

To verify CM4(ii), it suffices to show: if a map $X \to Y \in \mathbf{ab}/L$ is a trivial fibration, then so is the underlying map $UX \to UY \in \mathbf{S}/L$. By the small object argument [**6**, Ch.II, §3] the map $X \to Y$ admits a factorization $X \to X' \to Y$ in $\mathbf{ab}/L$ such that the underlying map $UX' \to UY \in \mathbf{S}/L$ is a trivial fibration and the map $X \to X' \in \mathbf{ab}/L$ is a cofibration. In view of 4.3, the map $X' \to Y \in \mathbf{ab}/L$ is a weak equivalence and, because the map $X \to Y \in \mathbf{ab}/L$ is a weak equivalence, it follows that the map $X \to X' \in \mathbf{ab}/L$ is a trivial cofibration. As the map $X \to Y \in \mathbf{ab}/L$ is also a fibration,one can apply CM4(i) and deduce that it is a retract of the map $X' \to Y \in \mathbf{ab}/L$ and that its underlying map therefore is also a trivial fibration.

Finally to verify CM5(ii) one uses the Bousfield argument [**1**, §11], i.e., one combines the small object argument [**6**, Ch. II, §3] with the observation that proposition 4.8 readily implies

5.8 LEMMA. *Let c be an infinite cardinal number at least as large as the number of simplices in L. Then a map in $\mathbf{ab}/L$ is a fibration iff it has the right lifting property with respect to all trivial cofibrations $A \to B \in \mathbf{ab}/L$ in which the number of simplices in B is at most c.*

§6 DEPENDENCE OF $\mathbf{ab}/L$ ON L

Our aim in this section is to show

6.1 PROPOSITION. *Let $g : L \to L' \in \mathbf{S}$ be a weak equivalence. Then g induces an equivalence of categories $ho(\mathbf{ab}/L) \cong ho(\mathbf{ab}/L')$ (2.6) and hence a 1–1 correspondence between the weak equivalence classes of the simplicial coefficient systems over L and the ones over L'.*

To prove this we consider

6.2 A pair of adjoint functors $g_* : \mathbf{ab}/L \leftrightarrow \mathbf{ab}/L' : g^*$. Given a map $g : L \to L' \in \mathbf{S}$, the *pull back functor* $g^* : \mathbf{ab}/L' \to \mathbf{ab}/L$ has as left adjoint the *push out functor* $g_* : \mathbf{ab}/L \to \mathbf{ab}/L'$ which "takes direct sums, over the simplices of L which have the same image in L', of their inverse images", and which clearly has the property that $\bigoplus_{L'} g_*A = \bigoplus_{L} A$ for every object $A \in \mathbf{ab}/L$.

Moreover, one readily verifies:

6.3 PROPOSITION. *The left adjoint $g_* : \mathbf{ab}/L \to \mathbf{ab}/L'$ preserves cofibrations and weak equivalences and the right adjoint $g^* : \mathbf{ab}/L' \to \mathbf{ab}/L$ preserves fibrations and weak equivalences between fibrant objects.*

The desired result now follows immediately from [6, Ch. I, §4, Th. 3] and

6.4 PROPOSITION. *Let $g : L \to L' \in \mathbf{S}$ be a trivial cofibration (i.e., a weak equivalence which is 1–1). Then, for every cofibrant object $A \in \mathbf{ab}/L$ and every fibrant object $A' \in \mathbf{ab}/L'$, a map $A \to g^*A' \in \mathbf{ab}/L$ is a weak equivalence iff its adjoint $g_*A \to A' \in \mathbf{ab}/L'$ is so.*

PROOF: As g is 1–1, for every object $A \in \mathbf{ab}/L$ the adjunction map $A \to g^*g_*A \in \mathbf{ab}/L$ is an isomorphism and hence a map $A \to B \in \mathbf{ab}/L$ is a weak equivalence iff the induced map $g_*A \to g_*B \in \mathbf{ab}/L'$ is so. Moreover, because g is a weak equivalence, the adjunction map $g_*g^*A' \to A' \in \mathbf{ab}/L'$ is (in view of 3.6(ii) and 6.2) a weak equivalence for every fibrant object $A' \in \mathbf{ab}/L'$. The proposition now readily follows.

§7. SIMPLICIAL MODULES OVER THE LOOP GROUP GL OF L

We end with showing that (7.4), for $L \in \mathbf{S}$ connected, *the weak equivalence classes of the simplicial coefficient systems over L are in a natural 1–1 correspondence with the weak equivalence classes of the simplicial modules over the loop group GL of L* (or more precisely, its integral group ring ZGL).

First we recall from [6] the existence of

7.1 A closed model category structure for simplicial modules over GL. Let $L \in \mathbf{S}$ be connected and have a base point, let GL be its *loop group* [4] (which is a free simplicial group which has the homotopy type of the loops on L) and let $\mathbf{M}_{GL}$ denote the category of *simplicial (left) modules over the integral group ring ZGL of GL*. Then [6, ch. II, §6] $\mathbf{M}_{GL}$ *admits a closed simplicial model category structure* in which the simplicial structure is the obvious one and in which a map is a weak equivalence or a fibration whenever the underlying map in $\mathbf{S}$ is a weak equivalence or a fibration.

Next we observe the existence of

7.2 A pair of adjoint functors $h : \mathbf{ab}/L \leftrightarrow \mathbf{M}_{GL} : k$. Let $EL \to L \in \mathbf{S}$ be the *path fibration* of [4], which is a principal fibration with group GL (i.e., [5, §18] GL acts freely on EL from the right and $EL/GL = L$). Then it is not difficult to see that the induced map $Z_L EL \to L \in \mathbf{S}$ is a principal fibration with group ZGL. Thus one can consider the functor $h : \mathbf{ab}/L \to \mathbf{M}_{GL}$ which sends an object $A \in \mathbf{ab}/L$ to the object $\bigoplus_L (Z_L EL \otimes A) \in \mathbf{M}_{GL}$ and the functor $k : \mathbf{M}_{GL} \to \mathbf{ab}/L$ which sends an object $M \in \mathbf{M}_{GL}$ to the object $M \otimes_{ZGL} Z_L EL \in \mathbf{ab}/L$. A straightforward calculation then yields that these functors form a pair of adjoint functors $h : \mathbf{ab}/L \leftrightarrow \mathbf{M}_{GL} : k$ and that

7.3 PROPOSITION. *Both of the functors h and k preserve weak equivalences and both adjunction maps $hk \to id$ and $id \to kh$ are natural weak equivalences.*

7.4 COROLLARY. *The functor h induces an equivalence of categories $ho(\mathbf{ab}/L) \cong ho(\mathbf{M}_{GL})$ (2.6) and hence a 1–1 correspondence between the weak equivalence classes of the simplicial coefficient systems over L and the weak equivalence classes of the simplicial modules over GL.*

7.5 REMARK: The weak equivalence classes of the simplicial coefficient systems over L are also in natural 1–1 correspondence with the *weak equivalence classes of the differential graded modules over the normalized chains on GL, which are trivial in negative dimensions.* This follows immediately from 7.4 and the fact that [5, §22 and §29] the normalization functor N gives rise to a functor $N : \mathbf{M}_{GL} \to \mathbf{m}_{GL}$ (where $\mathbf{m}_{GL}$ denotes the category of these differential graded modules), which induces an equivalence of categories $\text{ho}(\mathbf{M}GL) \cong \text{ho}(\mathbf{m}_{GL})$. One proves this by observing that

(1) the category $\mathbf{m}_{GL}$ admits a closed model category structure in which a map is a weak equivalence whenever it induces an isomorphism on homology and a fibration whenever it is onto in positive demensions, and

(2) the functor $N : \mathbf{M}_{GL} \to \mathbf{m}_{GL}$ and its right adjoint satisfy the conditions of [6, Ch. I, §4, Th.3].

REFERENCES

1. A.K. Bousfield, *The localization of spaces with respect to homology*, Topology **14** (1975), 133–150.
2. A.K. Bousfield and D.M. Kan, "Homotopy limits, completions and localizations," Lect. Notes in Math. 304, Springer, Berlin, 1972.
3. W.G. Dwyer and D.M. Kan, *Hochschild-Mitchell cohomology of simplicial categories and the cohomology of simplicial diagrams of simplicial sets*, Proc. Kon. Ned. Akad. van Wetensch. (Indag. Math.) (to appear).
4. D.M. Kan, *A combinatorial definition of homotopy groups*, Ann. of Math. **67** (1958), 282–312.
5. J.P. May, "Simplicial objects in algebraic topology," Van Nostrand, 1967.
6. D.G. Quillen, "Homotopical algebra," Lect. Notes in Math. 43, Springer, Berlin, 1967.

Department of Mathematics, University of Notre Dame, Notre Dame, Indiana

Department of Mathematics, Massachusetts Institute of Technology, Cambridge, Massachusetts

On the Double Suspension

by

Brayton Gray

1. In this note we show how to do the clutching construction in the category of quasi fiberings. Our main application is to make some general constructions related to $\Omega^2 S^2 A$ when A is a suspension, and to prove a conjecture made by Mahowald [M] when p = 2 and by Cohen, Neisendorfer, and Moore [CMN] when $p > 3$. Namely,

Theorem 1. There is a space B_n and a map $v_n : \Omega^2 S^{2n+1} \longrightarrow B_n$ such that the homotopy fiber of v_n is S^{2n-1}.

Note that this is not a localized statement and applies to all primes simultaneously. In section 2 we prove this theorem, and in section 3 draw out some of the consequences. In particular, we show that when localized at p, B_n is the fiber of a localized map $\varphi : \Omega^2 S^{2np+1} \longrightarrow S^{2np-1}$. A stable decomposition is also obtained. In section 4 we give a different application of Proposition 2, this time to H space bundles over spheres.

2. In this section we will give a general clutching construction for quasifiberings over mapping cones. We will use this, in our main application, to approximate Hurewicz fiberings. Let $X \cup CA$ be the unreduced mapping cone (this construction could also be done in the reduced case provided that the base point is nondegenerate).

Proposition 2. a) Let $E \xrightarrow{\pi} X$ be a Hurewics fibering with fiber $F = \pi^{-1}(*)$. Let $\theta : F \times A \to E$ be a trivialization of $\pi|_A$; i.e., $\pi\theta(f,a) = a$, and for each $a \in A$, $\theta_a : F \to \pi^{-1}(a)$ is a homotopy equivalence. Then there is a quasifibering $E' \xrightarrow{\pi'} X \cup CA$ such that

$\pi'|_X = \pi$; i.e., there is a pull back diagram:

$$\begin{array}{ccc} E & \longrightarrow & E' \\ \Big\downarrow{\scriptstyle \pi} & & \Big\downarrow{\scriptstyle \pi'} \\ X & \longrightarrow & X \cup CA, \end{array}$$

and E'/E is homeomorphic to $F \ltimes SA$.

b) Suppose* $\overline{E} \xrightarrow{\overline{\pi}} X \cup CA$ is a Hurewicz fibering with fiber $F = \overline{\pi}^{-1}$ (vertex) such that $\overline{\pi}|_X = \pi$. Then there is a suitable trivialization θ and a mapping $\Gamma : E' \longrightarrow \overline{E}$ over the identity which is a weak homotopy equivalence:

$$\begin{array}{ccc} E' & \xrightarrow{\quad \Gamma \quad} & \overline{E} \\ \Big\downarrow{\scriptstyle \pi'} & & \Big\downarrow{\scriptstyle \overline{\pi}} \\ X \cup CA & = & X \cup CA \end{array} .$$

Finally there is a weak homotopy equivalence between $\overline{E}/E$ and $F \ltimes SA$**.

<u>Proof</u>: We define E' as follows:

$$E' = E \coprod F \times CA \;/\; (f,a,0) \sim \theta(a,f);$$

the projection, $\pi' : E' \rightarrow X \cup CA$, is given by $\pi'|_E = \pi$, $\pi'(f.a.t) = (a,t) \in CA \subset X \cup CA$, putting 0 at the base of the cone. Write $U_1 = X \cup CA - X$ and $U_2 = X \cup CA - \{\text{vertex}\}$. These sets are open and cover $X \cup CA$. We will show that π' is a quasifibering over U_1, U_2 and $U_1 \cap U_2$. Applying [DT,2.2] we conclude that π' is a quasifibering.

* See [W;VII,1.1]

**Note added in proof: Our main application of Proposition 2 is via the map $\overline{E} \rightarrow F \ltimes SA$ defined by this equivalence. In the case that $\overline{\pi}$ is induced by a map $j: X \cup CA \rightarrow Y$, John Moore and Joe Neisendorfer have recently observed that such a map may also be obtained by considering the composition $X \cup CA \longrightarrow X \cup CA \vee SA \longrightarrow Y \vee SA$ and applying the results of [G4].

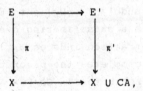

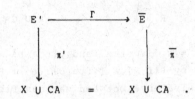

$(\pi')^{-1}(U_1)$ is compatibly homeomorphic with $U_1 \times F$, so π' is a quasifibering over U_1 and $U_1 \cap U_2$. There are compatible deformation retractions of U_2 onto X and $(\pi')^{-1}(U_2)$ onto E. We wish to apply [DT.2.10]. Clearly π' is a quasifibering over X since it restricts to π there. Finally the retraction induces $\theta_a : F \to \pi^{-1}(a)$ over each point $(a,t) \in U_2$. This completes the proof that π' is a quasifibering. Clearly E'/E is homeomorphic to $F \times CA/F \times A \equiv F \ltimes SA$.

To prove b) we use the homotopy lifting property to construct a map Γ in the diagram:

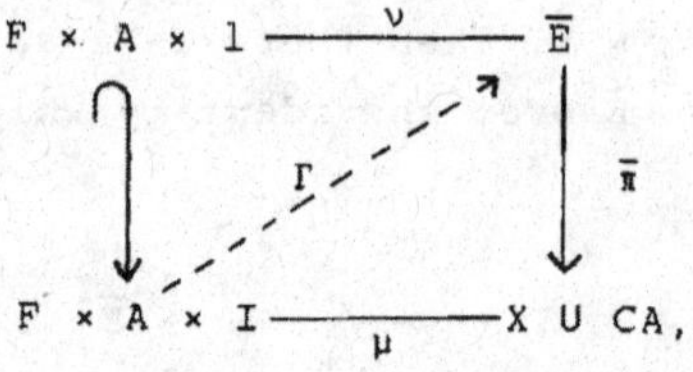

where $\gamma(f,a) = f \in F = \bar{\pi}^{-1}\{\text{vertex}\}$ and $\mu(f,a,t) = (a,t) \in CA \subset X \cup CA$. Define $\theta : F \times A \to E$ by $\theta(f,a) = \Gamma(f,a,0)$. $\theta_a : F \to \pi^{-1}(a)$ is just the standard homotopy equivalence between various fibers in a Hurewicz fibering. Thus part a) applies and Γ defines a map $\Gamma : E' \to \bar{E}$ over the identity, where $E \subset E'$ is mapped by the inclusion. Γ is clearly a weak homotopy equivalence by the 5 lemma. (E',E) has the homotopy extension property, as does $(\bar{E},E)$ since $(X \cup CA, X)$ does. Thus $\bar{E}/E \simeq \bar{E} \cup CE$ and $E'/E \simeq E' \cup CE$. Thus the induced map $E'/E \equiv F \ltimes SA \to \bar{E}/E$ is a weak homotopy equivalence by [Gl.16,17].

3. We now make some general constructions. Suppose that A is a suspension.* Consider the fibering:

$$\Omega^2 S^2 A \to X(A) \to SA$$

induced from the path space fibration over $\Omega S^2 A$. In case $A = S^{2n-1}$, $X(A)$ localized at 2 is just $\Omega^2 S^{4n+1}$ and the fibering represents the EHP sequence. We construct a map

$$\varphi : (X(A), \Omega^2 S^2 A) \to (SA \wedge A, *) \quad .$$

Applying Proposition 2 to the fibering, we have $X(A)/\Omega^2 S^2 A \simeq (\Omega^2 S^2 A) \ltimes SA$. Define φ to be the composition:

*See 4c

$$X(A) \to X(A)/\Omega^2 S^2 A \simeq (\Omega^2 S^2 A) \ltimes SA \to (\Omega^2 S^2 A) \wedge SA \longrightarrow SA \wedge A,$$

where the last map is a retraction using the fact that A is a suspension. In case $A = S^{2n-1}$, we write X_n and φ_n for X(A) and φ. We proceed without localizing.

Proposition 4. X_n is 4n-2 connected, $H_{4n-1}(X_n) \simeq Z$ and φ_n has degree 2.

Proof: We look at the integral and mod 2 cohomology Serre spectral sequences for the fibering $\Omega^2 S^{2n+1} \to X_n \to S^{2n}$ in dimensions less than 4n.

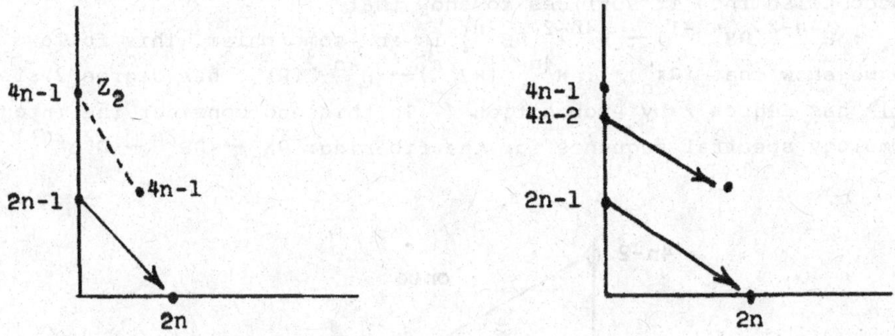

By comparing these we see that $H^{4n-1}(X_n) \simeq Z$, there is an extension in the left hand spectral sequence, and the map

$$Z \simeq H^{4n-1}(X_n) \to H^{4n-1}(\Omega^2 S^{2n+1}) \simeq Z_2$$

is onto. The conclusion follows since $X_n/\Omega^2 S^{2n+1} \xrightarrow{\varphi_n} S^{4n-1}$ induces an isomorphism in H^{4n-1}.

We now construct our first basic diagram. We do this for a general space A on the left giving the specific example with $A = S^{2n-1}$ on the right. Recall that we require A to be a suspension. All maps and spaces are functorial for maps which are suspensions; in these diagrams all sequences are fiber sequences. Thus, for example, B(A) is the fiber of φ and T(A) is the fiber of γ.

$$\begin{array}{ccccc}
\Omega B(A) & \longrightarrow & \Omega X(A) & \xrightarrow{\Omega\varphi} & \Omega S\wedge A \\
\downarrow & & \downarrow{\scriptstyle \Omega\pi} & & \| \\
T(A) & \longrightarrow & \Omega SA & \xrightarrow{\gamma} & \Omega SA\wedge A \\
\downarrow & & \downarrow & & \downarrow \\
\Omega^2S^2A & = & \Omega^2S^2A & \longrightarrow & * \\
\downarrow & & \downarrow & & \downarrow \\
B(A) & \longrightarrow & X(A) & \xrightarrow{\varphi} & SA\wedge A
\end{array}
\qquad
\begin{array}{ccccc}
\Omega B_n & \longrightarrow & \Omega X_n & \xrightarrow{\Omega\varphi_n} & \Omega S^{4n-1} \\
\downarrow & & \downarrow{\scriptstyle \Omega\pi_n} & & \| \\
S^{2n-1} & \longrightarrow & \Omega S^{2n} & \xrightarrow{\gamma_n} & \Omega S^{4n-1} \\
\downarrow & & \downarrow & & \downarrow \\
\Omega^2S^{2n+1} & = & \Omega^2S^{2n+1} & \longrightarrow & * \\
\downarrow & & \downarrow & & \downarrow \\
B_n & \longrightarrow & X_n & \xrightarrow{\varphi_n} & S^{4n-1}
\end{array}$$

The right hand lower square follows since $\varphi\big|_{\Omega^2S^2A} \sim *$. The proof of theorem 1 will be complete if we show that $T(S^{2n-1}) = S^{2n-1}$. In order to accomplish this it suffices to show that $(\gamma_n)_* : H^{4n-2}(\Omega S^{4n-1}) \to H^{4n-2}(\Omega S^{2n})$ is an isomorphism. This follows when we show that $(\Omega\pi_n)^* : H^{4n-2}(\Omega S^{2n}) \to H^{4n-2}(\Omega X_n)$ has degree 2 since $(\Omega\varphi_n)^*$ has degree 2 by Proposition 4. To this end consider the integral cohomology spectral sequence for the fibering: $\Omega X_n \to \Omega S^{2n} \to \Omega^2S^{2n+1}$

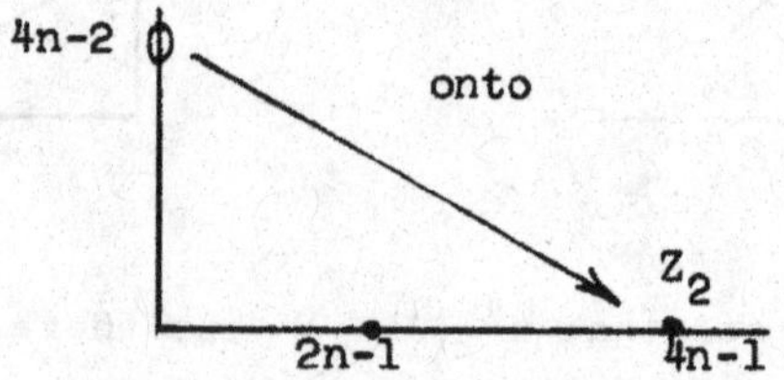

the differential is onto since there is no torsion in $H^{4n-1}(\Omega S^{2n})$. Thus twice the generator survives and the proof of theorem 1 is complete.

Of course there is no hope that $T(A) \simeq A$ is general, just as there is no hope that the EHP sequence should be exact for all spaces.

<u>Corollary 5</u>. The composition

$$\pi_{r+2}(S^{4n+1}) \xrightarrow{P} \pi_r(S^{2n}) \xrightarrow{H} \pi_r(S^{4n-1})$$

in the EHP sequence at p = 2 is induced by $\varphi_n : \Omega^2 S^{4n+1} \to S^{4n-1}$.

Proof: $\Omega\varphi_n \sim \gamma_n \sim \Omega\pi_n$. Here $\gamma_n = H$ is some Hopf invariant. This may not be the same one constructed by James [J].

We wish to construct a mod p analog for Corollary 5 for p > 2. To do so we consider the localized fibration:

$$\Omega^2 S^{2n+1} \to \Omega^2 S^{2np+1} \to S^{2n}_{(p-1)},$$

where $S^{2n}_{(p-1)}$ is the subspace of the James construction S^{2n}_{∞} of words of length $\leq p - 1$. In general we construct a fibration:

$$\Omega^2 S^2 A \to X_k(A) \to (SA)_{k-1}$$

ans as in the case p = 2, obtain a homotopy equivalence

$$X_k(A)/X_{k-1}(A) \simeq (\Omega^2 S^2 A) \ltimes (SA)^{k-1}$$

using Proposition 2. From this we construct a map:

$$\Omega : X_k(A) \to S^{k+1} A^k \quad .$$

In case $A = S^{2n-1}$ and $k = p$, $X_k \simeq \Omega^2 S^{2np+1}$ and $\varphi : \Omega^2 S^{2np+1} \to S^{2np-1}$ has degree p.

Proposition 6. There is a fibration sequence localized at p:

$$B_n \to \Omega^2 S^{2np+1} \xrightarrow{\varphi} S^{2np-1} \quad .$$

Let F_φ be the fiber of φ.

<u>Proof</u>: Comparing the fiberings:

$$\begin{array}{ccc}
\Omega^2 S^{2n+1} & = & \Omega^2 S^{2n+1} \\
\downarrow & & \downarrow \\
X_n & \xrightarrow{\alpha} & \Omega^2 S^{2np+1} \\
\downarrow & & \downarrow \\
S^{2n} & \longrightarrow & S^{2n}_{(p-1)};
\end{array}$$

for $p > 2$ we see that $X_n \to \Omega^2 S^{2np+1} \xrightarrow{\varphi} S^{2np-1}$ is null homotopic. Consequently we may factor α through F_φ. The composition $\Omega^2 S^{2n+1} \to B_n \to X_n \to F_\varphi \to \Omega^2 S^{2np+1}$ is onto in Z_p homology. Since $H_*(F_\varphi;Z_p) \cong H_*(\Omega^2 S^{2np+1};Z_p) \otimes H_*(\Omega S^{2np-1};Z_p)$ and the generators in dimensions 2np-1 and 2np-1 are connected by a Bockstein, the composition $B_n \to X_n \to F_\varphi$ is onto in homology. However these homology groups have the same rank, so $B_n \simeq F_\varphi$ and we are done.

We now record the mod p version of diagram I.

<u>Theorem 7</u>. For all primes p we have a commutative diagram of spaces localized at p:

$$\begin{array}{ccccc}
\Omega B_p(A) & \to & \Omega X_p(A) & \to & \Omega S^{p-1}A(p) \\
\downarrow & & \downarrow & & \| \\
T_p(A) & \to & \Omega(\Sigma A)_{(p-1)} & \to & \Omega S^{p-1}A(p) \\
\downarrow & & \downarrow & & \downarrow \\
\Omega^2 S^2 A & \to & \Omega^2 S^2 A & \to & * \\
\downarrow & & \downarrow & & \downarrow \\
B_p(A) & \to & X_p(A) & \to & S^{p-1}A(p)
\end{array}
\qquad
\begin{array}{ccccc}
\Omega B_n & \to & \Omega^2 S^{2np+1} & \to & \Omega S^{2np-1} \\
\downarrow & & \downarrow & & \| \\
S^{2n-1} & \to & \Omega S^{2n}_{(p-1)} & \to & \Omega S^{2np-1} \\
\downarrow & & \downarrow & & \downarrow \\
\Omega^2 S^{2n+1} & = & \Omega^2 S^{2n+1} & \to & * \\
\downarrow{\scriptstyle v_n} & & \downarrow & & \downarrow \\
B_n & \to & \Omega^2 S^{2np+1} & \xrightarrow{\varphi} & S^{2np-1}
\end{array}$$

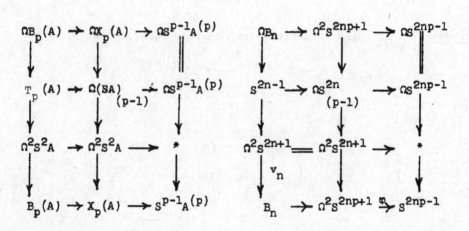

Remarks:

a) Here we have, as a biproduct, constructed a natural Toda-Hopf invariant $\Omega(SA)_{(p-1)} \rightarrow \Omega S^{p-1}A^{(p)}$ for any space A provided that $p > 2$ or A is a suspension. This is slightly stronger than the version constructed in [G3], where the target space is $\Omega^2 S^p A^{(p)}$.

b) An automatic consequence is that the composition $\Omega^3 S^{2np+1} \rightarrow \Omega S^{2n}_{(p-1)} \rightarrow \Omega S^{2np-1}$ from the EHP sequence is a loop map: it is $\Omega\varphi$.

c) We have also constructed a mod p version of the Whitehead product map in a functorial way for each A; namely $\Omega^2 S^{p-1}A^{(p)} \rightarrow T_p(A)$ which specializes to $\Omega^2 S^{2np-1} \rightarrow S^{2n-1}$ in case $A = S^{2n-1}$.

d) Another arrangement defines a useful diagram:

$$\begin{array}{ccc} \Omega^2 S^{2n+1} & \xrightarrow{v_n} & B_n \\ \uparrow & & \uparrow \\ \Omega S^{2n}_{(p-1)} & \longrightarrow & \Omega S^{2np-1} \end{array}$$

where all the maps are from the above diagram. This also has consequences for general spaces A.

e) The James Hopf invariant $\Omega S^2 A \xrightarrow{H_p} \Omega S^{p+1}A^{(p)}$ when restructed to $(SA)_{(p-1)}$ is null homotopic, so we may define a colifting:

$$\begin{array}{ccc} \Omega^2 S^2 A & \xrightarrow{\Omega H_p} & \Omega^p S^{p+1} A^{(p)} \\ \downarrow & \nearrow{\scriptstyle c} & \\ X_p(A) & & \end{array}$$

which is a homotopy equivalence when $A = S^{2n-1}$. Diagrams similar to the above can be constructed using the composition $B_p(A) \rightarrow X_p(A) \rightarrow \Omega^p S^{p+1} A^{(p)}$ which is null homotopic on $\Omega^2 S^2 A$.

f) Let $S^{2np-1}\{p\}$ be the fiber of the degree p map on S^{2np-1}. Then there is a commutative diagram of fiberings:

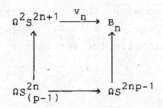

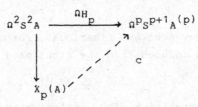

$$\begin{array}{ccccc}
S^{2np-1}\{p\} & \longrightarrow & S^{2np-1} & \xrightarrow{p} & S^{2np-1} \\
\downarrow & & \downarrow & & \| \\
B_n & \longrightarrow & \Omega^2 S^{2np+1} & \xrightarrow{\varphi} & S^{2np-1} \\
\downarrow & & \downarrow & & \downarrow \\
B_{np} & \longrightarrow & B_{np} & \longrightarrow & *
\end{array}$$

g) It is reasonable to ask whether the sequence

$$S^{2n-1} \to \Omega^2 S^{2n+1} \to B_n$$

can be extended one more stage to the right. A necessary condition, of course, is that there be a space BS^{2n-1} with $\Omega BS^{2n-1} \simeq S^{2n-1}$. Suppose, when localized at p, this is the case. It is well known that this happens if $n = 1$ or 2 or $n|(p-1)$. There is a natural inclusion $S^{2n} \subset BS^{2n-1}$. We obtain from the sequence $B_n \to X_n \to S^{2n} \to BS^{2n-1}$; a diagram:

$$\begin{array}{ccccccc}
S^{2n-1} & \longrightarrow & \Omega S^{2n} & = & \Omega S^{2n} & \longrightarrow & S^{2n-1} \\
\downarrow & & \downarrow & & \downarrow & & \downarrow \\
\Omega^2 S^{2n+1} & \longrightarrow & \Omega^2 S^{2n+1} & \longrightarrow & * & \longrightarrow & * \\
\downarrow & & \downarrow & & \downarrow & & \downarrow \\
B_n & \longrightarrow & X_n & \longrightarrow & S^{2n} & \longrightarrow & BS^{2n-1}
\end{array}$$

where the vertical sequences are fiberings. From this we see that the left hand fibering is induced from the right hand one and our task is accomplished.

We seek a stable decomposition of B_n analogous to [BP] and [C]. Using the notation of [C], Cohen proves that stably

$$\Omega^2 S^{2n+1} \sim \bigvee_{\substack{k \geq 1 \\ k \equiv 0,1 \pmod p}} S^{k(2n-2)} \; t(V_k)$$

where the $t(V_k)$ are Brown Gitler spectra.

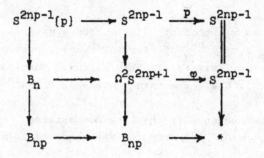

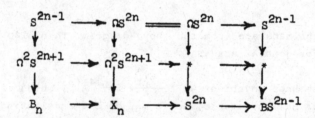

Theorem 8. Stably $B_n \sim \bigvee_{\substack{k\geq 1\\ k\equiv 0 (\mathrm{mod}\ p)}} S^{k(2n-2)} t(V_k)$

Proof: We have a stable map π given as the composition:

$$W = \bigvee_{\substack{k\geq 1\\ k\equiv 0 (\mathrm{mod}\ p)}} S^{k(2n-2)} t(V_k) \to \Omega^2 S^{2n+1} \to B_n$$

Each $t(V_k)$ is a cyclic $A_{(p)}$ module whose bottom cell is mapped to $x_1^k \in H_{(2np-2)k}(\Omega^2 S^{2n+1})$ where $x_1 \in H_{2np-2}(\Omega^2 S^{2n+1})$ is a polynomial generator. This is easy to see since these classes are in the image of $\Omega S^{2n}_{(p-1)} \to \Omega^2 S^{2n+1}$ and hence they are stably spherical. Thus $\pi^* : H^*(B_n;Z_p) \to H^*(W;Z_p)$ is onto. Comparing ranks we see that π^* is an isomorphism.

Proposition 9. $S^2(\Omega^2 S^{2n+1}) \sim S^2(S^{2n-1} \times B_n)$.

Proof: We define a homology isomorphism by first taking the double suspension of the composition:

$$\Omega^2 S^{2n+1} \xrightarrow{\Delta} \Omega^2 S^{2n+1} \times \Omega^2 S^{2n+1} \xrightarrow{1 \times v_n} \Omega^2 S^{2n+1} \times B_n$$

and then use the retraction $S^2\Omega^2 S^{2n+1} \to S^{2n+1}$ to obtain

$$S^2(\Omega^2 S^{2n+1} \times B_n) \xrightarrow{\sim} S^2\Omega^2 S^{2n+1} \vee S^2(\Omega^2 S^{2n+1}) \wedge B_n \vee S^2 B_n$$
$$\longrightarrow S^{2n+1} \vee S^{2n+1} \wedge B_n \vee S^2 B_n \simeq S^2(S^{2n-1} \times B_n).$$

This may well be the finest decomposition of $S^2\Omega^2 S^{2n+1}$. Several people have independently tried to produce a map to the Moore space Y^{2np+1}, but the problem seems at this point to be unsettled.

Corollary 10. The composition $B_n \to \Omega^2 S^{2np+1} \xrightarrow{p} \Omega^2 S^{2np+1}$ is null homotopic where p is the double loops on the degree p map for p > 2 or the loop space squaring map for p = 2.

<u>Proof</u>: The composition $\Omega^2 S^{2n+1} \xrightarrow{H_p} \Omega^2 S^{2np+1} \xrightarrow{P} \Omega^2 S^{2np+1}$ is null homotopic. (See [C2], [G2], or [S]). Since H_p factors through B_n, its double adjoint factors through $S^2(\Omega^2 S^{2n+1}) \to S^2 B_n$. However this map has a right inverse by Proposition 9.

As an immediate consequence we define a colifting using this homotopy to get:

<u>Corollary 11</u>. There is a commutative diagram:

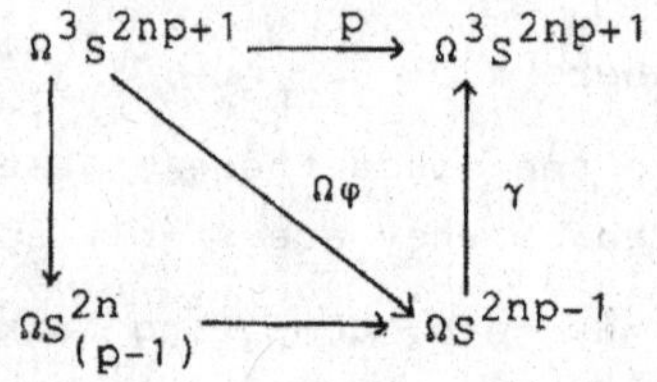

where γ is some map of degree 1.

This is almost Conjecture 15 of [G2]. Unfortunately it is not strong enough for the intended application.

We make one last observation before we close this section. In [S], Selick proved that if $2^r \pi_*(S^{4n-1}) = 0$ then $2^{r+1}\pi_*(S^{4n+1}) = 0$. His proof can be simplified using the map $\varphi : \Omega^2 S^{4n+1} \to S^{4n-1}$.

4. In this section we develop a totally different application of Proposition 2. This result was developed to analyze finite H spaces.

Suppose that X is an H space and that Y is an X module (i.e., we assume a map $X \times Y \to Y$). Suppose further that there is a base point $e \in Y$ such that the associativity formula $(x_1 \cdot x_2) \cdot e \sim x_1 \cdot (x_2 \cdot e)$ holds up to homotopy. Define a map $X \to Y$ by $x \to x \cdot e$. Finally, suppose that X and Y are simply connected.

<u>Proposition 12</u>. Suppose $H_*(Y)$ is free over $H_*(X)$ with one generator in dimension n. Then there is a fibration $X \to Y \to S^n$ in the homotopy category and $Y/X \simeq S^n \rtimes X$.

Proof: Let $\alpha: (CS^{n-1}, S^{n-1}) \to (X,Y)$ generate $\pi_n(Y,X) \cong H_n(Y,X) \cong Z$. Define $\theta: X \times S^{n-1} \to X$ by $\theta(x,u) = x \cdot \alpha(w)$. We apply proposition 2 with π the trivial fibration over a point. This constructs a quasi-fibering $X \to E' \xrightarrow{\pi'} S^n$. We define $\Gamma: E' \to Y$ as follows:

$$\begin{aligned} \Gamma(x) &= x \cdot e && \text{for } x \in X \\ \Gamma(x,u,t) &= h(x,\alpha(u),2t) && u \quad S^{n-1} \in t \le \tfrac{1}{2} \\ \Gamma(x,u,t) &= x \cdot \alpha(u,2t-1) && u \quad S^{n-1} \in t \ge \tfrac{1}{2} \end{aligned}$$

where h is a homotopy from $(x_1 \cdot x_2) \cdot e$ to $x_1 \cdot (x_2 \cdot e)$. Now define $\chi: (B^n, S^{n-1}) \to (E', X)$ by $\chi(u) = (*,u) \in X \times B^n$. Then $\Gamma \cdot \chi \simeq \alpha$ and $\pi' \cdot \chi$ generates $\pi_n(S^n)$. Thus $H_*(E') \cong H_*(X) \otimes H_*(S^n)$ and Γ_* induces isomorphisms in homology. Thus Γ is a weak homotopy equivalence.

5. We list here some questions we find interesting.

A. Is B_n an H space? Is $\varphi: \Omega^2 S^{2np+1} \to S^{2np-1}$ a H map when $p > 2$? Is $\Omega S^{2n}_{(p-1)} \to \Omega S^{2np-1}$ an H map when $p > 2$? (Compare [G3, Theorem 1c]). It is easy to see that any two H maps $\Omega S^{2n}_{(p-1)} \to \Omega S^{2np-1}$ with S^{2n-1} as fiber are homotopic.

B. In case $p = 2$, what can be said about the Hopf invariant $\Omega SA \to \Omega SA \wedge A$? Does it behave properly on $A \times A$? If $[\iota,\iota]: SA \wedge A \to SA$ is lifted to $X(A)$ what is its composite with φ?

C. How generally can φ be constructed? If $A = CP^2$, there is no retraction $(\Omega S^2 A) \wedge A \to SA \wedge A$ but there still may be a retraction $(\Omega^2 S^2 A) \wedge SA \to SA \wedge A$. What if A is a co H space?

D. Suppose $p > 2$. Is $\Omega^2 S^{2np+1} \xrightarrow{\varphi} S^{2np-1} \to \Omega^2 S^{2np+1}$ the double loops on the degree p map? If so we could strengthen the results from [G2]. (See [G2; Conjecture 14]).

REFERENCES

[BP] E.H. Brown and F.P. Peterson, On the stable decomposition of $\Omega^2 S^{r+2}$, Trans. Amer. Math. Soc. 243(1978), 287-298.

[C] R.L. Cohen, Odd primary infinite families in stable homotopy theory, Memoirs of the AMS Vol. 30, 242, (March 1981).

[CMN] F.R. Cohen, J.C. Moore, and J.A. Neisendorfer, Decompositions of loop spaces and applications to exponents. Algebraic Topology, Aarhus 1978 LMN 763 (1979), 1-12.

[DT] A. Dold and R. Thom, Quasifasserungen und Unendliche Symmetrische Produkte, Annals of Mathematics Vol. 67, No. 2 (1958), 239-281.

[G1] B.I. Gray, Homotopy Theory: An Introduction to algebraic topology. Academic Press (1975).

[G2] B.I. Gray, Unstable families related to the image of J. Math. Proc. Cambridge Philos. Soc. 96 (1984), 95-113.

[G3] B.I. Gray, On Toda's fibrations, Math. Proc. Camb. Philos. Soc. 97(1985), 289-298.

[G4] B.I. Gray, A note on the Hilton-Milnor Theorem, Topology Vol. 10 (1971), 199-201.

[J] I.M. James, The suspension triad of a sphere, Annals of Math, (2) 63 (1956), 407-429.

[M] J.P. May, The Geometry of iterated loop spaces, LNM 271(1972).

[M] M. Mahowald, On the double suspension homomorphism. Trans. Amer. Math. Soc. 214(1975), 169-178.

[S] P. Selick, private communication.

[T] H. Toda, On the double suspension E^2, Journal of Inst. Poly. Osaka City University 7(1956), 103-145.

[W] G. Whitehead, Elements of homotopy theory, Springer (1978).

A WHITEHEAD PRODUCT FOR TRACK GROUPS

K A Hardie and A V Jansen

Abstract: We study a pairing $\pi(\Sigma^m W,X) \times \pi(\Sigma^n W,X) \to \pi(\Sigma^{m+n-1} W,X)$ which specializes to the classical Whitehead product operation in the case $W = S^0$. If W is a suspension the operation is always trivial, but if W is not a suspension and a product vanishes then a version of the Hopf construction gives rise to an element that can be detected by a generalized Hopf invariant.

0. Introduction

Recall that the generalized Whitehead product in the sense of Arkowitz [1] is an operation

(0.1) $\pi(\Sigma A,X) \times \pi(\Sigma B,X) \to \pi(\Sigma(A \# B),X)$.

In the study of the unstable homotopy of track groups it is convenient to consider operations whose domain and codomain belong to the family of groups $\pi(\Sigma^n W,X)$, $n \geq 1$. Except in very special cases this condition is not satisfied by the operation 0.1.

However there is a bijection

(0.2) $\theta\colon \pi(\Sigma^m W,X) \to \pi_m(X^W)$,

where by X^W we denote the space of pointed maps $W \to X$ with compactly generated topology whose base point is the constant map $W \to X$.

The classical Whitehead product [5] of elements of $\pi_\cdot(X^W)$ gives rise then (under the bijection 0.2) to an operation

(0.3) $\pi(\Sigma^m W,X) \times \pi(\Sigma^n W,X) \to \pi(\Sigma^{m+n-1} W,X)$.

If $\alpha \in \pi(\Sigma^m W,X)$ and $\beta \in \pi(\Sigma^n W,X)$ we shall denote the element of $\pi(\Sigma^{m+n-1} W,X)$ obtained by $[\alpha,\beta]^W$. In the special case $W = S^0$ then the left and right sides of 0.2 can be identified by the bijection so that the new operation is also a generalization of the classical

Whitehead product. If $m = n = 1$, it follows that 0.3 coincides with the commutator construction. Similar arguments show that the operation 0.3 satisfies bilinearity properties and Jacobi identities corresponding to the classical Whitehead product.

Since X^W is an h-space whenever W is a suspension, it follows that in this situation the operation 0.3 will always be trivial. Moreover, since suspension Σ is equivalent to composition with a map $X^W \to \Sigma X^{\Sigma W}$ we recover the classical identity

$$\Sigma[\alpha,\beta]^W = 0 \quad \text{in} \quad \pi(\Sigma^{m+n}W,\Sigma X) . \tag{0.4}$$

Our main purpose is to obtain an analogue of the classical Hopf construction associated with a vanishing Whitehead product and to represent its generalized Hopf invariant as a type of join.

1. The Hopf-James invariant

Let Ω denote the pointed loop functor and let $S: X^W \to \Sigma X^{\Sigma W}$ be the (continuous) map such that $S(f) = \Sigma f: \Sigma W \to \Sigma X$, with Σ denoting reduced suspension. Then there is an exponential homeomorphism

$$E: \Sigma X^{\Sigma W} \to (\Omega\Sigma X)^W$$

given by

$$Eh(w)(t) = h(w,t) \qquad (h \in \Sigma X^{\Sigma W},\ w \in W,\ t \in I). \tag{1.1}$$

Let also $e: X \to \Omega\Sigma X$ be the unit of the $\Omega\Sigma$ adjunction, recalling that

$$e(x)(t) = (x,t) \qquad (x \in X,\ t \in I) .$$

Let X_∞ denote James' reduced product space [2] and $\alpha: X_\infty \to \Omega\Sigma X$ the map that James proves is a homotopy equivalence under mild restrictions on X.

1.2 Lemma: *There is a commutative diagram*

$$\begin{array}{ccc} X^W & \xrightarrow{\;i^W\;} & (X_\infty)^W = X_\infty^W \\ {\scriptstyle S}\downarrow & \overset{e^W}{\searrow} & \downarrow{\scriptstyle \alpha^W} \\ \Sigma X^{\Sigma W} & \xrightarrow{\;E\;} & (\Omega\Sigma X)^W \end{array} .$$

Proof: Only the commutativity of the lower triangle need be checked and we have :

$$e^W(f)(w)(t) = ef(w)(t) = (fw,t) = (Sf)(w,t) = (ESf)(w,t) .$$

We shall denote by $\phi = \theta^{-1}(E^{-1}\alpha^W)_*$ the following form of James' canonical isomorphism :

$$\pi_r(X_\infty^W) \xrightarrow{(E^{-1}\alpha^W)_*} \pi_r(\Sigma X^{\Sigma W}) \xrightarrow{\theta^{-1}} \pi(\Sigma^{r+1}W,\Sigma X), \qquad \phi : \pi_r(X_\infty^W) \to \pi(\Sigma^{r+1}W,\Sigma X)$$

(If $W = S^0$ we recover the case considered by James.)

Recall that James' map $h_2\colon X_\infty \to (X \# X)_\infty$ is the combinatorial extension [2] of the shrinking map

$$h_2\colon X_2, X \to X \# X, * .$$

We shall denote by $H = H_2$ the homomorphism given by the diagram

$$\begin{array}{ccc} \pi_r(X_\infty^W) & \xrightarrow{(h_2^W)_*} & \pi_r((X\#X)_\infty^W) \\ {\scriptstyle \phi}\downarrow{\scriptstyle \approx} & & {\scriptstyle \approx}\downarrow{\scriptstyle \phi} \\ \pi(\Sigma^{r+1}W,\Sigma X) & \xrightarrow{\;H\;} & \pi(\Sigma^{r+1}W,\Sigma(X\#X)) \end{array} .$$

2. The Hopf construction

Suppose that $[\alpha,\beta]^W = 0$ in $\pi(\Sigma^{m+n-1}W,X)$. Then there exists a map

(2.1) $$F\colon S^m \times S^n \to X^W$$

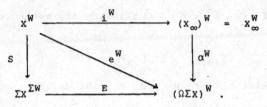

of type $(\theta\alpha,\theta\beta)$. The multiplication on X_∞ induces a multiplication on X_∞^W by means of which we can define another map $M: S^m \times S^n \to X_\infty^W$ which agrees with F on $S^m \vee S^n$. Following the method due to I.M. James [3], let

$$d(M,F) \in \pi_{m+n}(X_\infty^W)$$

be the "difference element" of M and F. We define

$$(2.2) \qquad c(\alpha,\beta) = \{\phi d(M,F) \mid F \text{ of type } (\theta\alpha,\theta\beta)\} \subseteq \pi(\Sigma^{m+n+1}W,\Sigma X)$$

to be the Hopf construction subset associated with the vanishing of $[\alpha,\beta]^W$. Applying an argument given in [3], we can prove that $c(\alpha,\beta)$ is a coset of the suspension subgroup of $\pi(\Sigma^{m+n+1}W,\Sigma X)$. Now let

$$(2.3) \qquad \chi: W \to W \# W$$

be the class of the diagonal map. We have the following theorem.

2.4 Theorem: *Given* $\alpha \in \pi(\Sigma^m W,X)$, $\beta \in \pi(\Sigma^n W,X)$ *with* $[\alpha,\beta]^W = 0$ *then* $Hc(\alpha,\beta) = \pm \Sigma(\alpha \# \beta) \circ \Sigma^{m+n+1}\chi$.

2.5 Remark: In the case $W = S^0$ we recover essentially [4; Theorem 5.1].

Proof of Theorem 2.4: Let $f: S^m \to X^W$, $g: S^n \to X^W$ denote representatives of $\theta\alpha$, $\theta\beta$. In the following commutative diagram let q denote the smash identification and let p be a map which agrees with the multiplication in X_∞.

$$\begin{array}{ccccc}
S^m \times S^n & \xrightarrow{f \times g} & X^W \times X^W & & \\
\downarrow q & & \downarrow p^W & & \\
 & & X_2^W & & \\
 & & \downarrow h_2^W & & \\
S^m \# S^n & \overset{\Gamma}{\dashrightarrow} & (X \# X)^W & \xrightarrow{i^W} & (X \# X)_\infty^W
\end{array}$$

It can be checked that the dotted arrow Γ is well defined since $h_2^W p^W(f \times g)$ maps $S^m \vee S^n$ to $*$. Moreover, since the composition

$$S^m \times S^n \xrightarrow{F} X^W \hookrightarrow X_2^W \xrightarrow{h_2^W} (X \# X)^W$$

is trivial, we have

$$Hc(\alpha, \beta) = \pm \phi\, i_*^W \Gamma .$$

To complete the proof of the theorem we analyse the class Γ. It may be checked that the arrow χ' in the following diagram is well defined and that the bottom composite is equal to Γ.

$$\begin{array}{ccccc}
S^m \times S^n & \xrightarrow{\theta^{-1} 1_{\Sigma^m W} \times \theta^{-1} 1_{\Sigma^n W}} & (\Sigma^m W \times \Sigma^n W)^W & \xrightarrow{(\alpha \times \beta)^W} & (X \times X)^W \\
\Big\downarrow q & & \Big\downarrow & & \Big\downarrow h_2^W p^W \\
S^m \# S^n & \overset{\chi'}{- - - - - \rightarrow} & (\Sigma^m W \# \Sigma^n W)^W & \xrightarrow{(\alpha \# \beta)^W} & (X \# X)^W
\end{array}$$

However it is straightforward to check that $\theta\{\chi'\} = \pm \Sigma^{m+n}\chi$, the sign depending on orientation conventions.

2.6 Remark: If W is a suspension then, as might be expected, Theorem 2.4 yields no information. However if W has, for example, a non-trivial cup square then it is easy to see that the theory is non-vacuous. Let ι denote the class of the identity $\Sigma W \rightarrow \Sigma W$. Then $[\iota,\iota]^W = 0$ since, as we have observed, the product here coincides with the commutator. Applying 2.4 we obtain $Hc(\iota,\iota) = \pm \Sigma^3 \chi$, which is non-zero.

2.7 Remark: If X is an H space then so is X^W and it follows that the elements $[\alpha,\beta]^W$ are trivial. We shall see that in this case also non-trivial Hopf construction elements can sometimes be detected.

3. Some computations

In this section we give a few illustrations of the possible applications of Theorem 2.4 in the context of unstable cohomotopy of the complex projective plane.

Let $W = S^2 \cup e^4$, where the 4-cell is attached by the Hopf class η. Since $\Sigma\eta$ is of order 2, a map $S^3 \to S^3$ of degree 2 can be extended over the 5-cell of ΣW giving rise to a class

(3.1) $$\alpha \in \pi(\Sigma W, S^3) .$$

Since S^3 is a topological group the Whitehead square $[\alpha,\alpha]^W$ vanishes.

3.2 Proposition: *The coset* $c(\alpha,\alpha) \subseteq \pi(\Sigma^3 W, S^4)$ *is of infinite order.*

Proof: Since $W \# W = S^4 \cup e^6 \cup e^6 \cup e^8$ the diagonal $\chi : W \to W \# W$ can be deformed into S^4 (i.e. to a cellular map). Using cohomology we can see that this can be regarded as the map $p : W \to S^4$ shrinking S^2. Then we have a diagram :

$$\begin{array}{ccccc} & & \Sigma^2 W \# W = \Sigma W \# \Sigma W & & \\ & \overset{\Sigma^2\chi}{\nearrow} & \uparrow & \overset{\alpha \# \alpha}{\searrow} & \\ \Sigma^2 W & \xrightarrow{\Sigma^2 p} & \Sigma^2 S^2 \# S^2 = S^3 \# S^3 & \xrightarrow{2 \# 2} & S^3 \# S^3 \end{array} .$$

The group $\pi(\Sigma^2 W, S^6)$ is already in the stable range and examining the stable Puppe sequence of $\eta_4 : S^5 \to S^4$ (of which class $\Sigma^2 W$ is the mapping cone) :

$$0 \to \{S^6, S^6\} \to \{\Sigma^2 W, S^6\} \to \{S^4, S^6\} \to \{S^5, S^6\}$$

we see that $\{\Sigma^2 W, S^6\}$ is infinite cyclic generated by $\{\Sigma^2 p\}$. Applying Theorem 2.4 we have $H\, c(\alpha,\alpha) = \Sigma(\alpha \# \alpha) \circ \Sigma^3\chi = 4\{\Sigma^3 p\}$ which completes the proof.

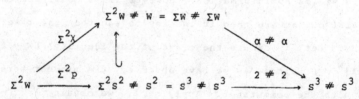

Theorem 2.4 can also be used sometimes to establish nontriviality of the Whitehead products as in the following argument. Let α denote the element in 3.1. Then we claim :

3.3 Proposition: $[\Sigma\alpha,\Sigma\alpha]^W$ *is of infinite order in* $\pi(\Sigma^3 W, S^4)$.

For if for some integer $m \neq 0$ we have $m[\Sigma\alpha,\Sigma\alpha]^W = [m\Sigma\alpha,\Sigma\alpha]^W = 0$ then, by Theorem 2.4,

$$H\, c(m\Sigma\alpha,\Sigma\alpha) = \pm\, \Sigma((m\Sigma\alpha)\ \#\ (\Sigma\alpha)) \circ \Sigma^5\chi = \pm\, 4m\{\Sigma^5 p\} .$$

Hence $c(m\Sigma\alpha,\Sigma\alpha)$ is a coset of infinite order in $\pi(\Sigma^5 W, S^5)$. But examining the Puppe sequence of η_7

$$\to \pi_9(S^5) \to \pi(\Sigma^5 W, S^5) \to \pi_7(S^5) \to \pi_8(S^5)$$

we arrive at a contradiction, for neither $\pi_9(S^5)$ nor $\pi_7(S^5)$ has an element of infinite order.

3.4 Remarks: The referee has pointed out the following direct method for obtaining the result in Proposition 3.3. There is a functional Samelson product adjoint to the Whitehead product. Given $\Sigma\alpha \in \pi(\Sigma^2 W, S^4)$ as above with adjoint $\bar{\alpha} \in \pi(\Sigma W, \Omega S^4)$ we obtain a Samelson product $\langle \bar{\alpha},\bar{\alpha} \rangle$ given by the composite

$$S^1 \wedge S^1 \wedge W \xrightarrow{1 \wedge 1 \wedge \chi} S^1 \wedge S^1 \wedge W \wedge W \xrightarrow{\text{twist}} S^1 \wedge W \wedge S^1 \wedge W$$

$\xrightarrow{\bar{\alpha} \wedge \bar{\alpha}} \Omega S^4 \wedge \Omega S^4 \xrightarrow{S} \Omega S^4$ where S is given by the commutator.

Then $H_6(S^1 \wedge S^1 \wedge W ; \mathbb{Z}) \approx \mathbb{Z} \approx H_6(\Omega S^4 ; \mathbb{Z})$ and $\langle \bar{\alpha},\bar{\alpha} \rangle_*$ (generator) = 2(generator) by inspecting the definitions and χ_*. Thus $\langle \bar{\alpha},\bar{\alpha} \rangle$ has infinite order because it has infinite order in homology and 3.3 follows. The argument suggests that the relations between the Arkowitz product, the $[-,-]^W$ product and the Samelson product deserve further study. It is easy to see that similar computations can be performed with the quaternionic and Cayley projective planes. The authors have also considered a related Whitehead product operation in homotopy pair theory; the results will be reported in due course.

References

[1] M. Arkowitz; The generalized Whitehead product. Pacific J.M. 12 (1962), 7-23.

[2] I.M. James; Reduced product spaces. Ann. of Math. 62 (1955), 170-197.

[3] I.M. James; On spaces with a multiplication. Pacific J.M. 7 (1957), 1083-1100.

[4] G.W. Whitehead; A generalization of the Hopf invariant. Ann. of Math. 51 (1950), 192-237.

[5] J.H.C. Whitehead; On adding relations to homotopy groups. Ann. of Math. 42 (1941), 409-429.

The authors acknowledge a grant to the Topology Research Group from the Foundation for Research Development of the South African Council for Scientific and Industrial Research.

Department of Mathematics
University of Cape Town
Rondebosch 7700
South Africa

Minimal Atlases of Real Projective Spaces

MICHAEL J. HOPKINS
University of Chicago

INTRODUCTION

Manifolds are usually described as the result of gluing together open subsets of a fixed Euclidean space. It is natural at the outset to ask how efficiently a given manifold can be constructed. In [**6**] Berstein introduced the following invariants. All manifolds, embeddings, and immersions are assumed to be differentiable of class C^∞.

Definition. *Let M be a closed n-manifold.*

(1) *The embedding covering number $N(M)$ is the least integer k such that M can be covered by k open subsets, each of which embeds in $\mathbb{R}^n$.*

(2) *The immersion covering number $n(M)$ is the least integer k such that M can be covered by k open sets, each of which immerses in $\mathbb{R}^n$.*

Berstein introduced upper and lower bounds for $n(M)$ and $N(M)$ when $M = \mathbb{RP}^n$. Unfortunately, they agree only in certain cases. This paper closes these gaps (mostly).

Theorem 1. *Write $n+1 = 2^k m$, with m odd.*

(1)

$$n(\mathbb{RP}^n) = \begin{cases} \max\{2, m\} & \text{if } k \le 3 \\ \text{the least integer} \ge \dfrac{n+1}{2(k+1)} & \text{if } k \ge 3. \end{cases}$$

(2) *$N(\mathbb{RP}^n) = n(\mathbb{RP}^n)$ with the possible exception of the values $n = 31$ and $n = 47$. There are inequalities:*

$$3 \le N(\mathbb{RP}^{31}) \le 4$$
$$5 \le N(\mathbb{RP}^{47}) \le 6.$$

One consequence of the proof of Theorem 1 is that $n(\mathbb{RP}^n)$ depends only on the homotopy type of $\mathbb{RP}^n$, and that $N(\mathbb{RP}^n)$ depends only on the topological type of $\mathbb{RP}^n$ (at least when $n \ne 31$ or 47). It would be interesting to know if this holds more generally.

The values of $n(M)$ and $N(M)$ when M is either $\mathbb{CP}^n$ or $\mathbb{HP}^n$ can be computed using Berstein's techniques. The values are:

$$n(\mathbb{CP}^n) = N(\mathbb{CP}^n) = \begin{cases} n+1 & \text{if } n \text{ is even} \\ \frac{n+1}{2} & \text{if } n \text{ is odd} \end{cases}$$

$$n(\mathbb{HP}^n) = N(\mathbb{HP}^n) = n+1.$$

Using the work of Hirsch and Poenaru ([**12**], and [**20**]), the problem of computing $n(M)$ reduces to determining the minimum number of parallelizable open subsets it takes to cover M, an invariant of the tangent bundle. A similar invariant can be defined for any vector bundle [**15**].

The author is currently supported by a Presidential Young Investigator award, and by the Sloan foundation.

Definition. *Let ξ be a vector bundle over a space X. A trivializing cover of ξ is a covering of X with open subsets over which ξ is stably trivial. The category of ξ,* Vecat(ξ), *is the cardinality of a minimal trivializing cover*

James produced upper and lower bounds for this invariant which were good enough to determine the category of all complex vector bundles over real and complex projective spaces. Unfortunately there are gaps between these bounds in the case of real vector bundles over real projective spaces. This paper closes these gaps by reducing the question to one in stable homotopy theory. A relationship between Vecat and geometric dimension will also be discussed. The main results are stated in Sections 1 and 2.

This paper is an extract of the author's 1984 Oxford thesis, written under the direction of Ioan James. I would like to thank Ioan James for suggesting this problem and Mark Mahowald for introducing his theory of *bo*-resolutions at a key moment.

§1 The main theorems

First a review of 'category' in the sense of Lusternik-Schnirelmann. For more details, see [**15**].

Definiton ([7], [8]). *Let $f : X \to Y$ be a map to a path connected space Y. Then*

$$\mathrm{cat}(f) \le n$$

means that X admits a covering by n open subsets U_i, with the property that the restrictions

$$f : U_i \to Y \qquad i = 1, \ldots, n$$

are nullhomotopic.

Set $\mathrm{cat}(f) = n$ if $\mathrm{cat}(f) \le n$, but $\mathrm{cat}(f) \not\le n-1$.

When $Y = X$ and f is the identity map then $\mathrm{cat}(f)$ is just the Lusternik-Schnirelmann category of X. When $Y = BO$, BU, or BSp and f classifies a vector bundle ξ, then $\mathrm{cat}(f)$ is just $\mathrm{Vecat}(\xi)$.

If $f : X \to Y$ and $g : Y \to Z$ then $\mathrm{cat}(g \circ f) \le \min\{\mathrm{cat}(f), \mathrm{cat}(g)\}$.

If $Y_1, \ldots, Y_n$ have basepoints $*$, let

$$T^n(Y_1, \ldots, Y_n) = \{(y_1, \ldots, y_n) \in Y_n \times \cdots \times Y_n \mid \text{some } y_i = *\}$$

denote the fat wedge. When all $Y_i = Y$ this will be abbreviated to $T^n(Y)$. There is the following analogue of Whitehead's definition of category [7]:

Proposition 1.1. *Let $f : X \to Y$ be a map of paracompact, locally contractible spaces. Suppose that Y is path connected. Then $\mathrm{cat}(f) \le n$ if and only if $\Delta \circ f$ factors (up to homotopy) through $T^n(Y)$, where $\Delta : Y \to Y^n$ is the iterated diagonal.*

$$\begin{array}{ccccc} & & & & T^n(Y) \\ & & & & \downarrow \\ X & \xrightarrow{f} & Y & \xrightarrow{\Delta} & Y^n \end{array}$$

From now on, all spaces will be assumed CW-complexes in order to guarantee the hypothesis of (1.1).

The cofibre of $T^n(Y) \to Y^n$ is the iterated smash product $\bigwedge^n Y$.

Definition. *The weak category of f, $\mathrm{wcat}(f)$, is less than or equal to n if the composite*

$$X \xrightarrow{f} Y \xrightarrow{\Delta} Y^n \to \bigwedge^n Y$$

is nullhomotopic.

Proposition 1.2.

(1) $\mathrm{wcat}(f) \leq \mathrm{cat}(f)$

(2) *If X has has dimension d and Y is $(m-1)$-connected, then*

$$\mathrm{wcat}(f) = \mathrm{cat}(f)$$

in the stable range

$$\mathrm{cat}(f) + 1 \geq (d+2)/m.$$

PROOF: Part (1) follows almost immediately from the definition. Part (2) follows from the following lemma of Ganea:

Lemma 1.3 [10]. *Let $A \to X \to X/A$ be a cofibration, $F \to X$ the fibre of the collapse, and $A \to F$ the canonical map. Then*

$$\Sigma(F/A) \approx A * \Omega(X/A).$$

Now let ξ be a vector bundle over X and let E be a multiplicative cohomology theory.

Definition. *An element $c \in \tilde{E}^*(X)$ is a characteristic class of ξ if $i^*(c) = 0$ for any $i : U \to X$ with $i^*(\xi)$ stably trivial.*

The basic examples of characteristic classes are the Chern, Pontryagin, and Stiefel-Whitney classes, and the reduced class $[\xi]$ of ξ itself in the appropriate K-group.

Given an element c in a ring R, let $\mathrm{nil}(c) = \min\{k \mid c^k = 0\}$.

Proposition 1.4 ([15], [6]). *For any characteristic class c of ξ, there is an inequality*

$$\mathrm{nil}(\xi) \leq \mathrm{Vecat}(\xi).$$

Proposition 1.4 provides an effective lower bound for Vecat. Upper bounds come from actual coverings of the base. Projective spaces can be covered with vector bundles over lower dimensional projective spaces. Let $P(V)$ be the projective space of a vector space V over the field K of real or complex numbers. Then

$$P(V \oplus W) = [P(V \oplus W) \setminus P(V)] \cup [P(V \oplus W) \setminus P(W)].$$

The space $P(V \oplus W) \setminus P(V)$ is diffeomorphic to the total space of the vector bundle

$$(W \setminus \{0\}) \times_{K^*} V \to P(W),$$

so up to homotopy, there is a decomposition

$$P^{n+m-1} \approx P^{n-1} \cup P^{m-1}.$$

A covering of projective space obtained by iterating the above construction will be called an *affine covering*. Using affine coverings as a lower bound, James made the following computation:

Theorem 1.5 [15]. *If ξ is a complex vector bundle over either real or complex projective space, then*

$$\mathrm{Vecat}(\xi) = \mathrm{nil}([\xi]) = \mathrm{nil}(\textit{first non-vanishing Chern class}).$$

Any vector bundle over real projective space $\mathbb{RP}^n$ is stably a multiple of the canonical line bundle h. The mail result of this paper is:

Theorem 1.6. *Let $\xi = 2^k(odd)h$ be a vector bundle over $\mathbb{RP}^n$. Then*

$$\mathrm{Vecat}(\xi) = \begin{cases} \mathrm{nil}(\xi) & \text{if } k \geq 3 \\ \mathrm{nil}(w_{2^k}(\xi)) & \text{if } k \leq 3. \end{cases}$$

Let τ and ν denote respectively the tangent and normal bundles of $\mathbb{RP}^n$ and write $n+1 = 2^k m$ with m odd.

Corollary 1.7. $\mathrm{Vecat}(\tau) = \mathrm{Vecat}(\nu)$ *is the least integer greater than or equal to*

$$\begin{cases} m & \text{if } k \leq 3 \\ \dfrac{n+1}{2(k+1)} & \text{if } k \geq 3. \end{cases}$$

For example, the (complex) tangent bundle of $\mathbb{CP}^{31}$ has category 32, whereas the (real) tangent bundle of $\mathbb{RP}^{31}$ has category 3.

§2 A Connection with Geometric Dimension

At first it seems as if Vecat of a tangent bundle should have some connection with immersions of the underlying manifold, or more generally that $\mathrm{Vecat}(\xi)$ should have something to do with the geometric dimension of ξ. The result at the end of §1 shows, however, that $\mathbb{RP}^{31}$ can be covered by 3 open sets, each of which immerses in $\mathbb{R}^{31}$. The fact that $\mathbb{RP}^{31}$ doesn't even immerse in $\mathbb{R}^{53}$ seems to discourage such a notion.

There is a variant of Vecat which is related to geometric dimension. For a real vector bundle ξ over a finite dimensional base X let $P(\xi)$ be the associated bundle of projective spaces, and let h_ξ be the canonical line bundle over $P(\xi)$.

Definition. $\bar{\nu}\mathrm{cat}(\xi) = \mathrm{Vecat}(h_\xi) - \dim(\xi)$.

Lemma 2.1. *If the Stiefel-Whitney class $w_j(-\xi)$ is non-zero, then $\bar{\nu}\mathrm{cat}(\xi) \geq j$. In particular $\bar{\nu}\mathrm{cat}(\xi) \geq 0$.*

PROOF: Let x be the first Stiefel-Whitney class of h_ξ. If ξ is n-dimensional, then the Stiefel-Whitney classes $w_i = w_i(\xi)$ are defined by the relation

$$x^n + w_1 x^{n-1} + \cdots + w_n = 0.$$

Let $1 + \bar{w}_1 + \cdots + \bar{w}_k$, $\bar{w}_k \neq 0$, be the total Stiefel-Whitney class of $-\xi$. Then

$$x^{n+j} = (x^n + w_1 x^{n-1} + \cdots + w_n)(x^j + \bar{w}_1 x^{j-1} + \cdots + \bar{w}_j) + x^{n+j}.$$

The right hand side is zero if and only if $j \geq k$. It follows that $\mathrm{nil}(x) = n + k$. This completes the proof.

Example. If n is the n-dimensional trivial bundle over X then $\bar{\nu}\mathrm{cat}(n) = 0$.

Lemma 2.2. *Let ξ and η be vector bundles over X. There is an inequality*

$$\bar{\nu}\mathrm{cat}(\xi \oplus \eta) \leq \bar{\nu}\mathrm{cat}(\xi) \oplus \bar{\nu}\mathrm{cat}(\eta).$$

In particular, $\bar{\nu}\mathrm{cat}(\xi \oplus 1) \leq \bar{\nu}\mathrm{cat}(\xi)$.

PROOF: The decomposition

$$P(V \oplus W) = P(V) \cup P(W),$$

applied fibrewise, gives

$$P(\xi \oplus \eta) = P(\xi) \cup P(\eta).$$

The lemma follows from this.

Lemma 2.2 allows for the following definition:

Definition. $\nu\mathrm{cat}(\xi) = \lim_{n \to \infty} \bar{\nu}\mathrm{cat}(\xi \oplus n)$.

Lemma 2.1 provides the inequality

$$j \leq \nu\mathrm{cat}(\xi) \leq \bar{\nu}\mathrm{cat}(\xi)$$

whenever $w_j(-\xi) \neq 0$. This suggests a relationship between $\nu\mathrm{cat}(\xi)$ and the geometric dimension, $\mathrm{gd}(-\xi)$.

Theorem 2.3. *Let ξ be a vector bundle over X. There is an inequality*

$$\nu\mathrm{cat}(\xi) \leq \mathrm{gd}(-\xi).$$

Equality holds if $\nu\mathrm{cat}(\xi) > \frac{1}{2}\dim(X)$.

Equality is guaranteed in (2.3) if, for example $w_j(-\xi) \neq 0$ with $j > \frac{1}{2}\dim X$. In this sense, $\nu\mathrm{cat}(\xi)$ and $\mathrm{gd}(-\xi)$ agree in the metastable range.

Lemma 2.4. *Let $f : X \to P^\infty$ classify a line bundle h. Then* $\mathrm{Vecat}(h) \leq n$ *if and only if f factors (up to homotopy) through P^{n-1}.*

PROOF: See **[22]** or **[15]**.

PROOF OF THEOREM 2.3: Suppose that ξ has dimension n. fix a riemannian metric on ξ and let $E \to X$ be the associated principal $O(n)$-bundle. Let $V_{n+k,k}$ be the Stiefel manifold of orthonormal n-frames in $\mathbb{R}^{n+k}$ and let $E_{n+k,k}$ be the space of $\mathbb{Z}/2$-equivariant maps from S^{n-1} to S^{n+k-1}. The inclusion $V_{n+k,k} \subset E_{n+k,k}$ is a $(2k-1)$-equivalence (**[11]**, **[14]**). Consider the following diagram of fibrations:

$$\begin{array}{ccc}
V_{n+k,k} & \longrightarrow & E_{n+k,k} \\
\downarrow & & \downarrow \\
E \times_{O(n)} V_{n+k,k} & \longrightarrow & E \times_{O(n)} E_{n+k,k} \\
{\scriptstyle p}\downarrow & & \downarrow{\scriptstyle q} \\
X & \xrightarrow{=} & X.
\end{array}$$

The map of total spaces is a $(2k-1)$-equivalence. By adding trivial bundles to ξ, n can be increased. For sufficiently large values of n, $\nu\mathrm{cat}(\xi) \leq k$, if and only if q admits a section (Lemma 2.4). For sufficiently large falues of n, $\mathrm{gd}(-\xi) \leq k$ if and only if p admits a section. The theorem follows.

The characterization of $\bar{\nu}\mathrm{cat}$ used in the above proof also provides a condition on $\dim(\xi)$ which gaurantees the equality $\bar{\nu}\mathrm{cat}(\xi) = \nu\mathrm{cat}(\xi)$.

Lemma 2.5. *The suspension map $E_{m,n} \to E_{m+1,n+1}$ is a $\min\{2n-m-2, n-1\}$-equivalence.*

PROOF: The proof is by induction on n using the fibrations $\Omega^{n-1}S^{n-1} \to E_{m,n} \to E_{m,n-1}$ [14].

Corollary 2.6. *The numbers $\nu\mathrm{cat}(\xi)$, and $\bar{\nu}\mathrm{cat}(\xi)$ are equal provided either*

(1) $\nu\mathrm{cat}(\xi) \neq 0$ *and* $\dim(\xi) > \dim(X) - \nu\mathrm{cat}(\xi)$, *or*
(2) $\nu\mathrm{cat}(\xi) = 0$ *and* $\dim(\xi) > \dim(X) + 1$.

Example. Consider $\xi = (m+1)h$ on P^n. For most values of m and n, $\bar{\nu}\mathrm{cat}(\xi) = \nu\mathrm{cat}(\xi)$ by (2.6). The space $P(\xi)$ is just $P^m \times P^m$ and the line bundle h_ξ is the tensor product of the line bundles classified by the two projections. In this case, the problem of calculating νcat is equivalent to determining the smallest projective space to which the map induced by the tensor product

$$P^m \times P^n \to P^\infty$$

can be deformed. This is the well-known "axial map" problem [5]. In this sense, the problem of calculating νcat can be thought of as an axial map problem for arbitrary vector bundles.

§3. Skeleton of the proof

The main idea behind the proof of Theorem 1.6 is easy, but the bookkeeping and a few exceptional cases make the general argument somewhat prolonged. This section contains a 'no numbers' rendition of the proof. First some notation.

For a space or a spectrum X let $X\langle n\rangle$ be its $(n-1)$-connected cover. The spectrum for connective K-theory will be donoted *bo*. Since we will be dealing only with real projective spaces we can safely abbreviate $\mathbb{RP}^n$ to P^n. The symbol P_k^n will denote the stunted projective space P^n/P^{k-1}, and when it will cause no confusion, a single symbol will be used to denote both a vector bundle and the map classifying its stable reduced class.

Let $\xi : P^m \to BO$ be a vector bundle. We are trying to factor the map

$$\Delta \circ \xi : P^m \to BO \times \cdots \times BO$$

through the fat wedge. The first step is to factor $\Delta \circ \xi$ through $BO\langle n_1\rangle \times \cdots \times BO\langle n_k\rangle$ for some fortunate choice of $n_1, \ldots, n_k$. With luck the integers n_i can be chosen so that

$$T^k(BO\langle n_1\rangle, \ldots, BO\langle n_k\rangle) \to BO\langle n_1\rangle \times \cdots \times BO\langle n_k\rangle \to BO\langle n_1\rangle \wedge \cdots \wedge BO\langle n_k\rangle$$

is a fibration through dimension m.

This leaves the problem of showing that the map

$$P^m \to BO\langle n_1\rangle \wedge \cdots \wedge BO\langle n_k\rangle$$

is null. With more luck the integers n_i can be chosen so that the map

$$BO\langle n_1\rangle \wedge \cdots \wedge BO\langle n_k\rangle \to \Omega^\infty bo\langle n_1\rangle \wedge \cdots \wedge bo\langle n_k\rangle$$

is an equivalence through dimension m. This leaves the problem of showing that the map in the category of spectra

$$f : P^m \to bo(n_1) \wedge \cdots \wedge bo(n_k)$$

is null.

In the category of spectra there is an equivalence

$$bo(n_1) \wedge \cdots \wedge bo(n_k) \approx bo(n) \vee Y,$$

for some spectrum Y and some integer n. The component of f in $bo(n)$ is a lift of the tensor power ξ^k and so is null by assumption. The component of f in Y can be muscled away, completing the proof.

§4 A REDUCTION

In this section we reduce (1.6) to a series of special cases. Let ξ be a vector bundle over P^{m-1}, and let h be the canonical line bundle.

The Atiyah-Hirzebruch spectral sequence

$$H^s(P^{m-1}; KO^t(\text{pt})) \Rightarrow KO^{s+t}(P^{m-1}),$$

has the property that no differentials enter or leave the line $s+t=0$. A vector bundle ξ over P^{m-1} therefore determines a unique, non-zero element

$$w(\xi) \in H^s(P^{m-1}; KO^s(\text{pt})),$$

representing the reduced class $[\xi]$ of ξ. Write

$$[\xi] = 2^{4k+r-1}\text{odd} \cdot h \qquad 1 \leq r \leq 4.$$

The integer 'odd' is actually inconsequential since both Vecat(ξ) and nil(c), c a characteristic class of ξ, depend only on the ideal generated by $[\xi]$. If $k=0$ then $w(\xi)$ is the Stiefel-Whitney class $w_{2^r-1}(\xi)$. When $k \neq 0$, $w(\xi)$ is not a characteristic class of ξ.

The smallest affine, trivializing cover of ξ, has nil($w(\xi)$) elements. It follows that (1.6) is true when $w(\xi)$ is a characteristic class. It also follows that (1.6) is true when

$$\text{nil}([\xi]) = \text{nil}(w(\xi)).$$

This happens if and only if $r=4$.

Corollary 4.1. *Let $\xi = 2^{4k+r-1}odd \cdot h$ be the reduced class of a vector bundle over P^{m-1}. Then Theorem (1.6) is true if either $k=0$ or $r=4$. In these cases, affine coverings provide minimal trivializing covers.*

This is the first reduction. We may now assume that $k \neq 0$ and $r \neq 4$. The next goal is to reduce (1.6) to the case $n \leq 4$.

First notice that if an integer $d < m$ has the property

$$\text{nil}([\xi]|_{P^{d-1}}) = \text{nil}([\xi]|_{P^{m-1}}),$$

then the assertion of (1.6) for $\xi|_{P^{d-1}}$ follows from the assertion of (1.6) for ξ. We may as well take m to be as large as possible, making m a *function* of n, k, and r:

$$m = m(n,k,r) = \min\{d \mid (2^{4k+r-1}h)^n = 0 \text{ on } P^{d-1}\}.$$

It is possible to write down an explicit formula for m. Write $rn = 4p + q$ with $1 \le q \le 4$. Then

$$(2^{4k+r-1}h)^n = \pm(2^{4kn+rn-1}h) = \pm 2^{4(kn+p)+q-1}h \tag{4.2}$$

and

$$m = m(n,k,r) = 8(kn+p) + 2^{q-1}. \tag{4.3}$$

The important thing to notice is that

$$m(n+4,k,r) = m(n,k,r) + m(4,k,r). \tag{4.4}$$

Proposition 4.5. *It suffices to prove (1.6) in case $n \le 4$.*

PROOF: The proof is by induction on n. Consider the affine covering

$$P^{m(n+4,k,r)-1} = P^{m(n,k,r)-1} \cup P^{m(4,k,r)-1}.$$

By the induction hypothesis

$$\mathrm{Vecat}([\xi]|_{P^{m(n,k,r)-1}}) = \mathrm{nil}([\xi]|_{P^{m(n,k,r)-1}}).$$

By assumption

$$\mathrm{Vecat}([\xi]|_{P^{m(4,k,r)-1}}) = 4,$$

so

$$n+4 = \mathrm{nil}([\xi]) \le \mathrm{Vecat}([\xi]) \le \mathrm{Vecat}([\xi]|_{P^{m(n,k,r)-1}}) + \mathrm{Vecat}([\xi]|_{P^{m(4,k,r)-1}}) = n+4.$$

This completes the proof.

The proof of (1.6) will take place in two passes, first when $k > 3$, and then case by case when $k \le 3$.

Finally, here are some estimates which will be needed later. The proofs are left to the reader.

Lemma 4.6.

(1)

$$m(n,k,r) - 1 \le 8k(n+1) - 2 \quad \text{if} \quad \begin{cases} k > 3 \text{ and } rn \le 12, \text{ or} \\ k = 3 \text{ and } rn < 12, \text{ or} \\ k = 2 \text{ and } r = 1; \end{cases}$$

(2) $m(n,k,r) - 1 \le 2(8kn - 1)$ *if* $nk > 3$;

(3) $m(n,k,r) - 1 - (8k+1)(n+1) \le 7$.

§5 Some connective K-theory

This section recalls several results from connective K-theory. Let $p : bo\langle n\rangle \to bo$ be projection from the $(n-1)$-connective cover, and let $i : S^0 \to bo$ be the unit.

Lemma 5.1.

(1) $i_* : [P^m, S^0] \to [P^m, bo]$ *is an epimorphism.*
(2) $p_* : [P^m, bo\langle n\rangle] \to [P^m, bo]$ *is a monomorphism.*

Proof:

(1) $[P^m, bo]$ is a cyclic group generated by the reduced class of the canonical line bundle. This generator factors through S^0 via, say, the transfer map.
(2) This follows from the Aityah-Hirzebruch spectral sequence.

Lemma 5.2. *In the category of spectra, the "evaluation" map $BO\langle n\rangle \to bo\langle n\rangle$ is a $2n$-equivalence.*

Proof: This is a consequence of the fibration $\Omega X * \Omega X \to \Sigma\Omega X \to X$.

Lemma 5.3.

(1) $bo \wedge bo \approx bo \vee Y$, *with Y 3-connected.*
(2) $bo\langle 8k\rangle \approx \Sigma^{8k} bo$ *and the following diagram commutes:*

$$\begin{array}{ccccc} \Sigma^{8k}bo \wedge \Sigma^{8k}bo \approx \Sigma^{16k}bo \wedge bo & \xrightarrow{\approx} & \Sigma^{16k}bo \vee \Sigma^{16k}Y & \xrightarrow{\text{pinch}} & bo\langle 16k\rangle \\ \approx\uparrow & & & & \downarrow \\ bo\langle 8k\rangle \wedge bo\langle 8k\rangle & \xrightarrow{p\wedge p} & bo \wedge bo & \xrightarrow{\mu} & bo \end{array}$$

Proof: The following composite is the identity map:

$$bo \wedge S^0 \to bo \wedge bo \xrightarrow{\mu} bo.$$

From this is follows that $bo \wedge bo \approx bo \vee Y$ for some spectrum Y, and that under this identification, the multiplication map is just the "pinch" map. This establishes commutativity of the diagram. The connectivity of Y is a consequence of the fact that the unit $i : S^0 \to bo$ is a 3-equivalence.

§6 Proof of Theorem 1.6, $\xi = 2^{4k+r-1}\text{ODD}\cdot h$, $k > 3$

We can now proceed with the proof of (1.6) in case $k > 3$. Let $m = m(n,k,r)$, and consider $[\xi] = [2^{4k+r-1}\text{odd}\cdot h]$ on P^{m-1}. We need to show that $\text{Vecat}(\xi) \le n$, and are reduced to the case $1 \le r \le 3$, $2 \le n \le 4$.

Since $m(1,k,r) = 8k + 2^{r-1} > 8k$ (4.3), the map ξ lifts through $\overline{\xi}$ to $BO\langle 8k\rangle$. Clearly, $\text{Vecat}(\xi) \le \text{cat}(\overline{\xi})$. By (4.6), $m - 1 \le 8k(n+1) - 2$, so $\text{cat}(\overline{\xi}) = \text{wcat}(\overline{\xi})$. We are therefore led to the following composite:

$$P^{m-1} \to \prod^{n} BO\langle 8k\rangle \to \bigwedge^{n} BO\langle 8k\rangle. \tag{6.1}$$

By (4.6), $m-1 \leq 2 \cdot 8kn - 2$, so we can consider this as a map of suspension spectra. By (5.2) and (4.6) the range can be replaced with $\bigwedge^n bo\langle 8k\rangle$. The composite (6.1) can then be rewritten

$$P^{m-1} \to \bigwedge^n P_{8k+1}^{(m-1)-(8k+1)n} \to \bigwedge^n bo\langle 8k\rangle. \tag{6.2}$$

The stunted projective space $P_{8k+1}^{(m-1)-(8k+1)n}$ is the Thom complex of the vector bundle $(8k+1)h$ over the projective space $P^{(m-1)-(8k+1)(n+1)}$. By (4.5), part (3),

$$(m-1)-(8k+1)(n+1) \leq 7.$$

Since $8k \cdot h$ is trivial on P^7 this Thom complex is just $\Sigma^{8k} P^{(m-1)-(8k+1)(n+1)+1}$. By (5.1) any map $\Sigma^{8k} P^{(m-1)-(8k+1)(n+1)+1} \to bo\langle 8k\rangle \approx \Sigma^{8k} bo$ factors through S^{8k}. It follows that the composite (6.2) factors through S^{8kn},

$$P^{m-1} \to S^{8kn} \to \bigwedge^n bo\langle 8k\rangle. \tag{6.3}$$

Iterating (5.3) splits $\bigwedge^n bo\langle 8k\rangle$ into $bo\langle 8kn\rangle \vee Y'$, with Y' $8kn+2$-connected. The factorization (6.3) shows that the component of (6.2) in Y' is null. The component in $bo\langle 8kn\rangle$ is detected (5.1) by the composite $P^{m-1} \to bo\langle 8kn\rangle \to bo$, which by (5.3) classifies the "0" bundle $[\xi]^n$.

§7 More connective K-theory

To complete the proof of (1.6) requires a deeper study of the spectrum for connective K-theory. The basic references for this section are **[2]**, [4] and **[17]**.

Until the statement of Lemma 7.10, all spectra will be connective, localized at 2, and have $Z_{(2)}$-cohomology of finite type. This ensures that mod 2 homology equivalences are stable homotopy equivalences.

Let $A_1 \subset A$ be the subalgebra of the mod 2 Steenrod algebra generated by Sq^1 and Sq^2. For $i = 0, 1, 2, 4$ let $M_i = A_1/J_i$, where,

$$\begin{aligned} J_0 &= (Sq^1, Sq^2) \\ J_1 &= (Sq^2) \\ J_2 &= (Sq^3) \\ J_4 &= (Sq^1, Sq^2Sq^3). \end{aligned}$$

Lemma 7.1. *There is an isomorphism of graded A-modules*

$$H^*(bo\langle i\rangle; \mathsf{F}_2) \approx \Sigma^i A \otimes_{A_1} M_i \qquad n = 0, 1, 2, 3, 4.$$

PROOF: Proofs of this can be found in [4] and **[21]**.

Lemma 7.2. *There exist stable complexes N_i with $H^*(N_i; \mathsf{F}_2) \approx M_i$.*

PROOF: For N_0 use the sphere S^0. The existence of N_1 and N_4 follows from the fact $2\eta = 0$. The existence of N_2 follows from the relation $\eta^2 \in \langle 2, \eta, 2\rangle$.

Lemma 7.3. *For $i = 0, 1, 2, 4$, the generator of $\pi_i bo$ extends to a map $\Sigma^i N_i \to bo$.*

PROOF: This follows easily from the Atiyah-Hirzebruch spectral sequence.

Since bo is a ring spectrum these maps extend to maps $\Sigma^i N_i \wedge bo \to bo$.

Corollary 7.4. $\Sigma^i N_i \wedge bo \approx bo\langle i\rangle$.

PROOF: The maps $\Sigma^i N_i \wedge bo \to bo$ lift, for reasons of connectivity, to $bo\langle i\rangle$. By (7.3) these lifts induce isomorphisms in π_i and hence in $H^i(\ ;\mathbb{F}_2)$. This proves (7.4) since the mod 2 cohomology of the spectra involved are isomorphic cyclic A-modules.

Let $\overline{N}_2$ denote the 4-fold suspension of the Spanier-Whitehead dual of N_2. This arranges matters to that both N_2 and $\overline{N}_2$ have their bottom cells in dimension zero. Observe also that N_1 and N_3 are Spanier-Whitehead 2-duals.

Corollary 7.5. $\Sigma^2 \overline{N}_2 \wedge bo \approx bo\langle 2\rangle$.

ROOF: The proof of Corollary 7.4 depended only on the cohomology of N_2 as an A_1-module.

Lemma 7.6. *The duality pairings $N_2 \wedge \overline{N}_2 \to S^4$ and $N_1 \wedge N_4 \to S^2$ are projections onto wedge summands.*

PROOF: There are two types of duality pairings. Their composites $S^4 \to N_2 \wedge \overline{N}_2 \to S^4$ and $S^2 \to N_1 \wedge N_4 \to S^2$ are multiplication by the Euler characteristics, which are 1.

Corollary 7.7. *$N_2 \wedge \overline{N}_2 \approx S^4 \vee Z$ where $H^*(Z;\mathbb{F}_2)$ is free as an A_1-module. $N_1 \wedge N_4 \approx S^2 \vee Z'$ where $H^*(Z';\mathbb{F}_2) \approx A_1$ as A_1-modules.*

PROOF: Let $Q_0 = Sq^1$ and $Q_1 = [Sq^1, Sq^2]$ be the Milnor elements. Since $Q_0^2 = Q_1^2 = 0$ one can form the homologies of an A_1-module with respect to Q_0 and Q_1. An A_1-module is free if and only if both its Q_0 and Q_1 homologies vanish [3], [19]. Since $H^*(N_2 \wedge \overline{N}_2;\mathbb{F}_2) \approx M_2 \otimes M_2$ the Q_0 and Q_1 homologies of $H^*(N_2 \wedge \overline{N}_2;\mathbb{F}_2)$ both have rank 1. It follows from (7.6) that $H^*(Z;\mathbb{F}_2)$ has neither Q_0 nor Q_1 homology. A similar argument works for $N_1 \wedge N_4$. A dimension count verifies the last assertion.

Corollary 7.8. *The products $\Sigma N_1 \wedge \Sigma^3 N_3 \to bo$ and $\Sigma^2 N_2 \wedge \Sigma^2 \overline{N}_2 \to bo$ of the maps of (7.3) factor through the composition of a generator of $\pi_8 bo$ with the duality pairings.*

PROOF: An elementary Adams spectral sequence argument shows that any map $\Sigma^4 Z \to bo$ or $\Sigma^4 Z' \to bo$ must be null homotopic. The maps therefore factor through the duality pairings. Since $\pi_8 bo$ is torsion free, a calculation using the rational Hurewicz homomorphisms will suffice to verify the statment about generation of $\pi_8 bo$. This is easily done, for example, by calculating the Chern characters of the maps of (7.3). Details are left to the reader.

The purpose of introducing the complexes N_i is to provide a calculus for decomposing the smash products of connective covers of bo.

Lemma 7.9.

(1) *There is are equivalences*

$$bo\langle 8k+2\rangle \wedge bo\langle 8\ell+2\rangle \approx bo\langle 8(k+\ell+1)\rangle \vee Y'$$
$$bo\langle 8k+1\rangle \wedge bo\langle 8\ell+4\rangle \approx bo\langle 8(k+\ell+1)\rangle \vee Y'.$$

(2) *The following diagrams commutes up to homotopy:*

$$\begin{array}{ccccc} bo\langle 8k+2\rangle \wedge bo\langle 8\ell+2\rangle & \xrightarrow{p\wedge p} & bo\wedge bo & \xrightarrow{\mu} & bo \\ \approx\big\downarrow & & & & \big\| \\ bo\langle 8(k+\ell+1)\rangle \vee Y' & \xrightarrow{\text{pinch}} & bo\langle 8(k+\ell+1)\rangle & \xrightarrow{p} & bo, \end{array}$$

$$\begin{array}{ccccc} bo\langle 8k+1\rangle \wedge bo\langle 8\ell+4\rangle & \xrightarrow{p\wedge p} & bo\wedge bo & \xrightarrow{\mu} & bo \\ \approx\big\downarrow & & & & \big\| \\ bo\langle 8(k+\ell+1)\rangle \vee Y' & \xrightarrow{\text{pinch}} & bo\langle 8(k+\ell+1)\rangle & \xrightarrow{p} & bo. \end{array}$$

PROOF: The first equivalence in part (1) follows from the sequence:

$$\begin{aligned} bo\langle 8k+2\rangle \wedge bo\langle 8\ell+2\rangle &\approx \Sigma^{8k} bo\langle 2\rangle \wedge \Sigma^{8\ell} bo\langle 2\rangle \approx \Sigma^{8(k+\ell)+4} N_2 \wedge \overline{N}_2 \wedge bo \wedge bo \\ &\approx \Sigma^{8(k+\ell)+4}(S^4 \vee Z) \wedge (bo \vee Y) \approx bo\langle 8(k+\ell+1)\rangle \vee Y' \end{aligned}$$

A similar sequence establishes the second. Commutativity of the diagrams in (2) is left to the reader.

Lemma 7.9 breaks the problem of determining a map into $bo\langle n_1\rangle \wedge bo\langle n_2\rangle$ into two parts. One part consists of a map to $bo\langle n_3\rangle$ and the other part involves a map to some other spectrum Y'. The notion of *simplicity* of a spectrum Y is useful for obtaining information about this other map. For the rest of this section all spectra will be localized at some prime p and have $\mathbf{Z}_{(p)}$ cohomology of finite type.

If A is an abelian group then HA will denote the Eilenberg-MacLane spectrum with $\pi_0 HA \approx A$.

Lemma 7.10. *For a spectrum Y, the following are equivalent:*

(1) *If* $\dim X \leq n$ *then any non-trivial map* $X \to Y$ *induces a non-zero homomorphism* $H^*(Y;M) \to H^*(X;M)$, *for some* $Z_{(p)}$ *module* M.
(2) *The Hurewicz homomorphism* $\pi_j(Y) \to H_j(Y;Z_{(p)})$ *is a split monomorphism for* $j \leq n$.
(3) *There is a map* $Y \to \bigvee_{j\leq n} \Sigma^j H\pi_j(Y)$ *which is an* $(n+1)$*-equivalence.*

PROOF: (1) $\Longrightarrow$ (2) Let F denote the fibre of $Y \to Y \wedge HZ_{(p)}$ and let $F^{(n)}$ denote the n-skeleton of F. Since $F^{(n)} \to Y \to Y \wedge HZ_{(p)}$ is null homotopic, we have by part (1) $F^{(n)} \to Y$ null. This results in the following diagram:

$$\begin{array}{ccccc} Y & \longrightarrow & Y \wedge HZ_{(p)} & \longrightarrow & \Sigma F \\ & & & & \uparrow \\ & & & & \Sigma F^{(n)}. \end{array}$$

Applying π_* yields (2).

(2) $\Longrightarrow$ (3) The splittings are elements of $\mathrm{Hom}[H_j(Y; Z_{(p)}), \Pi_j(Y)]$ for $j \leq n$. Thinking of these as cohomolgoy classes gives the required map.

(3) $\Longrightarrow$ (1) This is clear.

Definition. *A spectrum Y is n-simple if it satisfies the equivalent conditions of Lemma 7.10.*

Example. If Y is $(n-1)$-connected then Y is n-simple.

Lemma 7.11. *Let Y be n-simple and suppose Z is $(k-1)$-connected. Then $Y \wedge Z$ is $(n+k)$-simple.*

PROOF: This follows from condition (3).

The following is a useful criteria for n-simplicity.

Lemma 7.12. *Y is n-simple if the following are satisfied for $j \leq n$:*

(1) *The p-torsion in $\pi_j(Y)$ is of order p;*

(2) *After tensoring with $\mathbb{F}_p$, the Hurewicz homomorphism $\pi_j(Y) \otimes \mathbb{F}_p \to H_j(Y) \otimes \mathbb{F}_p$ is a monomorphism.*

PROOF: This is immdiate from the next lemma.

Lemma 7.13. *let M and N be finitely generated $Z_{(p)}$-modules with torsion of order at most p. Then $f : M \to N$ is a split monomorphism if and only if $f \otimes \mathbb{F}_p : M \otimes \mathbb{F}_p \to N \otimes \mathbb{F}_p$ is a monomorphism.*

PROOF: The forward implication is trivial. For the reverse implication, write $N = T \oplus F$ where T is the torsion subgroup and F is a free $Z_{(p)}$-module. There is clearly a map $T \to M$ splitting the restriction of f to the torsion subgroup of M. Extend this to a map $g : N \to M$ by requiring that $g \otimes \mathbb{F}_p : N \otimes \mathbb{F}_p \to M \otimes \mathbb{F}_p$ split $f \otimes \mathbb{F}_p$. Then $g \circ f : M \to M$ is surjective by Nakayama's lemma, hence an isomorphism since M is finitely generated.

§8 Completion of the proof of Theorem 1.6 (nearly)

The remaining cases are more difficult since the estimates of (4.6) fail. The argument for handling these is essentially as before, however, and can be outlined as follows.

We are given a map

(8.1) $$P^{m-1} \xrightarrow{\xi} BO \xrightarrow{\Delta} BO \times \cdots \times BO$$

and must show that it factors through the fat wedge. We know that ξ can be lifted through some connective cover of BO, so that $\Delta \circ \xi$ can be lifted (uniquely) through

$$BO\langle n_1 \rangle \times \cdots \times BO\langle n_k \rangle,$$

with $n_1 \le \cdots \le n_k$. Equality need not hold. It will suffice to factor this map through the fat wedge.

If the n_i are large enough, it will be enough to demonstrate that

$$P^{m-1} \to BO\langle n_1\rangle \times \cdots \times BO\langle n_k\rangle \to BO\langle n_1\rangle \wedge \cdots \wedge BO\langle n_k\rangle$$

is null homotopic, by (1.3). Again, if the n_i are large enough, (8.1) can be replaced by the stable map,

$$P^{m-1} \to bo\langle n_1\rangle \wedge \cdots \wedge bo\langle n_k\rangle. \tag{8.2}$$

The condition on the n_i which we require is:

(A) $m < n_1 + (n_1 + \cdots + n_k)$.

Remark. In case $m = n_1 + (n_1 + \cdots + n_k)$, the map $P^{m-1} \to BO\langle n_1\rangle \wedge \cdots \wedge BO\langle n_k\rangle$ can still be replaced by $P^{m-1} \to bo\langle n_1\rangle \wedge \cdots \wedge bo\langle n_k\rangle$. The problem is that the map from the fat wedge to the fibre of the collapse is only $(m-2)$-connected ((1.3), (5.2)). There is one more obstruction.

At odd primes and rationally (8.2) is null. At the prime 2 the range will split. With carefully chosen n_i this splitting will be in the form $bo\langle n\rangle \vee Y$. The component of (8.2) in $bo\langle n\rangle$ will be detected by the composite $P^{m-1} \to bo\langle n\rangle \to bo$, which vanishes since it classifies $[\xi]^k = 0$. The map into Y will vanish since we will be able to show that Y is at least $m-1$-simple and that the map is not detected in cohomology with coefficients in any $Z_{(2)}$ module.

We have imposed two more conditions on the sequence $\{n_i\}$:

(B) After excluding at most one n_i, the number of $n_j \equiv 2\ (8)$ must be even and the number of $n_j \equiv 1\ (8)$ must equal the number of $n_j \equiv 4\ (8)$.

(C) $m(1, k, r) > n_1$.

Remark. Condition (B) guarantees the splitting and condition (C) guarantees that the map (8.2) will induce the zero map in cohomology with coefficients in any $Z_{(2)}$-module.

PROOF OF REMARK: Condition (B) is the situation described by (7.9). Given (C), observe that (8.2) factors as

$$P^{m-1} \subset P \xrightarrow{\Delta} P \wedge \cdots \wedge P \xrightarrow{\xi \wedge \cdots \wedge \xi} bo\langle n_1\rangle \wedge \cdots \wedge bo\langle n_k\rangle,$$

where P denotes infinite dimensional real projective space. The map $\bar{\xi} : P \to bo\langle n_1\rangle$ induces the zero map in $H^{n_1}(\ ;\mathbf{F}_2)$ by condition (C), and hence in $H^*(\ ;\mathbf{F}_2)$, since $H^*(bo\langle n_1\rangle;\mathbf{F}_2)$ is a cyclic A-module. It follows that $P \to bo\langle n_1\rangle \wedge \cdot \wedge bo\langle n_k\rangle$ induces the zero map in mod 2 cohomology which is enough since reduction mod 2 of (untwisted) coefficients is a monomorphism in the cohomology of P.

We can now continue the computations of Vecat. For each of the remaining cases, the sequence $\{n_i\}$ will be described, and a lower bound for the simplicity of the resulting summand Y will be stated. The task of verifying conditions (A), (B), and (C) is left to the reader.

The techniques for estimating the simplicities are described in the appendix.

k = 3. The only case is $r = 3$, $n = 4$. Use the sequence (26, 26, 26, 26). The summand Y has simplicity 119.

k=2. The argument in §5 works for $k = 2, r = 1$ and in fact for $k = 2, r = 2, n = 2$. Since $m(3,2,2) = m(2,2,2) + m(1,2,2)$ and $m(4,2,2) = m(2,2,2) + m(2,2,2)$, this settles $k = 2, r = 2, n = 3, 4$. When $r = 3$ we must deal with $(2^{10}h)$ on P^{41}, P^{64} and P^{87}. The sequences to choose are $(18,18), (18,18,20)$ and $(18,18,18,18)$. The resulting summands Y have simplicities 47,67, and 87.

k=1. When $r = 1$ the argument of section 5 dispenses with $n = 1,2,3$. When $n = 4$ use the sequence $(8,8,8,9)$. The simplicity of Y is 40.

When $r = 2$ it suffices to consider $n = 2$. In this case use $(8,10)$. The summand Y has simplicity 25.

For $r = 2$, $n = 2,3$ use the sequences $(10,10)$ and $(10,10,12)$. The resulting summands have simplicities 31 and 43.

The case $k = 1$, $r = 3$, $n = 4$ is dealt with in the next section.

§9 2^6H ON P^{55}

It remains to consider the case of 2^6h on P^{55}. Using the sequence (10, 12, 12, 12) the argument of §8 nearly goes through since $bo\langle 10\rangle \wedge bo\langle 12\rangle \wedge bo\langle 12\rangle \wedge bo\langle 12\rangle$ is already 55-simple by Example 2 in the appendix. What fails is that the map from the fat wedge to the fibre of the collapse

$$BO\langle 10\rangle \times BO\langle 12\rangle^3 \to BO\langle 10\rangle \wedge \bigwedge^3 BO\langle 12\rangle$$

is only 54-connected (see the remark after condition (A)). There is one more obstruction.

Let $T_1 \subset BO\langle 10\rangle \times BO\langle 12\rangle^3$ and $T_2 \subset BO\langle 10\rangle^4$ denote the fat wedges. Consider the following diagram in which the vertical sequences are fibrations:

$$\begin{array}{ccccc} & & E_1 & \longrightarrow & E_2 & \longrightarrow & E_2/T_2 \\ & & \downarrow & & \downarrow & & \\ P^{55} & \longrightarrow & BO\langle 10\rangle \times BO\langle 12\rangle^3 & \longrightarrow & BO\langle 10\rangle^4 & & \\ & & \downarrow & & \downarrow & & \\ & & BO\langle 10\rangle \wedge \bigwedge^3 BO\langle 12\rangle & \longrightarrow & \bigwedge^4 BO\langle 10\rangle. & & \end{array}$$

An application of (1.3), with $A = T_2$, $X = E_2$, reduces the problem to showing that the composite $P^{55} \to E_2/T_2$ is null homotopic.

An elementary calculation using (1.3) again shows that E_1/T_1 and E_2/T_2 can be replaced with

$$\Sigma^{-1} \bigwedge^2 bo\langle 10\rangle \wedge \bigwedge^3 bo\langle 12\rangle,$$

and a wedge of copies of

$$\Sigma^{-1} \bigwedge^5 bo\langle 10\rangle,$$

through dimension 59. It follows that the map $E_1/T_1 \to E_2/T_2$ induces the zero map in cohomology through dimension 59. We are therefore left with a stable map

$$P^{55} \to \Sigma^{-1} \bigwedge^5 bo\langle 10\rangle$$

which is not detected in mod two cohomology.

At odd primes and rationally $bo\langle 10\rangle \approx bo\langle 12\rangle$ so tht the range is 58-connected. At the prime 2, the methods of the appendix show that the range is 58-equivalent to a wedge of suspensions of the Eilenberg-MacLane spectrum $H\mathbb{F}_2$. It follows that this last obstruction vanishes.

This completes the proof of Theorem 1.6.

§10 Berstein Covering Numbers

Using Corollary 1.7 we can now prove Theorem 1 of the introduction. The following lemma is a slight modificationof a result of Berstein ([6], Proposition2.3).

Lemma 10.1. *Let M be a closed n-manifold with tangent bundle τ. The immersion covering number $n(M)$ is equal to* $\max\{2, \operatorname{Vecat}(\tau)\}$.

Proof: The lemma is immediate from the theorem of Hirsch-Peonaru ([12], [20]) which states that a non-closed n-manifold can be immersed in $\mathbb{R}^n$ if and only if it is parallelizable, together with the observation that a non-closed manifold is parallelizable and only. if it is stably parallelizable.

Part (1) of Theorem 1 is an immediate consequence of (10.1) and (1.7).

Following Berstein and James ([6], [16]), some information about the embedding numbers $N(P^n)$ can be obtained from the affine coverings of projective space. Suppose $P^n = P^{n_1} \cup \cdots \cup P^{n_k}$ is a trivializing cover of the tangent bundle of P^{n_i}, with $2n_i < n$, $i = 1, \ldots, k$. For reasons of dimension, P^{n_i} embeds in $\mathbb{R}^n$. Let ν_1 be the normal bundle of an embedding. The total space of ν_1 also embeds in $\mathbb{R}^n$. The open subset $P^n - P^{n-n_i-1} \subset P^n$, of which P^{n_i} is a deformation retract, is diffeomorphic to the normal bundle ν_2 of the embedding $P^{n_i} \subset P^n$. Since the tangent bundle of P^n restricts to a stably trivial bundle on P^{n_i}, ν_1 is stably equivalent to ν_2. However, the dimensions of ν_1 and ν_2 are $n - n_i > n_i$, so they are already stable. It follows that the total space of ν_2 embeds in $\mathbb{R}^n$ and that $N(P^n) \leq k$.

The affine coverings of P^n therefore place an upper bound on $N(P^n)$. A lower bound is given by the inequality $n(P^n) \leq N(P^n)$. Write $n+1$ in the form $2^k m$ with m odd. According to (4.1) and the remark after (1.7), these bounds agree when either $k \leq 3$ or $k \equiv 3 \mod 4$. They also supply the bounds for $N(P^{31})$ and $N(P^{47})$.

Under certain circumstances this upper bound can be improved. Consider for example the case of P^{79} where $\operatorname{Vecat}(\tau) = 8$. There is the decomposition $P^{79} = P^{39} \cup P^{39}$. Let ν_1 be the normal bundle of an embedding $P^{39} \subset \mathbb{R}^{79}$ and let ν_2 be the normal bundle of the inclusion $P^{39} \subset P^{79}$. The tangent bundle of P^{79} does not restrict trivially to P^{39} so ν_1 and ν_2 are not isomorphic. However, $\operatorname{Vecat}(80h) = 4$ on P^{39}. One can therefore decompose P^{39} as a union of 4 open sets over which the tangent bundle of P^{79} does restrict to the trivial bundle. Over each of these open subsets ν_1 and ν_2 are therefore isomorphic. This results in a decomposition of P^{79} into $4+4=8$ open subsets, each of which embeds in $\mathbb{R}^{79}$. It follows that $N(P^{79}) = 8$.

In general, this situation will arise whenever a minimal trivializing cover of the tangent bundle of P^n can be obtained by refining a decomposition $P^n = P^{n_1} \cup \cdots \cup P^{n_k}$ with $2n_i < n$, $i = 1, \ldots, k$. It turns out that this is nearly always the case. Let ξ be a vector bundle over P^n. According to the proof of (4.5) minimal trivializing covers of ξ are obtained by refining a decomposition $P^n = P^{n_1} \cup \cdots \cup P^{n_k}$ with $\operatorname{Vecat}(\xi) \leq 4$ on each P^{n_i}. The condition $2n_i < n$ is automatically satisfied if $\operatorname{Vecat}(\xi) \geq 8$. Combined with the previous estimate this shows that $n(P^n) = N(P^n)$ provided $k \leq 3$, $k \equiv 3$ (4), or $\frac{(n+1)}{2(k+1)} \geq 8$. The only values of n which do not satisfy these inequalities are 15, 31, 47 and 63. The decompositions $P^{15} = p^7 \cup P^7$ is a minimal trivializing cover of the tangent bundle of P^{15}. The decomposition $P^{63} = P^{25} \cup P^{11}$ can be refined to a minimal trivializing cover of the tangent bundle. This completes the proof of Theorem 1.

Remark. The values of $n(M)$ and $N(M)$ when M is complex or quaternionic projective space can be easily computed using the methods of [6] and [16]. The lower bound nil$[\tau]$ agrees with the upper bound coming from the affine coverings.

APPENDIX. SMASHING THE CONNECTIVE COVERS OF *bo*

This appendix sets up computation of the simplicities of the sumands of smash products of connective covers of *bo*. The technique is to use the Adams spectral sequence. The calculations needed in §8 are all fairly similar, so rather than do each one individually, two illustrative examples are worked.

The mod 2 Adams spectral sequence for a spectrum of the form $X \wedge bo$ begins with $\mathrm{Ext}^{s,t}_{A_1}(H^*(X : \mathbb{F}_2), \mathbb{F}_2)$ and abuts to the 2-adic completion of $\pi_{t-s}X \wedge bo$. The Hopf algebra A_1 is the subalgebra of the mod 2 Steenrod algebra generated by Sq^1 and Sq^2.

Let $A_0 \subset A$ be the subalgebra of the mod 2 Steenrod algebra generated by Sq^1. It is an exterior algebra on one generator. The Hopf algebra cohomology ring $\mathrm{Ext}^{*,*}_{A_0}(\mathbb{F}_2, \mathbb{F}_2)$ is a polynomial ring $\mathbb{F}_2[h_0]$, $h_0 \in \mathrm{Ext}^{1,1}_{A_0}(\mathbb{F}_2, \mathbb{F}_2)$.

Definition A.1. *An A_1-module is n-simple if*

(1) *For $t-s \leq n$, the h_0-torsion in $\mathrm{Ext}^{s,t}_{A_1}(M, \mathbb{F}_2)$ is annihilated by h_0, and*
(2) *for $t-s \leq n$ the "reduced Hurewicz homomorphism"*

$$\mathrm{Ext}^{s,t}_{A_1}(M, \mathbb{F}_2) \otimes_{\mathbb{F}_2[h_0]} \mathbb{F}_2 \to \mathrm{Ext}^{s,t}_{A_0}(M, \mathbb{F}_2) \otimes_{\mathbb{F}_2[h_0]} \mathbb{F}_2$$

is a monomorphism.

Proposition A.2. *Suppose that the Adams spectral sequence for $X \wedge bo$ satisfies*

$$E_2^{s,t} \approx E_\infty^{s,t} \qquad \text{for } t-s \leq n.$$

If the A_1-module $H^*(X)$ is n-simple then $X \wedge bo$ is n-simple.

PROOF: Under these assumptions, the conditions of Lemma 7.12 are satisfied.

One advantage of the simplicity of a module M is that it depends only on the stable class of M.

Definition ([3], [4], [19].) *Two A_1-modules M, and N are stably equivalent (written $M \approx_S N$) if there exist projective modules P_1 and P_2 with $M \oplus P_1 \approx N \oplus P_2$.*

Direct sums and tensor products preserve stable equivalence.

Proposition A.3. *If $M \approx_S N$ and M is n-simple, then N is n-simple.*

PROOF: This is clear

Lemma A.4. *Let $R_1 \rightarrowtail P_n \to \cdots \to P_1 \twoheadrightarrow M$ and $R_1 \rightarrowtail P'_n \to \cdots \to P'_1 \twoheadrightarrow M$ be two exace sequences with the P_i and P'_i projective. Then $R_1 \approx_S R_2$.*

Lemma A.5. *Let α be the Yoneda class of the extension*

$$R_1 \rightarrowtail P_n \to \cdots \to P_1 \twoheadrightarrow M.$$

Multiplication by α induces an isomorphism

$$\mathrm{Ext}^{s}_{A_1}(R, \mathbb{F}_2) \approx \mathrm{Ext}^{s+n}_{A_1}(M, \mathbb{F}_2)$$

when $s > 0$, and an epimorphism when $s = 0$.

The simplicties of the iterated tensor products $M_1^i \otimes M_2^j \otimes M_4^k$, $i \leq k$ can now be detemined. First we will determine the stable calsses. The resulting "Ext-charts" can then be extracted from [4] – though the cases we consider are easily enough calculated.

Let I denote the desuspension of the augmentation ideal of A_1. I is (-1)-connected, but not 0-connected.

Lemma A.6. *There are stable equivalences*

$$\Sigma^2 M_4 \approx_s M_2 \otimes I$$
$$M_2 \otimes M_2 \approx_S \Sigma^4 \mathbb{F}_2$$
$$M_1 \otimes M_4 \approx_S \Sigma^2 \mathbb{F}_2.$$

PROOF: The first equivalence follows from the exact sequences $\Sigma^3 M_4 \rightarrowtail A_1 \twoheadrightarrow M_2$ and $\Sigma I \otimes M_2 \rightarrowtail A_1 \otimes M_2 \twoheadrightarrow M_2$. The other two equivalences are contained in Corollary 7.7.

It follows that the A_1-modules $M_1^i \otimes M_2^j \otimes M_4^k$ with $i \leq k$ are stably isomorphic to modules of the form $\Sigma^a I^b$ or $\Sigma^a I^b \otimes M_2$. We therefore need only determine the simplicities of I^n and $I^n \otimes M_2$.

Proposition A.7. *Write $n = 4k + r$ and $n + 2 = 4l + s$ with $0 \leq r, s \leq 3$. then I^n has simplicity $8k + 2^r - 1$ and $I^n \otimes M_2$ has simplicity $8l + 2^s - 3$.*

PROOF: By lemma A.3 there are isomorphisms

$$\mathrm{Ext}_{A_1}^{s,t}(I^n, \mathbb{F}_2) \approx \mathrm{Ext}_{A_1}^{s+n,t+n}(\mathbb{F}_2, \mathbb{F}_2)$$
$$\mathrm{Ext}_{A_1}^{s,t}(I^n \otimes M_2, \mathbb{F}_2) \approx \mathrm{Ext}_{A_1}^{s+n,t+n}(M_2, \mathbb{F}_2)$$

for $s > 0$. The result then follows from the known computation of $\mathrm{Ext}_{A_1}^{*,*}(\mathbb{F}_2, \mathbb{F}_2)$ and $\mathrm{Ext}_{A_1}^{*,*}(M_2, \mathbb{F}_2)$. These 'Ext-charts' are displayed on the following page.

Corollary A.8. *The spectrum $N_j \wedge N_4 \wedge bo$, $j = 0, 1, 2, 3, 4$, is 3-simple.*

The following result can be found in [**17**] or [**18**].

Proposition A.9. *There is a 2-local equivalence*

$$bo \wedge bo \approx (S^0 \vee \Sigma^4 N_4 \vee Z) \wedge bo.$$

The spectrum Z is 7-connected.

Lemma A.10. *For $n \equiv 0, 1, 2, 4 \pmod 8$ there is a splitting*

$$bo\langle n\rangle \wedge \bigwedge^{j} bo \approx bo\langle n\rangle \vee Y$$

where Y is $(n+7)$-simple.

PROOF: Use (7.4), (A.8), (A.9) and induction.

Finally we have the tools for making the computations needed in §8. Here are two examples. The second one is used in §9.

Example 1: $\bigwedge^4 bo\langle 26\rangle$. There is an equivalence

$$\bigwedge^4 bo\langle 26\rangle \approx \Sigma^{104} N_2 \wedge \bar{N}_2 \wedge N_2 \wedge \bar{N}_2 \bigwedge^4 bo.$$

From (7.7), there is a splitting

$$N_2 \wedge \bar{N}_2 \wedge N_2 \wedge \bar{N}_2 \approx S^8 \vee V,$$

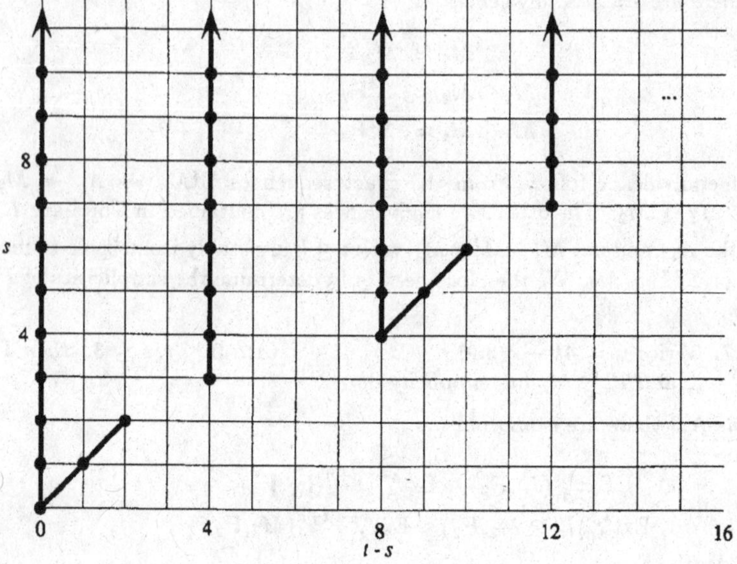

$\mathrm{Ext}^{s,t}_{A_1}(\mathbb{F}_2, \mathbb{F}_2)$

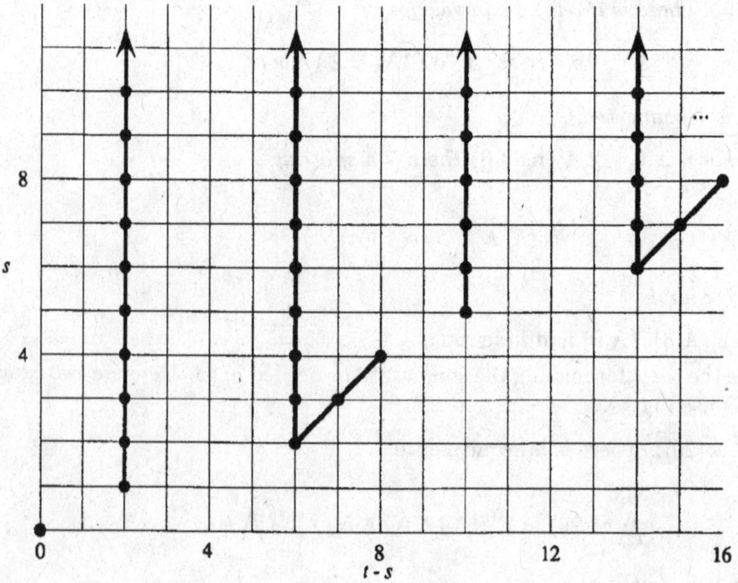

$\mathrm{Ext}^{s,t}_{A_1}(M_2, \mathbb{F}_2)$

with $H^*(V;\mathbb{F}_2)$ A_1-free. It follows that

$$\Sigma^{104} N_2 \wedge \bar{N}_2 \wedge N_2 \wedge \bar{N}_2 \bigwedge^4 bo \approx \Sigma^{104} S^8 \bigwedge^4 bo \vee HU,$$

where HU is a wedge of mod 2 Eilenberg-MacLane spectra, which cannot contribute to the simplicity. Appealing to (A.10) therefore yields a splitting

$$\bigwedge^4 bo\langle 26\rangle \approx bo\langle 112\rangle \vee Y$$

where Y is $112 + 7 = 119$-simple.

Example 2: $bo\langle 10\rangle \wedge bo\langle 12\rangle \wedge bo\langle 12\rangle \wedge bo\langle 12\rangle$. There is an equivalence

$$bo\langle 10\rangle \wedge bo\langle 12\rangle \wedge bo\langle 12\rangle \wedge bo\langle 12\rangle \approx \Sigma^{46} N_2 \wedge N_4 \wedge N_4 \wedge N_4 \bigwedge^4 bo.$$

First consider $\Sigma^{46} N_2 \wedge N_4 \wedge N_4 \wedge N_4 \wedge bo$. The module $M_2 \otimes M_4 \otimes M_4 \otimes M_4$ is stably isomorphic to

$$\Sigma^{-6} I^3 \otimes M_2^4 \approx_S \Sigma^2 I^3$$

which is $46+9 = 55$-simple. Smashing with further copies of bo doesn't change the simplicity by Lemma 7.11. It follows that $bo\langle 10\rangle \wedge bo\langle 12\rangle \wedge bo\langle 12\rangle \wedge bo\langle 12\rangle$ is 55-simple.

REFERENCES

1. J. F. Adams: On Chern characters and the structure of the unitary group, Proc. Cam. Phil. Soc. 57 (1961), 189–199.
2. J. F. Adams: Stable homotopy and generalized cohomology, University of Chicago Press 1974.
3. J.F. Adams and H. R. Margolis: Modules over the Steenrod algebra, Topology, 10(1971), 271–282.
4. J. F. Adams and S. B. Priddy: Uniqueness of BSO, Proc. Cam. Phil. Soc. 80 (1976), 475–509.
5. J. Adem, S. Gitler and I. M. James: On axial maps of a certain type, Bol. Soc. Mat. Mex. 17 91972), 59–62.
6. I. Berstein: On imbedding numbers of differentiable manifolds, Topology 7 (1968), 95–109.
7. I. Berstein and T. Ganea: The category of a map and of a cohomology class, Fundam. Math. (1962), 265–279.
8. R. H. Fox: On the Lusternik-Schnirelman Category, Ann. Math. 42 (1941), 333–370.
9. T. Ganea: A generalization of the homology and homotopy suspension, Comm. Math. Helv. 39 (1965), 295-322.
10. T. Ganea: On the homotopy suspension, Comm. Math. Helv. 43(1968), 225–234.
11. A Haefliger and M. W. Hirsch: Immersions in the stable range, Ann. Math. 75 (1962), 231–241.
12. M. W. Hirsch: Immersions of manifolds, Trans. Am. Math. Soc. 93 (1959), 242–276.
13. M. W. Hirsch: Smooth regular neighbourhoods, Ann. Mth. 79 (1962), 524–530.
14. I. M. James: The space of bundle maps, Topology 2 (1963), 45–59.
15. I. M. James: On category in the sense of Lusternik-Schnirelmann, Topology 17 (1978), 331–348.
16. I. M. James: On category and section category, Proceedings of the 1979 Topology Conference, Brazilian Math. Soc., APUCI, Rio de Janera.
17. M. Mahowald: bo resolutions, Pac. J. of Math. 92 (1981), 365–383.
18. R. J. Milgram: The Steenrod algebra and its dual for connective K-theory, Reunion Sobre Theorie de Homotopic Universidad de Northwestern 1974, Sociedad Matimatica Mexicana, 127–159.
19. F. P. Peterson: lectures on cobordism theory, Kinokuniya Book Store Co., Ltd. Tokyo, Japan, 1968.
20. V. Poenaru: Sur la theorie des immersions, Topology 1 (1962), 81–100.
21. R. E. Stong: Determination of $H^*(BO(k,\dots,\infty);\mathbb{Z}_{12})$ and $H^*(BU(k,\dots,\infty);\mathbb{Z}/2)$, Trans. Am. Math. Soc., 107 (1963), 526–544.
22. A. S. Svarc: The genus of a fibre space, Am. Math. Soc. Transl. 55 (1966), 49–140.
23. G. W. Whitehead: The homology suspension, Colloque de topologie algebrique, Louvain (1956), 89–95.

University of Chicago

HIGHER HOMOTOPY ASSOCIATIVITY

Norio IWASE and Mamoru MIMURA
Department of mathematics, Kyushu University, Hakozaki Fukuoka 812, Japan
Department of mathematics, Okayama University, Tushima-naka Okayama 700, Japan

§1. Introduction

Throughout the paper we work in the category of spaces having the homotopy type of 1-connected CW-complexes with base point. Let us recall some notions introduced by Stasheff [S].

Definition. An A_n-structure on a space X consists of a sequence of quasi-fibrations $p_i : E^i \to P^{i-1}$;

$$\begin{array}{ccccccccc} X = E^1 & \subset & E^2 & \subset & \cdots & \subset & E^n & & \\ \downarrow p_1 & & \downarrow p_2 & & & & \downarrow p_n & & \text{(commutative)} \\ * = P^0 & \subset & P^1 & \subset & \cdots & \subset & P^{n-1} & \subset & P^n \end{array}$$

with E^i contractible in E^{i+1} and $P^1 = SX$.

Stasheff then defines a special complex $K(i)$ such that

(1) $K(2) = *$, $K(i) \cong I^{i-2}$ (homeomorphic),

(2) the boundary $\partial K(i)$ of $K(i)$ is the union of $i(i-1)/2-1$ faces $K_k(r,s)$ for $2 \leqq r, s$; $1 \leqq k \leqq r$, $r+s = i+1$, where each face $K_k(r,s)$ is affine homeomorphic to $K(r)\times K(s)$ by the map $\partial_k(r,s) : K(r)\times K(s) \to K_k(r,s)$, a face operator,

(3) it has degeneracy operators $s_j : K(i) \to K(i-1)$ for $1 \leqq j \leqq i$.

Definition. An A_n-form on X consists of a family of maps

$$m_i : K(i)\times X^i \to X \quad \text{for} \quad 1 \leqq j \leqq i$$

such that

(1) m_2 is a multiplication with unit, $m_2(*,e,x) = m_2(x,*,e) = x$,

(2) $m_i(\partial_k(r,s)(\rho,\sigma);\ x_1,\ldots,x_i)$

$= m_r(\rho;x_1,\ldots,x_{k-1},m_s(\sigma;x_k,\ldots,x_{k+s-1}),x_{k+s},\ldots,x_i)$,

(3) $m_i(\tau;x_1,\ldots,x_{j-1},*,x_{j+1},\ldots,x_i)$

$= m_i(s_j(\tau);x_1,\ldots,x_{j-1},x_{j+1},\ldots,x_i)$.

Definition. The pair $(X,\{m_i\})$ is called an A_n-space.

Definition. If there exist m_i for any i, we call $\{m_i\}$ an A_∞-form and $(X,\{m_i\})$ an A_∞-space.

Remark. 1) A space is an A_2-space iff it is an H-space.

2) For $i = 3$, $K_3 = I$ the unit interval. The second condition (2) says that $m_2 \circ (Id \times m_2) \simeq m_2 \circ (m_2 \times Id)$, i.e. $x(yz) \sim (xy)z$. Thus m_3 is an associating homotopy (so that m_2 is a homotopy associative multiplication).

3) Any associative H-space admits n-forms for any n

$$m_n(\tau;x_1,\ldots,x_n) = x_1 \ldots x_n$$

so that it is an A_∞-space.

4) In the complex K_i, the symmetries are lost, because we do not assume strict associativity for the H-space. The faces are in the one to one correspondence with the variation of the (non-commutative) product and there are no inessential faces.

One of the main results by Stasheff in [S] is

Theorem. A space X has an A_n-form iff it admits an A_n-structure.

In the process of the proof he defined a space D^i such that there exists a relative homeomorphism $\sigma_{k+1} : (D^k,E^k) \to (P^k,P^{k-1})$

satisfying

(1) $$\begin{array}{ccccc} D^{k-1} & \subset & E^{k} & \subset & D^{k} \\ \downarrow{\scriptstyle \sigma_{k-1}} & & \downarrow{\scriptstyle p_k} & & \downarrow{\scriptstyle \sigma_k} \\ P^{k-1} & = & P^{k-1} & \subset & P^{k} \end{array} \qquad \text{(commutative)}$$

(2) $(CE^n, E^n) \simeq (D^k, E^k)$ (homotopy equivalence).

Notation (Convention). When we want to express the original space X explicitly (or to avoid ambiguity) we write:

$$E^n(X),\ P^{n-1}(X),\ D^n(X),\ p_n^X,\ m_i^X,\ \text{etc.}$$

It is quite natural to ask a functorial definition of an A_n-structure and A_n-form i.e. a definition of a map $f : X \to Y$ to be an A_n-map between A_n-spaces X and Y, which preserve, up to homotopy, A_n-structures of X and Y.

Before we give an explicit definition of an A_n-map, we state its fundamental properties:

P1) A map homotopic to an A_n-map is an A_n-map.

P2) A composition of A_n-maps is an A_n-map.

P3) An A_n-homomorphism is an A_n-map.

P4) The localization map is an A_n-map. The localization of an A_n-map is an A_n-map.

P5) If a homotopy equivalence is an A_n-map, so is its homotopy inverse.

P6) The homotopy fibre of an A_n-map admits an A_n-structure.

P7) The pull-back of two A_n-maps admits an A_n-structure.

P8) A map is an A_2-map, A_3-map or A_∞-map iff it is an H-map, an

H-map preserving homotopy associativity or a loop map, respectively.

P9) Let X be an A_n-space and G a monoid. A map $f : X \to G$ is an A_n-map iff its adjoint $ad(f) : SX \to BG$ is extendable over $P^n(X)$.

P10) Suppose that an A_n-space X dominates an A_{n-1}-space Y, namely, there are maps $f : X \to Y$, $g : Y \to X$ such that $f \circ g \simeq 1_Y$. If one of them is an A_{n-1}-map, Y has an A_n-form.

Historical remarks. 1) Stasheff [S] defined an A_n-homomorphism between A_n-spaces X and Y iff f commutes strictly with A_n-forms $(X,\{m_i\})$ and $(Y,\{n_i\})$: $f \circ m_i = n_i \circ (1_K \times f \times \ldots \times f)$ (a map homotopic to an A_n-homomorphism is not necessarily an A_n-homomorphism).

2) When the target space admits an A_∞-structure, he defined an A_n-form.

3) He also described a parametric complex for $n = 4$ giving an A_4-form of a map but did not give a unified construction of such complexes for all n.

4) Zabrodsky [Z1] defined an A_n-map for $n \leqq 3$ and mentioned the possibility for general n (but did not give an explicit definition).

The paper is organized as follows: In Section 2 we define a parameter complex $J(i)$. Then the notions, A_n-form and A_n-structure of a map, are defined by making use of the complex in Section 3. In Section 4 we give an outline of the proofs of the fundamental properties P1) ~ P10). In Section 5 we give some applications. In the last section, Section 6, we generalize the Zabrodsky theorem [Z1] and [Z2].

The details will appear somewhere.

§2. The construction of complex J(i)

In this section we define a complex $J(i)$ which will be needed to define an A_n-map in the later section.

<u>The definition of</u> $J(i)$

The definition of $J(i)$ is somewhat 'flexible'. Let n be any positive integer. Note that the $n-1$ cell $K(n+1)$ is homeomorphic with a complex $\overline{K}(n+1)$ whose boundary is a PL-manifold as follows:

$$\overline{K}(n+1) \subset \prod_{j=0}^{n}[0,j],$$
$$\overline{K}(n+1) \ni (u_0,\dots,u_n) \quad \text{if} \quad \sum_{i=0}^{j} u_i \leqq j \quad \text{and} \quad \sum_{i=0}^{n} u_i = n,$$
$$\overline{K}_{k+1}(r,s) \ni (u_0,\dots,u_n) \quad \text{if} \quad (u_k,\dots,u_{k+s-1}) \in \overline{K}(s).$$

From now on we identify $\overline{K}(n+1)$ with $K(n+1)$ if there is no misunderstanding.

The boundary of $K(n+1)$ is the union of $K_{k+1}(r,s)$'s. The face operators are described as

$$\partial_{k+1}(r+1,s+1)(\rho,\sigma) = (u_0,\dots,u_{k-1},v_0,\dots,v_s,u_k,\dots,u_r)$$

for $\rho = (u_0,\dots,u_r)$ in $K(r+1)$ and $\sigma = (v_0,\dots,v_s)$ in $K(s+1)$.

Now we define a complex $J(n+1)$ as follows:

$$J(n+1) \subset \prod_{j=0}^{n}[0,2j+1],$$

$$J(n+1) \ni (u_0,\dots,u_n) \quad \text{if} \quad \sum_{i=0}^{j} u_i \leqq 2j+1 \quad \text{and} \quad \sum_{i=0}^{n} u_i = 2n+1$$

and face operators are given by

$$\delta_{k+1}(r+1,s+1)(\rho,\sigma) = (u_0,\ldots,u_{k-1},2v_0,\ldots,2v_s,u_k,\ldots,u_r)$$

where $\rho = (u_0,\ldots,u_r)$ in $\bar{J}(r)$, $r+s = n$, and $\sigma = (v_0,\ldots,v_s)$ in $K(s+1)$ and

$$\delta(t+1;r_0,\ldots,r_t)(\tau;\rho_0,\ldots,\rho_t)$$
$$= (\rho_0',\ldots,\rho_t'),\ \rho_i' = (v_{i,0},\ldots,v_{i,r_i}+u_i)$$

where $\rho_i = (v_{i,0},\ldots,v_{i,r_i})$ in $J(r_i+1)$, $t+r_0+\ldots+r_t = n$ and $\tau = (u_0,\ldots,u_t)$ in $K(t+1)$ (see Figure 1).

Figure 1:

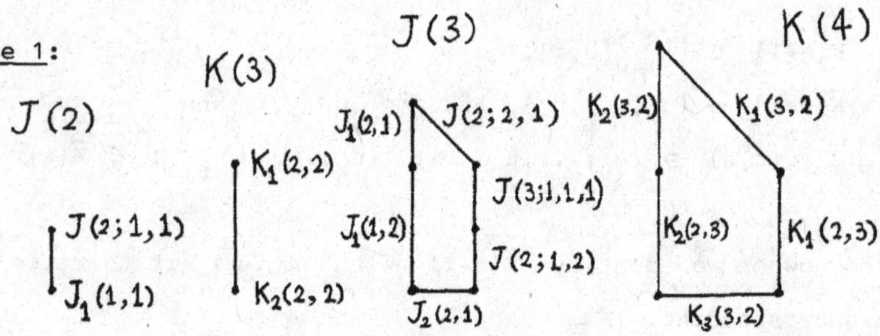

Next we will show the relationship of $J(n)$ with $K(n)$. For any given real number $a \geqq 0$, we can define a complex $\bar{J}(a)$ as follows:

$$\bar{J}(a+1) \subset \prod_{j=0}^{n}[0,j+1] \quad \text{with} \quad n = [a],$$
$$\bar{J}(a+1) \ni (u_0,\ldots,u_n) \quad \text{if} \quad \sum_{i=0}^{j} u_i \leqq a-n+j \quad \text{and} \quad \sum_{i=0}^{n} u_i = a;$$

and its face operators are given by

$$\bar{\delta}_{k+1}(b+1,s+1)(\rho,\sigma) = (u_0,\ldots,u_{k-1},v_0,\ldots,v_s,u_k,\ldots,u_r)$$

where $\rho = (u_0,\ldots,u_r)$ in $\bar{J}(b)$, $b+s = a$, $[b] = r$, $\sigma = (v_0,\ldots,v_s)$ in $K(s+1)$ and

$\bar{\delta}(t+1;a_0+1,\ldots,a_t+1)(\tau;\rho_0,\ldots,\rho_t)$

$= (\rho'_0,\ldots,\rho'_t)$, $\rho'_i = (v_{i,0},\ldots,v_{i,r_i}+u'_i)$ and $u'_i = u_i(n+1-a)$,

where $\rho_i = (v_{i,0},\ldots,v_{i,r_i})$ is in $\bar{J}(b_i+1)$, $[b_i] = r_i$, $b_i-[b_i] =$ $a-[a]$, $t(n+1-a)+a_0+\ldots+a_t = a$ and $\sigma = (u_0,\ldots,u_t)$ is in $K(t+1)$.

We denote the faces by $\bar{J}_k(r,s)$ which is the image of $\bar{\delta}_k(r,s)$ and by $\bar{J}(t+1;b_0+1,\ldots,b_t+1)$ which is the image of $\bar{\delta}(t+1;b_0+1,\ldots,b_t+1)$. We call the last ones upper faces and denote by $\partial_+\bar{J}(a+1)$, the union of all upper faces of $\bar{J}(a+1)$.

<u>Remark</u>. 1) The faces are all homeomorphisms if a is not an integer.

2) $J(n)$ is naturally equivalent to $\bar{J}(n+1/2)$ as complexes.

Under these notations, we have

Proposition 2.1. 1) $K(n+1) \supseteq \bar{J}(a) \supseteq K(n)$ <u>for</u> $[a] = n \geqq 1$ <u>and further</u> $\bar{J}(a) = K(n)$ <u>if</u> a <u>is an integer</u>.

2) <u>Let</u> a <u>be a non-integral real number and</u> $[a] = [b] = n \geqq 1$. <u>Then there is a map</u> $f_{a,b}$ <u>from</u> $\bar{J}(a)$ <u>to</u> $\bar{J}(b)$ <u>such that</u>

$$f_{a,b}\circ\bar{\delta}(t;a_1,\ldots,a_t) = \bar{\delta}(t;b_1,\ldots,b_t),$$
$$f_{a,b}\circ\bar{\delta}_k(a',s) = \bar{\delta}_k(b',s)\circ(f_{a',b'}\times \mathrm{Id}),$$

<u>where</u> b' <u>and</u> b_i <u>are given by</u> $a'-b' = a_i-b_i = a-b$. <u>Moreover the boundaries of</u> $\bar{J}(a)$ <u>and</u> $\bar{J}(b)$ <u>are equivalent as complexes, if</u> a <u>and</u> b <u>are not integers and</u> $[a] = [b] \geqq 1$.

3) <u>Let</u> a <u>be a non-integral real number and</u> $[a] = n \geqq 1$. <u>Then there is a projection</u> $\pi_n : J(n) \to K(n)$ <u>such that</u>

$$\pi_n\circ\delta(n;1,\ldots,1) = \mathrm{Id},$$
$$\pi_n\circ\delta_k(r,s) = \delta_k(r,s)\circ(\pi_r\times \mathrm{Id}).$$

Proof. 1) is clear by the above definitions of $K(n)$'s and $\bar{J}(n)$'s. We show 2). We have to show that there is a cellular map $f_{a,b} : \bar{J}(a) \to \bar{J}(b)$. Since $\bar{J}(a)$ is the union of the upper faces $\partial_+\bar{J}(a')$ for $n \leq a' \leq a$, it suffices to define the map on the upper faces of $\bar{J}(a)$. We define the map by induction on n. In the case $n = 1$, $\bar{J}(a) = \bar{J}(b) =$ one point set and the map is trivial. In general, we define $f_{a,b}$ by $f_{a,b}\cdot\bar{\delta}(t;a_1,\ldots,a_t) = \bar{\delta}(t;b_1,\ldots,b_t)$ on the upper faces and by the hypothesis. The latter identity obtained by the relation of the face operators. 3) is obvious, because $\pi_n = f_{n+1/2,n}$ satisfies the required properties by 1). This implies the proposition. Q.E.D.

Next, we define the degeneracies d_j on $J(n)$. It suffices to define degeneracies on $\partial_+\bar{J}(a)$. So we can define degeneracies by the relation with the upper faces (d-3) below and obtain

Proposition 2.2. $\pi_{n-1}\circ d_j = s_j\circ\pi_n$ for $1 \leq j \leq n$.

The properties of $J(i)$

The complexes $J(i)$ satisfy the following properties (2-a) ~ (2-e):

(2-a) $J(1) = \{*\}$, and $J(i)$ for $i \geq 2$ is affine homeomorphic with I^{i-1}.

(2-b) The boundary $\partial J(i)$ of the complex $J(i)$ is the union of $i(i-1)/2+2^{i-1}-1$ faces

$J_k(r,s)$ for $1 \leq k \leq r$, $1 \leq r \leq i-1$, $r+s = i+1$,

$J(t;r_1,\ldots,r_t)$ for $2 \leq t \leq i$, $r_j \geq 1$, $r_1+\ldots+r_t = i$.

(2-c) Let $r+s-1 = r_1+\ldots+r_t = i$. The faces $J_k(r,s)$ and

$J(t;r_1,\ldots,r_t)$ of $J(i)$ are affine and piecewise affine homeomorphic with $J(r)\times K(s)$ and $K(t)\times J(r_1)\times\ldots\times J(r_t)$ respectively through affine and piecewise affine homeomorphisms:

$$\delta_k(r,s) : J(r)\times K(s) \to J_k(r,s),$$

$$\delta(t;r_1,\ldots,r_t) : K(t)\times J(r_1)\times\ldots\times J(r_t) \to J(t;r_1,\ldots,r_t),$$

which are called face operators. The second face operators are called upper face operators by virtue of this property. The face operators satisfy the following four relations:

(c-1) $$\delta_k(r,s+t-1)\,(\mathrm{Id}\times\partial_j(s,t)) = \delta_{k+j-1}(r+s-1,t)\circ(\delta_k(r,s)\times\mathrm{Id}) : J(r)\times K(s)\times K(t) \to J(r+s+t-2),$$

(c-2) $$\delta_k(r+s-1,t)\circ(\delta_j(r,s)\times\mathrm{Id}) = \begin{cases} \delta_{j+s-1}(r+t-1)\circ(\delta_k(r,t)\times\mathrm{Id})\circ(\mathrm{Id}\times T) & \text{for } k < j \\ \delta_j(r,s+t-1)\circ(\mathrm{Id}\times\partial_{k-j+1}(s,t)) & \text{for } j \leqq k < j+s \\ \delta_j(r+t-1,s)\circ(\delta_{k-s+1}(r,t)\times\mathrm{Id})\circ(\mathrm{Id}\times T) & \text{for } j+s \leqq k \end{cases}$$
$$: J(r)\times K(s)\times K(t) \to J(r+s+t-2),$$

(c-3) for given $(r_1,\ldots,r_t)$ with $\sum_j r_j = i$ and k, let j be the index such that $r_1+\ldots+r_{j-1} < k \leqq r_1+\ldots+r_j$. Then

$$\delta_k(i,s)\circ(\delta(t;r_1,\ldots,r_t)\times\mathrm{Id}) = \delta(t;r_1,\ldots,r_{j-1},r_j+s-1,r_{j+1},\ldots,r_t)\circ(1_A\times\delta_{k-i'}(r_j,s)\times 1_B)\circ T' : K(t)\times J(r_1)\times\ldots\times J(r_t)\times K(s) \to J(i+s-1),$$

where 1_A and 1_B are the identity maps of $A = K(t)\times J(r_1)\times\ldots\times J(r_{j-1})$ and $B = J(r_{j+1})\times\ldots\times J(r_t)$ respectively, $i' = r_1+\ldots+r_{j-1}$ and $T' : A\times J(r_j)\times B\times K(s) \to A\times J(r_j)\times K(s)\times B$ is the map switching factors B and $K(s)$,

(c-4) $$\delta(t+u-1;r_1,\ldots,r_{t+u-1})\circ(\partial_k(t,u)\times\mathrm{Id}) = \delta(t;r_1,\ldots,r_{k-1},i'',r_{k+u},\ldots,r_{t+u-1})\circ(1_C\times$$

$$\delta(u;r_k,\dots,r_{t+u-1})\times 1_D)\circ T''$$

$$: K(t)\times K(u)\times J(r_1)\times\dots\times J(r_{t+u-1}) \to J(i),$$

where $i = r_1+\dots+r_{t+u-1}$, $i'' = r_k+\dots+r_{t+u-1}$, $C = K(t)\times C'$ with $C' = J(r_1)\times\dots\times J(r_{k-1})$, $D = J(r_{k+u})\times\dots\times J(r_{t+u-1})$ and T'' : $K(t)\times K(u)\times C'\times C''\times D \to K(t)\times C'\times K(u)\times C''\times D$ with $C'' = J(r_k)\times\dots\times J(r_{k+u-1})$ is the map switching factors $K(u)$ and C'.

(2-d) The complexes $J(i)$ have degeneracy operators d_j : $J_i \to J_{i-1}$, $1 \leqq j \leqq i$, satisfying the following three relations:

(d-1) $$d_k\circ d_j = \begin{cases} d_{j-1}\circ d_k & \text{for } k < j,\\ d_j\circ d_k & \text{for } k \geqq j,\end{cases}$$

(d-2) $$d_k\circ \delta_j(r,s) : J(r)\times K(s) \to J_i(r,s)$$
$$= \begin{cases} \delta_{j-1}(r-1,s)\circ(d_k\times Id) & \text{for } k < j,\\ \delta_j(r,s-1)\circ(Id\times s_{k-j+1}) & \text{for } j \leqq k < j+s,\ s > 2,\\ pr_1 & \text{for } j \leqq k < j+s,\ s = 2,\\ \delta_j(r-1,s)\circ(d_{k-s+1}\times Id) & \text{for } j+s \leqq k,\end{cases}$$

(d-3) for given k and $(r_1,\dots,r_t)$ with $\sum_a r_a = i$ let j be the index such that $r_1+\dots+r_{j-1} < k \leqq r_1+\dots+r_j$ and put $i' = r_1+\dots+r_{j-1}$. Then

$$d_k\circ\delta(t;r_1,\dots,r_t) : K(t)\times J(r_1)\times\dots\times J(r_t) \to J(t;r_1,\dots,r_t)$$
$$= \begin{cases} \delta(t;r_1,\dots,r_{j-1},r_j-1,r_{j+1},r_{j+1},\dots,r_t)\circ(1_E\times d_{k-i'}\times 1_F) \\ \qquad\qquad \text{for } r_j \geqq 2,\\ \delta(t-1;r_1,\dots,r_{j-1},r_{j+1},\dots,r_t)\circ(s_j\times 1_G)\circ\pi' & \text{for } r_j=1,\ t \geqq 3,\\ pr_2 & \text{for } t=2,\ k=r_1+1=i,\ j=2,\\ pr_3 & \text{for } t=2,\ k=r_1=1,\ j=1,\end{cases}$$

where $E = K(t)\times E'$ with $E' = J(r_1)\times\dots\times J(r_{j-1})$, $F = J(r_{j+1})\times\dots\times J(r_t)$, $G = E'\times F$, pr_t is the projection to the t-th factor and π' : $E\times J(r_j)\times F \to E\times F$ is the natural projection.

(2-e) There is the map $\omega_i : K(i) \to J(i)$ satisfying the following:

$$\omega_i \circ \partial_k(r,s) = \delta_k(r,s) \circ (\omega_r \times Id),$$

$$\begin{array}{ccc} K(r)\times K(s) & \xrightarrow{\omega_r \times Id} & J(r)\times K(s) \\ \partial_k(r,s) \downarrow & & \downarrow \delta_k(r,s) \\ K_k(r,s) & & J_k(r,s) \\ \cap & & \cap \\ K(i) & \xrightarrow{\omega_i} & J(i) \end{array}$$

$$d_j \circ \omega_i = \omega_{i-1} \circ s_j \quad \text{for } 2 \leqq j \leqq i-1,$$

$$\text{Image } \omega_i = \bigcup J(t; r_1, \ldots, r_t),$$

where the union runs over all the upper faces of $J(i)$.

§3. An A_n-map, an A_n-action and an A_n-equivariant map

In this section we introduce the notion of an A_n-map, an A_n-action (see [N]) and an A_n-equivariant map.

An A_n-map

Let X and Y be A_n-spaces.

Definition. An A_n-structure of $f : X \to Y$ consists of a sequence of maps $\{f_k^D\}$, $\{f_k^P\}$, $1 \leqq k \leqq n$ such that

1) $\sigma_{k+1}^Y \circ f_k^D = f_k^D \circ \sigma_{k+1}^X$

$$\begin{array}{ccc} (D^k X, E^k X) & \xrightarrow{f_k^D} & (D^k Y, E^k Y) \\ \sigma_{k+1}^X \downarrow & & \downarrow \sigma_{k+1}^Y \\ (P^k X, P^{k-1} X) & \xrightarrow{f_k^P} & (P^k Y, P^{k-1} Y) \end{array}$$

2) $f = f_1^D|_X$, $f_k^D = f_n^D|_{D^kX}$, $f_k^P = f_n^P|_{P^kX}$.

We have already introduced parameter complexes $J(i)$ for $i \geqq 1$ in the earlier section in order to define the notion "A_n-form" of a map.

Let X and Y be A_n-spaces with A_n-forms $\{M_i^X, i \leqq n\}$, $\{M_i^Y, i \leqq n\}$ respectively.

Definition. An A_n-form of a based map $f : X \to Y$ consists of a family of maps $\{F_i : J(i) \times X^n \to Y;\ i \leqq n\}$ satisfying

(1) $F_1 = f$

(2) $F_i(\delta_k(r,s)(\rho,\sigma);x_1,\ldots,x_i)$
$= F_r(\rho;x_1,\ldots,x_{k-1},m_s^X(\sigma;x_k,\ldots,x_{k+s-1}),x_{k+s},\ldots,x_i)$

(3) $F_i(\delta(t;r_1,\ldots,r_t)(\tau;\rho_1\ldots,\rho_t);x_1,\ldots,x_i)$
$= m_t^Y(\tau;F_{r_1}(\rho_1;x_1,\ldots,x_{r_1}),\ldots,F_{r_t}(\rho_t;x_{r_1+\ldots+r_{t-1}+1},\ldots,x_i))$

(4) $F_i(\gamma;x_1,\ldots,x_{j-1},*,x_{j+1},\ldots,x_i)$
$= F_{i-1}(d_j(\gamma);x_1,\ldots,x_{j-1},x_{j+1},\ldots,x_i)$

Definition. We call the pair $(f,\{F_i, i \leqq n\})$ an A_n-map. The pair $(f,\{F_i\})$ is called an A_∞-map if F_i exists for every i.

Theorem 3.1. <u>A map admits an A_n-structure iff it has an A_n-form</u>.

<u>(Outline of the proof)</u> Firstly we remark that $K(i)$ and the upper faces in $J(i+1)$ are homeomorphic and we can use the upper faces of $J(i+1)$ and d_j instead of $K(i)$ and s_j. Let f have an A_n-form. We define an A_n-structure as follows:

Definition. For σ in $K(i+1)$, $\omega_{i+1}(\sigma) = \delta(t;r_1,\ldots,r_t)(\tau;\rho_1,\ldots,\rho_t)$, we define $f_i^E : E^i(X) \to E^i(Y)$, $f_i^P : P^i(X) \to P^i(Y)$ and $f_i^D : D^i(X) \to D^i(Y)$ as follows:

$$(3.2.1)\quad f_{i-1}^E(\alpha_{i-1}^X(\sigma;x_1,\ldots,x_{i-1})) = \alpha_{t-1}^Y(\tau;F^f(r_1)(\rho_1;x_1,\ldots,x_{r_1}),\ldots,F^f(r_{t-1})(\rho_{t-1};x_{r_1+\cdots+r_{t-2}+1},\ldots,x_{r_1+\cdots+r_{t-1}}))$$

$$(3.2.2)\quad f_{i-1}^P(\beta_i^X(\sigma;x_2,\ldots,x_i)) = \beta_{t-1}^Y(\tau;F^f(r_2)(\rho_2;x_{r_1+1},\ldots,x_{r_1+r_2}),\ldots,F^f(r_{t-1})(\rho_{t-1};x_{r_1+\cdots+r_{t-2}+1},\ldots,x_{r_1+\cdots+r_{t-1}}))$$

$$(3.2.3)\quad f_{i-1}^D(\gamma_i^X(\sigma;x_2,\ldots,x_i)) = \begin{cases} \alpha_{t-1}^Y(\tau;F^f(r_1-1)(d_1(\rho_1);x_2,\ldots,x_{r_1}),\ldots,F^f(r_{t-1})(\rho_{t-1};x_{r_1+\cdots+r_{t-2}+1},\ldots,x_{r_1+\cdots+r_{t-1}})), & r_1 > 1, \\ \gamma_{t-1}^Y(\tau;F^f(r_2)(\rho_2;x_{r_1+1},\ldots,x_{r_1+r_2}),\ldots,F^f(r_{t-1})(\rho_{t-1};x_{r_1+\cdots+r_{t-2}+1},\ldots,x_{r_1+\cdots+r_{t-1}})), & r_1 = 1. \end{cases}$$

We leave the reader to check the well-definedness of this definition. The converse is similar to the proof of Theorem 5 of [S,I]. Q.E.D.

<u>An</u> A_n<u>-action</u> ([N])

We introduce here the idea of higher homotopy (<u>right</u>) action. A higher homotopy left action is defined similarly, so we omit it. Let G be an A_n-space with A_n-structure $\{p_i^G : E^i(G) \to P^{i-1}(G)$ for $i \leq n\}$ and h a based map from G to a space W, where $P^i(G)$ is the G-projective i-space.

Definition. An A_n-structure of a (right) action along the map $h : G \to W$ consists of a sequence of quasi-fibrations $p_i^h : E^i(h) \to P^{i-1}(G)$ with fibre W and relative homeomorphisms $\sigma_i^h : (D^i(h), E^i(h)) \simeq (P^i(G), P^{i-1}(G))$, where $E^1(h) = W$, and $E^i(h)$ and $D^i(h)$ are defined similarly to the definition of $E^i(G)$ and $D^i(G)$ respectively by the relative homeomorphisms

$$\alpha_{i+1}^h : (K(i+2)\times W\times G^i, \partial K(i+2)\times W\times G^i \cup K(i+2)\times W\times G^{[i]}) \to (E^{i+1}(h), E^i(h)),$$

$$\gamma_{i+1}^h : (K(i+2)\times G^i, \partial K(i+2)\times G^i \cup K(i+2)\times G^{[i]}) \to (D^i(G), E^i(G)),$$

and σ_i^h is obtained by the projection.

<u>Remark</u>. 1) $D^i(h)$ is not contractible in general.

2) The fibre of p_i^h is W.

3) Let $h : H \to G$ be a map between A_∞-spaces. If h is an A_∞-map, $E^\infty(h)$ can be identified with the homotopy fibre of the map $Bh : BH \to BG$. We may write $E^\infty(h)$ by G/H, if further h is an injective A_∞-homomorphism.

Let $\{m_i^G, i \leqq n\}$ be the A_n-form of G.

Definition. An A_n-form of a (right) action along $h : G \to W$ consists of a family of maps $\{N_i^f : K(i)\times W\times G^{i-1} \to W;\ i \leqq n\}$ satisfying

(1) $N_1 = Id$,

(2) $N_i(\partial_j(r,s)(\rho,\sigma);w;g_2,\dots,g_i)$
$= N_r(\rho;w;g_2,\dots,g_{j-1},m_s^G(\sigma;g_j,\dots,g_{j+s-1}),g_{j+s},\dots,g_i)$,

(2)' $N_i(\partial_1(r,s)(\rho,\sigma);w;g_2,\dots,g_i)$
$= N_r(\rho;N_s(\sigma;w;g_2,\dots,g_s),g_{s+1},\dots,g_i)$,

(3) $N_i(\tau;w;g_2,\dots,g_{j-1},*,g_{j+1},\dots,g_i)$
$= N_{i-1}(s_j(\tau);w;g_2,\dots,g_{j-1},g_{j+1},\dots,g_i)$,

(3)' $N_i(\tau;*;g_2,\dots,g_i) = h\circ m_{i-1}^G(s_1(\tau);g_2,\dots,g_i)$.

Definition. We call the pair $(h, \{N_i, i \leqq n\})$ a (right) A_n-action along h. The pair $(h, \{N_i\})$ is called a (right) A_∞-action along h if N_i exists for all i.

From a similar argument to [S,I], it follows

Theorem 3.3. <u>An</u> A_n<u>-structure of an action along</u> h <u>is equivalent to an</u> A_n<u>-action along</u> h.

<u>An</u> A_n<u>-equivariant map</u>

Let H and K be A_n-spaces and $f : H \to K$ be an A_n-map with the A_n-structure $\{f_i^P, f_i^D; i \leqq n\}$. We assume that there are A_n-structures of actions $\{p_i^h, \sigma_i^h\}$ on V along h and $\{p_i^k, \sigma_i^k\}$ on W along k. Let $F : V \to W$ be a based map with $F \circ h = k \circ f$.

Definition. An A_n-structure of $F : V \to W$ along f consists of a sequence of maps $\{F_i^D\}$, $1 \leqq i \leqq n$ such that

1) $f_i^P \circ \sigma_{i+1}^h = \sigma_{i+1}^k \circ F_i^D$

$$\begin{array}{ccc} D^i(h) & \xrightarrow{F_i^D} & D^i(k) \\ {\scriptstyle \sigma_{i+1}^h}\downarrow & & \downarrow{\scriptstyle \sigma_{i+1}^k} \\ P^i(H) & \xrightarrow{f_i^P} & P^i(K) \end{array}$$

2) $F = F_1^D|_V$, $F_i^D = F_n^D|_{D^i(h)}$.

Let $\{N_i^h, i \leqq n\}$ (and $\{N_i^k, i \leqq n\}$) be the A_n-action along h (and along k, resp.) and $\{F_i^f\}$ the A_n-form of f.

Definition. An A_n-form of the map $F : V \to W$ consists of a family of maps $\{R_i : J(i)\times V\times H^{i-1} \to W;\ i \leqq n\}$ satisfying

(1) $R_1 = F$

(2) $R_i(\delta_j(r,s)(\rho,\sigma);v;g_2,\ldots,g_i)$
$= R_r(\rho;v;g_2,\ldots,g_{j-1},m^H_s(\sigma;g_j,\ldots,g_{j+s-1}),g_{j+s},\ldots,g_i)$

(2)' $R_i(\delta_1(r,s)(\rho,\sigma);v;g_2,\ldots,g_i)$
$= R_r(\rho;N^h_s(\sigma;v;g_2,\ldots,g_s),g_{s+1},\ldots,g_i)$

(3) $R_i(\delta(t;r_1,\ldots,r_t)(\tau;\rho_1\ldots,\rho_t);v;g_2,\ldots,g_i)$
$= N^k_t(\tau;R_{r_1}(\rho_1;v;g_2,\ldots,g_{r_1}),F^f_{r_2}(\rho_2;g_{r_1+1},\ldots,g_{r_1+r_2}),\ldots$
$F^f_{r_t}(\rho_t;g_{r_1+\ldots+r_{t-1}+1},\ldots,g_i))$

(4) $R_i(\gamma;v;g_2,\ldots,g_{j-1},*,g_{j+1},\ldots,g_i)$
$= R_{i-1}(d_j(\gamma);v;g_2,\ldots,g_{j-1},g_{j+1},\ldots,g_i)$

(4)' $R_i(\gamma;*;g_2,\ldots,g_i) = k\circ F^f_{i-1}(d_1(\gamma);g_2,\ldots,g_i)$

Definition. We call the triple $(F,f,\{R_i,\ i \leqq n\})$ an A_n-equivariant map along f. The triple $(F,f,\{R_i\})$ is called an A_∞-equivariant map along f if R_i exists for all i.

By a same argument as above, we have

Theorem 3.4. <u>An A_n-structure of a map along f is equivalent to an A_n-equivariant map along it</u>.

§4. Fundamental properties

We indicate an outline of the proof of the properties.

P1) We deform the A_n-forms by the homotopy of maps.

P2) The composition of the A_n-structures gives the A_n-structure of the composition of the A_n-maps.

P3) By using π_n, we can define A_n-forms for an A_n-homomorphism.

P4) As the localization functor is continuous (by [I3]), the localization map is an A_n-homomorphism and the localization of an A_n-map is again an A_n-map.

P5) Let (f,g) be the homotopy equivalence pair with f an A_n-map, given homotopies $f \circ g \simeq \mathrm{Id}$ and $g \circ f \simeq \mathrm{Id}$. Through these homotopies, we can define an A_n-form for g.

P7) The homotopy fibre of the A_n-structure of an A_n-map gives the A_n-structure of the homotopy fibre of the A_n-map.

P6) This is a corollary of P7)

P8) We obtain this directly by the definition.

P9) It suffices to show the existence of the A_n-structure for a given map, if its adjoint has an extension to the X-projective n-space. We can construct it by the homotopy extension property of $(D^i(X), E^i(X))$ and by the homotopy lifting property of the principal fibration $p_i^X : E^i(X) \to P^{i-1}(X)$.

P10) Using the A_n-form of f (or g), we can define an n-form for Y similarly to the case when $n = 2$.

§5. Some applications

An A_n-primitive space

Let X be an A_n-space of finite type such that

$$H^*(X;\mathbb{Q}) \cong E(x_1,\ldots,x_r) \otimes P[y_1,\ldots,y_s]$$

with $\deg(x_i) = 2n_i - 1$, $n_1 \leqq \cdots \leqq n_r$ and $\deg(y_j) = 2m_j$, $m_1 \leqq \cdots \leqq m_s$. Then as is well known,

$$X \simeq_0 \prod S^{2n_i - 1} \times \prod K(\mathbb{Z}, 2m_j).$$

Recall from [I2, Theorem B]

(5.1) Let X be a finite CW-complex having an A_n-form. There is a homotopy equivalence $h : X_{(0)} \simeq \prod S^{2n_i-1}_{(0)}$ which is an A_{n-1}-map.

Definition. An A_n-space X is A_n-primitive if h is an A_n-map.

Definition. An element x in $H^*(X;Q)$ is A_n-primitive if there is an element y in $H^*(P^nX;Q)$ such that $s^*x = \iota_2^* \cdots \iota_n^* y$, where $\iota_i : P^{i-1} \hookrightarrow P^i$ is a natural inclusion and s^* is the suspension isomorphism.

Proposition 5.2. X <u>is an</u> A_n-<u>primitive iff</u> $H^*(X;Q)$ <u>is generated by</u> A_n-<u>primitive elements</u>.

<u>Notation</u>. $\mathfrak{A}_n$ = {finite A_n-spaces}

$\mathfrak{A}'_n$ = {finite A_n-primitive spaces}

By the definitions and by the above proposition we have the inclusions

(5.3) $\mathfrak{A}_2 \supseteq \mathfrak{A}'_2 \supseteq \mathfrak{A}_3 \supseteq \cdots \supseteq \mathfrak{A}'_{n-1} \supseteq \mathfrak{A}_n \supseteq \mathfrak{A}'_n \supseteq \cdots$

<u>Example</u>. The seven sphere S^7 is an example of $\mathfrak{A}'_2 \supsetneqq \mathfrak{A}_3$.

Recall that whether or not $\mathfrak{A}_2 \supsetneqq \mathfrak{A}'_2$ is still open problem.

Let $(X;\{m_i, i \leq n\})$ be a finite A_n-space. For the dimensional reasons, we have

Proposition 5.4. *If, for any* $i \leqq r$ *and any* r-*tuple* $\{\alpha_j; \emptyset \neq \alpha_j \subseteq \{1,\dots,r\}, 1 \leqq j \leqq r\}$, *we have* $2n_i \neq (n-1) + \sum_{j=1}^{r} |\alpha_j|$ *with* $|\alpha_j| = \sum(2n_t - 1)$ *where* t *ranges over* α_j, *then* $(X,\{m_i, i \leqq n\})$ *is in* $\mathfrak{A}'_n$.

Corollary 5.5. *Let* $G(m) = SU(m)$ *for* $d = 2$, $Sp(m)$ *for* $d = 4$. *If* $d(m-1) \leqq 4n$, *then* $G'(m) = (G(m), m_n)$ *is in* $\mathfrak{A}'_n$ *for any* n-*form* m_n. *So*, $P^nG'(m)$ *is rationally equivalent to* $P^nG(m)$.

Proposition 5.6 (Counter examples).

1) *If* $d(m-1) > 4n$, *then there is an* n-*form* m'_n *of* $G(m)$ *such that* $(G(m), m'_n)$ *is not in* $\mathfrak{A}'_n$

2) *If* $m \geqq 2n+1$, *then there is an* n-*form* m'_n *of* $S^3 \times S^3 \times SU(m)$ *such that* $(S^3 \times S^3 \times SU(m), m'_n)$ *is not in* $\mathfrak{A}'_n$.

In the proof of these proposition we need the following facts.

Let $(Y,\{m_i, i \leqq n\})$ be an A_n-space $(n \geqq 3)$.

Notation. $A_n(X;\{m_i\}) = \{m'_n : K_n \times X^n \to X \ ; \{m_i, i < n$ and $m'_n\}$ is an A_n-form of $X\}$

Definition. For any n-forms m'_n and m''_n in $A_n(X;\{m_i\})$, $m'_n \sim_{A_n} m''_n$ iff there is an n-form $F(n) : J_n \times X^n \to X$ of the identity map 1_X such that $\{m_i \circ (\pi_i \times 1_X \times \dots \times 1_X), i < n$ and $F(n)\}$ is an A_n-form of the identity 1_X.

By the fundamental property P5), we have

Proposition 5.7. *The relation* $\sim_{A_n}$ *is an equivalence relation*.

Lemma 5.8. $\mathbf{A}_n(X,\{m_i\})/\sim_{A_n}$ <u>is in the one to one correspondence with</u> $\pi_0\mathbf{A}_n(X,\{m_i\})$.

<u>(Outline of the proof)</u> Since 1_X is an A_{n-1}-homomorphism, the obstruction for the existence of an n-form of 1_X is deformed to be the obstruction for the existence of the homotopy between two n-forms m_n and m_n'.

The latter obstruction is classified by the set $[S^{n-2}\wedge X\wedge\ldots\wedge X,X]$. So we have

Theorem 5.9. $\mathbf{A}_n(X,\{m_i\})/\sim_{A_n} \cong [S^{n-2}\wedge X\wedge\ldots\wedge X,X]$.

Localization and Zabrodsky's theorem

Zabrodsky constructs in [Z2] an example of a finite A_{p-1}-space which does not have an A_p-form for each prime $p \geqq 3$. It seems, however, that his construction needs more precise arguments.

Let Π be the set of all primes and $\mathbf{P}$ be a subset of Π.

Definition. A space X or a map f admits a mod $\mathbf{P}$ A_n-form (or A_n-structure) iff $X_{\mathbf{P}}$ or $f_{\mathbf{P}}$ admits an A_n-form (or A_n-structure).

Let $\Pi = \coprod_i \mathbf{P}_i$ (a finite partition).

Proposition 5.10. 1) X <u>is an</u> A_n-<u>primitive iff</u> X <u>is a</u> mod $\mathbf{P}_i$ A_n-<u>primitive for all</u> i,

2) f <u>is an</u> A_n-map <u>iff</u> f <u>is a</u> mod $\mathbf{P}_i$ A_n-<u>map for all</u> i.

Proposition 5.11 (Mixing homotopy types) (see [MNT]). <u>If</u> X_i <u>are</u> mod $\mathbf{P}_i$ A_n-<u>space</u> $(i \geqq 1)$ <u>such that</u> $(X_i)_{(0)}$ <u>is</u> A_n-<u>equivalent to</u> $2n_i-1$

$\prod S_{(0)}$. Then there exists an A_n-space X such that X is mod P_i A_n-homotopy equivalent to X_i.

Proposition 5.12 (see [MNT]). In the category of connected complexes,

1) X is an A_n-space iff X is an A_n-space mod P_i for each i and the rationalization $X_{P_i} \to X_{(0)}$ induces an equivalent A_n-structure on $X_{(0)}$.

2) A map f between A_n-spaces is an A_n-map iff f is an A_n-map mod P_i for each i and the rationalization of f_{P_i} induces an equivalent A_n-structure on $f_{(0)}$.

Using these propositions, we have

Theorem 5.13 (Zabrodsky). For every prime $p \geq 3$ there exists a finite CW-complex which admits an A_{p-1}-structure but no A_p-structure.

§6. The sphere extension of a complex Lie group

A Lie group G often acts transitively on a sphere and is regarded as a total space of the fibre bundle over the sphere [B]. Then the Lie group G is called a sphere extension of the isotropy subgroup G_0. Let us consider a new sphere extension $\overline{G}$ of G_0 induced by a map f on spheres. Then $\overline{G}$ is no longer a Lie group, in general. Zabrodsky [Z] shows, however, that $\overline{G}$ is an H-space, provided that the map degree deg(f) of f is prime to 2. It is natural to ask when $\overline{G}$ is a group-like space, in other words, when it is an A_3-space. If the bundle projection $e : G \to S^{2n-1}$ were an A_3-map, it would be trivial. But it is not true in general. Y. Hemmi [H] shows the following

Theorem. Let n be a positive integer not dividing $2\cdot 3$* and let G be $U(n)$, $SU(n)$ or $Sp(m)$ $(n = 2m)$ acts on the odd sphere S^{2n-1}. Then $\deg(f)$ is prime to 6, provided that $\overline{G}$ is an A_3-space.

We discuss the existence problem of a higher homotopy associativity of $\overline{G}$ in this section. Firstly by a simple computation of the action of the mod p Steenrod algebra $\mathbf{A}(p)$ on the mod p cohomology algebra of the projective spaces, we have a generalization of the above theorem.

Theorem 6.1. Let p be a prime number and n a positive integer not dividing $(p-1)\cdot p$* and let G be $U(n)$, $SU(n)$ or $Sp(m)$ $(n = 2m)$ acting on the odd sphere S^{2n-1}. Then $\deg(f)$ is prime to p, provided that $\overline{G}$ is a mod p A_p-space and $\overline{f}$ is a mod p A_p-map.

Corollary 6.2. Let n be prime to $p!$. Then $\deg(f)$ is prime to $p!$, provided that $\overline{G}$ is an A_p-space and $\overline{f}$ is an A_p-map.

(Outline of the proof)

Assuming that $(\deg(f), p) \neq 1$, we deduce a contradiction. By the hypothesis on $\overline{G}$ and $\overline{f}$, the map $\overline{f}$ induces a homomorphism F between the spectral sequences $E_k(\overline{G})$ and $E_k(G)$ of the Stasheff type. We remark that $\overline{G}$ and G have torsion free integral cohomologies and $E_k(\overline{G})$ has also a torsion free integral cohomology for $k \leqq p$. By comparing the spectral sequences, we obtain

Proposition 6.3. $\overline{G}$ is A_p-primitive.

Then the mod p cohomology of $P^k(\overline{G})$ is the direct sum of a polynomial ring and a nilpotent ideal S_k (see [I2]). Let R be the quotient algebra of the mod p cohomology of $P^k(\overline{G})$ by the ideal generated by the image of $\overline{f}^*$ and the nilpotent ideal S_k. By the hypothesis, $R = \mathbf{Z}/p\mathbf{Z}[v]/(v^{p+1})$ with $\deg(v) = 2n$ must be an $\mathbf{A}(p)$-algebra. On the other hand, by the Adem relation, we have

Lemma 6.4. Let $R = \mathbf{Z}/p\mathbf{Z}[v]/(v^{p+1})$ with $\deg(v) = 2n$. Then R can not be an $\mathbf{A}(p)$-algebra unless n divides $(p-1)\cdot p^*$.

It is a contradiction and the proof of Theorem 6.1 is completed.

Q.E.D.

Our main goal of this section is the following

Theorem 6.5. Let G be a compact complex Lie group complex-linearly and transitively acting on the odd sphere S^{2n-1}. Then $\overline{G}$ is an A_k-space A_{k-1}-acting on the sphere and the map $\overline{f}$ covering f is an A_k-map, if the degree deg(f) of f is prime to k!. Moreover $\overline{f}$ preserves A_{k-1}-action in a homotopical sense (see Section 3 and also [N] for the definition of A_{k-1}-action).

Corollary 6.6. Let n be a positive integer not dividing $2\cdot 3^*$. Then $\overline{G}$ is a homotopy associative H-space iff deg(f) is prime to 6.

Remark that the conclusion is equivalent to that $\overline{G}$ is an A_4-space and $\overline{f}$ is an A_4-map.

We use the following method: For the unitary group U(n-1),

taking the equivariant localization of the base space S^{2n-1} and the map f, we get an equivariant A_k-space X and an equivariant A_k-map F. By the obstruction theory for an A_k-space and an A_k-map, we can show that this space has a higher homotopy associative equivariant structure for the total space G and the sphere map has also a higher homotopy associative equivariant structure. We often call the higher homotopy associative equivariant structure the A_k-equivariant structure for some k.

Decomposition of the equivariant A_n-action of $U(n)$

We work in the category of (strictly) $U(n-1)$-equivariant spaces and maps. Let p be an odd prime and $\mathbf{P}$ the set of primes $\geqq p$. Recall that the unitary group $U(n)$ acts complex linearly and transitively on the odd sphere $S^{2n-1} \subset \mathbf{C}^n$. Therefore S^{2n-1} is an equivariant based space whose fixed point set by any subgroup H is always a sphere of odd dimension $2n(H)-1$; $2n-1 \geqq 2n(H)-1 > 0$. So we can consider the equivariant localization X (and F) of S^{2n-1} (and a map $f : S^{2n-1} \to S^{2n-1}$ continuously, resp.) (see [I3]). Let $C_2(X)$ be the double (associative) loop space of the double reduced suspension of X. Then $C_2(X)$ is an equivariant A_∞-space by the first loop structure of the double loop.

Recall that the unitary group $U(n)$ is a left equivariant group by the conjugate action of $U(n-1)$ and is also a right equivariant space by the product from the right. We denote by $U(n)\times_{U(n-1)}S^{2n-1}$ the equivariant product of the right equivariant space $U(n)$ and the left equivariant space S^{2n-1}. Then S^{2n-1} admits an equivariant (strict) action of $U(n)$ by the equivariant map from $U(n)\times_{U(n-1)}S^{2n-1}$ to S^{2n-1} along the projection $e : U(n) \to S^{2n-1}$. On the other hand, X is equivariantly mod $\mathbf{P}$ equivalent to S^{2n-1}. Therefore $U(n)\times_{U(n-1)}X$ is mod $\mathbf{P}$

equivariantly equivalent to $U(n)\times_{U(n-1)}S^{2n-1}$ and X admits an A_∞-action of $U(n)$ (see [I1]) by the equivariant A_∞-form, a tuple of k-forms from $K_k\times U(n)\times_{U(n-1)}\cdots\times_{U(n-1)}U(n)\times_{U(n-1)}X$ to X.

Proposition 6.7. <u>There is an equivariant homotopy action</u> T <u>of</u> $U(n)$ <u>on</u> X <u>satisfying the following two conditions</u>:

(1) <u>The</u> A_∞<u>-action of</u> $U(n)$ <u>on</u> X <u>is</u> $U(n-1)$<u>-equivariantly equivalent to</u> $e'(g) \oplus T(g,x)$ <u>in</u> $C_2(X)$ <u>for</u> $x \in X$, $g \in U(n)$, <u>where</u> $\oplus$ <u>means the associative loop product</u>,

(2) $T(h,x) = hx$, $T(gh,x) = T(g,hx)$ <u>and</u> $T(hg,x) = hT(g,x)$, <u>for</u> $x \in X$, $g \in U(n)$, $h \in U(n-1)$,

<u>where</u> $e' = j\circ e$, j <u>is the inclusion of</u> X <u>into</u> $C_2(X)$.

We inductively construct the k-form of the homotopy action of $U(n)$ on $C_2(X)$ decomposed by the A_{k-1}-form of the homotopy action T. We may assume that the first loop structure of the double loop space $C_2(X)$ is equivariantly associative. We deform the A_k-form of the action in $C_2(X)$ to be decomposed by the A_k-form of T' given by $T'(g,w)(s) = T(g,w(s))$.

The action on X of $U(n)$ is homotopy equivalent to $e'(g) \oplus T'(g)(w)$ for all w in $C_2(X)$. The key lemma of the main theorem is described as follows:

Lemma 6.8. T <u>is an</u> A_{p-1}<u>-action with a k-form</u> N_k^T : $K_k\times U(n)\times_{U(n-1)}\cdots\times_{U(n-1)}U(n)\times_{U(n-1)}X \to X$ <u>of</u> $U(n)$ <u>on</u> X <u>for</u> $k < p$ <u>and there is an</u> A_∞<u>-action</u> $N_k^{e'}$: $K_k\times U(n)\times_{U(n-1)}\cdots\times_{U(n-1)} U(n)\times_{U(n-1)}C_2(X) \to C_2(X)$ <u>of</u> $U(n)$ <u>on</u> $C_2(X)$ <u>equivariantly equivalent to the usual action of</u> $U(n)$ <u>on</u> X <u>in</u> $C_2(X)$; <u>for given</u> $(\sigma;g_1,\ldots,g_{k-1};w)$ <u>in</u> $K_k\times\prod^{k-1}_{U(n-1)}U(n)\times_{U(n-1)}C_2(X)$, <u>the</u> A_∞<u>-action</u> $N_k^{e'}$ <u>has the form</u>:

$$(6.9)\quad N_k^{e'}(\sigma;g_1,\dots,g_{k-1};w)$$
$$=\bigoplus_{i=1}^{k-1} N_i^{T'}(s_{i+1}\dots s_{k+1}(\sigma);g_1,\dots,g_{i-1};e'(g_i))$$
$$\oplus\ N_k^{T'}(\sigma;g_1,\dots,g_{k-1};w).$$

Proof. We construct inductively the A_k-forms R_k^j and $N_k^{T'}$ of the map j along the identity and the action along T' respectively by the following formulae:

(1) $R_1^j(*;x) = j(x)$,

(2) $R_k^j(\delta_j(r,s)(\rho,\sigma);g_1,\dots,g_{k-1};x)$
$= R_r^j(\rho;g_1,\dots,g_{j-1},g_j\dots g_{j+s-1},g_{j+s},\dots,g_{k-1};x)$,

(2)' $R_k^j(\delta_r(r,s)(\rho,\sigma);g_1,\dots,g_{k-1};x)$
$= R_r^j(\rho;g_1,\dots,g_{r-1};g_r\dots g_{k-1}x)$,

(3) $R_k^j(\delta(t;r_1,\dots,r_t)(\tau;\rho_1,\dots,\rho_t);g_1,\dots,g_{k-1};x)$
$= \bigoplus_{i=1}^{t-1} N_i^{T'}(s_{i+1}\dots s_t(\tau);g_1\dots g_{r_1},\dots,g_{n_{i-2}+1}\dots g_{n_{i-1}};$
$R_{r_i}^j(\rho_i;g_{n_{i-1}+1},\dots,g_{n_i}))$
$\oplus\ N_t^{T'}(\tau;g_1\dots g_{r_1},\dots,g_{n_{i-2}+1}\dots g_{n_{i-1}};$
$R_{r_t}^j(\rho_t;g_{n_{t-1}+1},\dots,g_{k-1};x))$, where $n_i = r_1 + \dots + r_i$,

(4) $R_k^j(\gamma;g_1,\dots,g_{j-1},*,g_{j+1},\dots,g_{k-1};x)$
$= R_{k-1}^j(d_j(\gamma);g_1,\dots,g_{j-1},g_{j+1},\dots,g_{k-1};x)$,

(4)' $R_k^j(\gamma;g_1,\dots,g_{k-1};*) = e'(g_1\dots g_{k-1})$.

(Outline of the proof of Theorem 6.5)

Since X is mod $\mathbf{P}$ equivariantly equivalent to S^{2n-1}, we may identify X with S^{2n-1}. Recall that $e : G \to X$ is a fibration and $e' = j\circ e : G \to C_2(X)$. Let $\bar{e}$ be the induced fibration of e by f and let $f' = j\circ f$. Then $e\circ\bar{f} = f\circ\bar{e}$ and $e'\circ\bar{f} = f'\circ\bar{e}$.

We prove Theorem 6.5 through the inverse process of Lemma 6.8. Let $Y = C_2(X)\times U(n)$ and $\bar{E} = \langle f'\circ\bar{e},\bar{f}\rangle : \bar{G} \to Y$ which we may assume to be an inclusion. Then Y is the A_∞-space with the following A_k-form m_k^Y:

$$(6.10)\quad pr_1\circ m_k^Y(\tau;y_1,\dots,y_k) = \bigoplus_{i=1}^{k} N_i^{T'}(s_{i+1}\dots s_k(\tau);g_1,\dots,g_{i-1});x_i),$$

$$pr_2\circ m_k^Y(\tau;y_1,\dots,y_k) = g_1\dots g_k \quad \text{for} \quad y_j = (x_j,g_j).$$

By using the equivariant obstruction theory, we obtain that $\overline{G}$ is an A_k-space A_k-acting on X and the map f along $\overline{f}$ preserves A_k-actions by induction on $k < p$, namely,

Proposition 6.11. <u>There are</u> A_{p-1}<u>-forms</u> $m_k^{\overline{G}}$, $F_k^{\overline{E}}$, $N_k^{\overline{e}}$ <u>and</u> $R_k^{f'}$ <u>of the space</u> $\overline{G}$, <u>the map</u> $\overline{E}$, <u>the action along</u> $\overline{e}$ <u>and the map</u> f' <u>preserving action along the homomorphism</u> $\overline{f}$ <u>respectively satisfying the following formulae:</u>

(1) $R_1^{f'}(*;x) = j(x)$,

(2) $R_k^{f'}(\delta_j(r,s)(\rho,\sigma);\overline{g}_1,\dots,\overline{g}_{k-1};x)$
$= R_r^{f'}(\rho;\overline{g}_1,\dots,\overline{g}_{j-1},m_s^{\overline{G}}(\sigma;\overline{g}_j,\dots,\overline{g}_{j+s-1}),\overline{g}_{j+s},\dots,\overline{g}_{k-1};x)$,

(2)' $R_k^{f'}(\delta_r(r,s)(\rho,\sigma);\overline{g}_1,\dots,\overline{g}_{k-1};x)$
$= R_r^{f'}(\rho;\overline{g}_1,\dots,\overline{g}_{r-1};N_s^{\overline{e}}(\sigma;\overline{g}_r,\dots,\overline{g}_{k-1};x))$,

(3) $R_k^{f'}(\delta(t;r_1,\dots,r_t)(\tau;\rho_1,\dots,\rho_t);\overline{g}_1,\dots,\overline{g}_{k-1};x)$
$= \bigoplus_{i=1}^{t-1} N_i^{T'}(s_{i+1}\dots s_t(\tau);m_{r_1}^{\overline{G}}(\rho_1;\overline{g}_1,\dots,\overline{g}_{r_1}),\dots;$
$R_{r_i}^{f'}(\rho_i;\overline{g}_{n_{i-1}+1},\dots,\overline{g}_{n_i}))$
$\oplus N_t^{T'}(\tau;m_{r_1}^{\overline{G}}(\rho_1;\overline{g}_1,\dots,\overline{g}_{r_1}),\dots;R_{r_t}^{f'}(\rho_t;\overline{g}_{n_{t-1}+1},\dots,\overline{g}_{k-1};x))$,
where $n_i = r_1+\dots+r_i$,

(4) $R_k^{f'}(\gamma;\overline{g}_1,\dots,\overline{g}_{j-1},*,\overline{g}_{j+1},\dots,\overline{g}_{k-1};x)$
$= R_{k-1}^{f'}(d_j(\gamma);\overline{g}_1,\dots,\overline{g}_{j-1},\overline{g}_{j+1},\dots,\overline{g}_{k-1};x)$,

(4)' $R_k^{f'}(\gamma;\overline{g}_1,\dots,\overline{g}_{k-1};*) = e'\circ F_{k-1}^{\overline{f}}(\gamma;g_1,\dots,g_{k-1})$,

(5) $pr_1\circ F_k^{\overline{E}}(\gamma;\overline{g}_1,\dots,\overline{g}_k) = R_k^{f'}(\gamma;\overline{g}_1,\dots,\overline{g}_{k-1};x_k)$

(6) $pr_2\circ F_k^{\overline{E}}(\gamma;\overline{g}_1,\dots,\overline{g}_k) = g_1\dots g_k$ for $\overline{g}_j = (x_j,g_j)$.

Remark. 1) The A_k-form $R_k^{f'}$ gives the A_k-form R_k^f of A_k-equivariant map f (along $\overline{f}$) by using the equivariant compression theory.

2) These homotopy actions are left homotopy actions.

This completes the proof of Theorem 6.5.

REFERENCES

[B] Borel, A.: Some remarks about Lie groups transitive on spheres and tori. Bull. of A. M. S. 55, 580-587(1949)

[H] Hemmi, Y.: Mod 3 homotopy associative finite H-spaces and sphere extensions of classical groups, preprint

[HM] Hubbuck, J., Mimura, M.: Certain p-regular H-spaces. Arch. Math. 49, 79-82(1987)

[I1] Iwase, N.: On the ring structure of $K^*(XP^n)$. Master thesis in Kyushu Univ., (1982) (in Japanese)

[I2] Iwase, N.: On the K-ring structure of X-projective n-space. Mem. Fac. Sci. Kyushu Univ. Ser. 2 38, 285-297(1984)

[I3] Iwase, N.: Equivariant localization and completion as a continuous functor. preprint

[MNT] Mimura, M., Nishida, G.,Toda, H.: Localization of CW-complexes and its applications. J. Math. Soc. Japan 23, 593-624(1971)

[N] Norlan, R. A.: A_n-actions on fibre spaces. Indiana Univ. Math. J. 21, 285-313(1971)

[S] Stasheff, J. D.: Homotopy associativity of H-spaces, I, II. Trans. Amer. Math. Soc. 108, 275-292, 293-312(1963)

[Z1] Zabrodsky, A.: On construction of new finite CW H-spaces. Inventiones. Math. 16, 260-266(1972)

[Z2] Zabrodsky, A.: Homotopy associativity and finite CW complexes. Topology, 9, 121-128(1970)

HOMOTOPY APPROXIMATIONS FOR CLASSIFYING SPACES OF COMPACT LIE GROUPS

Stefan Jackowski and James E. McClure*

Instytut Matematyki
Uniwersytet Warszawski
PKiN, IX p.
00-901 Warszawa, Poland

Department of Mathematics
University of Kentucky
Lexington, KY 40506
U.S.A.

1. Introduction.

Let G be a compact Lie group. It is a well-known and useful fact that the map $BNT \to BG$, where NT is the normalizer of a maximal torus, is a mod p cohomology isomorphism whenever p does not divide the order of the Weyl group. This paper arose from a desire to know what happens when p does divide $|W|$. The first example of a satisfactory answer to this question was the remarkable observation, due to Dwyer, that the diagram

$$\begin{array}{ccc} B(N\Gamma \cap NT) & \rightarrow & BNT \\ \downarrow & & \downarrow \\ BN\Gamma & \rightarrow & BSO(3) \end{array} \tag{1.1}$$

is a homotopy pushout at the prime 2 (see [5, section 4]). Here Γ is the diagonal copy of $Z_2 \oplus Z_2$ in $SO(3)$, its normalizer $N\Gamma$ is the "octahedral" subgroup of order 24, and $N\Gamma \cap NT$ is dihedral of order 8. The same argument gives similar diagrams for BS^3 and $BU(2)$ in which one replaces the subgroups of $SO(3)$ by their inverse images under the quotient maps $S^3 \to SO(3)$ and $U(2) \to SO(3)$. These diagrams have been applied by Dwyer, Miller and Wilkerson [5] to prove the uniqueness of BS^3 and $BSO(3)$ at $p=2$, i.e., to show that the 2-completions of these spaces are uniquely determined by their mod 2 cohomology algebras over the Steenrod algebra. One can also use the diagrams to get a new proof of Mislin's celebrated classification [10] of the self-maps of BS^3 and a generalization to $BSO(3)$. Unfortunately, there is no general method for obtaining such a diagram for a given G, or even of guessing what it should look like. We shall give a condition (Corollary 3.4, or more generally Theorem 3.1) which when it is satisfied allows one to write BG at the prime p as an explicit homotopy colimit of classifying spaces of subgroups of G. This condition is unfortunately quite difficult to verify in practice, but we are able to get one new example, namely the homotopy pushout diagram

---

*Partially supported by NSF grant.

$$\begin{array}{ccc} B(NT\cap U(2)) & \rightarrow & BU(2) \\ \downarrow & & \downarrow \\ BNT & \rightarrow & BSU(3) \end{array} \qquad (1.2)$$

at $p=2$. Here $U(2)$ is the subgroup of $SU(3)$ consisting of matrices with a 2 by 2 block in the upper left positions and a single element in the lower right. This diagram can be applied to the uniqueness and classification problems for $BSU(3)$, as we shall show in a later paper, where we shall also give a pushout decomposition for $BSU(3)$ at the prime 3 which apparently cannot be obtained by our present method.

The first author would like to thank the University of Virginia for its hospitality.

2. A general point of view and some examples.

The reader may have concluded from what we have said so far that the "diagrams" we seek will always be homotopy pushouts. On the contrary, we think that the use of more general types of homotopy colimits will turn out to be both necessary (for most choices of G) and natural. The existence of homotopy pushout diagrams in the cases considered so far is due to special simplifying features of these cases (see section 5). Our objective, roughly speaking, is to approximate BG by gluing together copies of BK for certain subgroups K of G while taking into account the relations of conjugacy and inclusion between these subgroups. Formally, this means we shall consider homotopy colimits over subcategories of the "orbit category" of G.*

For the reader unfamiliar with general homotopy colimits, perhaps the simplest definition is as follows. Let C be a category with object set O and let $F: C \longrightarrow \mathrm{Top}$ be a covariant functor (so that F gives a "diagram with shape C" in the category of topological spaces). Then the homotopy colimit $\operatorname{hocolim}_C F$ is the bar construction $B(O,C,F)$ as defined in [9, section 12]. Explicitly, $\operatorname{hocolim}_C F$ is the geometric realization of the simplicial space $B_*(O,C,F)$ whose n-simplices have the form $[f_1,\ldots,f_n]x$, where $f_1, \ldots, f_n$ is a composable sequence of arrows in C and $x \in F(\text{domain } f_n)$, and for which the face and degeneracy maps have the usual form (cf. [9, section 7]). If C is a topological category, one takes this structure into account in topologizing the space of n-simplices. A good example to consider is the following: if C has a single object and the endomorphisms of this object form a group G then a covariant functor $F: C \longrightarrow \mathrm{Top}$ is essentially the same thing as a left G-space X and $\operatorname{hocolim}_C F$ is homeomorphic to $EG \times_G X$.

The *orbit category* $O(G)$ is a topological category defined as follows: its objects are the orbits G/H, and the space of morphisms $G/H \rightarrow G/K$ is the set of equivariant maps with the compact-open topology. The space of morphisms may be described more explicitly by using the fact that a G-map is determined by its effect on the identity coset: there is a homeomorphism

*At this point we should mention the related program of Friedlander and Mislin (see [6] and its sequels). Their construction allows one to write BG at a given prime as an explicit telescope of classifying spaces of finite groups. The main difficulty which arises in their work is in relating these finite groups back to G itself, a difficulty we seek to avoid by working entirely with subgroups of G.

$$\mathrm{Hom}(G/H, G/K) \to \{g \in G \mid g^{-1}Hg \subset K\}/K$$

taking f to $f(eH)$. This description confirms the claim made earlier that $O(G)$ encodes the relations of conjugacy and inclusion between the subgroups of G. We shall make use of two functors on $O(G)$: the inclusion functor I: $O(G) \subset \mathrm{Top}$ and the functor $EG \times_G I$. Note that $EG \times_G (G/G) = BG$ and that $EG \times_G (G/H) \simeq BH$.

By an *approximation to BG at the prime p* we mean a mod p cohomology isomorphism

$$\operatorname*{hocolim}_{C} EG \times_G I \to BG$$

for some subcategory C of $O(G)$. In looking for such approximations it is often convenient to use the following definition and lemma.

Definition 2.1. A subcategory C of $O(G)$ is *mod p dense* if the space $\operatorname*{hocolim}_{C} I$ is mod p acyclic.

Lemma 2.2. If C is mod p dense then the natural map

$$\operatorname*{hocolim}_{C} EG \times_G I \to EG \times_G * = BG$$

is a mod p cohomology isomorphism.

Proof. The functor $EG \times_G$ - is easily seen to commute with homotopy colimits, and it is well-known that it preserves mod p cohomology isomorphisms.

An obvious question is whether $O(G)$ itself is mod p dense. This is a special case of the following (cf. [15, Theorem 3.1]).

Lemma 2.3. Any full subcategory C of $O(G)$ containing the object G/e is mod p dense for all p.

Proof. There is a simplicial homotopy from the identity map of $B_*(O,C,I)$ to a constant map defined as follows:

$$h_i[f_1, \ldots, f_n]gH = [f_1, \ldots, f_i, f_{i+1} \cdot \cdots \cdot f_n \cdot \overline{g}, 1, \ldots, 1]eH$$

for $0 \le i \le n$, where $\overline{g}: G/e \to G/H$ is the map taking a to agH.

Attractive as this result is at first sight, we shall make use of it only by pointing out its deficiencies. For example, if C has the single object G/e then $\operatorname*{hocolim}_{C} I$ is just EG and the approximation obtained from 2.2 and 2.3 is $EG \times_G EG$, which does not differ from BG itself in any useful way. This is the simplest case of 2.3 and adding more objects to C does not improve the situation. Another way of saying what is wrong is that too much of the structure of BG is showing up in the morphism spaces of C (note that $\mathrm{Hom}(G/e, G/e) \cong G$) and not enough in the pieces $EG \times_G G/H$ (note that $EG \times_G G/e \simeq *$). A third objection, related to the second, is that the cohomological analysis of a homotopy colimit over C via the Bousfield-Kan spectral sequence (see section 4) is available only when the morphism sets of C are discrete. Motivated by these considerations, we shall confine our attention from now on to subcategories C which are both mod p dense and *discrete*. Of course, we would really like to have a general way of deducing what C should be from the structure of G (e.g. from its root system), but we have no idea how to do this at present so we merely list some examples.

Example 2.4. If p does not divide $|W|$ let C be the full subcategory of $O(G)$ with the single object G/T. Then C is discrete: indeed, $\mathrm{Hom}(G/T, G/T)$ can be identified with W by letting $w(gT) = gn^{-1}T$, where $w \in W$ is represented by $n \in NT$. Then $\underset{C}{\mathrm{hocolim}}\, I$ is homeomorphic to $EW \times_W G/T$, and this in turn is homotopic to $(G/T)/W = G/NT$ since W acts freely on G/T. It is well-known that G/NT is mod p acyclic, and so C is mod p dense. Note that the approximation to BG obtained from this example is

$$\underset{C}{\mathrm{hocolim}}\, EG \times_G I = EG \times_G (EW \times_W G/T) \simeq EG \times_G G/NT \simeq BNT,$$

so we have essentially recovered the "well-known fact" mentioned in the first sentence of the introduction.

Example 2.5. Let $G = SO(3)$ and $p = 2$. Let C be the full subcategory of $O(G)$ with objects G/Γ and G/NT (notations as in the introduction). Then $\mathrm{Hom}(G/NT, G/NT)$ contains only the identity map, while matrix calculations show that $\mathrm{Hom}(G/\Gamma, G/\Gamma)$ and $\mathrm{Hom}(G/\Gamma, G/NT)$ may be identified respectively with Σ_3 and $Z_2 \backslash \Sigma_3$ in such a way as to be compatible with compositions. In particular C is discrete. We shall give two different proofs, in sections 3 and 5, that C is mod 2 dense, and we shall see in section 5 that this example implies diagram 1.1.

We can generate further examples from example 2.5 by using the quotient maps $S^3 \to SO(3)$ and $U(2) \to SO(3)$ and the following elementary observation, whose proof is immediate.

Lemma 2.6. An epimorphism $G \to G'$ induces an imbedding of $O(G')$ as a full subcategory of $O(G)$. If $C \subset O(G')$ is mod p dense then so is its image in $O(G)$.

For example, if we let $\overline{\Gamma} \subset S^3$ be the *binary octahedral* group which covers $\Gamma \subset SO(3)$ then the full subcategory of $O(S^3)$ with the objects $S^3/\overline{\Gamma}$ and S^3/NT is discrete and mod 2 dense.

Example 2.7. Let $G = SU(3)$ and $p = 2$. Let C be the full subcategory of $O(G)$ with objects G/T and $G/U(2)$. Matrix calculation shows that C is discrete and in fact that it is abstractly isomorphic to the category of example 2.5. We shall show in section 3 that C is mod 2 dense, and we shall see in section 5 that this example implies diagram 1.2.

One way of obtaining explicit approximations to BG.

We begin by stating a sufficient condition for a full subcategory C of $O(G)$ to be mod p dense. Fix a prime p and a maximal torus T of G, and for any G-space X let $C(X)$ denote the full subcategory of $O(G)$ whose objects are the orbits of X.

Theorem 3.1. Let X be a G-space such that

(i) X is finite dimensional,

(ii) X^H is mod p acyclic whenever H is an isotropy subgroup of X,

(iii) X^H is mod p acyclic whenever H is a subgroup of T, and

(iv) $C(X)$ is discrete.

Then $C(X)$ is mod p dense.

The proof will be given in section 4. For the moment we concentrate on the problem of finding a suitable X.

Proposition 3.2. Let V be an orthogonal representation of G and let SV be its unit sphere. Let $f : SV \to SV$ be an equivariant map such that p divides the degree of f^H whenever H is either the trivial subgroup or an isotropy subgroup of SV. Let X be the telescope of the system

$$SV \xrightarrow{f} SV \xrightarrow{f} SV \to \cdots$$

Then X satisfies hypotheses (i), (ii), and (iii) of 3.1 and $C(X) = C(SV)$.

Proof. First observe that X^H is the telescope of the system

$$SV^H \xrightarrow{f^H} SV^H \to \cdots.$$

It follows at once that $C(X) = C(SV)$ and that X^H is mod p acyclic if H is either the trivial subgroup or an isotropy subgroup. It remains to show that X^H is mod p acyclic for all $H < T$. First recall that $\deg f = \deg f^T$ (this is an easy and standard consequence of Smith theory). But T/H is also a torus and it acts on SV^H with fixed-point set SV^T so we have $\deg f^H = \deg (f^H)^{T/H} = \deg f^T = \deg f$. This completes the proof.

In order to apply 3.2 we need to be able to know when equivariant self-maps of SV exist with specified degrees on the fixed-point sets. Fortunately, this problem has been completely solved by Tornehave. In order to explain his result, we must first recall some facts from [4] concerning the Burnside ring $A(G)$. Additively, $A(G)$ is free abelian on generators $[G/H]$ as H ranges over any set of representatives for the conjugacy classes of subgroups which have finite index in their normalizers. For each $K < G$ there is a map $\phi_K : A(G) \to Z$ taking $[G/H]$ to the Euler characteristic $\chi((G/H)^K)$.. $A(G)$ is isomorphic to the ring of stable equivariant self-maps of S^0 by an isomorphism which has the property that if one stabilizes a map $f: SV \to SV$ then the corresponding element a of $A(G)$ has $\phi_K a = \deg f^K$ for all $K < G$. Tornehave's theorem is an unstable analogue of this isomorphism. (Although Tornehave's paper is written with the assumption that G is finite, the obstruction-theoretic proof of the result in question works in general; see page 290 of [13]).

Theorem 3.3. [13, Theorem A]. Let V be an orthogonal representation of G and let $a \in A(G)$. There is an equivariant map $f: SV \to SV$ with $\deg f^H = \phi_H a$ for all H if and only if a satisfies

(i) $\phi_H a = 1$ whenever SV^H is empty

(ii) $\phi_H a \in \{-1,0,1\}$ whenever $SV^H = S^0$

(iii) $\phi_H a = \phi_K a$ whenever $SV^H = SV^K$

Combining what has been said so far, we have

Corollary 3.4. Let V be an orthogonal representation of G with $C(SV)$ discrete. Suppose that there is an element a of $A(G)$ such that

(i) $\phi_H a = 1$ whenever SV^H is empty

(ii) $\phi_H a \in \{-1,0,1\}$ whenever $SV^H = S^0$

(iii) $\phi_H a = \phi_K a$ whenever $SV^H = SV^K$, and

(iv) $\phi_H a$ is a multiple of p whenever H is either the trivial subgroup or an isotropy subgroup of SV.

Then $C(SV)$ is mod p dense.

We do not have a general method for finding a suitable choice for V and a. The following examples have been found by trial and error. In verifying them, we shall need to know that the map

$$\mathrm{Hom}(G/H, G/K) \to (G/K)^H$$

taking f to $f(eH)$ is a bijection.

Example 3.5. Let $G = SO(3)$ and let V be the space of 3×3 symmetric real valued matrices with trace zero. Let $SO(3)$ act on V by conjugation. A standard theorem of linear algebra says that every orbit of this $SO(3)$ action contains a diagonal matrix. The isotropy group of a diagonal matrix in SV is either Γ (if all three diagonal entries are distinct) or a conjugate of NT (if two of them are equal). Thus $C(SV)$ has, up to isomorphism, only the objects G/Γ and G/NT. Let $a = [G/G] - [G/NT]$. We claim that V and a satisfy the hypotheses of Corollary 2.4 with $p = 2$. First observe that $|(G/NT)^{NT}| = |\mathrm{Hom}(G/NT, G/NT)| = 1$ and that $|(G/NT)^{\Gamma}| = |\mathrm{Hom}(G/\Gamma, G/NT)| = 3$ (see example 2.5). Thus (iv) is satisfied. Next observe that $(SV)^H$ is the set of all points of SV whose isotropy subgroup contains H. In particular, if

$(SV)^H = (SV)^K$ then the sets $\{g \mid H \subset gNTg^{-1}\}$ and $\{g \mid K \subset gNTg^{-1}\}$ are the same, from which it follows at once that $\mathrm{Hom}(G/H, G/NT) = \mathrm{Hom}(G/K, G/NT)$, and hence that $\phi_H a = \phi_K a$. Thus (iii) is satisfied. Similarly, if $(SV)^H$ is empty then so is $\mathrm{Hom}(G/H, G/NT)$, showing (i), and if $(SV)^H = S^0$ then $(SV)^H = (SV)^K$ for some conjugate K of NT, showing (ii). We can now conclude from Corollary 3.4 that $C(SV)$ is mod 2 dense, as required for example 2.5.

We remark that the example just given does not actually require the use of Theorem 3.3, since the necessary G-map $SV \to SV$ in this case been explicitly constructed by Floyd (see [3]).

Example 3.6. Let $G = SU(3)$ and let V be the space of 3×3 Hermitian matrices with zero trace. Let $SU(3)$ act by conjugation. Again, every orbit contains a diagonal matrix, and in this case we conclude that every object of $C(SV)$ is isomorphism to either G/T or $G/U(2)$. Let $a = [G/G] - [G/U(2)]$. We claim that V and a satisfy the hypotheses of Corollary 3.4. The verification is exactly parallel to that of the previous example, and we conclude that $C(SV)$ is mod 2 dense as required for example 2.6.

4. Lim ^{i} and Bredon cohomology

In this section we give the proof of Theorem 3.1. In order to calculate $H^*(\operatorname{hocolim}_C I; Z_p)$ we shall use the Bousfield-Kan spectral sequence [1, XII.4.5]

$$\lim_C{}^i H^j(I(\ );Z_p) \Rightarrow H^{i+j}(\operatorname{hocolim}_C I; Z_p);$$

note that C must be discrete for this to be applicable. Theorem 3.1 will be an immediate consequence of this spectral sequence and our next result.

Proposition 4.1. Let X be as in Theorem 3.1. Let $C = C(X)$. Then

a) $\lim_C{}^i H^j(I(\ );Z_p) = 0$ for $i > 0$.

b) $\lim_C H^j(I(\ );Z_p) = H^j(*;Z_p)$.

We begin the proof of the proposition by translating it into a result about Bredon cohomology (cf. section 2 of Mislin's paper [10]). First we need some basic facts which may be found in Willson's article [14]. A *Bredon cohomology theory* is an equivariant cohomology theory for which the cohomology of every orbit G/H is concentrated in dimension zero; this is the equivariant dimension axiom. The restriction of the theory to the orbit category $O(G)$ is a functor $M: O(G) \to Ab$ and the theory is determined by M; this is the equivariant uniqueness theorem. The homotopy axiom for the theory implies that M must factor through the homotopy category $hO(G)$, i.e. the category whose morphism sets are the connected components of those of $O(G)$. Conversely, any functor $M: hO(G) \to R$-mod extends uniquely to a Bredon cohomology theory, denoted $H_G^*(X;M)$, which takes its values in R-mod.

If C is any subcategory of $O(G)$ then the functors from hC to R-mod form an abelian category in which one can define Ext groups $\operatorname{Ext}^i_{hC}(M,N)$. These groups can be used to calculate $H^*_G(X;M)$ in the following way.

Theorem 4.2. [14, Theorem 6.3 and Theorem 7.3]. Let X be a G-space. For each $j \geq 0$ define a functor $\underline{X}_j : hO(G) \to R$-mod by

$$\underline{X}_j(G/K) = H_j(X^K/W_0K;Z_p)$$

where W_0K is the unit component of NK/K. If C is any full subcategory of $O(G)$ containing $C(X)$ there is a spectral sequence

$$\operatorname{Ext}^i_{hC}(\underline{X}_j \mid hC, M \mid hC) \Longrightarrow H^{i+j}_G(X;M).$$

Theorem 4.2 gives a relation between $\lim^i$ and Bredon cohomology. From now on we fix $R = Z_p$.

Corollary 4.3. Let X be a finite-dimensional G-space and let C be a full subcategory of $O(G)$ containing $C(X)$. Suppose that X^K is mod p acyclic whenever $G/K \in C$. Then

$$\lim_{hC}{}^i M \cong H^i_G(X;M)$$

for every functor $M : hC \to Z_p$-mod.

Proof. Let $G/K \in C$. Since X^K is finite dimensional and mod p acyclic we may apply the Conner conjecture [12] to see that $H_*(X^K/W_0K;Z_p) = H_*(*;Z_p)$. In particular $\underline{X}_j = 0$ for $j > 0$ and the spectral sequence of 4.2 collapses to give $\operatorname{Ext}^i_{hC}(\underline{X}_0,M) \cong H^i_G(X;M)$. But $\underline{X}_0$ is the constant functor with value Z_p, so that $\operatorname{Hom}_{hC}(\underline{X}_0,M) = \lim_{hC} M$, and thus the derived functors $\operatorname{Ext}^i_{hC}(\underline{X}_0,M)$ are the same as $\lim_{hC}{}^i M$. This completes the proof.

Note that the assumptions that X is finite dimensional and that R is Z_p were used only to show that $\underline{X}_j(G/K) = H^j(*;R)$. If one has some other way of showing this, for example if G is finite, then these assumptions may be avoided.

We can now give the promised translation of Proposition 4.1 into Bredon cohomology: it is equivalent to our next result.

Proposition 4.4. Let X be as in Theorem 3.1. Let $M^j : hO(G) \to Z_p$-mod be the functor taking G/K to $H^j(G/K;Z_p)$. Then

a) $H^i_G(X;M^j) = 0$ for $i > 0$

b) $H^0_G(X;M^j) = H^j(*;Z_p)$.

The proof of 4.4 will occupy the rest of this section. First we observe that part (a) implies part (b). For there is an equivariant "Atiyah-Hirzebruch" spectral sequence

$$H^i_G(X;h^j(-)) \Rightarrow h^{i+j}(X)$$

for any equivariant theory h (see [14, Theorem 3.1]). Now ordinary nonequivariant mod p cohomology $H^*(X;Z_p)$ satisfies the axioms for an equivariant cohomology theory (but not, of course, the dimension axiom unless G is finite). Letting $h^*(X)=H^*(X;Z_p)$, we have $h^j(\ )=M^j$ and the spectral sequence just mentioned specializes to give

$$H^i_G(X;M^j) \Rightarrow H^{i+j}(X;Z_p).$$

If 4.4 (a) holds then the spectral sequence collapses to give $H^0_G(X;M^j)\cong H^j(X;Z_p)$. But $H^j(X;Z_p)\cong H^j(*;Z_p)$ by hypothesis (iii) of 3.1, showing part (b).

To prove 4.4(a) we need two lemmas. Let N_pT be the inverse image in NT of a p-Sylow subgroup of the Weyl group.

Lemma 4.5. Let X be any G-space and define $\bar{M}^j : hO(N_pT) \to Z_p$-mod by $\bar{M}^j(-)=M^j(G\times_{N_pT} -)$. Then there is a natural inclusion $H^*_G(X;M^j)\subset H^*_{N_pT}(X;\bar{M}^j)$.

Lemma 4.6. Let X be as in 3.1 and let $M : hO(N_pT) \to Z_p$-mod be any contravariant functor. Then $H^i_{N_pT}(X;M)=0$ for $i>0$.

Taken together, these clearly imply 4.4(a). The proof of 4.5 is essentially a standard transfer argument, but to carry it out we need some facts from [8] and [7].

Fix an orbit G/H and observe that the Pontrjagin-Thom construction gives rise to a map $S^V \to S^V \wedge G/H^+$ for any V in which G/H imbeds (see [8, Lemma IV.2.2]). This in turn gives rise to a map τ: $S^0 \to \Sigma^\infty(G/H^+)$ in the equivariant stable category $\bar{h}GS$. (The reader who has not yet read all of [8] may feel the urge to do so now). Composing this with the projection π: $G/H \to *$ we get an equivariant stable map $S^0 \to S^0$ which we shall denote by $\chi_G(G/H)$. If X is a G-space we get a map

$$\tau\wedge 1: \Sigma^\infty_G X^+ \to \Sigma^\infty_G(G/H^+\wedge X^+).$$

Now if h is any equivariant cohomology theory *defined on* $\bar{h}GS$ we obtain a natural transformation, called the transfer, from $h^*(G/H\times X)$ to $h^*(X)$. The composite of this map with the projection $(\pi\times 1)^*$ is a natural transformation $h^*X \to h^*X$ of equivariant cohomology theories, and in particular it is an isomorphism for all X if it is an isomorphism for all orbits G/K, that is, if the map

$$\chi_G(G/H)^*: h^*(G/K) \to h^*(G/K)$$

is an isomorphism for all K.

By the *stable orbit category*, which we shall denote by $O^s(G)$, we mean the full subcategory of $\bar{h}GS$ whose objects are the orbits G/K (or rather their stabilizations $\Sigma^\infty_G(G/K^+)$). A

Mackey functor is a functor $M: O^s(G) \to Ab$. It is shown in [7] that any such functor extends uniquely to a cohomology theory on $\bar{h}GS$ which satisfies the dimension axiom; this theory is denoted $H^*_G(X;M)$. One therefore has a transfer $H^*_G(G/H \times X;M) \to H^*_G(X;M)$, and the composite of transfer with projection is an isomorphism of $H^*_G(X;M)$ for all X if and only if the map $\chi_G(G/H)^*: M(G/K) \to M(G/K)$ is an isomorphism for all K.

Next observe that each functor M^j defined in Proposition 4.4 is a Mackey functor, since it is the composite

$$O^s(G) \to \bar{h}GS \to \bar{h}S \to Z_p\text{—mod};$$

here the first map is the inclusion, the second is the forgetful functor from the equivariant to the nonequivariant stable category, and the third map is $H^j(\ ;Z_p)$. In this case the map $\chi_G(G/H)^*$ is the composite

$$H^j(G/K;Z_p) \to H^j(G/H \times G/K;Z_p) \to H^j(G/K;Z_p)$$

of π^* and τ^*. Now an inspection of the definitions shows that τ^* is the standard Becker-Gottlieb transfer for the trivial bundle $G/H \times G/K \to G/K$, and thus the endomorphism $\chi_G(G/H)^*$ of $M^j(G/K)$ is just multiplication by the ordinary, nonequivariant Euler characteristic $\chi(G/H)$. In particular, if $H = N_pT$ then $\chi(G/H)$ is prime to p and we can conclude that

$$(\pi \times 1)^*: H^*_G(X;M^j) \to H^*_G(G/N_pT \times X;M^j)$$

is a monomorphism for all X. We have now finished with the equivariant stable category and may sink back to the level of ordinary G-spaces.

To complete the proof of Lemma 4.5 we need only recall the natural isomorphism

$$H^*_G(G \times_H X;M) \cong H^*_H(X;M(G \times_H -))$$

which holds for any H-space X and any functor $M: hO(G) \to Ab$. This isomorphism is a special case of the uniqueness theorem for Bredon cohomology mentioned earlier, since both sides are H-equivariant Bredon theories and they have the same effect on orbits H/K. If X is in fact a G-space then $G \times_H X$ is G-homeomorphic to $G/H \times X$. In particular, we have

$$H^*_G(G/N_pT \times X;M^j) \cong H^*_{N_pT}(X;\bar{M}^j).$$

Combining this with the conclusion of the preceding paragraph gives Lemma 4.5.

We now turn to the proof of Lemma 4.6. Let X be as in Theorem 3.1 and let K be any closed subgroup of N_pT. Then K is an extension

$$0 \to H \to K \to Q \to 1,$$

where $H < T$ and Q is a finite p-group. By hypothesis (iii) of 3.1 we have $H^*(X^H;Z_p) \cong H^*(*;Z_p)$. Since X^H is finite dimensional we may apply Smith theory [2, III.5.2] to see that $H^*((X^H)^Q;Z_p) \cong H^*(*;Z_p)$. But $(X^H)^Q = X^K$ so we conclude that each X^K is mod p acyclic. We can now apply Corollary 4.3 with $G = N_pT$ to conclude that

$$\lim_{hC}{}^i M \cong H^i_{N_pT}(X;M)$$

with C equal to all of $O(N_p T)$. But $O(N_p T)$ contains a terminal object $N_p T/N_p T$. Hence the functor $\lim_{hC}$ is exact, since $\lim_{hC} M = M(N_p T/N_p T)$. But then $\lim_{hC}^i M = 0$ for $i > 0$, and so $H^i_{N_p T}(X;M) = 0$ for $i > 0$ as required for Lemma 4.6.

5. Converting certain homotopy colimits to homotopy pushouts

There is obviously a close connection between examples 2.5 and 2.6 respectively and diagrams 1.1 and 1.2. In this section we make this connection precise. For this purpose let C be a category with the following properties.

(i) C has two objects a and b.

(ii) Hom (a,a) is a group A and Hom (b,b) is a group B.

(iii) Hom (b,a) is empty.

Let $F: C \to \text{Top}$ be any contravariant functor. Observe that $F(a)$ is a left A-space and $F(b)$ is a left B-space. We define an $A \times B$-action on the set $S = \text{Hom}(a,b)$ by letting $(\phi,\psi)s = \phi s \psi^{-1}$. The next result and its proof were shown to us by Bill Dwyer.

Proposition 5.1. The space $\operatorname{hocolim}_C F$ is weakly homotopy equivalent to the homotopy pushout of the diagram

$$\begin{array}{ccc} E(A\times B)\times_{A\times B}(S\times F(a)) & \xrightarrow{\pi_b} & EB\times_B F(b) \\ \downarrow \pi_a & & \\ EA\times_A F(a) & & \end{array}$$

where the map π_b is defined by the projection $A \times B \to B$ and the evaluation map $S \times F(a) \to F(b)$, and π_a is defined by the projections $A \times B \to A$ and $S \times F(a) \to F(a)$.

We begin with a general construction which allows an arbitrary category C to be replaced by a more complicated one. The *twisted arrow category* of C, denoted $Ar^t(C)$, is the category in which the objects are the arrows $X \to Y$ in C and the morphisms from $X \to Y$ to $X' \to Y'$ are the diagrams

$$\begin{array}{ccc} X & \to & Y \\ \downarrow & & \uparrow \\ X' & \to & Y'. \end{array}$$

(Note the direction of the map $Y' \to Y$). There is an obvious functor $f\colon Ar^t(C) \to C$ which assigns to every arrow $X \to Y$ its source X.

Lemma 5.2. For any functor $F\colon C \to \mathrm{Top}$ there is a homotopy equivalence

$$\operatorname*{hocolim}_{Ar^t(C)} F \cdot f \xrightarrow{\simeq} \operatorname*{hocolim}_{C} F$$

Proof. By [16, 9.4] it suffices to show that the functor f is "right cofinal", i.e. that for every object $c \in C$ the category $c \downarrow f$ is contractible. For this purpose it suffices to construct a functor $g\colon c \downarrow f \to c \downarrow f$ and two natural transformations $g \to I$ and $g \to c$, where I is the identity functor and c is the evident constant functor. The objects of the category $c \downarrow f$ can be identified with the diagrams $c \to X \to Y$ and the morphisms in $c \downarrow f$ can be identified with the commutative diagrams

$$\begin{array}{ccccc} & \nearrow & X & \to & Y \\ c & & \downarrow & & \uparrow \\ & \searrow & X' & \to & Y'. \end{array}$$

We define the functor g by $g(c \xrightarrow{\alpha} X \xrightarrow{\beta} Y) = c \xrightarrow{1} c \xrightarrow{\beta\alpha} Y$. The natural transformation $g \to I$ and $g \to c$ are given by the following diagram.

$$\begin{array}{ccccc} & \nearrow & c & \xrightarrow{1} & c \\ & & \uparrow 1 & & \downarrow \beta\alpha \\ c & \xrightarrow{1} & c & \xrightarrow{\beta\alpha} & Y \\ & & \downarrow \alpha & & \uparrow 1 \\ & \searrow & X & \xrightarrow{\beta} & Y \end{array}$$

This concludes the proof.

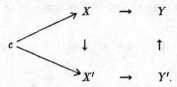

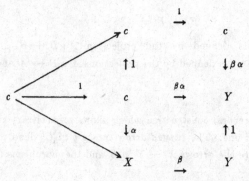

Next let P be the category with three objects v_0, v_1, and v_2 and two nontrivial morphisms, one from v_0 to v_1 and one from v_0 to v_2 (thus a homotopy colimit over P is a homotopy pushout). Let C be a category satisfying assumptions (i), (ii) and (iii) at the beginning of this section. Then there is a functor $G: Ar^t(C) \to P$ defined on objects by

$$G(a \to a) = v_1,\ G(a \to b) = v_0,\ G(b \to b) = v_2.$$

By [16, 9.8] there is a homotopy equivalence

$$\operatorname*{hocolim}_{Ar^t(C)} \bar{F} \cong \operatorname*{hocolim}_{P} G_* \bar{F}$$

for any functor $\bar{F}: Ar^t(C) \to \text{Top}$. Here $G_* \bar{F}$ denotes the homotopy Kan extension of $\bar{F}$ along G, so that $G_* \bar{F}(v_i) = \operatorname*{hocolim}_{G \downarrow v_i} (\bar{F} |(G \downarrow v_i))$.

We can now complete the proof of 5.1. It only remains to identify the diagram $G_* \bar{F}$ on P, where $\bar{F} = F \cdot f$, with the diagram in the statement of 5.1. We shall show that these diagrams agree on the objects v_0, v_1 and v_2 and leave it to the reader to worry about their effect on morphisms.

First of all, the category $G \downarrow v_0$ is isomorphic to the full subcategory of $Ar^t(C)$ whose set of objects is the set S of morphisms $a \to b$ in C. The functor $\bar{F} |(G \downarrow v_0)$ associates to each map $a \to b$ the space $F(a)$. Since all morphisms in $G \downarrow v_0$ are isomorphisms, it is easy to see that there is an equivalence

$$\operatorname*{hocolim}_{G \downarrow v_0} \bar{F} |(G \downarrow v_0) \simeq E(A \times B) \times_{A \times B} (S \times F(a))$$

Next, the category $G \downarrow v_2$ is isomorphic to the full subcategory of $Ar^t(C)$ whose objects are the morphisms $a \to b$ and $b \to b$. This subcategory, in turn, has a right cofinal full subcategory whose objects are the arrows $b \to b$. It follows that there is an equivalence

$$\operatorname*{hocolim}_{G \downarrow v_2} \bar{F} |(G \downarrow v_2) \simeq EB \times_B F(b).$$

Finally, the category $G \downarrow v_1$ is isomorphic to the full subcategory of $Ar^t(C)$ whose objects are the morphisms $a \to b$ and $a \to a$. As in the previous case, the subcategory whose objects are the arrows $a \to a$ is right cofinal and we obtain an equivalence

$$\operatorname*{hocolim}_{G \downarrow v_1} \bar{F} |(G \downarrow v_1) \simeq EA \times_A F(a).$$

References

1. Bousfield, A.K. and Kan, D.M., *Homotopy limits, completions and localizations,* Lecture Notes in Math. Vol. 304, Springer 1972.
2. Bredon, G., *Introduction to compact transformation groups,* Academic Press, 1972.

3. Conner, P.E. and Montgomery, D., *An example for SO*(3), Proc. Nat. Acad. Sci. U.S.A. **48** (1962), 1918-1922.

4. tom Dieck, T., *Transformation Groups and Representation Theory,* Lecture Notes in Math. Vol 766, Springer 1979.

5. Dwyer, W., Miller, H., and Wilkerson, C., *The homotopic uniqueness of BS^3, Preprint.*

6. Friedlander, E. and Mislin, G., *Locally finite approximations of Lie groups I.,* Inventiones Math. **83** (1986), 425-436.

7. Lewis, L.G., May, J.P., and McClure, J.E., *Ordinary RO(G)-graded cohomology,* Bull. Amer. Math. Soc. **4** (1981), 208-212.

8. Lewis, L.G., May, J.P., and Steinberger, M., *Equivariant Stable Homotopy Theory,* Lecture Notes in Math. Vol. 1213, Springer, 1986.

9. May, J.P., *Classifying spaces and fibrations,* Memoirs Amer. Math. Soc. **155** (1986).

10. Mislin, G., *The homotopy classification of self-maps of infinite quaterionic projective space,* Quarterly J. of Math. Oxford **38** (1987), 245-257.

11. Oliver, R., *Fixed-point sets of group actions on finite acyclic complexes,* Comment. Math. Helv. **50** (1975), 155-177.

12. Oliver, R., *A proof of the Conner conjecture,* Annals of Math. **103** (1976), 637-644.

13. Tornehave, J., *Equivariant maps of spheres with conjugate orthogonal actions,* Canadian Math. Soc. Conference Proceedings Vol. 2, Part 2, 1982, 275-301.

14. Willson, S.J., *Equivariant homology theories on G-complexes,* Trans. Amer. Math. Soc. **212** (1975), 155-171.

[15] Miller, H., *The Sullivan conjecture,* Bull. Amer. Math. Soc. **9** (1983), 75-79.

[16] Dwyer, W., and Kan, D.M., *A classification theorem for diagrams of simplicial sets,* Topology **23** (1984), 139-155.

Cyclic Homology and Characteristic Classes of Bundles with Additional Structures

MAX KAROUBI

This short paper is just a report on some work relating cyclic homology and characteristic classes. The proofs and more details may be found in the references at the end of this paper. I hope that the informal style adopted here will help a better understanding.

0. Dictionary. The functor which associates to any C^∞ vector bundle V over a manifold X its space of sections $E = \Gamma(V)$ is in fact an equivalence of categories

$$\mathcal{E}(X) \longrightarrow \mathcal{P}(A)$$

Here $\mathcal{E}(X)$ is the category of vector bundles over X and $\mathcal{P}(A)$ is the category of finitely generated projective A-modules with $A = C^\infty(X)$ (Serre-Swan's theorem). To a large extent, the classical Chern-Weil theory on $\mathcal{E}(X)$ can be extended in a Chern-Weil theory on $\mathcal{P}(A)$ for any k-algebra A where k is an arbitrary commutative ring(*) (but A not necessarily commutative). One of the purposes of cyclic homology is to accomplish this goal [C] [K1]. In particular, the generalization of the classical Chern character

$$K(X) \longrightarrow H^{\text{even}}(X)$$

will be

$$K_0(A) \longrightarrow H_{\text{even}}(A)$$

where H_{even} is essentially cyclic homology. The following dictionary gives more examples of translating a geometrical concept into an algebraic one.

Space X	Ring A
Locally compact space X	C^*-algebra
Manifold $X+$ ordinary differential calculus in $\Omega^* X$	Ring $A+$ differential graded algebra $\Omega_*(A)$ with $A = \Omega_0 A$ and A dense subalgebra of a C^*-algebra
Integration of smooth forms $\int$	Graded trace on $\Omega_*(A)$
Differential forms $\Omega^* X$	Hochschild homology $H_*(A, A)$
De Rham cohomology of forms $H^*(X)$	Cyclic homology $H_*(A)$ or $HC_*(A)$
Vector bundle V	Finitely generated projective module E

(*) We are assuming $\mathbf{Q} \subset k$ for sake of simplicity—see [K3] in general.

Connection $D : \Gamma(V) \longrightarrow \Gamma(V \otimes T^*X)$	Connection $D : E \longrightarrow E \underset{A}{\otimes} \Omega_1 A$
Matrix $\Gamma_{ij}^k = < D_{e_i} e_j,\ dx^k >$	Matrix $\Gamma_{ij} \in M_n(\Omega_1 A)$
Curvature $R = D^2 = d\Gamma + \frac{1}{2}[\Gamma, \Gamma] = d\Gamma + \Gamma^2$	same
Topological K-theory $K^0(X) = K(X)$	Algebraic K-theory $K_0(A) = K(A)$
Topological K-theory $K^{-n}(X) = \widetilde{K}(S^n X^+)$	Topological K-theory $K_n^{\mathrm{top}}(A)$ $(= \pi_{n-1}(GL(A)))$ if A is a Fréchet algebra
Bundles with additional structures	Modules with additional structures
Multiplicative K-theory $MK(X)^{(*)}$	Multiplicative K-theory $MK(A)$

1. The Chern character for the group K_0

Let A be a ring with a unit. We want to define a "Chern character" (*)

$$K_0(A) \longrightarrow H_*(A)$$

where $H_*(A)$ is some kind of "homology theory" associated to A of a De Rham type. We start with a differential graded algebra $\Omega_*(A)$

$$0 \longrightarrow A = \Omega_0 A \xrightarrow{d} \Omega_1 A \xrightarrow{d} \Omega_2 A \longrightarrow \ldots$$

(think of $\Omega_* A = \Omega^* X$, usual De Rham complex, if $A = C^\infty X$). If $w_n \in \Omega_n A$, $w_p \in \Omega_p A$, define their graded commutator $[w_n, w_p] = w_n w_p - (-1)^{np} w_p w_n$. Denote by $\overline{\Omega}_* A$ the quotient of $\Omega_* A$ by the k-module generated by graded commutators. The *non commutative De Rham homology* of A (or rather $\Omega_* A$) is the homology of the complex (cf. [K1])

$$0 \longrightarrow \overline{\Omega}_0 A \xrightarrow{d} \overline{\Omega}_1 A \xrightarrow{d} \overline{\Omega}_2 A \xrightarrow{d} \ldots$$

which we shall denote by $H_*(A)$. Now the Chern character

$$\mathrm{ch} : K_0(A) \longrightarrow H_{2*}(A)$$

can be defined in two ways:

a) If $E = \mathrm{Im}(e)$ with $e^2 = e \in M_r(A)$, then we put $\mathrm{ch}_n(E) = \frac{1}{n!}\mathrm{Trace}(e(de)^{2n}) \in H_{2n}(A)$. This is the simplest definition ([K1], [K3], [C]).

b) A *connection* on E is a k-linear map

$$D : E \underset{A}{\otimes} \Omega_* A \longrightarrow E \underset{A}{\otimes} \Omega_{*+1}(A)$$

(*) The letter M stands for "multiplicative". We give up the notation used in [K3] for typographical reasons.

with the Leibnitz rule

$$D(s.w) = D(s).w + (-1)^{\deg(s)} s.dw \ .$$

Its *curvature* $R = D^2$ is $\Omega_* A$-linear and we put $\mathrm{ch}_n(E)$ (also denoted by $\mathrm{ch}_n(D)$) $= \frac{1}{n!}\mathrm{Trace}(R^n) \in H_{2n}(A)$ because the homology class is independent of the choice of D. In this algebraic setting there is a *universal* example $\Omega_0 A = A, \Omega_1 A = \mathrm{Ker}(A \otimes A \longrightarrow A)$ which is a A-bimodule, $\Omega_n A = \Omega_1 A \underset{A}{\otimes} \Omega_1 A \underset{A}{\otimes} \ldots \underset{A}{\otimes} \Omega_1 A$ (n factors); $d : A \longrightarrow \Omega_1 A$ defined by $d(x) = 1 \otimes x - x \otimes 1$, etc. ...

Theorem: $H_n A = \mathrm{Ker}\left(HC_n(A) \xrightarrow{B} H_{n+1}(A, A)\right)$ *where* $HC_n(A)$ *is the cyclic homology of A. Connes and* $H_n(A, A)$ *is the Hochschild homology*(*).

2. The Chern character for the groups K_1 and K_i.

If $a \subset GL_r(A)$, define

$$\mathrm{ch}_n(\alpha) = \frac{(n-1)!}{(2n-1)!} \mathrm{Trace}(\alpha^{-1} d\alpha)^{2n-1}$$

This induces a character ($l = n - 1$)

$$K_1(A) \longrightarrow H_{1+2l}(A) \subset HC_{1+2l}(A)$$

More generally, one can define character maps (for $l \geq 0$)

$$K_i(A) \longrightarrow H_{i+2l}(A)$$

where K_i are the Quillen K-groups. This connects the homology of the group $GL(A)$ and the Lie algebra homology of $gl(A)$. In fact, one has $\mathrm{Prim}(H_*(gl(A))) = HC_{*-1}(A)$ (Loday-Quillen-Feigan-Tsygan theorem: cf. [LQ]).

If A is a Banach (or Fréchet) algebra, these higher Chern characters can be extended to topological K-theory through commutative diagrams

$$\begin{array}{ccc} K_i(A) & \longrightarrow & H_{i+2l}(A) \\ \downarrow & \nearrow & \\ K_i^{\mathrm{top}}(A) & & \end{array} \qquad \begin{array}{ccc} K_i^{\mathrm{top}}(A) & \longrightarrow & H_{i+2l}(A) \\ \beta \downarrow & & \downarrow S \\ K_{i+2}^{\mathrm{top}}(A) & \longrightarrow & H_{i+2l-2}(A) \end{array}$$

where β is Bott periodicity and S is the periodicity map of cyclic homology (coming from the periodicity of the homology of finite cyclic groups).

(*) Strictly speaking, one has to consider *reduced* cyclic homology in the statement of this theorem (cf. [C], [K3]).

3. "Relative K-theory"—Borel regulators.

Although the characters defined in §2 detect some part of algebraic or topological K-theory, this is not the full story!

Examples.

a) The simplest case of failure is the determinant map $K_1(\mathbf{C}) \longrightarrow \mathbf{C}^*$. This is not covered completely by §2.

b) If $A = \mathbf{Z}$, all characters in $H_*(A \underset{\mathbf{Z}}{\otimes} \mathbf{Q})$ are trivial. We know (by Borel and Quillen) that $K_5(\mathbf{Z}) = (\mathbf{Z} \oplus$ finite group) for example. This is not covered by §2 again.

c) If $A = C^\infty(X)$, A-modules correspond to C^∞-vector bundles. One would like to construct characteristic classes for bundles with additional structures: bundles which are flat, or foliated (X a foliated manifold), or analytic (X a complex analytic manifolds), or algebraic (X a complex algebraic manifolds).

1st approach. If A is a Fréchet algebra, denote by $\mathcal{F}(A)$ the homotopy fiber of the obvious map $(BGL(A)^+ \longrightarrow BGL(A)^{\text{top}}$ and define $K_n^{\text{rel}}(A) - \pi_n(\mathcal{F}(A))$

Theorem. (cf. [K3] and [CK]) *One has a commutative diagram*

$$\begin{array}{ccccccccc} K_{n+1}(A) & \longrightarrow & K_{n+1}^{\text{top}}(A) & \longrightarrow & K_n^{\text{rel}}(A) & \longrightarrow & K_n(A) & \longrightarrow & K_n^{\text{top}}(A) \\ \downarrow & & \downarrow & & \downarrow & & \downarrow & & \downarrow \\ H_{n+1}(A,A) & \longrightarrow & HC_{n+1}(A) & \longrightarrow & HC_{n-1}(A) & \longrightarrow & H_n(A,A) & \longrightarrow & HC_n(A) \end{array}$$

where 1st row = homotopy exact sequence of the fibration, 2nd row = fundamental sequence of A. Connes, columns = various Chern characters; the map $K_n(A) \longrightarrow H_n(A,A)$ is the Dennis trace map.

Examples.

a) $A = \mathbf{C}$—Diagram chasing gives an homomorphism $\alpha : K_{2i+1}(\mathbf{C}) \longrightarrow \mathbf{C}^*$ which is an isomorphism on the torsion part (Suslin). The composition $K_{4i+1}(\mathbf{Z}) \longrightarrow K_{4i+1}(\mathbf{C}) \longrightarrow \mathbf{C}^* \xrightarrow{\log|\,|} \mathbf{R}$ is the Borel regulator, therefore not trivial.

b) $A = C^\infty(S^1)$. One gets an homomorphism

$$K_2(C^\infty(S^1)) \longrightarrow \mathbf{C}^*$$

It is associated to the Kac-Moody extension of $SL(C^\infty(S^1))$ by $\mathbf{C}^*$ (cf. [CK]).

4. Multiplicative K-Theory.

The simplest case of such a "theory" is when we are considering flat complex vector bundles. Denote by $BGL_n(\mathbf{C})$ (resp. $BGL_n(\mathbf{C})^\delta$) the classifying space of $GL_n(\mathbf{C})$ with the usual (resp. discrete) topology. The Chern character is interpreted as a map of spaces $BGL_n(\mathbf{C}) \longrightarrow \Pi K(\mathbf{C}, 2i)$ (product of Eilenberg-Mac Lane spaces) which homotopy fiber we denote by $\mathcal{F}_n$.

Theorem. *There is a* **canonical** *map* $BGL_n(\mathbf{C})^\delta \longrightarrow \mathcal{F}_n$ *included in a commutative diagram*

$$\begin{array}{ccc} & & \mathcal{F}_n \\ & \nearrow & \downarrow \\ BGL_n(\mathbf{C})^\delta & \longrightarrow & BGL_n(\mathbf{C}) \\ & \overset{0}{\searrow} & \downarrow \\ & & \Pi K(\mathbf{C}, 2i) \end{array}$$

Making $n = \infty$, $\mathcal{F}_\infty$ can be interpreted as a classifying space of "multiplicative K-theory" $[X, \mathcal{F}_\infty] = K^{-1\mathrm{top}}(X; \mathbf{C}^*)$: it fits into an exact sequence

$$K^{-1\mathrm{top}}(X;\mathbf{Z}) \longrightarrow K^{-1\mathrm{top}}(X;\mathbf{C}) \longrightarrow K^{-1\mathrm{top}}(X;\mathbf{C}^*) \xrightarrow{\partial} K^{\mathrm{top}}(X;\mathbf{Z}) \longrightarrow$$

where $K^{*\mathrm{top}}(X;\mathbf{Z})$ is topological K-theory and $K^*(X;\mathbf{C})$ is De Rham cohomology (this is of course the origin of the terminology). If we put $K_\mathbf{C}(X) = [X, BGL(\mathbf{C})^+]$ (algebraic K-theory of Quillen of the space X), one has a characteristic map

$$K_\mathbf{C}(X) \longrightarrow K^{-1\mathrm{top}}(X;\mathbf{C}^*)$$

Composed with the usual "Chern classes"

$$K^{-1\mathrm{top}}(X;\mathbf{C}^*) \xrightarrow{c_k} H^{\mathrm{odd}}(X;\mathbf{C}^*)$$

it gives essentially the Chern-Cheeger-Simons characteristic classes for flat complex vector bundles.

5. General setup for multiplicative K-theory.

It seems to be the following in the algebraic situation of §1: we give ourselves an extra data which is a decreasing filtration F^r of the differential graded algebra $\Omega_* A$ with $F^0 = \Omega_* A$, $F^r F^s \subset F^{r+s}$ and $d(F^r) \subset F^r$.

Examples from geometry ($A = C^\infty(X)$)

1. X complex analytic manifold. The Hodge filtration $F^r\Omega^*(X) = \bigoplus_{p \geq r} \Omega^{p,q}(X)$. If X is compact Kähler, Hodge theory tells us that $H^*(F^r)$ injects in $H^*(X)$ (as $\bigoplus_{p\geq r} H^{p,q}(X)$).

2. X complex algebraic manifold. There is a refinement of the Hodge filtration called the Hodge-Deligne filtration F^r; $H^*(F^r)$ injects in $H^*(X)$.

3. X foliated manifold; F^r is then generated by products of differentials of r functions transversal to the leaves, we have $F^r = 0$ for r larger than the codimension of the leaves.

4. X any manifold and put $F^r = 0$ for $r > 0$: this is the *trivial* filtration. It is of interest for characteristic classes of flat bundles (as in §4).

5. X any manifold and put $F^r\Omega^n = \Omega^n$ if $r \leq n$, and 0 otherwise. This is the filtration "bête" (Deligne terminology). It will be of interest in §6.

Grothendieck K-theory.

Define the following category $\mathcal{P}^F(A)$:

Objects are finitely generated projective A-modules provided with a connection $D : E \otimes \Omega_* \longrightarrow E \otimes \Omega_{*+1}$ such that the curvature R factorizes through $E \otimes F^1\Omega_{*+2}$:

$$\begin{array}{ccc} E \otimes \Omega_* & \xrightarrow{R} & E \otimes \Omega_{*+2} \\ & \searrow \quad \nearrow & \\ & E \otimes F^1\Omega_{*+2} & \end{array}$$

Morphisms α: $(E, D) \longrightarrow (E', D')$ are A-module maps making the following diagram commutative

$$\begin{array}{ccc} E \otimes \Omega_* & \longrightarrow & E' \otimes \Omega_* \\ \downarrow & & \downarrow \\ E \otimes \Omega_{*+1} & \longrightarrow & E' \otimes \Omega_{*+1}/E' \otimes F^1\Omega_{*+1} \end{array}$$

(they commute with the connections mod F^1). This category is an *exact* category for the obvious definition of exact sequences. Call $K^F(A)$ its Grothendieck group.

Comment. We are really coming back to the origin of K-theory. If X is a complex projective algebraic manifold, $\Omega_*(A) = \Omega^*(X)$, F^r the Hodge filtration, this is the original definition of Grothendieck (use the Newlander-Nirenberg theorem and the paper of Atiyah-Hitchin-Singer). We call it $K(X)$ as Grothendieck did.

But, as we know, $K(X)$ is difficult to compute in general, much more than its topological counterpart $K^{\text{top}}(X)$ due to Atiyah and Hirzebruch (which is $K(A)$ in this context). So we need an intermediary group $MK^F(A)$ which can be inserted between $K^F(A)$ and $K(A)$

$$K^F(A) \longrightarrow MK^F(A) \longrightarrow K(A)$$

with two properties:

It is computable;

It is keeping part of the geometry.

How to define it?

The category $\mathcal{P}^F(A)$ is too rigid: asking the curvature R to be in F^1 is very strong. Instead, we may assume Trace $(R^r) \in F^r$. Precisely, we take triples (E, D, w) where D is a connection on E and $ch_r(D) \equiv dw_r \operatorname{mod} F^r$ with $w = \Sigma w_r$. $MK^F(A)$ is the Grothendieck group built out of these triples.

Theorem: *There is an exact sequence*

$$K_1(A) \longrightarrow \oplus H_{2r-1}(\Omega_* A/F^r) \longrightarrow MK^F(A) \longrightarrow K(A) \longrightarrow \oplus H_{2r}(\Omega_* A/F^r)\,.$$

Moreover, the map $K^F(A) \longrightarrow K(A)$ factorizes through $MK^F(A)$.

Example 1. X complex projective algebraic manifold with the Hodge filtration on $\Omega_* A = \Omega^*(X)$. Then the group $MK^F(A)$ denoted by $MK(X)$ can be computed using Hodge theory. We have an exact sequence

$$0 \longrightarrow T^\beta \longrightarrow MK(X) \longrightarrow G \longrightarrow 0$$

where T^β is a compact torus of dimension $\beta = \mathrm{rank}(H^{\mathrm{odd}}(X))$ and G a finitely generated abelian group(*). The map $K(X) \longrightarrow MK(X)$ is far to be trivial.

Example 2. X foliated manifold with the canonical filtration on its De Rham complex. The normal bundle of the foliation has a natural class in $MK^F(A)$. All the known characteristic classes of the foliation (like Godbillon-Vey) can be simply deduced from this class using the fact that $F^r = 0$ for $r >$ codimension of the foliation.

6. Higher and lower multiplicative K-theory.

If A is a (real or complex) Banach algebra one can define groups $MK_n^F(A)$ (with $MK_0^F(A) = MK^F(A)$) with a natural periodicity map $MK_n^F(A) \longrightarrow MK_{n-8}^F(A)$: they fit into an exact sequence

$$K_{n+1}^{\mathrm{top}}(A) \longrightarrow \oplus H_{2r-n-1}(\Omega_* A/F^r) \longrightarrow MK_n^F(A) \longrightarrow K_n^{\mathrm{top}}(A) \longrightarrow \oplus H_{2r-n}(\Omega_* A/F^r)$$

Example 1. With a good choice of the filtration, we have a nice group $MK_n(A)$ (suggested by G. Segal) in the diagram:

$$\begin{array}{ccccccccc} K_{n+1}^{\mathrm{top}}(A) & \longrightarrow & K_n^{\mathrm{rel}}(A) & \longrightarrow & K_n(A) & \longrightarrow & K_n^{\mathrm{top}}(A) & \longrightarrow & K_{n-1}^{\mathrm{rel}}(A) \\ \| & & \downarrow & & \downarrow & & \| & & \downarrow \\ K_{n+1}^{\mathrm{top}}(A) & \longrightarrow & HC_{n-1}(A) & \longrightarrow & MK_n(A) & \longrightarrow & K_n^{\mathrm{top}}(A) & \longrightarrow & HC_{n-2}(A) \end{array}$$

(Compare with §3.)

Example 2: Higher multiplicative K-theory seems to be the good set up for regulators in algebraic geometry. With the Hodge filtration, one gets exact sequences of the type

$$K_{n+1}^{\mathrm{top}}(X) \longrightarrow \bigoplus_{\substack{p+q=2r-n-1 \\ p<r}} H^{p,q}(X) \longrightarrow MK_n(X) \longrightarrow K_n^{\mathrm{top}}(X) \longrightarrow \bigoplus_{\substack{p+q=2r-n \\ p<r}} H^{p,q}(X)$$

with a "regulator map" $K_n(X) \longrightarrow MK_n(X)$. The groups $MK_n(X)$ are related with Deligne cohomology and Beilinson work. Some work has also been done by Soulé in this direction.

(*) we have $\mathrm{rank}(G) \leq \sum_p \dim H^{p,p}(X)$

References

[Be] A.A. Beilinson, Higher regulators and values of L-functions of curves. *Funk. Analysis* **14** (1980), p. 116–117.

[Bo] A. Borel, Stable real cohomology of arithmetic groups. *Annales Sci. Ec. Norm. Sup.* **4** (1974), p. 253–272.

[BS] A. Borel and J.P. Serre, Le théorème de Riemann-Roch (d'après Grothendieck). *Bull. Soc. Math. France* **86** (1958), p. 97–136.

[C] A. Connes, Non commutative differential geometry. *Publ. Math. IHES* No. 62 (1985), p. 257–360.

[CK] A. Connes and M. Karoubi, Caractère multiplicatif d'un module de Fredholm. *C.R. Acad. Sci. Paris* **299** (1984), p. 963–968.

[CS] S.S. Chern and J. Simons, Characteristic forms and geometric invariants. *Annals of Math.* **99** (1974), p. 48.

[D] P. Deligne, Théorie de Hodge. *Publ. Math. IHES* **40** (1972), p. 5–57.

[KT] F.W. Kamber and P. Tondeur, Characteristic invariants of foliated bundles. *Manuscripta Math.* **11** (1974), p. 48–69.

[K1] M. Karoubi, Connexions, courbures et classes caractéristiques en K-théorie algébrique *Canadian Math. Soc. Conference Proceedings*, **2** Part I (1982), p. 19–27.

[K2] ———, Four Notes in the *Comptes Rendus Acad. Sci. Paris* **297** (1983), p. 381–384, 447–450, 513–516, 557–560.

[K3] ———, Homologie cyclique et K-théorie. *Astérisque* No. 149, Société Mathématique de France (1987).

K4] ———, Sur une théorie générale des classes caractéristiques secondaires (submitted to *K-theory*).

[LQ] J.L. Loday and D. Quillen, Cyclic homology and the Lie algebra homology of matrices. *Commentarii math. Helvetici* **59** (1984), p. 565–591.

[Q] D. Quillen, Higher algebraic K-theory. *Lecture Notes in Math.* No. 341 (1973).

[S] J. P. Serre, Géométrie algébrique et géométrie analytique. *Ann. Inst. Fourier* No. 6 (1955), p. 1–42.

[So] C. Soule, Connexions et classes caractéristiques de Beilinson. Prépublication IHES (1985).

[Su] A. Suslin, On the K-theory of local fields. *J. Pure Appl. Algebra* **34** (1984), p. 301–318.

Université de Paris VII
Département de Mathématiques
2 place Jussieu
75251 Paris
France

MORAVA K-THEORIES AND INFINITE LOOP SPACES

Nicholas J. Kuhn*
Department of Mathematics
University of Virginia
Charlottesville, VA 22903
and
Department of Pure Mathematics
Cambridge University
Cambridge, England

§1 Introduction and main results

For a fixed prime p, the Morava K-theories $K(0)_*$, $K(1)_*$, $K(2)_*$, ... are a sequence of p-local periodic homology theories generalizing complex K-theory: $K(0)_*$ is ordinary rational homology and $K(1)_*$ is one of the (p-1) isomorphic summands of K-theory with mod p coefficients. These theories have various nice properties - for example, the associated spectra are ring spectra, the coefficients form a graded field, and thus there is a Kunneth formula (see [R2, Chapter 4, §2]). Recently, their central role in stable homotopy has been demonstrated by the work of M. Hopkins and J. Smith on maps between finite complexes [HS], using the remarkable nilpotence theorem of [DHS].

In [B3], A.K. Bousfield proved a beautiful theorem - K(1)-localization factors through the 0th space functor - and used this to reprove and strengthen the various delooping results of Adams-Priddy [AP] and Madsen-Snaith-Tornehave [MST]. In this paper, I show how the Hopkins-Smith work allows one to generalize Bousfield's argument to all n.

To state our main theorem, we need some notation. Let "Spaces" and "Spectra" respectively denote the homotopy categories of p-local spaces and spectra (as in, say, [A2]), and let Ω^∞: Spectra → Spaces be the 0th space functor, right adjoint to the suspension Σ^∞. Let $L_{K(n)}$: Spectra → Spectra be K(n)-localization [B1]. We will often write $E_{K(n)}$ for $L_{K(n)}(E)$. Recall the characterizing properties: $E \to E_{K(n)}$ is a $K(n)_*$-equivalence, and $[X, E_{K(n)}] = 0$ if $K(n)_*(X) = 0$.

Theorem 1.1 For each $n \geq 1$, there exists a functor ϕ_n: Spaces → Spectra such that $\phi_n \circ \Omega^\infty = L_{K(n)}$. Furthermore, ϕ_n preserves fibration

* Research partially supported by the NSF, SERC, and the Sloan Foundation.

sequences.

We note two pleasant corollaries.

Corollary 1.2 If $\Omega^\infty X \simeq \Omega^\infty Y$ then $K(n)_*(X) \simeq K(n)_*(Y)$ for all $n \geq 1$.

Proof Applying Φ_n to the homotopy equivalence $\Omega^\infty X \simeq \Omega^\infty Y$ shows that $X_{K(n)} \simeq Y_{K(n)}$.

Corollary 1.3 Let $f: X \to Y$ be a map between spectra. If $\Omega^\infty f$ has a section (i.e. a right inverse) then so does $K(n)_*(f)$ for $n \geq 1$. In particular, the $K(n)$-homology suspension $K(n)_*(\Omega^\infty E) \to K(n)_*(E)$ is onto for all E.

Proof Applying Φ_n to the section of $\Omega^\infty f$ shows that $f_{K(n)}$ has a section, and the first statement follows. For the second, note that the homology suspension is induced by the evaluation map $\varepsilon: \Sigma^\infty\Omega^\infty E \to E$, and $\Omega^\infty\varepsilon$ has a section.

Note that these corollaries would be false with $K(n)_*$ replaced by ordinary homology, while they are essentially tautologically true (when restricted to (-1)-connected spectra) with $K(n)_*$ replaced by π_*^S. The failure of 1.3 for HZ/p is the source of the "unstable" condition for A-modules, thus 1.3 implies that there is no analogous condition for $K(n)_*$.

The key to all of these results is the existence of interesting self maps of finite complexes inducing isomorphisms in $K(n)_*$. An introduction to the use of such maps in our context occurs in §2. This contains an elementary proof of corollary 1.3 that is independent of both the theory of localizations (and thus Theorem 1.1) and the nilpotence conjecture. The only nontrivial input needed here is

(1.4) There exists a finite complex Z with $K(n)_*(Z) \neq 0$ and a $K(n)_*$-equivalence $v: \Sigma^d Z \to Z$ with $d > 0$.

For small n, this has been known for quite awhile. Adams constructed a $K(1)_*$-self equivalence of the Moore space in [A1], while, for $n = 2$ or 3, Toda's spaces V(n) do the job [T]. For arbitrary n, a Z satisfying (1.4) is constructed in [HS], independent of the main theorems of [DHS]. (To paraphrase Mike Hopkins, this Z is the *first* kid on the block with an ice cream cone.)

Section 3 has the proof of the main theorem - a streamlined version of Bousfield's argument in [B3]. The proof is basically formal, except that (1.4) must be strengthened to

(1.5) There exists a commutative diagram of finite complexes

$$\begin{array}{ccccccc} Z_1 & \longrightarrow & Z_2 & \longrightarrow & Z_3 & \longrightarrow & \cdots \\ \downarrow & \swarrow & & \swarrow & & & \\ S^0 & & & & & & \end{array}$$

such that

(i) $\varinjlim K(n)_*(Z_i) \simeq K(n)_*(S^0)$.

(ii) each Z_i has a $K(n)_*$-equivalence $v_i: \Sigma^{d_i} Z_i \to Z_i$ with $d_i > 0$, and these self maps are compatible, for different i, after suitable finite iteration.

This is proved in §4 using the whole strength of Hopkins and Smith's work. It should perhaps be pointed out that the length of [B3] is partly due to the fact that Bousfield had to prove a version of (1.5) pre Devanitz-Hopkins-Smith.

Section 5 contains some questions, conjectures, and examples, e.g., a computation of $BP^*(g/p\ell)$ at the prime 2.

Finally, I wish to thank Pete Bousfield and Mike Hopkins for their help in this project. I came across [B2] after having already discovered Corollary 1.3 and the argument of §2. An exchange of letters (and preprint [B3]) with Pete, and subsequent conversations with Mike, led to Theorem 1.1. It is only excessive modesty that caused each of them to decline joint authorship.

§2 A proof that (1.4) => (1.3)

We begin with the following elementary observation.

<u>Lemma 2.1</u> Let $f: X \to Y$ be a map between spectra. The following are equivalent:

(1) $\Omega^\infty f$ has a section.

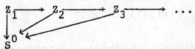

(2) Any map $g: \Sigma^\infty W \to Y$ lifts

$$\begin{array}{ccc} & \nearrow & X \\ & & \downarrow f \\ \Sigma^\infty W & \xrightarrow{g} & Y \end{array} .$$

Now we assume that $f: X \to Y$ is a map satisfying the conditions of this last lemma. We wish to show that $K(n)_*(f)$ is onto.

Since Y is a direct limit of its finite subspectra, it suffices to show that given a map $g: F \to Y$, where F is a finite complex, there exists an algebraic lifting

$$(2.2) \qquad \begin{array}{ccc} & \nearrow & K(n)_*(X) \\ & & \downarrow f_* \\ K(n)_*(F) & \xrightarrow{g_*} & K(n)_*(Y) \end{array} .$$

Now let Z be as in (1.4). Because $K(n)_*$ is a field, $K(n)_*(Z) \neq 0$, and $K(n)_*$ satisfies a Kunneth formula, to show that the lifting exists in (2.2), it suffices to show that it exists in the diagram

$$(2.3) \qquad \begin{array}{ccc} & \nearrow & K(n)_*(Z \wedge X) \\ & & \downarrow (1 \wedge f)_* \\ K(n)_*(Z \wedge F) & \xrightarrow{(1 \wedge g)_*} & K(n)_*(Z \wedge Y) \end{array} .$$

Now choose N so large that $\Sigma^{dN} Z \wedge F \wedge DZ$ is a suspension spectrum, where DZ denotes the Spanier-Whitehead dual of Z. By our assumption on f, there is a lifting

$$(2.4) \qquad \begin{array}{ccc} & \nearrow & X \\ & & \downarrow f \\ \Sigma^{dN} Z \wedge F \wedge DZ & \xrightarrow{g_N} & Y \end{array} ,$$

where g_N is dual to $v^N \wedge g: \Sigma^{dN} Z \wedge F \to Z \wedge Y$. ($v^N$ is the Nth iterate of v.)

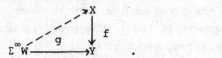

Adjointing yields a diagram

$$(2.5)\qquad \begin{array}{ccc} & & Z \wedge X \\ & \nearrow & \downarrow{\scriptstyle 1 \wedge f} \\ \Sigma^{dN}Z \wedge F & \xrightarrow{v^N \wedge g} & Z \wedge Y \end{array} \quad .$$

Since $K(n)_*(v)$ is an isomorphism, applying $K(n)_*(\quad)$ to (2.5) yields the lifting in (2.3).

§3 Proof of Theorem 1.1

We start with our basic construction.

Construction 3.1 Suppose that Z is a *space*, and $v: \Sigma^d Z \to Z$ is a self map with $d > 0$. We construct a functor

$$\Phi_Z' : \text{Spaces} \to \text{Spectra}$$

as follows: $\Phi_Z'(X)$ is the spectrum with (md)th-space

$$\Phi_Z'(X)_{md} = \mathrm{Map}(Z,X)$$

with structure maps

$$\Phi_Z'(X)_{md} = \mathrm{Map}(Z,X) \xrightarrow{v^*} \Omega^d\ \mathrm{Map}(Z,X) = \Omega^d \Phi_Z'(X)_{(m+1)d} \quad .$$

Note that a commutative diagram

$$\begin{array}{ccc} \Sigma^{dN}Z & \xrightarrow{\Sigma^{dN}\alpha} & \Sigma^{dN}Z' \\ \downarrow{\scriptstyle v^N} & & \downarrow{\scriptstyle v'^N} \\ Z & \xrightarrow{\alpha} & Z' \end{array} \quad ,$$

with $N \in \mathbb{N}$, induces a natural transformation.

$$\alpha^* : \Phi_{Z'}' \longrightarrow \Phi_Z' \quad .$$

We list some basic properties of Φ'.

Proposition 3.2

(1) $v^*: \Phi'_Z(X) \simeq \Phi'_{\Sigma^d Z}(X)$, naturally in both Z and X.

(2) $\Phi'_Z(X)$ preserves fibrations in the variable X, and cofibrations in the variable Z.

(3) $\Phi'_Z(X)$ is periodic with period d.

(4) $\Phi'_Z(\Omega^\infty E) \simeq v^{-1}F(Z,E)$, naturally in both Z and X.

In (4), F(Z,E) denotes the function spectrum defined, by Brown Representability, so that $[Y \wedge Z, E] \simeq [Y, F(Z,E)]$, and $v^{-1}F(Z,E)$ is the direct limit $\lim_{\to} \{F(Z,E) \xrightarrow{v} F(\Sigma^d Z,E) \xrightarrow{v} F(\Sigma^{2d} Z,E) \longrightarrow \dots\}$.

Proof of Proposition 3.2 Properties (1), (2) and (3) are clear by inspection. For (4), note that $\mathrm{Map}(Z, \Omega^\infty E) \simeq \Omega^\infty F(Z,E)$. Then (4) follows from the next lemma by letting $E(m) = F(\Sigma^{md} Z,E)$ and $d_m = md$.

Lemma 3.3 Suppose given a sequence of spectra $E(0) \xrightarrow{f(0)} E(1) \xrightarrow{f(1)} E(2) \longrightarrow \dots$, and an increasing sequence of natural numbers, $d_0 < d_1 < d_2 < \dots$. Define a new spectrum E by letting $E_{d_m} = \Omega^\infty \Sigma^{d_m} E(m)$, with structure maps

$\Omega^\infty \Sigma^{d_m} f_m : E_{d_m} \to \Omega^{d_{m+1} - d_m} E_{d_{m+1}}$. Then $E \simeq \lim_{\to} E(m)$.

Proof We can assume $E(m)_i = \Omega^\infty \Sigma^i E(m)$. Thus, for $n \geq m$, the f_m induce maps $E(m)_{d_n} \to E_{d_n}$. These fit together to give a map $\lim_{\to} E(m) \to E$, which is easily checked to be an isomorphism on homotopy groups.

Note that property (1) of Proposition 3.2 allows us to extend the construction Φ' to any *finite spectrum* Z with self map v: one simply replaces the pair (Z,v) by $(\Sigma^{dN} Z, \Sigma^{dN} v)$ with N large.

Now let $\Phi_Z = L_{K(n)} \circ \Phi'_Z$: Spaces → Spectra.

Proposition 3.4 If Z is finite and $K(n)_*(v)$ is an isomorphism, there is a natural equivalence $\Phi_Z(\Omega^\infty E) \simeq F(Z, E_{K(n)})$.

Proof By Proposition 3.2 (4), we need to show that there is an equivalence

$$F(Z,E_{K(n)}) \simeq (v^{-1}F(Z,E))_{K(n)} \quad .$$

First note that $F(Z, E_{K(n)})$ is $K(n)$-local, since $E_{K(n)}$ is. Thus, to finish the proof, it suffices to show that the natural maps

$$F(Z,E_{K(n)}) \longleftarrow F(Z,E) \longrightarrow v^{-1} F(Z,E)$$

induce isomorphisms in $K(n)_*$. But, since Z is finite, these maps are equivalent to the maps

$$DZ \wedge E_{K(n)} \longleftarrow DZ \wedge E \longrightarrow v^{-1}(DZ \wedge E) \ .$$

Both of these maps are clearly $K(n)_*$-equivalences, the second because $K(n)_*(v)$ is an isomorphism.

The definition of Φ_n, and the proof of theorem 1.1 are now remarkably easy, assuming (1.5).

Definition 3.5 With $Z_1 \to Z_2 \to Z_3 \to \ldots$ as in (1.5), let

$\Phi_n = \lim\limits_{\leftarrow} \Phi_{Z_i}$: Spaces $\to$ Spectra.

Proof of Theorem 1.1 There are natural equivalence

$$\begin{aligned}\Phi_n(\Omega^\infty E) &\simeq \lim_{\leftarrow} \Phi_{Z_i}(\Omega^\infty E)\\ &\simeq \lim_{\leftarrow} F(Z_i, E_{K(n)})\\ &\simeq F(\lim_{\rightarrow} Z_i, E_{K(n)})\\ &\simeq F(S^0, E_{K(n)})\\ &\simeq E_{K(n)} \ .\end{aligned}$$

Here the second equivalence follows from Proposition 3.4. The fourth equivalence holds because $\lim\limits_{\rightarrow} Z_i \to S^0$ is a $K(n)_*$-equivalence ((1.5) (i)), and $E_{K(n)}$ is $K(n)$-local.

Finally, that Φ_n preserves fibrations is a direct consequence of Proposition 3.2 (2).

<u>Exercise</u> Prove Corollary 1.2 (without using Theorem 1.1!), by just using the Φ_Z' construction applied to the pair (Z,v) of (1.4).

§4 C_n-resolutions

In this section we develop the theory of what we dub "C_n-resolutions": approximations to a fixed finite complex by complexes admitting $K(n)_*$-equivalences. A special case will be (1.5).

We need some notation and definitions from [HS]. Let C be the p-local, stable homotopy category of finite complexes, and let C_n be the full subcategory consisting of the $K(n-1)_*$-acyclic complexes.

<u>Definition 4.1</u> For $X \in C$, a map $v: \Sigma^d X \to X$ is a *v_n-self map* if $K(n)_*(v)$ is an isomorphism, and $K(m)_*(v)$ is nilpotent for $m \neq n$.

For the rest of this section we will repress suspensions "Σ^d" (i.e. view morphisms as having possibly nonzero degrees).

The main theorem of [HS] is

<u>Theorem 4.2</u>

(1) $X \in C_n$ if and only if X has a v_n-self map.

(2) Given $X,Y \in C_n$, with respective v_n-self maps v_X, v_Y, and $f: X \to Y$, there exist integers i, j such that

$$\begin{array}{ccc} X & \xrightarrow{\ f\ } & Y \\ \downarrow{\scriptstyle v_X^i} & & \downarrow{\scriptstyle v_Y^j} \\ X & \xrightarrow{\ f\ } & Y \end{array}$$

commutes.

Note that this has the following consequence.

<u>Corollary 4.3</u> Given $X(1) \xrightarrow{f(1)} X(2) \xrightarrow{f(2)} X(3) \xrightarrow{f(3)} \ldots$ with $X(i) \in C_n$, there exist $k_i \in \mathbb{N}$ and v_n-self maps $v(i): X(i) \to X(i)$

such that

$$\begin{array}{ccc} X(i) & \xrightarrow{f(i)} & X(i+1) \\ \downarrow{\scriptstyle v(i)^{k_i}} & & \downarrow{\scriptstyle v(i+1)} \\ X(i) & \xrightarrow{f(i)} & X(i+1) \end{array}$$

commutes for all i.

We now define our approximations.

Definition 4.4 For $X \in \mathcal{C}$, a *$\mathcal{C}_n$-resolution of X* (written $X_* \to X$) is a commutative diagram

$$\begin{array}{ccccccc} X_1 & \longrightarrow & X_2 & \longrightarrow & X_3 & \longrightarrow & \cdots \\ \downarrow & \swarrow & & \swarrow & & & \\ X & & & & & & \end{array}$$

such that

(1) $X_i \in \mathcal{C}_n$

(2) $\lim\limits_{\substack{\to \\ i}} K(m)_*(X_i) \to K(m)_*(X)$ is an isomorphism for $m \geq n$.

In light of Corollary 4.3, (1.5) is essentially the case $X = S^0$ of the following theorem.

Theorem 4.5 Every $X \in \mathcal{C}$ has a $\mathcal{C}_n$-resolution.

It is handy to have the following notion: A *map* $f_*: X_* \to Y_*$ of $\mathcal{C}_n$-resolutions, over $f: X \to Y$, is a collection of maps $f_n: X_n \to Y_n$ making all the obvious diagrams commute. Similarly, given a commutative diagram D of spectra in $\mathcal{C}$, there is an obvious notion of a commutative diagram of $\mathcal{C}_n$-resolutions over D.

As a first step towards proving Theorem 4.5, we prove

Proposition 4.6 Given $X(1) \xrightarrow{f(1)} X(2) \xrightarrow{f(2)} X(3) \xrightarrow{f(3)} \cdots$ with

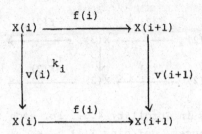

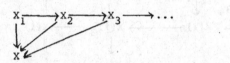

$X(i) \in C_n$, there exists a commutative diagram of C_{n+1}-resolutions

$$\begin{array}{ccccccc} X(1)_* & \xrightarrow{f(1)_*} & X(2)_* & \xrightarrow{f(2)_*} & X(3)_* & \xrightarrow{f(3)_*} & \cdots \\ \downarrow & & \downarrow & & \downarrow & & \\ X(1) & \xrightarrow{f(1)} & X(2) & \xrightarrow{f(2)} & X(3) & \xrightarrow{f(3)} & \cdots \end{array}$$

<u>Proof</u> Let $v(i)$ and k_i be as in Corollary 4.3. For each i, we have a diagram of cofibration sequences, defining $X(i)_j$:

$$\begin{array}{ccccccc} X(i) & \xrightarrow{v(i)} & X(i) & \longrightarrow & X(i)_1 & \longrightarrow & X(i) \\ \| & & \downarrow{\scriptstyle v(i)} & & \downarrow & & \| \\ X(i) & \xrightarrow{v(i)^2} & X(i) & \longrightarrow & X(i)_2 & \longrightarrow & X(i) \\ \| & & \downarrow{\scriptstyle v(i)} & & \downarrow & & \| \\ X(i) & \xrightarrow{v(i)^3} & X(i) & \longrightarrow & X(i)_3 & \longrightarrow & X(i) \\ \| & & \downarrow & & \downarrow & & \| \\ \vdots & & \vdots & & \vdots & & \vdots \end{array}$$

Furthermore, we then get a commutative diagram

$$\begin{array}{ccccccc} X(1)_* & \xrightarrow{f(1)_*} & X(2)_* & \xrightarrow{f(2)_*} & X(3)_* & \xrightarrow{f(3)_*} & \cdots \\ \downarrow & & \downarrow & & \downarrow & & \\ X(1) & \xrightarrow{f(1)} & X(2) & \xrightarrow{f(2)} & X(3) & \xrightarrow{f(3)} & \cdots \end{array}$$

by letting $f(i)_j$ be the composite $X(i)_j \to X(i)_{k_i j} \to X(i+1)_j$.

We claim that $X(i)_* \to X(i)$ is a C_{n+1}-resolution. By construction $X(i)_j \in C_{n+1}$ for all j. For $m \geq n+1$, $K(m)_*(v(i))$ is nilpotent. Thus applying $K(m)_*$ to (4.7) yields an isomorphism

$$\lim_{\substack{\to \\ j}} K(m)_*(X(i)_j) \xrightarrow{\sim} K(m)_*(X(i)).$$

<u>Proof of Theorem 4.5</u> We prove this by induction on n. Assuming that

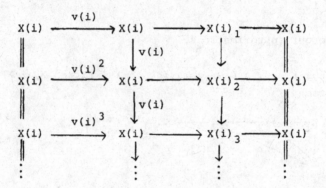

there is a C_n-resolution of X,

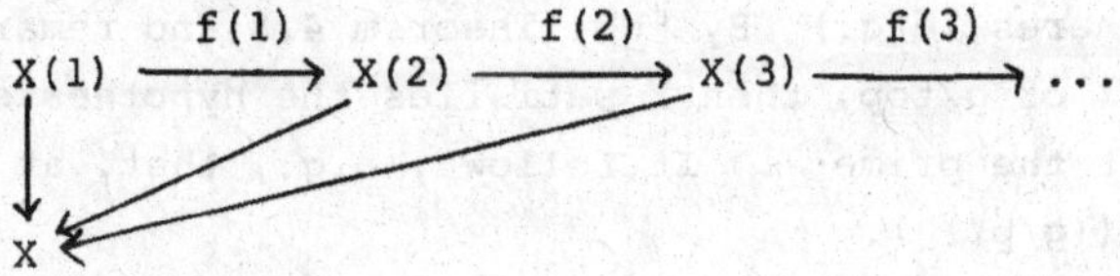

,

we will construct a C_{n+1}-resolution.

Find C_{n+1}-resolutions of the $X(i)$ as in the last proposition. Then let $Y_i = X(i)_i$, and let $Y_i \to Y_{i+1}$ be the composite $X(i)_i \to X(i+1)_i \to X(i+1)_{i+1}$. We claim that $Y_* \to X$ is a C_{n+1}-resolution. This is easy to check: $\{Y_i\}$ are cofinal in $\{X(i)_j\}$, so that, for $m \geq n+1$,

$$\lim_{\substack{\to \\ i}} K(m)_*(Y_i) \simeq \lim_{\substack{\to \\ i}} \lim_{\substack{\to \\ j}} K(m)_*(X(i)_j) \simeq \lim_{\substack{\to \\ i}} K(m)_*(X(i))$$
$$\simeq K(m)_*(X) \qquad .$$

§5 Examples and Conjectures

Our first example uses a consequence of Corollary 1.2. Recall a definition from [R1]: *Harmonic* localization is localization with respect to $\bigvee_{n\geq 0} K(n)$. Ravenel shows that BP is harmonic, as is any finite complex.

Proposition 5.1 Suppose that $\Omega^\infty X$ is homotopic to a weak product of spaces $\Omega^\infty Y_i$, where each Y_i has only finitely many non zero homotopy groups. Then

$$X \longrightarrow X_Q$$

is harmonic localization.

Proof Eilenberg-MacLane spectra are $K(n)_*$-acyclic for $n \geq 1$[R1], thus so are spectra with only a finite number of homotopy groups. By hypothesis, $\Omega^\infty X \simeq \Omega^\infty Y$ where Y is a wedge of such spectra. It follows that $K(n)_*(X) = 0$ for $n \geq 1$, so that $X \to X_Q$ is a $K(n)_*$-equivalence for all $n \geq 0$.

Example 5.2 Let g, $p\ell$, and top be the usual spectra with 0th spaces G, PL and Top, as in [MM]. (G is the group of stable homotopy equivalences of spheres, etc.) By [MM, Theorem 4.8 and remark 4.36], if X is either $g/p\ell$ or g/top, then X satisfies the hypothesis of the last proposition at the prime 2. It follows, e.g., that, at the prime 2, $BP^*(g/p\ell) \simeq BP^*((g/p\ell)_Q)$.

Remark 5.3 We would like to thank Frank Adams for explaining to us how easy it is to calculate [HQ,E]. In particular, $BP^*(\Sigma^{-1} HQ) \simeq Z_{\hat{p}} / Z_{(p)} \otimes BP^*$, from which one can calculate $BP^*(X_Q)$.

Example 5.4 $K(n)_*(g) \simeq K(n)_*(S^0)$ for $n \geq 1$.

Proof Let Q_0S^0 be the basepoint component of $\Omega^\infty\Sigma^\infty S^0$, so that $Q_0S^0 = \Omega^\infty(S\langle 0\rangle)$ where $S\langle 0\rangle$ is the 0-connected cover of $\Sigma^\infty S^0$. Then $\Omega^\infty g = G \simeq Q_0S^0 \times Z/2 = \Omega^\infty(S\langle 0\rangle \vee HZ/2)$. Thus 1.2 implies that $K(n)_*(g) \simeq K(n)_*(S\langle 0\rangle \vee HZ/2)$. But $HZ/2$ and HZ are both $K(n)_*$-acyclic, so $K(n)_*(S\langle 0\rangle \vee HZ/2) \simeq K(n)_*(S^0)$.

Our next examples are applications of Corollary 1.3. As in [K2], call a sequence of spectra $\ldots \to X_2 \to X_1 \to X_0 \to E_0$ *exact* if it is formed by splicing together cofibration sequences $E_{i+1} \to X_i \xrightarrow{f_i} E_i$ where each $\Omega^\infty f_i$ has a section. Corollary 1.3 implies

Proposition 5.4 If $\ldots \to X_2 \to X_1 \to X_0 \to E$ is an exact sequence of spectra, then, for all $n \geq 1$, there is a long exact sequence

$$\ldots \to K(n)_*(X_1) \to K(n)_*(X_0) \to K(n)_*(E) \to 0.$$

Example 5.5 The K(n)-homology suspension epimorphism can be extended, using the canonical resolution based on the adjoint pair $(\Sigma^\infty, \Omega^\infty)$:

$$\ldots \to K(n).(Q\Omega^\infty E) \to K(n).(\Omega^\infty E) \xrightarrow{\varepsilon_*} K(n).E \to 0.$$

(Here QX denotes $\Omega^\infty\Sigma^\infty X$.)

Example 5.6 For a "smaller" bound on the kernel of ε_*, we use the main theorem of [K4]: If E is 0-connected, at the prime 2 there is an exact sequence

$$\Sigma^\infty D_2\Omega^\infty E \xrightarrow{f_2} \Sigma^\infty\Omega^\infty E \xrightarrow{\varepsilon} E\,.$$

Here $D_2(X)$ is the quadratic construction on X, and f_2 is the composite

$$\Sigma^\infty D_2(\Omega^\infty E) \hookrightarrow \Sigma^\infty Q(\Omega^\infty E) \xrightarrow{\Sigma^\infty\Omega^\infty\varepsilon} \Sigma^\infty\Omega^\infty E.$$

(See [K4] for the odd primary analogue.) It follows that there is an exact sequence:

$$K(n)_*(D_2(\Omega^\infty E)) \xrightarrow{f_{2*}} K(n)_*(\Omega^\infty E) \xrightarrow{\varepsilon_*} K(n)_*(E) \to 0$$

Example 5.7 In [K1, KP1], we constructed a "minimal spacelike resolution" of $HZ_{(p)}$ extending the Kahn-Priddy epimorphism:

$$\cdots \to L(2) \to L(1) \to L(0) \to HZ_{(p)} .$$

Here $L(0) = \Sigma^\infty S^0$, $L(1) = \Sigma^\infty B\Sigma_p$, and, in general, $L(m)$ is an indecomposable stable wedge summand of $B(Z/p)^m_+$. $L(m)$ is $K(n)_*$-acyclic if $m > n$ [W, K5]. It follows that, for all $n \geq 1$, there is an exact sequence

(5.8) $\quad 0 \to K(n)_*(L(n)) \to \cdots \to K(n)_*(L(1)) \to K(n)_*(L(0)) \to 0 .$

This generalizes the well known isomorphism $K(1)_*(B\Sigma_p) \xrightarrow{\sim} K(1)_*(S^0)$ (see e.g. [K3]).

Note that (5.8) implies that

$$\sum_{m=0}^{n} (-1)^m \dim_{K(n)_*} K(n)_*(L(m)) = 0 .$$

This was observed computationally in [K5], and first caused us to try to prove Corollary 1.3.

With end this section with some questions and conjectures, aimed at making stronger use of Theorem 1.1.

Question 5.8 How faithful is the functor $\bigvee_{n\geq 0} L_{K(n)}$: Spectra $\to$ Spectra?

Clearly, one should begin by restricting to the harmonic subcategory. Bousfield [B4] has pointed out that the cofibers of maps $S_K^{-2} \to S_K^0$ provide infinitely many distinct K-local spectra all having identical K(0) and K(1) localizations.

With Mike Hopkins, we conjecture

Conjecture 5.9 If X and Y are finite spectra, then

$$X_{K(n)} \simeq Y_{K(n)} \text{ for all } n \Rightarrow X \simeq Y.$$

A consequence of Theorem 1.1 and the validity of this conjecture would be:

$$\Omega^{\infty}X \simeq \Omega^{\infty}Y \Rightarrow X \simeq Y \text{ for all finite } X \text{ and } Y.$$

We repeat a question from [R1].

Question 5.10 Is every suspension spectrum harmonic?

Conjecture 5.11 $QX \simeq QY \Rightarrow \Sigma^{\infty}X \simeq \Sigma^{\infty}Y$.

References

[A1] J.F. Adams, On the groups J(X), IV, Topology 5 (1966), 21-71.

[A2] __________, *Stable Homotopy and Generalized Homology*, University of Chicago Press, 1974.

[AP] J.F. Adams and S.B. Priddy, Uniqueness of BSO, Math. Proc. Camb. Phil. Soc. 80 (1978), 475-509.

[B1] A.K. Bousfield, The localization of spectra with respect to homology, Topology 18 (1979), 257-281.

[B2] __________, On the homotopy classification of K-theoretic spectra and infinite loop spaces, Symposium on Algebraic Topology in honor of Peter Hilton, A.M.S. Cont. Math. Series 37 (1985), 15-24.

[B3] __________, Uniqueness of infinite deloopings for K-theoetic spaces, preprint, 1986.

[B4] __________, private correspondance.

[DHS] E.S. Devinatz, M.J. Hopkins, and J.H. Smith, Nilpotence and Stable Homotopy Theory I, preprint, 1987.

[HS] M.J. Hopkins and J.H. Smith, in preparation.

[K1] N.J. Kuhn, A Kahn-Priddy sequence and a conjecture of G.W. Whitehead, Math. Proc. Camb. Phil. Soc. 92 (1982), 467-483.

[K2] __________, Spacelike resolutions of spectra, Northwestern Homotopy Theory Conference, A.M.S. Cont. Math. Series 19 (1983), 153-165.

[K3] __________, Suspension spectra and homology equivalences, Trans. A.M.S. 283 (1984), 303-313.

[K4] __________, Extended powers of spectra and a generalized Kahn-Priddy theorem, Topology 23 (1985), 473-480.

[K5] __________, The Morava K-theories of some classifying spaces, Trans. A.M.S., to appear.

[KP] N.J. Kuhn and S.B. Priddy, The transfer and Whitehead's conjecture, Math. Proc. Camb. Phil. Soc. 98 (1985), 459-480.

[MM] I. Madsen and R.J. Milgram, *The Classifying Spaces for Surgery and Cobordism of Manifolds*, Annals of Math. Study 92, Princeton University Press, 1979.

[MST] I. Madsen, V. Snaith, and J. Tornehave, Infinite loop maps in geometric topology, Math. Proc. Camb. Phil. Soc. 81 (1977), 399-430.

[R1] D.C. Ravenel, Localization with respect to certain periodic homology theories, Amer. J. Math. 106 (1984), 351-414.

[R2] __________, *Complex Cobordism and Stable Homotopy Groups of Spheres*, Academic Press, 1986.

[T] H. Toda, On spectra realizing exterior parts of the Steenrod algebra, Topology 10 (1971), 53-65.

[W] P.J. Welcher, Symmetric fiber spectra and K(n)-homology acyclicity, Indiana J. Math. 30 (1981), 801-812.

LIE GROUPS FROM A HOMOTOPY POINT OF VIEW

James P. Lin
Department of Mathematics
University of California, San Diego
La Jolla, California 92093 USA

Dedicated to the memory of
Alex Zabrodsky

In this talk, I will try to survey some recent results in the theory of finite H-spaces. But first let me give you a little historical perspective by describing some results about Lie groups and some elementary homotopy theory.

First, what is a Lie group? A Lie group is first of all a group G, secondly, it's an analytic manifold, (so the transition functions can be expanded in terms of power series), and thirdly, if I take the map

$$\theta: G \times G \longrightarrow G$$

that sends (x, y) to xy^{-1}, I want θ to be an analytic map. This means $G \times G$ endowed with the product manifold structure becomes an analytic manifold, so it makes sense to say that θ is an analytic map.

Examples of Lie groups are the circle S^1, $\mathbb{R}^n$, the general linear group $GL(n, \mathbb{R})$ of $n \times n$ matrices of nonzero determinant, the unitary group $U(n)$, and the orthogonal group $O(n)$. So there are lots of examples of Lie groups.

One of the first things one learns about Lie groups is that they have been all classified. The classification can be described roughly as follows: For simplicity let us assume the Lie groups we are talking about are simply connected. So take a Lie group G and pass to the Lie algebra g. The Lie algebra g is the set of left invariant vector fields on the Lie group. That is, given a tangent vector at the identity of G, it can be uniquely extended to a left invariant vector field on G. The set of left invariant vector fields are closed under a "bracket" operation which is anticommutative.

1980 Mathematics Subject Classification (1985 Revision): 55P45, 55S35, 55S45, 55U99.
Partially supported by the National Science Foundation.

What is remarkable about the passage from the group G to the Lie algebra g is that if two groups are isomorphic then their Lie algebras are isomorphic [10]. And conversely, if two Lie algebras are isomorphic, the corresponding Lie groups are "locally isomorphic". That is, in a neighborhood of the identity there is a "local" isomorphism, from the neighborhood of one Lie group to a neighborhood of the identity of another Lie group. In the simply connected case, this local isomorphism can be extended to an isomorphism of the first Lie group with the second Lie group. So in the case of simply connected Lie groups, there is a one to one correspondence between Lie groups and their Lie algebras.

Therefore, the classification of simply connected Lie groups is reduced to the classification of Lie algebras. That was done mainly by Cartan and Killing [21]. They classified all the "simple" Lie algebras. Then one can 'pass' back to the corresponding "simple" Lie groups. It turns out that there are four infinite families of simple Lie groups A_n, B_n, C_n, and D_n and five exceptional Lie groups called G_2, F_4, E_6, E_7 and E_8.

So this completes the classification of simple Lie groups. Then if you want to compute their cohomology, historically, this is done by using case by case computations via this classification theorem. The whole scheme of computation can be summarized by the following diagram:

(A)

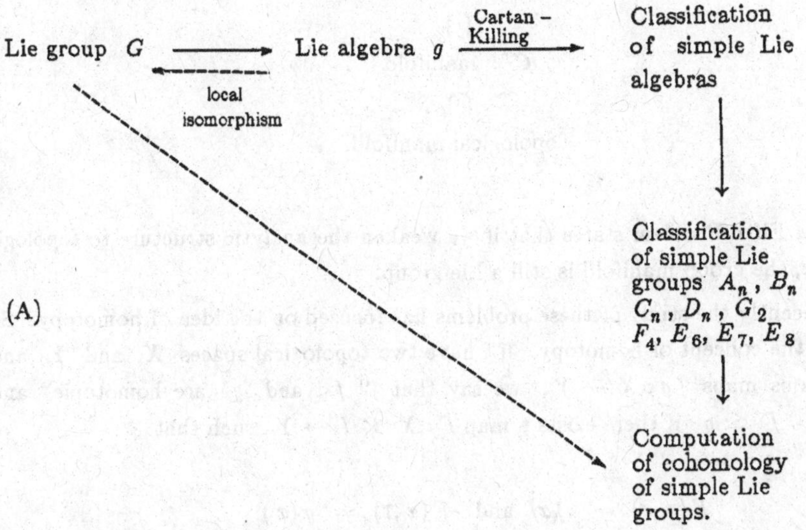

What I want to talk about today, is the question of whether or not there is some *other* method to compute the cohomology of Lie groups without resorting to the complicated classification of Lie algebras done by Cartan and Killing. So now I would like to introduce some topological properties of Lie groups.

One of the first observations is due to Iwasawa [10]. He observed that every Lie group splits as the product of a compact Lie group and a Euclidean space. So in fact, every Lie group has the homotopy type of a compact Lie group; hence it has the homotopy type of a finite complex. I will use the term "finite" to denote any space that has the homotopy type of a finite complex.

The second fact I would like to mention is of historical interest. It is called Hilbert's Fifth Problem. Hilbert's Fifth Problem characterizes Lie groups in the following way:

Theorem 1. (Montgomery, Gleason, Zippin) [20]. *Every locally Euclidean topological group is a Lie group.*

So let me remind you that a topological group is just a topological space that is a group together with the fact that the multiplication and inverse maps are continuous. In Hilbert's Fifth Problem, one can see that the topology of the Lie group is altered, but the group structure is held constant. That is, there is the stratification of structures:

$$\begin{array}{c} \text{Analytic manifold} \\ \cap \\ C^{\infty} \text{ manifold} \\ \cap \\ C^{n} \text{ manifold} \\ \cap \\ \text{Topological manifold.} \end{array}$$

Hilbert's Fifth Problem states that if we weaken the analytic structure to topological structure, the group manifold is still a Lie group.

More recently the study of these problems has focused on the idea of homotopy. So let's discuss the concept of homotopy. If I have two topological spaces X and Y and two continuous maps $f\, g : X \to Y$, we say that " f and g are homotopic" and denote it by $f \simeq g$ if there exists a map $F : X \times I \to Y$ such that

$$F(x, \ ) = f(x) \text{ and } F(x, 1) = g(x) \quad .$$

Pictorially, this means that for a given x, I can connect the points $f(x)$ and $g(x)$ by a path:

and this path varies continuously as a function of x.

In a basic course in topology, one finds that homotopic maps induce the same map on cohomology. Similarly, we say two spaces X and Y have the same "homotopy type" and write $X \simeq Y$ if there are maps

$$f : X \longrightarrow Y \quad \text{and} \quad g : Y \longrightarrow X$$

such that $gf \simeq \mathrm{id}_X$ and $fg \simeq \mathrm{id}_Y$. In this case, $H^*(X)$ is isomorphic to $H^*(Y)$. So if I want to compute the cohomology of a Lie group, it suffices to consider spaces that have the same homotopy type as the Lie group.

Now recall in Hilbert's Fifth Problem, the manifold structure was changed, but the group structure was held fixed. Why don't we change the group structure also? By this I mean we can introduce such concepts as "homotopy identity". This means I have a topological space X together with some binary operation

$$X \times X \xrightarrow{\mu} X, \quad \mu(x_1, x_2) = x_1 x_2$$

and a distinguished point e in X that I want to behave like an identity.

A homotopy identity e should have the property that the maps

$$\begin{array}{l} \mu(e, \ ): X \longrightarrow X \\ \mu(\ , e): \end{array}$$

be homotopic to the identity, 1_X. Schematically, for each x in X, there are paths

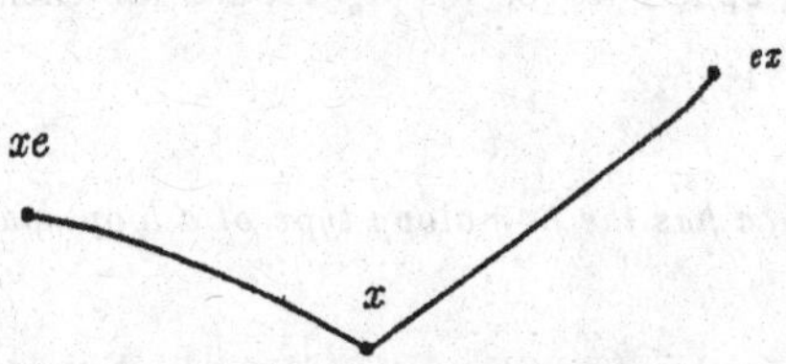

These paths should vary continuously as functions of the variable x.

Such a pairing $\mu: X \times X \to X$ together with a homotopy identity will be called an "H-space," in honor of Heinz Hopf who invented the concept [11].

A second concept can be added to this. Recall a group is always associative. We introduce the concept of "homotopy associativity." Given an H-space (X, e, μ), and three points x, y, z in X we can pair them in two different ways:

$$\begin{aligned} (xy)z &= \mu(\mu(x, y), z) \\ x(yz) &= \mu(x, \mu(y, z)) \end{aligned}$$

If there is a path from $(xy)z$ to $x(yz)$

that varies continuously as a function of the three variables x, y and z then we say (X, e, μ) is "homotopy associative."

We can continue this process with four variables x, y, z, w. In this case we obtain a picture like this:

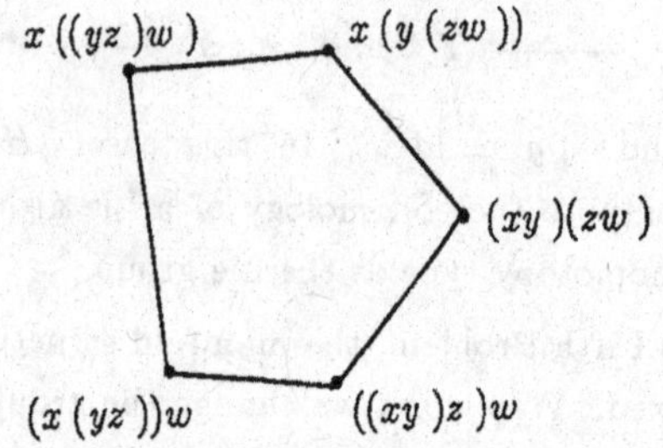

But given that the H-space is homotopy associative, there is a further obstruction to shrinking the pentagon to a point. Roughly speaking, if this can be done continuously as a function of the four variables x, y, z and w then (X, e, μ) is called an "a_4-space." In fact there are examples of H-spaces that are not homotopy associative or homotopy associative H-spaces that are not a_4-spaces.

These concepts were studied by Sugawara and Stasheff [22,23]. In fact, there is a stratification of H-spaces. If we call an H-space an a_2-space, a homotopy associative H-space an a_3-space, then Stasheff shows there are corresponding definitions for a_n-spaces for each positive integer $n \geq 2$.

Let's define an "a_∞ space" to be an a_n space for each $n \geq 2$. The following theorem is due to Stasheff [22]:

Theorem 2. *An a_∞ space has the homotopy type of a loop space ΩB.*

The following theorem due to Milnor describes the homotopy type of a topological group:

Theorem 3. *A topological group has the homotopy type of a loop space ΩB.*

So there is a stratification of spaces.

$$\text{Lie groups} \subset \text{finite topological groups} \subset \text{finite loop spaces}$$
$$\subset \text{finite } a_n\text{-spaces} \subset \text{finite } H\text{-spaces} \quad .$$

A reasonable question to ask is: What are the cohomological properties of finite H-spaces?

Given a finite H-space (X, e, μ), we can apply cohomology to the map

$$\mu : X \times X \longrightarrow X .$$

We get a map

$$\Delta = \mu^* : H^*(X) \longrightarrow H^*(X \times X) .$$

With coefficients in a field, the Kunneth Theorem tells us that $H^*(X \times X) \cong H^*(X) \otimes H^*(X)$. In addition $H^*(X)$ is a ring with respect to cup product. So we obtain maps

$$\Delta : H^*(X) \longrightarrow H^*(X) \otimes H^*(X)$$

$$v : H^*(X) \otimes H^*(X) \longrightarrow H^*(X) .$$

Δ is called a "coalgebra structure." When Δ is a map of algebras, we say $H^*(X)$ is a Hopf algebra.

The following theorems are Hopf algebra theorems [19]:

Theorem 4. (a) (Hopf) $H^*(X;\mathbb{Q}) = \Lambda(x_1, \ldots, x_l)$ *where degree* x_i *are odd as algebras*

(b) (Borel) $H^*(X;\mathbb{Z}_2) = \otimes \Lambda(x_i) \otimes \mathbb{Z}_2 \dfrac{[y_j]}{y_j^{2^{f_j}}}$ *as algebras.*

One sees immediately that certain spaces cannot admit Lie group structures. For example an even sphere does not have the correct rational cohomology, so it cannot be a Lie group, (or a finite H-space).

The second tool is a sharpening of the Borel structure theorem. The Borel structure theorem is a statement about the ring structure of $H^*(X;\mathbb{Z}_2)$. It makes no statement about the possible coalgebra structures. If we assume $H_*(X;\mathbb{Z}_2)$ is an associative ring, we can say quite a bit about the coalgebra structure.

Define $\overline{\Delta} : IH^*(X;\mathbb{Z}_2) \to IH^*(X;\mathbb{Z}_2) \otimes IH^*(X;\mathbb{Z}_2)$ to be $\overline{\Delta} x = \Delta x - x \otimes 1 - 1 \otimes x$. This is called the "reduced coproduct." Let $\xi : H^*(X;\mathbb{Z}_2) \to H^*(X;\mathbb{Z}_2)$ be the squaring map. Let

$$R = \{x \in IH^*(X;\mathbb{Z}_2) \mid \overline{\Delta} x \in \xi H^*(X;\mathbb{Z}_2) \otimes H^*(X;\mathbb{Z}_2)\} .$$

Note that R itself is a coalgebra. We use the fact that $H_*(X;\mathbb{Z}_2)$ is associative.

Let $S(R)$ be the free commutative Hopf algebra generated by the coalgebra R. Define $I \subset S(R)$ to be the ideal generated by terms of the form $\xi x - x \otimes x$ for $x \in R$. Then $I \subset S(R)$ becomes a Hopf ideal and

Theorem 5. (Lin [14]) $H^*(X;\mathbb{Z}_2) \cong S(R)/I$ *as Hopf algebras.*

This theorem shows that the coalgebra structure is determined by the submodule R and that all the generators in a Borel decomposition may be chosen to lie in the module R. Furthermore, it may be shown that there is an exact sequence

$$0 \longrightarrow \xi H^*(X;\mathbb{Z}_2) \longrightarrow R \longrightarrow QH^*(X;\mathbb{Z}_2) \longrightarrow 0 \quad .$$

Here $QH^*(X;\mathbb{Z}_2) = IH^*(X;\mathbb{Z}_2)/IH^*(X;\mathbb{Z}_2)^2$ is the module of "indecomposables."

The third tool is the Steenrod algebra $\mathcal{A}(2)$. There is a sequence of natural transformations of the function $H^*(\ ;\mathbb{Z}_2)$

$$Sq^i : H^*(\ ;\mathbb{Z}_2) \longrightarrow H^{*+i}(\ ;\mathbb{Z}_2), \quad Sq^i \in \mathcal{A}(2) \ .$$

Given a continuous map $f : X \to Y$, the following diagram commutes:

$$\begin{array}{ccc} H^l(X;\mathbb{Z}_2) & \xrightarrow{Sq^i} & H^{l+i}(X;\mathbb{Z}_2) \\ \uparrow f^* & & \uparrow f^* \\ H^l(Y;\mathbb{Z}_2) & \xrightarrow{Sq^i} & H^{l+i}(Y;\mathbb{Z}_2) \end{array}$$

I'll list a few more facts that will be needed for this talk.

Cartan Formula. $Sq^i(x \cup y) = \sum_{l=0}^{i} Sq^l x \cup Sq^{i-l} y$.

If degree $x = k$, then $Sq^k x = x^2$.

If $QH^*(X;\mathbb{Z}_2)$ denotes the module of indecomposables; i.e., $QH^*(X;\mathbb{Z}_2) = IH^*(X;\mathbb{Z}_2)/IH^*(X;\mathbb{Z}_2)^2$, we note that Sq^i induces a natural transformation of the indecomposables

$$Sq^i : QH^*(\ ;\mathbb{Z}_2) \longrightarrow QH^{*+i}(\ ;\mathbb{Z}_2) \ .$$

We can ask

Question 1. What is the action of the Sq^i on the module $QH^*(X;\mathbb{Z}_2)$ for X a finite H-space?

We have the following results. Assume $H_*(X;\mathbb{Z}_2)$ is an associative ring. Then given any integer n, I can expand it dyadically and look for the first power of two missing from the expansion.

$$\begin{aligned} n &= 1 + 2 + \cdots + 2^{r-1} + 2^{r+1} + \cdots + 2^s \\ &= 2^r + 2^{r+1}k - 1 \quad \text{for some} \quad k \geq 0 \quad . \end{aligned}$$

Theorem 6. (Lin [17]) *For* $k > 0$

(a) $QH^{2^r+2^{r+1}k-1}(X;\mathbb{Z}_2) = Sq^{2^r k}\, QH^{2^r+2^r k-1}(X;\mathbb{Z}_2)$.

(b) $Sq^{2^r}\, QH^{2^r+2^{r+1}k-1}(X;\mathbb{Z}_2) = 0$.

(c) $\sigma^* QH^{2^r+2^{r+1}k-1}(X;\mathbb{Z}_2) \subseteq Sq^{2^r} H^*(\Omega X;\mathbb{Z}_2)$ *where* $\sigma^* : H^*(X;\mathbb{Z}_2) \to H^*(\Omega X;\mathbb{Z}_2)$ *is the suspension map.*

(d) *If* $t = 2^{i_1} + 2^{i_2} + \cdots + 2^{i_l}$ *where* $i_1 < i_2 < \cdots < i_l$ *then*

$$Sq^t\, QH^*(X;\mathbb{Z}_2) = Sq^{2^{i_1}} Sq^{2^{i_2}} \cdots Sq^{2^{i_l}}\, QH^*(X;\mathbb{Z}_2) \quad .$$

Let's give an example of how Theorems 5 and 6 can be used.

The mod 2 cohomology of E_8 has the following form

$$H^*(E_8:\mathbb{Z}_2) = \mathbb{Z}_2 \frac{[x_3, x_5, x_9, x_{15}]}{x_3^{16}, x_5^{8}, x_9^{4}, x_{15}^{4}} \otimes \Lambda(x_{17}, x_{23}, x_{27}, x_{29}) \quad .$$

By Theorem 5 all the x_i's may be chosen to lie in R; hence, they have

$$\overline{\Delta} x_i \in \xi H^*(X;\mathbb{Z}_2) \otimes R \quad .$$

Theorem 6, parts (a) and (d), yield

$$\begin{aligned} r=1: \quad & x_5 = Sq^2 x_3 \\ & x_9 = Sq^4 x_5 \\ & x_{17} = Sq^8 x_9 \\ & x_{29} = Sq^2 Sq^4 Sq^8 x_{15} \end{aligned}$$

$$r=2: \quad x_{27} = Sq^4 Sq^8 x_{15}$$

$$r=3: \quad x_{23} = Sq^8 x_{15}$$

A simple analysis using Theorem 6, part (c) shows

$$x_{17} = Sq^2 x_{15} .$$

This is the complete action of the Steenrod algebra on the generators of $H^*(E_8;\mathbb{Z}_2)$. It can be shown that x_{15} is not primitive [13].

Theorem 6 has many applications. Part (a) yields the following corollary.

Corollary 7. (Lin [17]) *The first nonvanishing* mod 2 *cohomology group occurs in a degree a power of two minus one. In fact, it occurs in a degree of the form* 1, 3, 7 *or* 15.

Since the first nonvanishing mod 2 cohomology group occurs in the same degree as the first nonvanishing homotopy group we have

Corollary 8. (Lin [17]) *The first nonvanishing homotopy group of a finite H - space occurs in degree* 1, 3, 7 *or* 15. *If $H^*(X;\mathbb{Z})$ has no two torsion then the first nonvanishing homotopy group lies in degree* 1, 3 *or* 7.

Corollary 8 is a generalization of Adams' Hopf Invariant One Theorem [1] which states that the only spheres that are H-spaces are S^1, S^3 or S^7. Thomas [24] proved theorems of this type in the case that $H^*(X;\mathbb{Z}_2)$ is primitively generated.

Theorem 6 in the case that $r = 0$ states that

$$QH^{2k}(X;\mathbb{Z}_2) = Sq^k QH^k(X;\mathbb{Z}_2) .$$

Recall that $Sq^k x = x^2$ if $\deg x = k$, so this implies $QH^{even}(X;\mathbb{Z}_2) = 0$. This turns out to be equivalent to

Corollary 9. (Lin [18]) *$H_*(\Omega X;\mathbb{Z})$ has no two torsion.*

R. Bott [6] gave a geometric proof of Corollary 9 when X is a Lie group. His proofs involve Morse theory which of course is not applicable to finite H-spaces.

Another simple application yields

Corollary 10. (Lin [17]) *The* mod 2 *Hurewicz map*

$$h_n \otimes \mathbb{Z}_2 \colon \pi_n(X) \otimes \mathbb{Z}_2 \longrightarrow H_n(X;\mathbb{Z}_2)$$

is zero unless n is a power of two minus one.

This result is best possible because the mod 2 Hurewicz map is nontrivial in degrees a power of two minus one for certain spin groups.

Corollary 11 (Lin [13]) *The* mod 2 *cohomology of the exceptional groups can be determined from the rational cohomology of the exceptional groups. In fact, the rational cohomology of the exceptional groups determines the* mod 2 *cohomology of the exceptional groups as algebras over the Steenrod algebra.*

Corollary 11 comes from a Bockstein spectral sequence argument. Historically, the computation of the mod 2 cohomology of the exceptional groups was quite difficult. The result for E_8 was announced by Araki and Shikata in 1961 [4], but the details of the calculation used Lie group properties and did not appear until 1984 [12].

I would like to turn now to some recent questions posed by Adams and Wilkerson. In broad terms the question might be posed as follows:

Question 2. Given a finite loop space ΩB, is the mod 2 cohomology of ΩB isomorphic to the mod 2 cohomology of a Lie group?

There are certain restrictions that are placed on $H^*(B;\mathbb{Z}_2)$ in order for $H^*(\Omega B;\mathbb{Z}_2)$ to be a finite dimensional vector space. In recent years, two Hopf algebras have emerged as possible non-Lie candidates for $H^*(\Omega B;\mathbb{Z}_2)$. They are

$$A_1 = \mathbb{Z}_2 \frac{[x_7]}{x_7^4} \otimes \Lambda(x_{11}, x_{13})$$

$$A_2 = \mathbb{Z}_2 \frac{[x_{15}]}{x_{15}^4} \otimes \Lambda(x_{23}, x_{27}, x_{29})$$

By Theorem 6 the following generators must be connected by Steenrod operations

$$x_{11} = Sq^4 x_7, \quad x_{13} = Sq^2 x_{11}$$

$$x_{23} = Sq^8 x_{15}, \quad x_{27} = Sq^4 x_{23}, \quad x_{29} = Sq^2 x_{27}$$

Several false proofs using K-theory have been attempted to show A_1 cannot be the mod 2 cohomology of an H-space. U. Suter (unpublished) has a proof that A_2 is not the mod 2 cohomology of a finite H-space. Recently, we have found

Theorem 12. (Lin [15]) A_1 *cannot be the* mod 2 *cohomology of an* H *- space.*

(Lin-Williams), (Suter) A_2 *cannot be the* mod 2 *cohomology of an* H *- space.*

I want to shift gears to look at another question, that is,

Question 3. Given a finite loop space ΩB is the rational cohomology of ΩB isomorphic to the rational cohomology of a Lie group?

Hopf's theorem states

$$H^*(\Omega B;\mathbb{Q}) \cong \Lambda(y_{n_1-1}, \ldots, y_{n_r-1})$$

as Hopf algebras where $n_1 \leq n_2 \leq \cdots \leq n_r$ n_i are even.

A theorem of Borel shows the rational cohomology of B is a polynomial algebra on generators $x_{n_1}, \ldots, x_{n_r}$

$$H^*(B;\mathbb{Q}) = \mathbb{Q}[x_{n_1}, \ldots, x_{n_r}] \quad .$$

We call the sequence of numbers $[n_1, \ldots, n_r]$ the "type" of ΩB and r is called the "rank" of ΩB. So Question 3 reduces to checking that the type of a finite loop space is the same as the type of a Lie group.

Some other facts relate the rational cohomology to the mod p cohomology.

Theorem 13. (Borel, Browder [5,7]). *If* $p \nmid n_i$ *for each* i*, then*

$$H^*(B;\mathbb{Z}_p) = \mathbb{Z}_p[x_{n_1}, \ldots, x_{n_r}] \quad .$$

There have been two recent major achievements that assist us in answering Question 3. The first is a result due to Clark and Ewing [9]. Essentially they create a list of types that are shown to be realizable at various primes. The second major advance is due to Adams and Wilkerson [2].

Theorem 14. *If* $p \nmid n_i$ *for every* i*, then* $[n_1, \ldots, n_r]$ *must be a union of types that occur in the Clark-Ewing list.*

J. Aguadé observes that most of the types occurring in the Clark-Ewing list occur at primes p where $p \equiv 1 \mod m$ for some m that divides n_i for some i. Using Dirichlet's theorem to find primes $p \not\equiv 1 \mod m$ for all $m \mid n_i$. is possible to reduce the size of the list and obtain the following theorems.

Theorem 15. (Aguadé [3]). *If* $H^*(\Omega B;\mathbb{Z})$ *has no torsion and has type* $[n_1, \ldots, n_r]$ *with* $n_1 < n_2 < \cdots < n_r$ *then* $H^*(\Omega B;\mathbb{Q})$ *is isomorphic to the rational cohomology of a Lie group.*

Theorem 16. (Lin [16]). *Let ΩB have type $[n_1, \ldots, n_r]$.*

(a) *Then $[n_1, \ldots, n_r]$ is a union of simple Lie types and sets of the form* [4,24] *and* [12,16].

(b) $[n_1, \ldots, n_r]$ *is a union of simple Lie types and* [4,16], [4,24] *and* [4,48].

Theorem 16 shows us that rationally, finite loop spaces and Lie groups are almost the same. There are, in some sense, four exceptional simple loop types [4,24], [12,16], [4,16] and [4,48].

There is a homotopy theoretic way to characterize simple Lie groups.

We can define a "simple loop space" to be a finite loop space ΩB with $\pi_3(\Omega B) = \mathbb{Z}$. Note that this definition, when applied to Lie groups, may be used to define a simple Lie group. We have the following theorem which follows from Theorem 16(b):

Theorem 17. (Lin [16]). *Every simple loop space has the rational cohomology of a simple Lie group.*

Let's return now to the concepts I mentioned at the beginning of my talk and summarize the basic strategy used to obtain cohomological calculations for Lie groups.

The original method is to begin with a Lie group, pass to its Lie algebra, then apply the Cartan-Killing classification to obtain all simple Lie algebras and then go back to obtain the corresponding simple Lie groups. Then apply case by case computations to compute the cohomology of all simple Lie groups. This is described by diagram (A).

Hilbert's Fifth Problem states that we may alter the analytic structure on the manifold and still obtain a Lie group. This schematically is described like this:

Lie group G

$\cap$

Topological group with locally Euclidean structure

If we have finite loop spaces, their rational cohomology can be computed using results of Adams, Wilkerson, Clark and Ewing. So we have

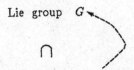

$$
\begin{array}{lllll}
\text{Lie groups} & & & \searrow & \\
\cap & & & & \\
\text{Topological groups} & & & & \\
\cap & & & & \\
\text{Finite loop spaces } \Omega B & \longrightarrow & H^*(\Omega B;\mathbb{Q}) & \underset{\text{simple loop space}}{\rightleftarrows} & H^*(\text{Lie group};\mathbb{Q})
\end{array}
$$

Finally if we have just a finite H-space X, we can pass to its mod 2 cohomology $H^*(X;\mathbb{Z}_2)$ which is a Hopf algebra over the Steenrod algebra to obtain formulae that restrict the indecomposables, the coalgebra structure and the connectivity. We have

$$
\begin{array}{lllll}
\text{Finite loop spaces} & & & & \\
\cap & & & & \\
\text{Finite } a_n\text{-spaces} & & & & \\
\cap & & & & \\
\text{Finite } H\text{-spaces } X & \longrightarrow & H^*(X;\mathbb{Z}_2) & \longrightarrow & H^*(\text{Lie group};\mathbb{Z}_2)
\end{array}
$$

Putting these diagrams all together, we obtain a picture like this:

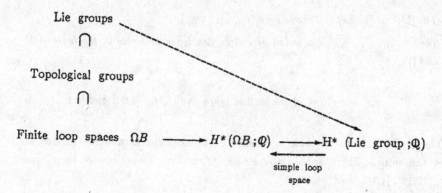

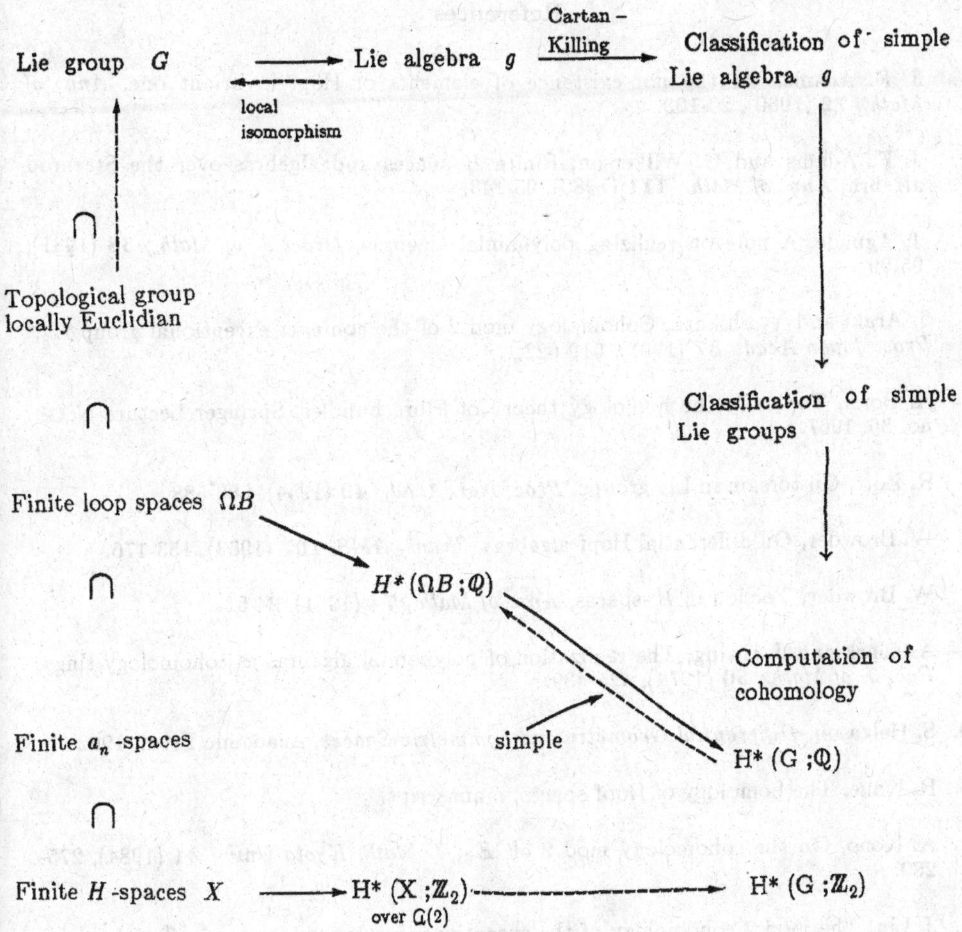

I hope that what I've been able to show is that even though Lie groups have a very rigid analytic manifold and group structure, there is another more homotopy theoretic viewpoint, and that even without the rigid group structure one can still obtain many topological results about Lie groups.

References

1. J. F. Adams, On the non-existence of elements of Hopf invariant one, *Ann. of Math.*, **72** (1960), 20-103.

2. J. F. Adams and C. Wilkerson, Finite H-spaces and algebras over the Steenrod algebra, *Ann. of Math.*, **111** (1980), 95-143.

3. J. Aguadé, A note on realizing polynomial algebras, *Israel J. of Math.*, **38** (1981), 95-99.

4. S. Araki and Y. Shikata, Cohomology mod 2 of the compact exceptional group E_8, *Proc. Japan Acad.*, **37** (1961), 619-622.

5. A. Borel, Topics in the homology theory of Fibre bundles, Springer Lecture Notes no. 36, 1967.

6. R. Bott, On torsion in Lie groups, *Proc. Nat. Acad.*, **40** (1954), 586-588.

7. W. Browder, On differential Hopf algebras, *Trans. AMS*, **107** (1963), 153-176.

8. W. Browder, Torsion in H-spaces, *Ann. of Math.*, **74** (1961), 24-51.

9. A. Clark and J. Ewing, The realization of polynomial algebras as cohomology rings, *Pac. J. of Math.*, **50** (1974), 425-439.

10. S. Helgason, *Differential Geometry and Symmetric Spaces*, Academic Press, 1962.

11. R. Kane, The homology of Hopf spaces, manuscript.

12. A. Kono, On the cohomology mod 2 of E_8, *J. Math. Kyoto Univ.*, **24** (1984), 275-280.

13. J. Lin, The mod 2 cohomology of the exceptional groups, to appear in *Topology and Its Applications*.

14. J. Lin, On the Hopf algebra structure of the mod 2 cohomology of a finite H-space, *Pub. RIMS, Kyoto Univ.*, **20** (1984), 877-892.

15. J. Lin and F. Williams, On 6-connected finite H-spaces with two torsion, to appear in *Topology.*

16. ______. The rational cohomology of finite loop spaces, *J. of Pure and Applied Algebra*, **53** (1988), 71-73.

17. ______. Steenrod connections and connectivity in H-spaces, *Memoirs of AMS*, **68** (1987), no. 329.

18. ______. Two torsion and the loop space conjecture, *Ann. of Math.*, **115** (1982), 35-91.

19. J. Milnor and J. C. Moore, On the structure of Hopf algebras, *Ann. of Math.*, **81** (1965), 211-264.

20. D. Montgomery and L. Zippin, Small subgroups of finite dimensional groups, *Proc. Nat. Acad. Sci., USA,* **38** (1952), 440-442.

21. H. Samelson, *Notes on Lie Algebras,* Van Nostrand, 1969.

22. J. Stasheff, Homotopy associativity of H-spaces, I, II, *Trans. AMS,* **108** (1963), 275-292; 293-312.

23. M. Sugawara, A condition that a space is group-like, *Math. J. Okayama Univ.,* **7** (1957), 123-149.

24. E. Thomas, Steenrod squares and H-spaces I, *Ann. of Math.,* **77** (1963), 306-317.

Order of the identity map of the Brown-Gitler spectrum

By Wen-Hsiung Lin

Let $B(k)$ be the 2-primary k^{th} Brown-Gitler spectrum ([3]). Let $\nu(k)$ be the integer defined by $k = 2^{\nu(k)}(2\ell+1)$. In this note we prove

Theorem 1. In the stable homotopy group $[B(k),B(k)]$ the order of the identity map $1_k: B(k) \to B(k)$ is $2^{\nu(k)+2}$ for $k \geq 1$.

Cohen and Goerss prove in [6] that for any odd prime p the identity map 1_k of the p-primary Brown-Gitler spectrum $B(k)$ ([5]) is of order p for $k \geq 1$, that is, the order is independent of k.

The proof of Theorem 1 consists of two parts.

Proposition 2. $2^{\nu(k)+2} \geq$ order of 1_k.

Proposition 3. $2^{\nu(k)+2} \leq$ order of 1_k.

For the first part it suffices to consider the cases $k = 2^i$ by the following result of Cohen and Goerss.

Proposition 4. ([6]) Suppose $k = 2^{i_1} + 2^{i_2} + 2^{i_m}$ with $i_1 < i_2 < \ldots < i_m$ and $m \geq 2$. So $\nu(k) = i_1$. Then $B(k)$ is a wedge summand of $B(2^{i_1}) \wedge B(k-2^{i_1})$.

To prove Proposition 2 for $k = 2^i$ we recall that $H^*(B(2^i)) \cong M(2^i) = A/A\{\chi(Sq^\ell) \mid \ell > 2^i\}$ where A is the mod 2 Steenrod algebra and $\chi : A \to A$ is the canonical auti-automorphism.

Lemma 5. $Ext_A^{s,s}(M(2^i),M(2^i)) = 0$ for $s \geq i + 2$.

This implies that, in the mod 2 Adams spectral sequence $\{E_r^{s,s}\}$ for the stable groups $[B(k),B(k)]_*$, $E_\infty^{s,s} = 0$ for $s \geq i + 2$. Since $B(2^i)$ is 2-local it follows that each element in $[B(2^i),B(2^i)]$, in particular 1_{2^i}, has order at most 2^{i+2}, and this is Proposition 2 for $k = 2^i$.

To prove Lemma 5 consider the lambda algebra Λ of Kan et al. ([1]). Recall that Λ is a bigraded algebra over $\mathbb{Z}/2$ generated by $\lambda_i \in \Lambda^{1,i}$ for $i \geq 0$ and has the set admissible monomials $\{\lambda_{i_1}\lambda_{i_2} \cdots \lambda_{i_s} \mid 2i_j \geq i_{j+1}\}$ as a $\mathbb{Z}/2$-base. Let $\Lambda_k \subset \Lambda$ be the left ideal generated by $\lambda_0, \lambda_1, \ldots, \lambda_{k-1}$. This ideal is a subcomplex, and we let $\Lambda(k)$ denote the quotient Λ/Λ_k. Then $\Lambda(k)$ has as a base the images of the admissible monomials $\lambda_I=\lambda_i \cdots \lambda_{i_s}$ for which $i_s \geq k$. Mahowald proves in [8] the following.

Proposition 6. ([10]) $Ext_A^{s,t}(M(k),\mathbb{Z}/2) \cong \Lambda(k)^{s,t-s}$ for $t-s \leq 2k$.

This result is actually implicit in [3].

Lemma 7. $Ext_A^{s,t}(M(2^i),\mathbb{Z}/2) \cong \Lambda(2^i)^{s,t-s} = 0$ for $s \geq i+2$ and $t-s \leq 2^{i+1}-1$.

Proof. $\{\lambda_I \mid I = (i_1,\ldots,i_s)$ admissible and $i_s \geq 2^i\}$ is a $\mathbb{Z}/2$-base for $\Lambda(2^i)$. It is easy to see, then, that for $s \geq i$ the smallest integer j_s for which $\Lambda(2^i)^{s,j_s} \neq 0$ is $j_s = 2^{i+1} + s - (i+2)$ and $\Lambda(2^i)^{s,j_s} \cong \mathbb{Z}/2$ is generated by

$$\underbrace{\lambda_1 \cdots \lambda_1}_{\leftarrow s-i \rightarrow}\lambda_2\lambda_4\lambda_8 \cdots \lambda_{2^{i-1}}\lambda_{2^i}.$$

If $s \geq i+2$ and $j \leq 2^{i+1} - 1$ then

$$j \leq 2^{i+1}-1 \leq 2^{i+1}+s-(i+2)-1 = j_s-1 < j_s$$

and so $\Lambda(2^i)^{s,j} = 0$. This proves Lemma 7.

Proof of Lemma 5: Filter the A-module $M(2^i)$ by $F_p = F_pM(2^i) = \bigoplus_{q \geq p} M(2^i)_q$. The resulting spectral sequence $\{E_r^{p,s,t}\}_{r \geq 1}$ obtained by applying $Ext_A^{*,*}(M(2^i),-)$ to this filtration with $E_1^{p,s,t} = Ext_A^{s,t}(M(2^i),F_p/F_{p+1})$ has $\bigoplus_p E_\infty^{p,s,t} \cong Ext_A^{s,t}(M(2^i),M(2^i))$. In particular,

$$(1) \qquad Ext_A^{s,s}(M(2^i),M(2^i)) \cong \bigoplus_p E_\infty^{p,s,s}.$$

The conclusion of the lemma follows from (1) together with the following (2)(a) and (2)(b).

$$(2)(a) \qquad E_\infty^{p,s,s} = E_1^{p,s,s} = Ext_A^{s,s}(M(2^i),F_p/F_{p+1}) = 0 \quad \text{for } p > 2^{i+1}-1.$$

$$(2)(b) \qquad E_\infty^{p,s,s} = E_1^{p,s,s} = Ext_A^{s,s}(M(2^i),F_p/F_{p+1}) = 0 \quad \text{for } s \geq i+2 \text{ and } p \leq 2^{i+1}-1.$$

(2)(a) follows from the fact that $M(2^i)_q = 0$ for $q > 2^{i+1}-1$ ([3]). To see (2)(b) note that if $F_p/F_{p+1} \neq 0$ then $\mathrm{Ext}_A^{s,s}(M(2^i),F_p/F_{p+1})$ is a direct sum of Ext groups each of which is

$$\mathrm{Ext}_A^{s,s}(M(2^i),\Sigma^p\mathbb{Z}/2) = \mathrm{Ext}_A^{s,s+p}(M(2^i),\mathbb{Z}/2).$$

By Lemma 7, for $s \geq i+2$ and $p \leq 2^{i+1}-1$, $\mathrm{Ext}_A^{s,s+p}(M(2^i),\mathbb{Z}/2) = 0$. Thus $\mathrm{Ext}_A^{s,s}(M(2^i),F_p/F_{p+1}) = 0$ for $s \geq i+2$ and $p \leq 2^{i+1}-1$. This proves Lemma 5.

The second part follows the following.

Proposition 8. For each $k \geq 1$ there is a homotopy element in $\pi_{2k}(B(k))$ having order at least $2^{\nu(k)+2}$.

Proof. Brown and Cohen prove in [2] that in the mod 2 Adams spectral sequence for $B(k)$

$$E_\infty^{s,t} = E_2^{s,t} \cong \Lambda(k)^{s,t-s} \quad \text{for} \quad t-s \leq 2k.$$

Let $\{\lambda_{2k}\} \in \pi_{2k}(B(k))$ be a homotopy element detected by $\lambda_{2k} \in E_\infty^{1,2k+1}$. Let $k = 2^m\bar{\ell}$, $\bar{\ell}$ = odd. One calculates by induction that

$$\lambda_0^i\lambda 2k = \lambda_{2^{m-i+1}\bar{\ell}}\,\lambda_{2^{m-i+1}\ell}\,\lambda_{2^{m-i+2}\bar{\ell}}\,\lambda_{2^{m-i+3}\bar{\ell}}\cdots\lambda_k$$

for $1 \leq i \leq m+1$. Since these are non-zero classes in $E_\infty^{*,2k+*}$ and multiplication by λ_0 correspond to multiplication by 2 in homotopy it follows that $\{\lambda_{2k}\}$ has order at least $2^{m+2} = 2^{\nu(k)+2}$. Q.E.D.

The above argument is first given by Barrat and Peterson for $k = 2^i$ ([4]).

Using a recent result of P. Goerss ([8]) one can give a direct proof of Theorem 1 which avoids the Adams spectral sequence argument. We sketch the proof as follows.

Let $T(2k)$ be the Spanier-Whitehead dual of $B(k)$ with top class $\iota_{2k} \in H_{2k}(T(2k))$. $T(2k)$ is first considered by H. Miller in [11]. He observes that $H_*(T(2k))$ is a projective in the category of unstable modules over the mod 2 Steenrod algebra A. Here the notion of an unstable module M over A is dual to that of the usual one, that is, A acts on M from the right and the Steenrod operations

$$Sq^n : M_{2n} \to M_n$$

lower the degrees. Then $H_*(T(2k)) \cong \Sigma^{2k}A/\{Sq^i : i > k\}A$ as right modules over A.

Theorem 1 is equivalent to

Theorem 1'. $1_{2k} : T(2k) \to T(2k)$ has order $2^{\nu(k)+2}$.

P. Goerss proves in [8] the following.

Theorem 9. ([8]) For any CW-complex X, the map

$$\Phi : [T(2k), \Sigma^\infty X] \to \mathrm{Hom}_R(H_*\Omega^\infty T(2k), H_*\Omega^\infty\Sigma^\infty X)$$

given by $\Phi(f) = \Omega^\infty f_*$ is an isomorphism.

Here R is the abelian category of commutative, cocommutative Hopf algebras that are left modules over the Dyer-Lashof algebra and right modules over A with the two actions being mixed by the Nishida relations. The group multiplication in $\mathrm{Hom}_R(H,K)$ is given by

$$H \xrightarrow{\Delta} H \otimes H \xrightarrow{f\otimes g} K \otimes K \xrightarrow{m} K.$$

J. Lannes has shown ([9]) that T(2k) is a wedge summand of a suspension spectrum. (Another proof of this result is given by Goerss in [7]). This implies Goerss' result 9 is still true when $\Sigma^\infty X$ is replaced by T(2k). So to prove Theorem 1' it suffices to show

$$(3) \qquad 2^{\nu(k)+2}\Phi(1_{2k}) = 0 \quad \text{and} \quad 2^{\nu(k)+1}\Phi(1_{2k}) \neq 0.$$

Since T(2k) is a retract of a suspension spectrum there is a map $\tau : T(2k) \to \Sigma^\infty\Omega^\infty T(2k)$ so that $\sigma\tau = 1$ where $\sigma : \Sigma^\infty\Omega^\infty T(2k) \to T(2k)$ is the evaluation map. Let $j_{2k} \in H_{2k}(\Omega^\infty T(2k))$ be defined by $j_{2k} = \tau_*\iota_{2k}$ (identify $H_*(\Omega^\infty T(2k))$ with $H_*(\Sigma^\infty\Omega^\infty T(2k))$). Goerss also shows in [8] that

$$f \in \mathrm{Hom}_R(H_*(\Omega^\infty T(2k)), H_*(\Omega^\infty T(2k))) \text{ is } 0 \iff f(j_{2k}) = 0.$$

So to prove (3) is to prove

$$(4) \qquad [2_\nu^{\nu(k)+2}(\Omega^\infty 1_{2k})_*](j_{2k}) = 0 \quad \text{and}$$

$$[2^{\nu(k)+1}(\Omega^\infty 1_{2k})_*](j_{2k}) \neq 0.$$

It is easy to see that

$$\Delta(j_{2k}) = 1 \otimes j_{2k} + j_{2k} \otimes 1 + \sum_i x_i' \otimes x_i'' + \xi j_{2k} \otimes \xi j_{2k}$$

where Δ is the coproduct of $H_*(\Omega^\infty T(2k))$, $x_i' \neq x_i''$ and $\xi = Sq^n : H_{2n}(\Omega^\infty T(2k)) \to H_n(\Omega^\infty T(2k))$. From the definition of the multiplication in $\mathrm{Hom}_R(\ ,\ )$ and the fact that Δ is cocommutative we see

(5) For $f \in [T(2k),T(2k)]$ if $\Omega^\infty f_*(j_{2k}) = x$ then $[2(\Omega^\infty f_*)](j_{2k}) = \Omega^\infty(2f)_*(j_{2k}) = (\xi x)^2$.

Since $H_*(\Omega^\infty T(2k))$ is a polynomial algebra (this also follows from Lannes' result) and $\xi(y^2) = (\xi y)^2$ for $y \in H_*(\Omega^\infty T(2k))$ it follows from (5) that (4) is equivalent to

(6) $\xi^{\nu(k)+2}(j_{2k}) = 0$ and $\xi^{\nu(k)+1}(j_{2k}) \neq 0$.

Since τ_* is a monomorphism, (6) is equivalent to

(7) $\xi^{\nu(k)+2}(\iota_{2k}) = 0$ and $\xi^{\nu(k)+1}(\iota_{2k}) \neq 0$.

Now since $H_*(T(2k)) = \Sigma^{2k}A/\{Sq^i : i > k\}A$ one easily calculates that

$$\xi^{\nu(k)+1}(\iota_{2k}) = \iota_{2k} Sq^k Sq^{k/2} Sq^{k/2^2} \dots Sq^{(2\ell+1)} \neq 0 \text{ and}$$

$$\xi^{\nu(k)+2}(\iota_{2k}) = 0$$

where $k = 2^{\nu(k)}(2\ell+1)$. This proves (7) and therefore Theorem 1'.

Proposition 8 is suggested by the referee which replaces the original lengthy proof of a result of the same kind. He also suggests Theorem 1' together with the above simple and elegant argument. I would like to thank the referee for taking efforts in making these suggestions and for his allowing me to record his arguments in the paper.

This research is supported by the National Science Council of the Republic of China and also a China Foundation Fellowship.

References

[1] A.K. Bousfield, E.B. Curtis, D.M. Kan, D.G. Quillen, D.L. Rector and J.W. Schlessinger, The mod-p lower central series and the Adams spectral sequence II, Topology 9(1970), 309-316.

[2] E.H. Brown Jr. and R.L. Cohen, The Adams spectral sequence of Ω^2S^3 and Brown-Gitler spectra, To appear on the conference in celebration of J.C. Moore 60th birthday (1983).

[3] E.H. Brown Jr. and S. Gitler, A spectrum whose cohomology is certain cyclic module over the Steenrod algebra, Topology 12(1973), 283-296.

[4] E.H. Brown Jr. and F.P. Peterson, The Brown-Gitler spectrum, Ω^2S^3 and $\eta_j \in \pi_{2^j}(s)$, Proc. Steklov. Inst. Math. Issue 4(1984), 41-45.

[5] R.L. Cohen, Odd primary infinite families in stable homotopy theory, Memo. of AMS, No. 242 (1981).

[6] R.L. Cohen and P. Goerss, Secondary Cohomology Operations that detect homotopy classes, Topology Vol. 23, No. 2 (1984), 177-194.

[7] P.G. Goerss, A direct construction for the duals of Brown-Gitler spectra, Indiane J. Math. 34(1985), 733-751.

[8] P.G. Goerss, Unstable projectives and stable Ext: with applications, Proc. London Math. Soc. (3) 53 (1986) 539-561.

[9] J. Lannes, Sur le n-dual du n-eme spectre de Brown-Gitler', preprint, Ecole Polytechnique, 1984.

[10] M.E. Mahowald, A new infinite family in ${}_2\pi_*^S$, Topology 16(1977), 249-256.

[11] H. Miller, The Sullivan conjecture on maps from classifying spaces, Ann. of Math. 120(1984), 449-458.

Department of Mathematics
National Tsing Hua University
Hsinchu, Taiwan 30043
Republic of China

TOPOLOGY OF THE INTERSECTION OF QUADRICS IN $\mathbb{R}^n$

Santiago López de Medrano
Instituto de Matemáticas and Facultad de Ciencias
Universidad Nacional Autónoma de México
Ciudad Universitaria, 04510 México, D. F. MEXICO

1. The results. Consider the set of points in $\mathbb{R}^n$ that satisfy the following 2 quadratic homogeneous equations:

$$a_1x_1^2 + \ldots + a_nx_n^2 = 0 \qquad (1)$$

$$b_1x_1^2 + \ldots + b_nx_n^2 = 0 \qquad (2)$$

This set is a cone from the origin and we let M be its intersection with the unit sphere. M is determined by the configuration of n points $A_i = (a_i, b_i)$ in $\mathbb{R}^2$. In particular, it is the transverse intersection of the sets defined by its three defining equations if and only if, the following generic condition is satisfied:

(*) *The origin in* $\mathbb{R}^2$ *is not a convex combination of any pair of vectors* A_i .

Under condition (*), M is a smooth manifold of dimension $n-3$, whose diffeomorphism type will be described below.

First we have to describe all essentially different configurations of n points in $\mathbb{R}^2$ satisfying condition (*).

We consider a configuration as an unordered set of n vectors in $\mathbb{R}^2$, some of which may coincide. If one configuration satisfying condition (*) is deformed continuously without breaking this condition, then the diffeomorphism type of the manifold M does not change, so we have to describe the connected components of the space of configurations satisfying (*).

To do this, define an odd cyclic partition of the positive integer n to be an equivalence class of partitions

$$n = n_1 + n_2 + \ldots + n_k$$

of n into an odd number k of positive integers, where two such partitions are equivalent if they differ by a cyclic permutation of the n_i.

To an odd cyclic partition we associate a canonical configuration as follows: consider a regular k-gon with center at the origin, and the configuration consisting of the vertices of this polygon with succesive multiplicities n_i (in the positive direction).

Theorem 1. *This association establishes a one-to-one correspondence between the set of connected components of the space of configurations satisfying (*) and the set of odd cyclic partitions of n.*

The partition $n = n_1 + \ldots + n_k$ plays the same role as the Morse signature $n = p + q$ of a single non-degenerate quadratic form, so it can be thought of as the "signature" of a non-degenerate pair of quadratic forms. (It can be defined also when they are not simultaneously diagonizable).

The number of odd cyclic partitions of n can be computed for a given n, and a formula can be produced using Möbius inversion. It is roughly equal to $2^{n-1}/n$, this being a lower bound which is attained if, and only if, n is a power of 2. If n is prime it is the integer following $2^{n-1}/n$. For $n = 4,5,6$ it is $2,4,6$ corresponding to the types of regions into which the plane is divided by a generic n-gon. For $n = 23$ it is $182,362$.

To describe the diffeomorphism type of M, let $k = 2\ell + 1$ and define d_i to be the sum of the numbers from n_i to $n_{i+\ell-1}$ in the cyclic order given by the partition $n = n_1 + \ldots + n_k$. Then we have

Theorem 2. *If M is the manifold corresponding to the partition $n = n_1 + \ldots + n_k$, then*

i) *if* $k = 1$, $M = \emptyset$

ii) *if* $k = 3$, $M = S^{n_1-1} \times S^{n_2-1} \times S^{n_3-1}$

iii) *if* a) $k = 5$ and $d_i > 2$ *or*

b) $k > 5$ and $n \neq 7$

then M *is diffeomorphic to the connected sum of the manifolds*

$$S^{d_i-1} \times S^{n-d_i-2} \qquad i = 1,\ldots,k.$$

Remarks 1. The restrictions in case iii) come from the fact that our proof uses the h-cobordism theorem, so we have to assume that M is simply connected and of dimension different from 3 and 4. Other cases can be dealt with using the 2-dimensional ones and an open book construction. In all known cases the conclusion of iii) holds, so it really must be true without restriction for $k > 3$.

As stated, Theorem 2 covers most of the cases and certainly all those used in the applications (cf. #3). Still, it is clear that a different proof is needed to cover all the cases.

2. Most of the steps in the proofs work also for the intersection of 3 or more quadrics, a question that arises naturally in the study of $\mathbb{R}^k$-actions ([Ch]). The main unsolved question is the determination of the basic configurations (analogous to the odd regular polygons). It is easy to see that one can obtain (when one intersects again with the unit sphere) all products of spheres, products of connected sums of the type iii) with spheres, etc., but also new connected sums (which are clearly not products).

3. At the Berkeley International Congress, professor F. Hirzebruch kindly explained to us his construction of quadrics from arrangements of lines ([H]). From this construction all 2-dimensional cases of intersections of any number of quadrics can be described. In fact, it complements our approach to give a perfect correspondence between a certain class of polytopes and intersections of quadrics, and with $\mathbb{Z}_2$, S^1 and torus actions having a polytope as quotient (cf. #3).

4. After the Berkeley Congress we learned from professor C.T.C. Wall that Theorem 1 and some of the homology computations leading to Theorem 2 had been previously discovered by him when solving a problem coming from singularity theory ([W]). In this paper a connection is established with the theory of Galė diagrams, coming from the theory of convex polytopes, and closely related with the construction mentioned in the previous remark (Cf. [G]). (The numbers given in [G] and [W] differ from ours, due to the fact that they correspond to equivalence classes, and not to connected components).

Our slightly different approach to this problem, coming from questions in dynamical systems, led us to Theorem 2, which is not contained in [W] and seems to be new in its essential part (case iii).

2. Proofs of Theorems 1 and 2.

To prove Theorem 1 one must show that any configuration S can be deformed to a unique canonical one. To do that one defines an equivalence relation on the elements of S as follows: $A \sim B$ if for no element C of S the convex hull of A, B, C contains the origin. This is the same as saying that A can be deformed to B, all the points of $S \setminus \{A\}$ remaining fixed, without breaking condition (*). It follows that we really have an equivalence relation and that each equivalence class can be deformed to a multiple point. Also each class can be deformed radially to the unit circle. Let $A_1, \ldots, A_k$ be the points in the unit circle corresponding to these classes, in their cyclic order.

<u>Lemma 1</u>. *k is odd: $k = 2\ell + 1$. If $k > 1$ and A_i, A_{i+1} are two contiguous points, then $j = i + \ell + 1$ is the unique (mod. k) value of j such that 0 is in the convex hull of A_i, A_{i+1}, A_j.*

<u>Proof</u>. If $k > 1$, consider the points A_k, A_1. Since they correspond to different classes there must be (at least) a point $A_{\ell+1}$ such that $0 \in \operatorname{Conv}\{A_k, A_1, A_{\ell+1}\}$. For $i = 1, \ldots, \ell$ there is also (at least) one point $s(i)$ such that $0 \in \operatorname{Conv}\{A_i, A_{i+1}, A_{s(i)}\}$. Since all $s(i)$ must be different for different i, and $s(i)$ must be in the interval $[\ell + 2, k]$, it follows that $\ell \leqslant k - \ell - 1$. Applying the same argument to $i = \ell + 1, \ldots, k - 1$ one obtains $\ell \geqslant k - \ell - 1$, so $k = 2\ell + 1$. It follows also that there is only one choice for $s(i)$, and that it is $i + \ell + 1$, so lemma 1 is proved.

No two i's can have the same $s(i)$

If α_i is the angle $A_i\, O\, A_{i+1}$, then lemma 1 says that $\alpha_i + \cdots + \alpha_{i+\ell-1} < \pi < \alpha_i + \cdots + \alpha_{i+\ell}$. Fixing A_1, the set of points $\{A_2, \ldots, A_k\}$ is determined by the k-ad of angles $(\alpha_1, \ldots, \alpha_{k-1})$ satisfying theses inequalities, and since the collection of such k-ads is convex, given any two sets of points $\{A_1, \ldots, A_k\}$ as in lemma 1 one can be deformed into the other one. In particular any one can be deformed to the vertices of a regular polygon with $2\ell + 1$ sides. Thus S can be deformed to a canonical form which is a unique up to a rotation, and Theorem 1 is proved. (Actually, this shows that each component of the set of configurations satisfying (*) has the homotopy type of S^1).

<u>Proof</u> <u>of</u> <u>Theorem</u> <u>2</u>. From Theorem 1 we can assume the configuration S is in normal form, having points $V_1, \ldots, V_k$ at the vertices

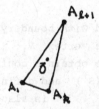

of a regular polygon of $k = 2\ell + 1$ sides, with multiplicities $n_1, \dots, n_k$, $n_1 + \cdots + n_k = n$.

Let $\mathcal{O} = (\mathbb{R}^+)^n$ be the closed first orthant of $\mathbb{R}^n$ and $K = M \cap \mathcal{O}$. The homeomorphism of $\mathcal{O}$ to itself given by $x_i \to x_i^2$ sends K to the set K' defined by

$$\begin{aligned} a_1x_1 + \cdots + a_nx_n &= 0 \\ b_1x_1 + \cdots + b_nx_n &= 0 \qquad x_i \geqslant 0 \\ x_1 + \cdots + x_n &= 1 \end{aligned}$$

which is a convex polytope, whose faces are its intersections with the coordinate subspaces of $\mathbb{R}^n$. M is the union of 2^n copies of K obtained by reflection on the hyperplanes $x_i = 0$ and this gives a cell decomposition of M.

K' is also the set of ways of expressing $\bar{0}$ as a convex combination of the points A_i, so K^1, and M, are empty if, and only if, $\bar{0}$ is not in the convex hull of the A_i. In this case S is called a *Poincaré configuration*, and is in the case $k = 1$ of Theorem 2. Otherwise S called a *Siegel configuration*, $k > 1$, and M is non-empty.

Together with M we will consider the manifold with boundary $Q \subset \mathbb{R}^{n+1}$ defined by adding a new point A_o at the vertex V_1 (i.e. increasing the multiplicity of V_1 to $n_1 + 1$) to obtain a configuration S^+, forming the corresponding manifold $M(S^+)$ and defining $Q = M(S^+) \cap \{x_o \geqslant 0\}$. Then clearly $M = \partial Q = M \cap \mathbb{R}^n$. Q is also the union of 2^n cells obtained by reflecting $K(S^+)$ on the hyperplanes $x_i = 0$, $i = 1,..,n$. (Reflecting also on $x_o = 0$ would give the double of Q, which is $M(S^+)$).

(Projecting Q to the hyperplane $x_o = 0$ gives a diffeomorphic Q' of the form $\Sigma(A_i - A_o)x_i^2 = -A_o$, $\Sigma x_i^2 \leqslant 1$, so Q can be considered as an unfolding of the cone singularity $\Sigma A_i x_i^2 = 0$ at the origin).

For every subset $R \subset S$ denote by $K(R)$ the corresponding face of $K = K(S)$. We say that R is *indispensable* if $K(S \setminus R)$ is empty (i.e., if $S \setminus R$ is Poincaré). We will see that the topology of M and Q is determined by the indispensable subsets of S. In particular we have

<u>Lemma 2</u>. M *is* r*-connected iff every indispensable subset of* S *has at least* $r + 2$ *elements.*

This is clear if we look at dual cell decomposition of M, which can be thought of as a subcomplex of the n-cube, a cell being included in the subcomplex if and only if the dual cell $K(R)$ is non-empty. If there are no indispensable subsets of S with $r+1$ elements or less, then the $(r+1)$-skeleton of the dual cell complex is the whole $(r+1)$-skeleton of the cube. The *minimal* indispensable subsets produce cycles, a fact that will be made clear in the following lemmas.

Let g_i be the reflection on the hyperplane $x_i = 0$, and for $R \subset S$, g_R be the product of the g_i such that $A_i \in R$. The cells of M are then of the form $g_R K(R')$ where $R \subset R' \subset S$, so the chains are

$$C_* = \{ \Sigma\, n(R,R') g_R K(R') \mid n(R,R') \in \mathbb{Z},\ \ R \subset R' \}$$

Let now $G(R)$ be the group generated by the g_i with $A_i \in R$ and $\mathbb{Z}G(R)$ the corresponding group ring. Let also $h_i = 1 - g_i \in \mathbb{Z}G(S)$ and h_R the product of the h_i for which $A_i \in R$.

Lemma 3. $\mathbb{Z}G(R)$ *is freely generated (as a group) by the elements* $h_{R'}$, $R' \subset R$.

The proof is immediate by induction on the number of elements of R.

By lemma 3, $C_*(M)$ decomposes as the direct sum of the *subcomplexes* $h_R C_I(K)$, which have a geometric meaning: let $L(S,A_i)$ be the face $K(S \setminus \{A_i\})$ and $L(S,R)$ the union of the faces $L(S,A_i)$ for $A_i \in R$. Since on the face $L(S,A_i)$ g_i acts as the identity, h_R annihilates all the chains in $L(S,R)$ so we can identify $h_R C_*(K)$ with $C_*(K,L(S,R))$ so we have

Lemma 4.

$$H_*(M) \approx \bigoplus_{R \subset S} H_*(K,L(S,R))$$

New let U_i be the set of points concentrated on the vertex V_i, (i.e., U_i is one of the equivalence classes considered in the proof of Lemma 1). The *minimal indispensable* (m.i.) subsets of S are precisely the unions of ℓ consecutive classes U_i, and their complements, the *maximal Poincaré* (M.P.) subsets of S, are unions of $\ell + 1$ consecutive classes. If R is m.i. let R^* be R union one point from each class contiguous to R. If R is P.M. then R^* will be R union one point from one of the contiguous classes:

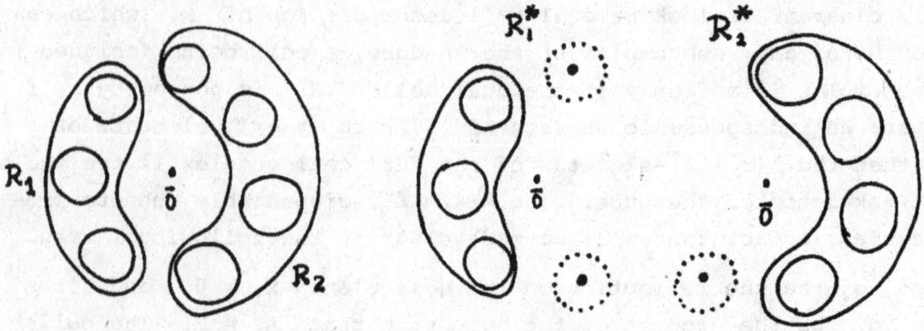

R_1 minimal indispensable, $R_2 = S \setminus R_1$ maximal Poincaré.

If R is m.i. or P.M. then $\partial K(R^*) \subset L(S,R)$.

<u>Proposition 1</u>. *If* $R = \phi$ *then* $L(S,R) = \phi$
If $R = S$ *then* $L(S,R) = \partial K \sim S^{n-4}$

If R *is minimal indispensable or maximal Poincaré then*

$$\partial K(R^*) \subset L(S,R)$$

is a homotopy equivalence.

In all other cases $L(S,R)$ *is contractible.*

The proof follows from the next 4 lemmas, where it is more convenient to work in the homeomorphic linear model K'.

<u>Lemma 5</u>. *If* R *is not a union of classes* U_i *then* $L'(S,R)$ *is contractible.*

<u>Proof</u>. Assume $A_1, A_2 \in U_1$, $A_1 \in R$ but $A_2 \notin R$. Then $F: K\,(S) \times I \to K'(S)$ given by $F(x_1, x_2, x_3, \ldots, n, t) =$ $= ((1-t)x_1, x_2 + tx_1, x_3, \ldots, x_n)$ defines a deformation of $K'(S)$ onto $K'(S \setminus A_1) \approx L'(S,A_1)$ where every $L'(S,A_i)$ except $L'(S,A_2)$ is deformed inside itself. Therefore $L'(S,A_1)$ is a deformation retract of $L'(S,R)$.

<u>Lemma 6</u>. *If* R *or* $S \setminus R$ *is Siegel then* $L'(S,R)$ *is a* $(n-4)$-*cell.*

<u>Proof</u>. If $S \setminus R$ is Siegel then $K'(S \setminus R)$ is not empty so $L'(S,R)$ retracts to any point of $K'(S \setminus R)$. But actually $L'(S,R)$ is a union of cells glued along cells in their boundaries, so inductively is also a cell. If R is Siegel then $L'(S,R)$ is the complementary cell of $L'(S, S \setminus R)$.

<u>Lemma 7</u>. *If* R *is m.i. then* $\partial K'(R^*) \subset L'(S,R)$ *is a homotopy*

equivalence.

Proof. Assume $R^* = R \cup \{P_1, P_2\}$ and let $(u_1, u_2, \bar{x}, \bar{y})$ represent a point of K' where $\bar{x}$ denotes coordinates corresponding to points in R, $\bar{y}$ denotes those corresponding to points in $S \setminus R^*$ and u_i the one corresponding to P_i. Then given $(u_1, u_2, \bar{x}, \bar{y})$ there is a unique point $\frac{1}{v}(u_1', u_2', \bar{x}, 0) \in K'(R^*)$ where $v = u_1' + u_2' + \Sigma\bar{x}$. (Since one can find unique u_1', u_2' such that $(u_1', u_2', \bar{x}, 0)$ satisfies the first two equations defining K'; then u_1', u_1' must be positive since K' becomes Poincaré if we change R_1 to $-P_1$ or P_2 to $-P_2$, and we can normalize by dividing by $v > 0$). This gives a retraction of $L'(S,R)$ to $\partial K'(R^*)$ which deforms linearly to the identity keeping $L'(S,R)$ inside itself.

Lemma 8. *If* R *is* P.M. *then* $\partial K'(R^*) \subset L'(S,R)$ *is a homotopy equivalence.*

Proof. Let $R^* = R \cup \{P_2\}$ and P_1 be a point in the class of R opposite P_2:

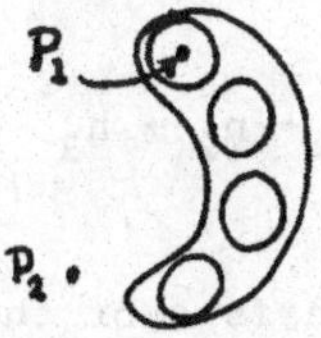

Let $S' = S \setminus \{P_1\}$, $R' = R \setminus \{P_1\}$. Then $L'(S,R) = L'(S,R') \cup \cup L'(S,P_1)$ and $L'(S,R') \cap L'(S,P_1) = L'(S',R')$. If P_1 is the only element in its class then in S', P_2 becomes equivalent to the points in the class of R near it, so by lemma 5 $L'(S',R')$ is contractible and $L'(S,R)$ retracts to $L'(S,R')$. But in that case R' is m.i. in S, $R'^* = R^*$ and $\partial K'(R^*) = \partial K'(R'^*) \subset L'(S,R')$ is a homotopy equivalence by lemma 7.

If there is more than one element in the class of P_1, then we proceed by induction on their number. In that case R' is still M.P. in S', $R'^* = R' \cup P_2$ and from the Mayer-Vietoris sequences of the triples $(L'(S,R); L'(S,R'), L'(S,P_1))$ and $(L'(R^*,R); L'(R^*,R'), L'(R^*,P_1))$ one gets

$$\begin{array}{ccc} H_{i+1}(L'(S,R)) & \overset{\approx}{\rightarrow} & H_i(L'(S',R')) \\ \uparrow & & \uparrow \\ H_{i+1}(L'(R^*,R)) & \overset{\approx}{\rightarrow} & H_i(L'(R'^*,R')) \end{array}$$

(since the second terms of each triple are contractible by lemma 5 and the third ones are cells). Now, because $L'(R^*,R) = \partial K'(R^*)$ and $L'(R'^*,R') = \partial K'(R'^*)$, the vertical arrow on the right is an isomorphism for all i (by induction hypothesis) and so is the one on the left, and the lemma is proved.

Going back to Proposition 1, the first two assertions are immediate, the second one follows from lemmas 7 and 8 and all the "other cases" follow from lemmas 5 and 6.

We have now computed all the homology groups of M: they are all free, and beside the top and bottom classes there is one generator in dimension $s - 1$ for every m.i. with s elements, and one in dimension $r - 2$ for every P.M. with r elements. This follows from the proposition, since $H_{i+1}(K,L) \to H_i(L)$ is always an isomorphism. Notice the shift in dimension of the relative classes.

(Notice also that complementarity of subset of S corresponds to Alexander duality in ∂K and to Poincaré duality in M).

This shows that the homology of M corresponds with the conclusions of Theorem 2. To finish the proof we have to distinguish the cases $k = 3$ and $k > 3$.

<u>Lemma 9</u>. *If* $k = 3$, $n = n_1 + n_2 + n_3$ *then* $M = S^{n_1-1} \times S^{n_2-1} \times S^{n_3-1}$.

<u>Proof</u>. We can assume S consists of the points $(2,0)$, $(-1,1)$, $(-1,-1)$ with multiplicities n_1, n_2, n_3. Then the equations defining M are equivalent to three equations of the form $\Sigma x_i^2 = 1/3$ with no common variables, so M is *equal* to the product of those three spheres.

In this case it is not true that all homology classes below the top dimension are always spherical. Surprisingly we have

<u>Lemma 10</u>. *If* $k > 3$ *then all the homology classes of* M *in dimensions* $\neq 0, n - 3$ *can be represented by embedded spheres with trivial normal bundle.*

<u>Proof</u>. If $R \subset S$ then $M(R) \subset M(S)$ with trivial normal bundle. If R is m.i. then R^* is a configuration with three classes, one is R and the others consist of one point. Therefore $M(R) = S^{r-1} \times S^0 \times S^0$ and it is clear from Proposition 1 that the corresponding class in $M(S)$ is represented by one of these spheres.

If R is M.P. then R^* is again a configuration with three classes; one is the extra point, another one is the class U of R opposite to this point (with u elements), and the third one V is the rest of R (with v elements).

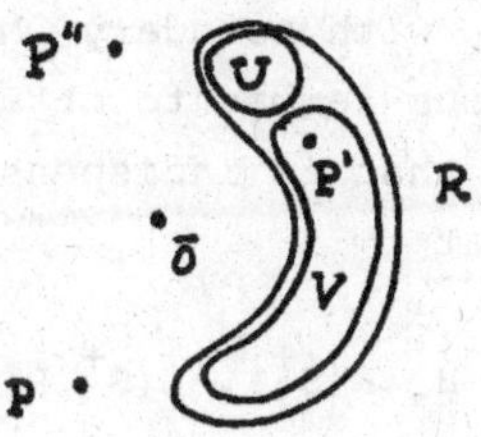

Then it follows that $M(R^*) = S^{u-1} \times S^{v-1} \times S^0$ and that the class corresponding to R in $M(S)$ is represented by an embedded $S^{n-1} \times S^{v-1}$ with trivial normal bundle. Now V is an m.i. in S, so S^{v-1} is non trivial in $M(S)$.

But now the hypothesis $k > 3$ *comes in: it implies that* U *cannot be* m.i. *nor* M.P. *in* S, *so it cannot carry any homology and therefore the sphere* S^{u-1} *has to be trivial in* $M(S)$.

To be more precise, let $P' \in V$ be a point in a class of S contiguous to U, and $P'' \notin V$ a point in the other class contiguous to U. Let $U^* = U \cup \{P,P'\}$ and $\bar{U} = U^* \cup \{P''\}$. Then, since $k > 3$, $\bar{U}$ has three classes: $U \cup \{P''\}$, $\{P\}$ and $\{P'\}$, so $M(\bar{U}) \approx$ $\approx S^u \times S^0 \times S^0$. (If $k = 3$, P and P'' would be in the same class). Now we have

$$\begin{array}{l} M(R^*) \approx S^{u-1} \times S^{v-1} \times S^0 \\ \quad \cup \\ M(U^*) = S^{u-1} \times S^0 \times S^0 \\ \quad \cap \\ M(\bar{U}) \approx S^u \times S^0 \times S^0 \end{array}$$

and $M(R^*) \cap M(\bar{U}) = M(R^* \cap \bar{U}) = M(U^*)$, so the sphere S^{u-1} in $S^{u-1} \times S^{v-1}$ bounds a disk D^u in $M(S)$ such that $S^{u-1} \times S^{v-1} \cap D^u$ $= S^{u-1}$. So we can perform surgery on S^{u-1} to get an embedded $(u+v-2)$-sphere representing the same homology class as $S^{u-1} \times S^{v-1}$. To be sure that one can extend the framing on S^{u-1} to D^u (so that the $(u+v-2)$-sphere has trivial normal bundle) one needs, as usual, that $u - 1 < \frac{1}{2}(n-3)$. If this is not the case start all over again interchanging the roles of P and P'', and considering a class U'' on the other side of R. Now we cannot have also $u'' - 1 \geqslant \frac{1}{2}(n-3)$, because $u + u'' + 3 \leqslant n$, as there are at least three more classes in S. Lemma 10 is now proved.

To conclude the case $k > 3$ of Theorem 2 from lemma 10, one would need that all spheres be disjoint, except for those representing dual classes, but this is not always true. To finish the proof one has to use the manifold Q with boundary M. One can easily modify the proofs of the previous lemmas to obtain that Q is r-connected iff every subset of S that is indispensable in S^+ has at least $r+2$ elements, and that

$$H_*(Q) \approx \bigoplus_{R\subset S} H_*(K(S^+), L(S^+,R)).$$

It follows that non-trivial homology classes in Q correspond with subsets of S that are m.i. or P.M. in S^+. This shows that out of each pair of complementary classes in M exactly one goes to 0 in Q, and that $H_*(M) \to H_*(Q)$ is onto. (If one adds a point to a configuration in such a way that the number of classes increases, then this is not true). It follows that all the homology of Q can be represented by embedded spheres with trivial normal bundle coming from $\partial Q = M$, which means they can be made disjoint in the interior of Q. Taking the closed tubular neighborhoods of these spheres and joining them by thin tubes one gets a manifold Q' with boundary M' which is a connected sum of products of spheres. If the conditions of Theorem 2 part (iii) are fulfilled, then $Q - \mathrm{int}Q'$ is a simply connected h-cobordism between M and M', which are then diffeomorphic by the h-cobordism theorem. This completes the proof of Theorem 2.

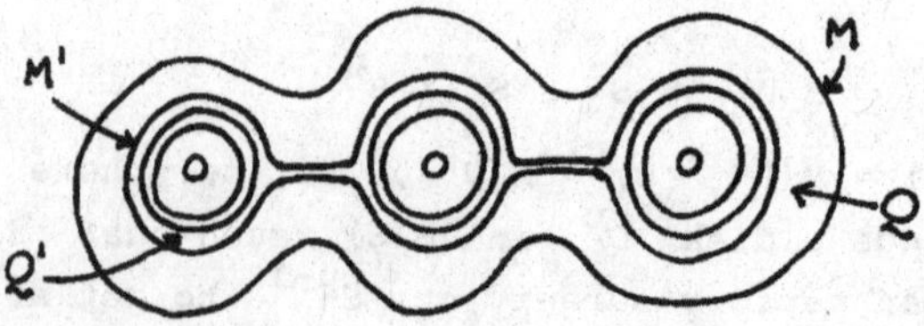

3. Applications

a) *Holomorphic dynamical sistems.* This was the original motivation for this work. Camacho, Kuiper and Palis study in [CKP] the dynamical system in $\mathbb{C}^n$:

$$\dot{Z} = AZ$$

In this case the natural generic condition is

(**) *Any two eigenvalues of* A *are linearly independent over* $\mathbb{R}$.
This condition on the configuration of eigenvalues is stronger than (*), but it is clear that the connected components are again given by Theorem 1. Canonical representatives must consist in this case of clouds of points on the unit circle in the neighborhood of a regular polygon. This solves a question posed in [CKP].

The odd cyclic partition is then *the* discrete invariant of operators A satisfying (**), all the rest being moduli.

In this case the relevant variety is

$$\lambda_1 z_1 \bar{z}_1 + \lambda_2 z_2 \bar{z}_2 + \ldots + \lambda_n z_n \bar{z}_n = 0$$

where λ_i are the eigenvalues of A. This is the intersection of two quadrics in $\mathbb{R}^{2n}$ whose configuration consists of all the λ_i with multiplicity 2, so one gets the cones on the manifolds of Theorem 2 with all the n_i even. This is the variety of "Siegel leaves" of the dynamical system, i.e., the space of its solutions that do not have the origin on their closure. It is from the classical distinction in dynamical systems between the Poincaré and Siegel domains that we have adopted our terminology.

b) *Group actions*. In the manifold M corresponding to $n = n_1 + \ldots + n_k$ we clearly have a natural action of $O(n_1) \times \ldots \times O(n_k)$. In particular the following examples look interesting:

Theorem a) *For* $\ell > 1$ *there is an action of* $\mathbb{Z}_2^{2\ell+1}$ *on the connected sum of* $2\ell+1$ *copies of* $S^{\ell-1} \times S^{\ell-1}$.

b) *For* $\ell > 1$ *there is an action of* $(S^1)^{2\ell+1}$ *on the connected sum of* $2\ell+1$ *copies of* $S^{2\ell-1} \times S^{2\ell}$.

Part b) generalizes in one direction an example of McGavran [Mc] of an action of the 5-torus on the connected sum of 5 copies of $S^3 \times S^4$, and we can therefore give a description of the first of McGavran's actions by simple algebraic equations:

$$z_0\bar{z}_0 + \rho z_1\bar{z}_1 + \rho^2 z_2\bar{z}_2 + \rho^3 z_3\bar{z}_3 + \rho^4 z_4\bar{z}_4 = 0$$

$$z_0\bar{z}_0 + z_1\bar{z}_1 + z_2\bar{z}_2 + z_3\bar{z}_3 + z_4\bar{z}_4 = 1$$

where ρ is a fifth root of unity. It should be acknowledged that McGavran's result was a very important element in the process of finding Theorem 2.

When all the n_i are equal, as in the above example, we have an additional action of $\mathbb{Z}_k$ by cyclic permutation of coordinates. This even gives an action of $\mathbb{Z}_k$ on the polytope K'. When all n_i are 1 this action is fixed point free of $\partial K'$, from which we get interesting cell decompositions of some lens spaces.

4. Acknowledgments. Central parts of this work were done while the author enjoyed a visit to I.H.E.S., for which he is grateful to N. Kuiper, O. Lanford and D. Sullivan. Conversations with W. Browder, M. Chaperon, X. Gómez-Mont, F. González-Acuña, F. Hirzebruch, N. Kuiper and D. Sullivan were very helpful. Thanks are due also to miss Lourdes Arceo for her kind and efficient typing of several versions of this manuscript.

The problem itself was suggested by Alberto Verjovsky. This work is largely a consequence of Alberto's insistence, enthusiasm and crucial orientations.

References

[CKP] C. Camacho, N. Kuiper and J. Palis. The topology of holomorphic flows with singularities. Publications Mathematiques I.H.E.S., 48 (1978), 5-38.

[Ch] M. Chaperon, Géometrie différentielle et singularités de systèmes dynamiques. Asterisque (1986), 138-139.

[G] B. Grünbaum, Convex Polytopes, Wiley, 1967.

[H] F. Hirzebruch, Arrangements of lines and algebraic surfaces, in M. Artin, J. Tate, eds., Arithmetic and Geometry, Papers dedicated to I.R. Shafarevich, Vol. II, Birkhäuser, 1983, 114-140.

[Mc] D. McGavran. Adjacent connected sums and torus actions, Trans. Amer. Math. Soc. 251 (1979), 235-254.

[W] C.T.C. Wall, Stability, pencils and polytopes, Bull. London Math. Soc., 12 (1980), 401-421.

ORIENTATIONS FOR POINCARÉ DUALITY SPACES AND APPLICATIONS

R. James Milgram[†]

In the first section of this paper we associate to any compact 2n-dimensional manifold or Poincaré duality complex X with fundamental group π, a $(-1)^n$-symmetric non-singular form $(\mathbf{Z}[s|\pi)^v \longrightarrow (\mathbf{Z}[s|\pi)^{v*}$, where v is approximately the number of n-cells in X. Likewise, to any compact $2n+1$-dimensional Poincaré complex we associate a self-orthogonal direct summand of the $(-1)^n$-symmetric hyperbolic form on $(\mathbf{Z}[s]\pi)^v \bigoplus (\mathbf{Z}[s]\pi)^{v*}$, v as before.

Complete examples are given in dimensions 1 and 2. Further examples can be found in [MII].

The invariance of these constructions under bordism is investigated in §2, and we arrive at a direct derivation of the definition of the surgery L-groups, together with a natural homomorphism

$$\Omega_*^{P.D}(B_\pi) \xrightarrow{e_\pi} L_*^h(\mathbf{Z}[s]\pi).$$

This is an example of an orientation theory for Poincaré duality bordism, and when π is a finite 2-group the structure of the orientation groups $L_*^h(\mathbf{Z}[s]\pi)$ is completely determined in all dimensions except those where $* \equiv 1 (\mathrm{mod} 4)$, where it is still determined at least up to some minor indeterminacy.

In §4 we study the lifted Steenrod construction

$$\mathbf{RP}^\infty \times X \xrightarrow{S_\pi} E_{\mathbf{Z}/2} \times_{\mathbf{Z}/2} (\tilde{X} \times \tilde{X}/\Delta\pi)$$

where $\tilde{X}$ is the π-cover of X, and from its properties, modeled on Ranicki's definition of the Mischenko-Ranicki symmetric L-groups $L_h^*(\mathbf{Z}\pi)$, we define a new orientation theory $L_{R.L}^*(\mathbf{Z}\pi)$. These orientations fit together as follows

$$\Omega_*^{P.D}(B_\pi) \xrightarrow{E_\pi} L_{R.L}^*(\mathbf{Z}\pi) \longrightarrow L_h^*(\mathbf{Z}\pi) \longrightarrow L_*^h(\mathbf{Z}[s]\pi).$$

where the composite is exactly e_π.

These 2 extreme orientation theories turn out to be within range of effective calculation, and, as I show in the remaining 3 sections, acting together they seem to carry the invariants needed to answer all questions related to classification (up to h-cobordism, but see §7 for extensions), and most questions related to detecting surgery obstructions.

This last is rather unexpected. (In particular, the orientation into the Mischenko-Ranicki L-groups detects almost none of the torsion information relating to surgery obstructions.) It

---

[†] *Research partially supported by the NSF.*

comes about as follows. There is the Levitt exact sequence which is vital for studying Poincaré duality bordism

$$\cdots \longrightarrow \pi^s_*(B_{\pi+} \wedge MSG) \xrightarrow{\partial} L^h_{*-1}(\mathbf{Z}\pi) \xrightarrow{v} \Omega^{P.D}_{*-1}(B_\pi) \longrightarrow \pi^s_{*-1}(B_{\pi+} \wedge MSG) \longrightarrow \cdots$$

and the composition $E_\pi v$ is shown in §5 to be essentially an isomorphism away from the contributions from the 2-Sylow subgroup of π (at least when π is 2-hyperelementary). But, given a surgery problem over π, $M^n \longrightarrow X^n$ with surgery obstruction $\omega \in L^h_n(\mathbf{Z}\pi)$, then the difference $(\{M^n\} - \{X^n\}) = v(\omega) \in \Omega^{P.D}_n(B_\pi)$.

If one compares these remarks with the ideas in e.g. [L], [M3], [H-Mad], one sees that the E_π-orientation provides a uniform method for detecting the non-triviality of all the surgery problems involved in the compact space-form problem. In fact, it seems that with very few exceptions it detects all the non-trivial obstructions previously seen.

On the other hand, the orientation into $L^h_*(\mathbf{Z}[s]\pi)$ that we constructed originally is seen in §6 to exactly carry the information needed to evaluate surgery obstructions on products of the form {Poincaré duality complex over π} $\times$ {simply connected surgery problem}. These are the key surgery problems which must be solved to evaluate surgery obstrcutions over closed topological manifolds. Indeed, there is a map, [Wa]

$$\sigma : \Omega^{TOP}_*(G/TOP \times B_\pi) \longrightarrow L^h_*(\mathbf{Z}\pi)$$

which takes a surgery problem over M^n (thought of as corresponding to a homotopy class of maps $f : M^n \longrightarrow G/TOP$, see e.g. [M-M] chapters 2 and 3) and, if $e : M^n \longrightarrow B_{\pi=\pi_1(M)}$ classifies the universal comvering of M, then σ of the bordism class $\{M^n, f \times e\}$ is the surgery obstruction for the original surgery problem.

Now, by looking explicitly at the 2-local bordism groups of $G/TOP \times B_\pi$, which can be written as $H_*(G/TOP \times B_\pi; \Omega^{TOP}_*(pt))$, we can choose representatives for the bordism classes which enter have trivial surgery obstructions, are products, as above, or are twisted products of the form

$$I \times K \times N \cup_{\{0,1\}\times K\times N} K \times W \longrightarrow I \times S \times N \cup_{\{0,1\}\times S\times N} S \times W$$

where W is an oriented manifold with boundary equal to two copies of N. (A discussion of most of the ideas relevant to this can be found in [M5].

Thus, the results of §3 give, as they stand, almost the complete determination of the characteristic class formulae for determining the surgery obstructions over any manifold with finite fundamental group. Moreover, when we extend them to $\mathbf{Z}/2$-coefficients, then complete results are indeed obtainable.

This, in turn, enables us to completely evaluate the ∂-map in the Levitt sequence above.

Finally, in §7 we indicate recent extensions of some of these ideas, based mostly on work of R. Oliver, leading toward classification up to s-cobordism.

In late 1984 I proved the main structure theorems, (1.9 and 6.1 of [MI], and, as applications, theorems B, C, D, E, of [MII]), for classifying the possible surgery obstructions over closed manifolds with finite fundamental groups. This work is recounted in [MI], [MII], and an extension of these techniques gives further results in [H-M-T-W].

However, the point of view of those papers was, I felt, too restricted, making the theorems appear almost accidental. In this paper I feel that they have been put into proper perspective – if you have a problem in this area, and want to solve it, the first step is to use an existing orientation theory, or if they prove insufficient, invent a new one!

In particular Ranicki introduced his L-groups by clarifying previous work of Mischenko. They certainly provide an orientation, but (1) they are extremely difficult to calculate, and (2) they do not detect even the simplest of the invariants useful in the surgery approach to studying group actions on manifolds (the R. Lee semicharacteristic classes†).

On the other hand, theoretically, the L_h^*-groups are very useful for analyzing product formulae for surgery obstructions. Of course, the difficulty with calculations limits their application when specific information is needed. Thus, it is reasonable to look for a *weakening* of the definition of the L_h^*-groups with better calculational properties to actually study specific product formulae.

The first hint that other theories could replace the L_h^*-orientation appeared in the paper [Cl] of F. Clauwens, and provided the initial motivation for me to first complete his work to a full orientation, and second, begin to look for other theories, which might also be useful.

§1. Poincaré complexes, the orientation matrix.

Let A be a ring with involution $\tau : A \longrightarrow A$ $(\tau(ab) = \tau(b)\tau(a),\ \tau^2 = 1)$. Then, if M is a left A-module $Hom_A(M, A) = M^*$ is naturally a *right* A-module, $(f \circ a(m) = f(m) \circ a)$ and τ converts M^* into a left A-module by $(af)(m) = f(m) \circ \tau(a)$. Clearly $A^* \approx A$, so, if P is a finitely generated free or projective, then so is P^*. Now, suppose

$$C = \{C_0, C_1, \cdots, C_n, \cdots, \partial : C_i \longrightarrow C_{i-1}, \partial^2 = 0\}$$

is a finitely generated complex of free or projective left A-modules. An n-dimensional *Poincaré duality structure* on C is a chain map

$$\phi : C^* = \{\cdots C_n^* \cdots C_0^*,\ \delta\} \longrightarrow C_*$$

satisfying $\phi(C^{n-i}) \subset C_i$, for each i, which is first an A-map, second, a chain homotopy equivalence, and third, has the property that there is a chain homotopy

$$\Phi : C^* \longrightarrow C_* \quad (\Phi : C^{n-i} \longrightarrow C_{i+1})$$

so that

$$\Phi\partial + \partial\Phi = \phi^* - (-1)^{n(n+1)/2}\phi.$$

In other words, ϕ and $\pm\phi^*$ are chain homotopic.

The standard proof of Poincaré duality for a combinatorial manifold M^n gives rise to such a structure on the chain complex of its universal covering. (Here, we regard the first Stiefel-Whitney class of M^n as an element $w \in Hom(\pi_1(M^n), \mathbf{Z}/2)$, so the involution on $A = \mathbf{Z}(\pi_i(M^n))$ is given by $\tau(\sum n_i g_i) = \sum n_i w(g) g_i^{-1}$.)

So far everything is standard. Now we introduce the orientation of (C_*, ϕ, Φ) in a way similar in many respects to the construction of the torsion for an h-cobordism. The key thing to note is that just as the mapping cone of an h-cobordism is acyclic, so, too, is the mapping cone of ϕ. However, this modification of the construction of the torsion of an h-cobordism is not quite direct since ϕ is not $\pm$ symmetric directly, but only so up to a homotopy. To remedy this difficulty we begin by enlarging A.

† I am greatly indebted here to J. Davis, who pointed this out to me. Once he convinced me of this the development of the R.L-L-groups was almost immediate.

Let $\mathbf{Z}[s]$ denote the ring of polynomials in the free variable s with integer coefficients.

1.1 $$B = \mathbf{Z}[s] \otimes_{\mathbf{Z}} \mathbf{Z}(\pi_1(M^n))$$

is given an involution by the rule $\tau(s) = 1 - s$, $\tau(g) = w(g)g^{-1}$. Let

$$C'_* = B \otimes_{\mathbf{Z}(\pi_1(M^n))} C_*$$

and define $\mu : C'^* \longrightarrow C'_*$ as $s \cdot \phi$. Then we have

Lemma 1.2: $\mu + (-1)^{n(n+1)/2}\mu^* : C'^{(n-*)} \longrightarrow C_*$ *is a chain homotopy equivalence, chain homotopic to* $\pm\phi^*$.

Proof: $\mu + (-1)^{n(n+1)/2}\mu^* = (-1)^{n(n+1)/2}\phi^* + s(\phi - (-1)^{n(n+1)/2}\phi^*)$ so

$$\mu + (-1)^{n(n+1)/2}\mu^* \pm s \cdot (\partial\Phi + \Phi\delta) = \pm\phi^*,$$

and the result follows.

Remark 1.3: Since $\mu + (-1)^{n(n+1)/2}\mu^*$ is directly $\pm$ symmetric, we can assume the chain homotopy is identically zero.

Definition 1.4: *Let* $\lambda : C_* \longrightarrow D_*$ *be a chain map, then the mapping cone of* λ, *written* $MC(\lambda)$ *is the complex* $\sum C_* \bigoplus D_*$ *with* ∂ *given by the rule*

$$\partial(c, d) = (-\partial c, \lambda(c) + \partial(d)).$$

In particular,

$$MC(\lambda)^* = MC(\lambda^*) = (c^j, d^{j+1}), \ \delta(c^j, d^{j+1}) = (-\delta(c^j), \lambda^*(c^j) + \delta(d^{j+1})).$$

It is direct to verify that there is an exact sequence of chain complexes

$$0 \longrightarrow D_* \xrightarrow{I} MC(\lambda) \xrightarrow{J} \Sigma C_* \longrightarrow 0$$

where

$$I(d) = (0, d), \quad J(c, d) = c,$$

and the connecting homomorphism in the resulting homology exact sequence is λ_*. Consequently

1.5
(1) $MC(\mu + (-1)^{n(n+1)/2}\mu^*)$ is acyclic.
(2) $MC(\mu + (-1)^{n(n+1)/2}\mu^*)$ is self-dual.

In particular 1.5.1 implies $\text{im}(\partial_i) = \ker(\partial_{i-1})$ is projective (or stably free if C_* is free) for each i . 1.5.2 implies, if $n = 2k$ that

$$\text{coker}(\partial_{k+2}) \cong \ker(\partial_k)^*$$

and that

$$\partial = (-1)^{n(n+1)/2}\partial^* : \text{coker}(\partial_{k+2}) \longrightarrow \ker(\partial_k)$$

is an isomorphism. Thus we associate a non-singular $(-1)^k$ symmetric quadratic form on the B-free or projective module $\text{coker}(\partial_{k+2})$ to any Poincaré duality complex over A .

Remark 1.6: Generally, $\text{coker}(\partial_{k+2})$ is free over B on a number of generators only slightly larger than the number of generators for C_k. In the simplest case, where $\pi_1(M) = 1$ this matrix basically gives the middle dimensional intersection form. However, it seems to contain a great deal of more recondite information when the fundamental group is non-trivial.

When $n = 2k+1$, in the middle dimensions of $MC(\mu + (-1)^{n(n+1)/2}\mu^*)$ we have the sequence

1.7 $$C^{k-1} \oplus C_{k+1} \xrightarrow{\partial_1} C^k \oplus C_k \xrightarrow{\partial_0} C^{k+1} \oplus C_{k-1}$$

with

$$\partial = \begin{pmatrix} -\delta, & 0 \\ \mu \pm \mu^*, & \partial \end{pmatrix}.$$

Since the mapping cone is self dual we have $im(\partial_1) \cong ker(\partial_0)^*$. On the other hand, if we define a (hyperbolic) form on $C^k \bigoplus C_k$ by

$$< (A,\ B),\ (C,\ D) > = < A,\ D > - \pm\tau(< C,\ B >)$$

we have

$$< (-\delta x,\ (\mu \pm \mu^*)x + \partial y),\ (-\delta z,\ (\mu \pm \mu^* z + \partial w) > =$$
$$< -\delta x,\ (\mu \pm \mu^*)x + \partial w > \pm - \tau(< -\delta z,\ (\mu \pm \mu^*)x + \partial y >$$
$$= < -\delta x,\ (\mu \pm \mu^*)z > \pm - \tau(< -\delta z,\ (\mu \pm \mu^*)x >) = 0$$

and we have that $im(\partial_1)$ is a direct summand of $C^k \bigoplus C_k$, and satisfies the properties

(1) $< \partial_1 r,\ \partial_1 s > = 0$, all $r,\ s$.

(2) $< \delta x,\ (\mu \pm \mu^*)x + \partial y > = < \delta x,\ \mu(x) > \pm\tau(< \delta x,\ \mu(x) >)$.

Definition 1.8: *Let* $\mathbf{H}^{\mp} = C^* \bigoplus C$, *with* C *projective, and hyperbolic form as above. A KERNEL* K *in* $\mathbf{H}^{\mp}$ *is a projective direct summand* $K \subset \mathbf{H}^{\mp}$ *with* $K^{\perp} = K$. *Moreover, writing* $k \in K$ *as* $(k_1,\ k_2)$, *and setting*

$$\phi(k) = k_1(k_2),\ \textit{there is}\ \mu : C^* \longrightarrow C$$

so that $\phi(k) = (1 \rightleftharpoons \tau)k_1(\mu(k_2))$. *A triple* $(\mathbf{H}^*,\ K,\ \mu)$ *with* K *a kernel is called a* $\mp$ *quadratic formation.*

Thus, the construction above associates to every $2k+1$ dimensional Poincaré complex over A a $(-1)^{k+1}$-quadratic formation over B.

Example 1.9: Suppose M is a point. Then ϕ is the identity $\mathbf{Z} \longrightarrow \mathbf{Z}$, and the resulting quadratic form is simply $< 1 >$.

Example 1.10: Suppose M is the circle S^1, then $\pi_1(M) = \mathbf{Z}$. The group ring of $\mathbf{Z}$ is $A = \mathbf{Z}[t, t^{-1}]$, with involution $\tau(t) = t^{-1}$. A resolution of $\mathbf{Z}$ over A is given as follows

$$A \xrightarrow{t-1} A \longrightarrow \mathbf{Z},$$

with dual resolution

$$A \xrightarrow{t^{-1}-1} A \longrightarrow \mathbf{Z}.$$

A chain equivalence from the dual resolution to the original resolution is constructed using $\phi_0 = 1$, $\phi_1 = -t^{-1}$, and a homotopy of ϕ and $-\phi^*$ is easily constructed. Using this, we obtain the following formation

1.11 $\mathbf{H}^- = A^* \bigoplus A$, $K = A$ with embedding $1 \mapsto (1 - t^{-1}, -1 + (1 - t^{-1})s)$,

$$\text{and } \mu = -1 + (1 - t^{-1})s.$$

Example 1.12: Let M_g^2 be a closed surface of genus g . It has fundamental group

$$\Gamma = \{a_1, \cdots a_h, b_1, \cdots b_g \mid [a_1, b_1][a_2, b_2] \cdots [a_g, b_g] = 1\}$$

where $[a, b] = aba^{-1}b^{-1}$ so $[a, b]^{-1} = [b, a]$.

A resolution of Γ is given by

1.13 $$0 \longrightarrow \mathbf{Z}(\Gamma)(R) \longrightarrow (\mathbf{Z}(\Gamma))^{2n} \longrightarrow \mathbf{Z}(\Gamma)(1) \xrightarrow{\epsilon} \mathbf{Z} \longrightarrow 0$$

where the $2n$ generators of C_1 are

$$[a_1], \cdots, [a_n], [b_1], \cdots, [b_n]$$

and $\partial[a_i] = (a_i - 1)(1)$, $\partial[b_i] = (b_i - 1)(1)$.

Let

$$N_i = (1 - [a_i, b_i]b_i)[a_i]$$
$$M_i = (a_i - [a_i, b_i])[b_i],$$

then

$$\partial(N_i + M_i) = ([a_i, b_i] - 1)(1),$$

and we can set

$$\partial(R) = [a_1, b_1] \cdots [a_{n-1}, b_{n-1}](N_n + M_n) +$$
$$[a_1, b_1] \cdots [a_{n-2}, b_{n-2}](N_{n-1} + M_{n-1}) + \cdots + [a_1, b_1](N_2 + M_2) + (N_1 + M_1).$$

The dual resolution is given by

$$\mathbf{Z}(\Gamma)(1^*) \longrightarrow \Sigma\mathbf{Z}(\Gamma)[a_i]^* \bigoplus \Sigma\mathbf{Z}(\Gamma)[b_i]^* \longrightarrow \mathbf{Z}(\Gamma)(R^*) \longrightarrow \mathbf{Z} \longrightarrow 0$$

where $\delta(1^*) = \Sigma(a_i^{-1} - 1)[a_i]^* + \Sigma(b_i^{-1} - 1)[b_i]^*$, and

$$\delta[a_i]^* = (1 - b_i^{-1}[b_i, a_i])[b_{i-1}, a_{i-1}] \cdots [b_1, a_1](R^*)$$
$$\delta[b_i]^* = (a_i^{-1} - [b_i, a_i])[b_{i-1}, a_{i-1}] \cdots [b_1, a_1](R^*)$$

Definition 1.14:

$$S_i = ([b_i, a_i] - a_i^{-1} - b_i)[a_i] + ([b_i, a_i]a_i - 1)[b_i]$$
$$T_i = (b_i^{-1}[b_i, a_i] - 1)[a_i] + a_i b_i^{-1}[b_i]$$
$$W_i = -(S_i + a_i^{-1}[a_i])$$
$$= (b_i - [b_i, a_i])[a_i] + (1 - [b_i, a_i]a_i)[b_i].$$

One verifies directly the formulae

$$\partial S_i = (a_i^{-1} - [b_i, a_i])(1)$$
$$\partial T_i = (1 - b_i^{-1}[b_i, a_i])(1)$$
$$\partial W_i = ([b_i, a_i] - 1)(1).$$

Also,

$$W_i = -[b_i, a_i](N_i + M_i)$$

and

$$(a_i^{-1} - 1)T_i + (b_i^{-1} - 1)S_i = W_i.$$

We now construct a chain homotopy equivalence

$$\phi : C^{2-*} \longrightarrow C_*$$

by setting $\phi_0(R^*) = (1)$. Note that if we set

$$V_i = W_i + [b_i, a_i]W_{i-1} + [b_i, a_i][b_{i-1}, a_{i-1}]W_{i-2} + \cdots +$$

$$[b_i, a_i] \cdots [b_1, a_1]W_1$$

then

$$\partial V_i = ([b_i, a_i] \cdots [b_1, a_1] - 1)(1)$$

so a suitable lifting of ϕ_0 is defined on generators by

$$\phi_1([a_i]^*) = (1 - b_i^{-1}[b_i, a_i])V_{i-1} + T_i$$

$$\phi_1([g_i]^*) = (a_i^{-1} - [b_i, a_i])V_{i-1} + S_i$$

We have as a consequence of the expansion of W in terms of S_i, T_i given above that

$$(a_i^{-1} - 1)(1 - b_i^{-1}[b_i, a_i]) + (b_i^{-1} - 1)(a_i^{-1} - [b_i, a_i]) = [b_i, a_i] - 1$$

Thus

$$\phi_1(\Sigma(a_i^{-1} - 1)[a_i]^* + \Sigma(b_i^{-1} - 1)[b_i]^*) = V_n$$

and this in turn is

$$-(\Sigma[b_n, a_n] \cdots [b_{i+1}, a_{i+1}](N_i + M_i)) = -\partial(R)$$

since $[a_1, b_1] \cdots [a_n, b_n] = 1$. In particular we must set $\phi_2((1)^*) = -(R)$, and it is easily verified that ϕ_* is a chain isomorphism. A matrix representation for ϕ_1 is given by a (2×2 block) lower triangular matrix if we choose our ordered basis as $[a_1]$, $[b_1]$, $[a_2]$, $[b_2]$, $\cdots$, $[a_n]$, $[b_n]$, and the corresponding dual basis $[a_1]^*$, $[b_1]^*$, $\cdots$, $[a_n]^*$, $[b_n]^*$. Here the diagonal blocks look like

1.15
$$\begin{pmatrix} a_i b_i^{-1} a_i^{-1} - 1 & a_i b_i^{-1} \\ [b_i, a_i] - a_i^{-1} - b_i & b_i a_i^{-1} b_i^{-1} - 1 \end{pmatrix}$$

and the other terms have the forms

$$(1 - a_i b_i^{-1} a_i^{-1})[b_{i-1}, a_{i-1}] \cdots [b_{j+1}, a_{j+1}](b_j - [b_j, a_j])[a_i]^* \longrightarrow [a_j]$$

$$(a_i^{-1} - [b_i, a_i])[b_{i-1}, a_{i-1}] \cdots [b_{j+1}, a_{j+1}](b_j - [b_j, a_j])[b_i]^* \longrightarrow [a_j]$$

$$(1 - a_i b_i^{-1} a_i^{-1})[b_{i-1}, a_{i-1}] \cdots [b_{j+1}, a_{j+1}](1 - b_j a_j^{-1} b_j^{-1})[a_i]^* \longrightarrow [b_j]$$

$$(a_i^{-1} - [b_i, a_i])[b_{i-1}, a_{i-1}] \cdots [b_{j+1}, a_{j+1}](1 - b_j a_j^{-1} b_j^{-1})[b_i]^* \longrightarrow [b_j]$$

Remark 1.16: Note that ϕ^{2-*} is chain homotopic over $\mathbf{Z}(\Gamma)$ to $-\phi$.

In particular, the matrix for the torus $M_1^2 = T^2$ is

$$1.17 \qquad \begin{pmatrix} 1-b, & a+b^{-1}-1 \\ -a^{-1}b, & 1-a^{-1} \end{pmatrix} + s\begin{pmatrix} b+b^{-1}-2, & 1-a-b^{-1}+ab^{-1} \\ 1-a^{-1}-b+a^{-1}b, & a+a^{-1}-2 \end{pmatrix}$$

§2. Poincaré pairs and cobordism

In this section we introduce an equivalence relation on the orientation matrices and formations constructed in the last section, and show that the assignment of a form or formation to a Poincaré duality structure on a manifold gives rise to a well defined homomorphism of bordism groups into the resulting groups of equivalence classes.

Definition 2.1: *An $n+1$ dimensional Poincaré pair is a 5-tuple $\{D, C, f, \phi_*, \Phi\}$ where $\{C, \phi_*\}$ is a Poincaré complex of dimension n, D_* is a finitely generated projective A chain complex, $f : C_* \longrightarrow D_*$ is an A-chain map, and $\Phi : D^{N+1-*} \longrightarrow D_*$ is an A-map, satisfying*

(1) $\partial\Phi + \Phi\delta = (-1)^{*-1} f_* \phi_0 f^*$ where $*$ denotes the dimension of the class being mapped.

(2) the chain map

$$p : D^{n+1-*} \longrightarrow MC(f)$$

defined by

$$p(d^i) = (-1)^i (\phi_0 f^*(d^i), -\Phi(d^i))$$

is a chain homotopy equivalence.

Remark 2.2: If (M, W) is a finite CW manifold pair with

$$\partial W = M, \text{ and } \pi_1(M) \longrightarrow \pi_1(W) \text{ an isomorphism,}$$

then geometric Poincaré duality gives rise to a Poincaré pair structure as above on the chain complexes of the universal covers of M and W. Here $A = \mathbf{Z}(\pi_1(M))$.

Remark 2.3: As in the last section, after we pass to $B = \mathbf{Z}[s] \otimes_{\mathbf{Z}} A$, we can assume that

$$\phi^* = (-1)^{n(n+1)/2}\phi, \text{ and } \Phi^* = (-1)^{(n+1)(n+2)/2}\Phi.$$

If (C, ϕ) is the boundary of a Poincaré pair over A, then the same is true on tensoring with B. Moreover, the next theorem shows that being a boundary over B severely restricts the kinds of forms and formations which arise via the construction of the last section.

Theorem 2.4: *Assume (C, ϕ_*) is the boundary in a Poincaré pair over B, then*

a. *If $n = 2k$ the associated form is hyperbolic,*

b. *If $n = 2k+1$, then we can write $C^* \oplus C$ in the form $C_1^* \oplus C_1 \perp C_2^* \oplus C_2$, with $C_1 \perp C_2 = C$, $C_1^* \perp C_2^* = C^*$, and $K = K_1 \perp K_2$ a corresponding splitting, with the projections*

$$p_1 : K_1 \longrightarrow C_1^*$$
$$p_2 : K_2 \longrightarrow C_2$$

both isomorphisms.

Proof: Define $DMC(\Phi, \phi)$ as

$$DMC(\Phi, \phi)_j = D^{n-j+2} \oplus C_{j-1} \oplus D_j$$

with boundary given by the matrix

2.5
$$\partial = \begin{pmatrix} \delta, & 0, & 0 \\ \phi f^*, & -\partial, & 0 \\ -\Phi, & f, & \partial \end{pmatrix}.$$

The coboundary δ for the dual complex is given by

2.6
$$\delta = \begin{pmatrix} \partial, & f\phi^*, & -\Phi^* \\ 0, & -\delta, & f^* \\ 0, & 0, & \delta \end{pmatrix}$$

Our assumptions imply that both $DMC(\Phi,\phi)$ and $DMC(\Phi, \phi)^*$ are acyclic. Let $\epsilon = (-1)^{n(n+1)/2}$, and define

2.7
$$\nu = \begin{pmatrix} 0 & 0 & 1 \\ 0 & \phi & 0 \\ \epsilon & 0 & 0 \end{pmatrix} : DMC(\Phi,\phi)^{n-j+2} \longrightarrow DMC(\Phi,\phi)_j,$$

then we easily verify that ν is a chain map, and we can define a surjective chain map

$$\theta : MC(\nu) \longrightarrow MC(\phi_0)$$

by the matrix

2.8
$$\theta = \begin{pmatrix} 0, & (-1)^{i+1}, & 0, & (-1)^i f^*, & 0, & 0 \\ 0, & 0 & ,0, & 0, & (-1)^{i+1}, & 0 \end{pmatrix}.$$

The construction of the matrix above is a direct consequence of the result

Lemma 2.9: *Suppose given $m : C \longrightarrow D$, and $n : \wp \longrightarrow C$ so that $m \circ n = \partial h + h\partial$, with $h : \wp \longrightarrow D$, then $\mu : MC(n) \longrightarrow D$ defined by $\mu(b, r) = (h(b) + m(r))$ is a chain map.* (The proof is direct.)

In order to use the above lemma set

2.10
$$(-1)^i \begin{pmatrix} f^*, & 0, & 0 \\ 0, & -id, & 0 \end{pmatrix} : DMC(\Phi,\phi) \longrightarrow M(\phi).$$

It is routine to verify that this is a chain map, and

2.11
$$(-1)^{i+1} \begin{pmatrix} 0, & 1, & 0 \\ 0, & 0, & 0 \end{pmatrix} : DMC(\Phi,\phi)^* \longrightarrow MC(\phi)$$

defines a chain homotopy for the composite

$$(-1)^i \begin{pmatrix} f^*, & 0, & 0 \\ 0, & -id, & 0 \end{pmatrix} \begin{pmatrix} 0, & 0, & 1 \\ 0, & \phi, & 0 \\ \epsilon, & 0, & 0 \end{pmatrix}.$$

The kernel of θ is the chain complex with generators of 4 kinds.

I.	d_*,	0,	0,	0,	0,	0
II.	0,	$\mp f^* d^*$,	0,	d^*,	0,	0
III.	0,	0,	d^*,	0,	0,	0
IV.	0,	0,	0,	0,	0,	d

The boundary operator connects types I to IV, and types III to II via projections. Specifically, one obtains the matrix for the boundary on the kernel as

$$\begin{pmatrix} -\partial, & 0, & 0, & 0 \\ 0, & \delta, & \epsilon, & 0 \\ 0, & 0, & -\partial, & 0 \\ 1, & -\Phi, & 0, & \partial \end{pmatrix}.$$

From this it is direct to see, in the even dimensional case that the kernel quadratic form is $D^{k+1} \bigoplus D_{k+1}$ with matrix

$$\begin{pmatrix} 0 & 1 \\ \epsilon & 0 \end{pmatrix}.$$

So the kernel form is hyperbolic.

Also, if we embed $C^k \ \pi \ C_{k+1}$ into $MC(\theta)$ in the evident way, we see directly that the form on the subspace is identical to that on $MC(\phi)$. Thus, the total form splits as the form on $MC(\phi) \bigoplus \mathbf{H}$ where $\mathbf{H}$ is hyperbolic.

On the other hand, since both $DMC(\Phi,\ \phi)$ and $DMC(\Phi,\ \phi)^*$ are acyclic, it is direct to check that the form on $MC(\theta)$ is hyperbolic (the cokernel of ∂ is isomorphic to the direct sum of (coker δ, coker ∂), and the latter is a kernel.) But this is the result in the even case.

The odd case is completely similar.

Thus, we introduce an equivalence relation in the set of isomorphism classes of non-singular quadratic forms and formations under orthogonal direct sum using the relations specified in the theorem. The resulting groups are the quadratic L-groups

2.12 $$L^d_n(\mathbf{Z}[s]\pi,\ \tau)$$

where $d = h$ if we restrict attention to finitely generated free modules, and $d = p$ when we allow finitely generated projectives as well.

Remark 2.13: From the definitions $L^d_* = L^d_{*+4}$ for all $* \geq 0$.

Clearly, the construction of §1 when applied to any CW decomposition of a manifold with fundamental group π gives the same element in $L^h_n(\mathbf{Z}[s]\pi,\ \tau)$ by applying theorem 2.4 to a decomposition of a homotopy equivalence. In fact, more generally, we can apply the construction to a bordism, and now 2.4 shows that the construction in §1 gives rise to a well defined homomorphism

2.14 $$e_\pi :\ \Omega_*(B_\pi) \longrightarrow L^h_*(\mathbf{Z}[s]\pi, \tau)_0)$$

where $\tau_0(\sum n_i g_i) = \sum \tau(n_i) \circ g_i^{-1}$.

This orientation and its obvious extension to $\mathbf{Z}/2$ manifolds is an essential ingredient in our main results since it factors all product formulae for surgery problems [MI] and also §6. As we will see in the next section, it is possible to calculate the groups $L^*_h(\mathbf{Z}[s]\pi)$ when π is a finite 2-group. This calculation then gives the solution of the "oozing conjecture" for manifolds with finite fundamental group, as well as an effective determination of the groups $\Omega^{P.D}_*(B_\pi)$ when π is finite. However, in another sense, this orientation is too weak. In the fourth section we will introduce a more delicate (symmetric) L-theory $L^*_{R.L}(\mathbf{Z}\pi)$, together with an orientation homomorphism

2.15 $$E_\pi :\ \Omega^{P.D\cdot}_*(B_\pi) \longrightarrow L^*_{R.L.}(\mathbf{Z}\pi)$$

which factors e_π in the sense that there is a natural transformation

$$p: L^*_{R.L}() \longrightarrow L^h_*()$$

and $e_\pi = p \circ E_\pi$.

It will turn out (Remark 5.15), that this more delicate orientation enables us to detect surgery obstructions. In fact it provides a clear explanation of the results a few years back of the current author and Ib Madsen on the structure of the solutions to the compact space form problem for the final class of periodic groups which had not been handled previously. These considerations essentially give us complete control of the basic classification questions in the category of manifolds of dimension ≥ 5, with finite fundamental group. However, they are not of too much help in low dimensions, and we are far from any real understanding in the case of infinite π_1.

§3. The groups $L^p_*(\mathbf{Z}[s]\pi)$ and $L^h_*(\mathbf{Z}[s]\pi)$

In this section π is always a finite 2-group. Under this assumption the groups $L^p_h(\mathbf{Z}[s]\pi)$ and $L^h_*(\mathbf{Z}[s]\pi)$ are studied and largely determined in [M-O] and [D-M]. In fact, the results given in these references are complete up to a hard (but purely algebraic) problem of computing the Witt ring of the *ordinary quaternion algebra* over the quotient field of $\mathbf{Q}[s]$, with involution given by $s \leftrightarrow 1-s$. The key idea in these references as in most previous papers concerned with the structure of L-groups is to reduce the calculational questions to questions about a much smaller class of groups, called model groups.

The Fontaine model groups [F] are the ones which we need in the oriented case. For a discussion of the groups needed in the non-oriented case see [D-M] or [H-T-W].

Table 3.1: *The Fontaine model groups.*

Symbol	Definition	Group
$\mathbf{Z}/2^n$	$\{x \mid x^{2^n} = 1\}$	cyclic
$D_{2^{n+1}}$	$\{x, y \mid x^{2^n} = y^2 = (xy)^2 = 1\}$	generalized dihedral
$SD_{2^{n+1}}$	$\{x, y \mid x^{2^n} = y^2 = 1,\ (xy)^2 = x^{2^{n-1}}\}$	semi-dihedral
$Q_{2^{n+1}}$	$\{x, y \mid x^{2^{n-1}} = y^2 = (xy)^2\}$	generalized quaternion

They have the property that, given any irreducible rational representation

$$r: \pi \longrightarrow M_n(D)$$

then there is a subgroup $H \subset \pi$, a model group M_r in the Fontaine list, together with a surjection

$$p: H \longrightarrow M_r$$

so that $tr(p^*(f_M) = r$ where f_M is the faithful irreducible rational representation of M_r. Moreover, the center of f_M is equal to the center of $M_n(D)$, as is the class of f_M in the Brauer group of the center.

Using naturality we have

3.2 $$res^d = \coprod_r p_* \cdot tr^h_\pi : L^d_*(\mathbf{Z}[s]\pi) \longrightarrow \coprod_r L^d_*(\mathbf{Z}[s](M_r))$$

and we say that $\alpha \in L^d_*(\mathbf{Z}[s]\pi)$ is *detected on models* if and only if $res^d(\alpha) \neq 0$.

res^p is always injective, hence, by naturality, we are able to easily calculate

$$e_\pi : \Omega^{PL}_*(B_\pi) \longrightarrow L^p_*(\mathbf{Z}[s]\pi)$$

reducing it to the determination of e_π on the Fontaine model groups.

res^h tends to have a large and uncontrolled kernel for the groups $L^h_*(\mathbf{Z}\pi)$ and $L^s_*(\mathbf{Z}\pi)$, and this had been a major stumbling block in previous work on the general problem of surgery on closed manifolds. But, thanks to a miracle, ultimately derivable from work of R. Oliver on p-adic logarithms, res^h on the groups $L^h_*(\mathbf{Z}[s]\pi)$ has an extremely small and well controlled kernel.

The main structure result for these groups is

Theorem 3.3:

(1) $L^h_*(\mathbf{Z}[s]\pi)$ is detected on models except (possibly) when $* = 1$, where there are a small number of possible, but well controlled classes, which, if they are non-zero, must be in the kernel of res^h.

(2) The map e_M is known for every Fontaine model group, and the groups $L^h_*(\mathbf{Z}[s]M)$ are known except when $M = Q_8$, the ordinary quaternion group.

We now turn to a brief description of the techniques used in the proof of 3.3. We begin by discussing the methods used to determine the projective L-groups, $L^p_*(\mathbf{Z}[s]\pi)$.

Let $R = \varprojlim(\mathbf{Z}/2^i[s]) = (\mathbf{Z}[s]^\wedge)_2$ be the completion of $\mathbf{Z}[s]$ at 2. R is an integral domain, in fact, but it is not a P.I.D. It is not hard to check, by projection onto the residue class ring that

3.4 $$L^h_*(R\pi) = L^p_*(R\pi) = \begin{cases} \mathbf{Z}/2 & * \text{ even} \\ 0 & * \text{ odd} \end{cases}$$

when π is a finite 2-group. Likewise, we write $S = R(1/2)$. Then, using a result of Clauwens (See [Cl] for the original result, and [M-O; 3.12] for the required sharpening) which shows, in particular, that

3.5 $$L^h_*(\mathbf{Z}(1/2)\pi) = L^h_*(\mathbf{Z}(1/2)[s]\pi)$$

for π a finite 2-group, we obtain a Meyer-Vietoris sequence

3.6 $$\cdots \longrightarrow L^p_*(\mathbf{Z}[s]\pi) \longrightarrow L^h_*(R\pi) \oplus L^h_*(\mathbf{Z}(1/2)\pi) \longrightarrow L^h_*(S\pi) \longrightarrow L^p_{*-1}(\mathbf{Z}[s]\pi) \longrightarrow \cdots$$

which reduces the calculation of $L^p_*(\mathbf{Z}[s]\pi)$ essentially to that of $L^h_*(S\pi)$, since $L^h_*(\mathbf{Z}(1/2)\pi)$ is well understood.

Theorem [M-O] 3.7: *Let K be a finite extension of $\hat{\mathbf{Q}}_2$ with involution either trivial or a Galois automorphism, then*

$$L^p_*(M_n(S \otimes K)) = \begin{cases} \mathbf{Z}/2 \oplus \mathbf{Z}/2 & * \text{ even} \\ 0 & * \text{ odd} \end{cases}$$

with generators $< 1 >, < \wp >$ when $* = 0$ where $\wp$ is a uniformizing parameter in $(S \otimes K)^\tau$ over 2.

Using this, the theorem of Fontaine [F], and the further result

Theorem [D-M] 3.8: *$L^h_{2*+1}(S\pi) = 0$ when π is a finite 2-group.*

we are easily able to calculate the groups $L^p_*(\mathbf{Z}[s]\pi)$.

Typical of the results obtained are

Theorem 3.9: $L^h_*(\mathbf{Z}[s]) = \begin{cases} \mathbf{Z} & * \equiv 0 (\mathrm{mod} 4) \\ \mathbf{Z}/2 & * \equiv 1 (\mathrm{mod} 4) \\ 0 & * \equiv 2,3 (\mathrm{mod} 4) \end{cases}$ *with the* $\mathbf{Z}$ *representing the signature and the* $\mathbf{Z}/2$ *the DeRham invariant.*

Remark 3.10: In particular, $e_1 : \Omega_*(pt) \longrightarrow L^h_*(\mathbf{Z}[s])$ is determined by the closed formula

3.11 $$\{M\} \longrightarrow \langle L_{4i}(M), [M] \rangle e_{4i} + \langle VSq^1V, [M] \rangle e_{4i+1}$$

where $L_{4i}(M)$ is the L-genus of M and V is the total Wu class.

Similarly we have

3.12 $$L^p_*(\mathbf{Z}[s]\mathbf{Z}/2) = \begin{cases} \mathbf{Z}^2 & * \equiv 0 (\mathrm{mod} 4) \\ (\mathbf{Z}/2)^3 & * \equiv 1 (\mathrm{mod} 4) \\ 0 & * \equiv 2,3 (\mathrm{mod} 4) \end{cases}.$$

When M is a manifold, the image of $\{M\}$ is in a single copy of $\mathbf{Z}$ in $\mathbf{Z}^2$ and is represented by the pair (Index(M), Index(M)). But the situation is considerably different for $\{M\} \in \Omega^{P.D}_{4k}(B_{\mathbf{Z}/2})$. Here the image actually is a finite index subgroup of $\mathbf{Z}^2$, indeed the index is at most 8. In dimension $4k+1$ the first $\mathbf{Z}/2$ gives the DeRham invariant, while the second $\mathbf{Z}/2$ has generator written d_{4k+1} and is associated to a middle dimensional torsion symmetric form. Finally the third $\mathbf{Z}/2$ has generator $\chi_{1/2,4k+1}$ which is detected in $L_{4k+1}(\mathbf{Q}(\mathbf{Z}/2))$ by a semi-characteristic class, as in e.g. [S] or [D-M]. The formulae defining

$$e_{\mathbf{Z}/2} : \Omega^{TOP}_*(B_{\mathbf{Z}/2}) \longrightarrow L^h_*(\mathbf{Z}[s]\mathbf{Z}/2)$$

are given by

$$\{M^{4i}\} \longrightarrow \langle L_*(M), [M] \rangle (e, e),$$

$$\{M^{4i+1}, f\} \longrightarrow \langle VSq^1V, [M] \rangle r_{4i+1} + \langle V^2 \cup f^*(e_1), [M] \rangle (d + \chi)$$

where $f : M \longrightarrow B_{\mathbf{Z}/2}$ classifies the $\mathbf{Z}/2$-covering, and e_1 is the non-trivial class in $H^1(B_{\mathbf{Z}/2}; \mathbf{Z}/2)$.

Remark 3.13: In general the calculations give

$$\begin{aligned} L^p_0(\mathbf{Z}[s]\pi) &= \mathbf{Z}^{c+N_1}, \\ L^p_2(\mathbf{Z}[s]\pi) &= \mathbf{Z}^c, \\ L^p_1(\mathbf{Z}[s]\pi) &= \mathbf{Z}/2 \oplus \coprod(\mathbf{Z}/2)d_i \oplus \coprod(\mathbf{Z}/2)\chi_i \oplus A_1 \\ L^p_3(\mathbf{Z}[s]\pi) &= \coprod(\mathbf{Z}/2) \oplus A_3 \end{aligned}$$

Here, the d_i, χ_i classes occur in a manner analogous to that for the group $\mathbf{Z}/2$, the $\mathbf{Z}/2$'s in L_3 are similar to the d_i classes, and A_1, A_3 are groups depending only on the number of "Q_8" summands

$$M_{n_i}(\langle {}^{-1}\,{}_{Q}\,{}^{-1} \rangle) \in \mathbf{Q}\pi.$$

We don't know A_1, A_3 at this time, but they seem to play no role in the image of $\Omega(B_\pi)$ so can be safely ignored.

Now we discuss the groups $L^h_*(\mathbf{Z}[s]\pi)$ when π is a finite 2-group. The method used in [M-O] to analyze these groups is to use the Ranicki-Rothenberg sequence

$$\cdots \tilde{H}^i(\mathbf{Z}/2; \tilde{K}_0(\mathbf{Z}[s]\pi)) \longrightarrow L^h_i(\mathbf{Z}[s]\pi) \longrightarrow L^p_i(\mathbf{Z}[s]\pi) \longrightarrow \tilde{H}^{i-1}(\mathbf{Z}/2; \tilde{K}_0(\mathbf{Z}[s]\pi)) \longrightarrow \cdots$$

together with the Meyer-Vietoris sequence for calculating the group $K_0(\mathbf{Z}\pi)$

3.14 $K_1(\mathbf{Z}(1/2)[s]\pi) \oplus K_1(R\pi) \longrightarrow K_1(S\pi) \longrightarrow K_0(\mathbf{Z}[s]\pi) \longrightarrow K_0(\mathbf{Z}(1/2)[s]\pi) \oplus K_0(R\pi)$.

Since $\mathbf{Z}(1/2)\pi$, $S\pi$ are both regular rings, K_0 and K_1 for these functors are easily studied. Moreover, in [M-O] we extend the logarithm techniques of [O] to obtain enough information about $K_1(R\pi)$ to understand its image in $K_1(S\pi)$.

What happens is that the d_i's in L_3^p map non-trivially to $\tilde{H}^2(\mathbf{Z}/2;\ \tilde{K}_0)$ and all the d_i classes in L_1^p except those associated with π^{AB} map non-trivially to $\tilde{H}^0(\mathbf{Z}/2;\ \tilde{K}_0(\mathbf{Z}[s]\pi))$, while the semicharacteristic classes and DeRham invariant classes map to 0 .

The result is summarized via the extensions

3.15
$$0 \longrightarrow (\mathbf{Z}/2)^{D-D(1)} \longrightarrow L_3^h(\mathbf{Z}[s]\pi) \longrightarrow A_3 \longrightarrow 0$$
$$0 \rightarrow (\mathbf{Z}/2)^{D-D(1)+L} \rightarrow L_1^h(\mathbf{Z}[s]\pi) \oplus A_1 \rightarrow (\mathbf{Z}/2)^{rk(\pi^{AB})-1} \oplus (\mathbf{Z}/2)^{SC} \rightarrow 0$$

where SC is the number of semi-characteristic terms. Similarly, in even dimensions we have extensions

3.16
$$0 \longrightarrow (\mathbf{Z}/2)^{l(0)} \longrightarrow L_0^h \longrightarrow \mathbf{Z}^{C+N} \longrightarrow 0$$
$$0 \longrightarrow (\mathbf{Z}/2)^{l(2)} \longrightarrow L_2^h \longrightarrow \mathbf{Z}^{C} \longrightarrow 0$$

where $l(0)$ and $l(2)$ depend only on the rational represention ring of π .

In particular, the new classes in $L_3^h(\mathbf{Z}[s]\pi)$ are important. Indeed we have

$$L_{4i+3}^h(\mathbf{Z}[s](Q_{2^i})) = \begin{cases} \mathbf{Z}/2 & i > 3 \\ \mathbf{Z}/2 \oplus A_3', & i = 3 \end{cases}$$

and the $\mathbf{Z}/2$ summand q_{4i+3} evaluates on $\Omega_{4i+3}(B_{Q_{2^i}})$ according to the formula

$$(M^{4i+3},\ f) \longrightarrow \ < V^2 \cup f^*(e_3),\ [M^{4i+3}] > q_{4i+3}\ +\ \epsilon$$

where $\epsilon \in A_3'$ so $\epsilon = 0$ if $g\ >\ 3$. Moreover e_3 is the non-trivial element in

$$H^3(B_{Q_{2^i}};\ \mathbf{Z}/2)\ =\ \mathbf{Z}/2.$$

Similarly, the new classes in $L_2^h(\mathbf{Z}[s]\pi)$ are important. There is one for each "dihedral" representaton.

§4. Steenrod maps and R.L-L-theory

Let $\tilde{X}$ be the univeral cover of the finite simplicial complex X with $\pi\ =\ \pi_1(X)$ as group of covering transformations. Then $\tilde{X}\ \times\ \tilde{X}$ is acted on freely by $\pi\ \times\ \pi$, and the diagonal map

$$\tilde{\Delta} : \tilde{X} \longrightarrow \tilde{X}\ \times\ \tilde{X},\ \ g \mapsto (x,\ x)$$

covers the diagonal embedding homomorphism

$$\Delta : \pi \longrightarrow \pi\ \times\ \pi,\ \ g \mapsto (g,\ g).$$

Thus there is a lifting Δ^π of the ordinary diagonal on X to give the diagram

$$\begin{array}{ccc} X & \xrightarrow{\Delta^\pi} & \tilde{X} \times \tilde{X}/\Delta\pi \\ \| \downarrow & & \downarrow \\ X & \xrightarrow{\Delta} & X \times X \end{array}$$

There is an involution on $\tilde{X} \times \tilde{X}/\Delta\pi$ $(T(\{x, y\}) = \{y, x\})$, and Δ^π extends to give a map

4.1 $$\mathbf{RP}^\infty \times X \xrightarrow{S(\pi)} E_{\mathbf{Z}/2} \times_{\mathbf{Z}/2} (\tilde{X} \times \tilde{X}/\Delta\pi)$$

which is a lifting of the usual Steenrod map [St], [S-E],

$$\mathbf{RP}^\infty \times X \xrightarrow{S(1)} E_{\mathbf{Z}/2} \times_{\mathbf{Z}/2} (X \times X).$$

Example 4.2: Let $\Sigma^i \mathbf{Z}\pi$ be the chain complex (with trivial ∂)

$$(\Sigma^i \mathbf{Z}\pi)_j = \begin{cases} \mathbf{Z}\pi & j = i \\ 0 & j \neq i \end{cases}$$

$\mathbf{Z}\pi$ acts from both the left and the right on $\Sigma^i \mathbf{Z}\pi$, so, via the diagonal action, it acts from both the left and the right on $\Sigma^i \mathbf{Z}\pi \otimes \Sigma^i \mathbf{Z}\pi$, as well. Under the usual left action we see that $\Sigma^i \mathbf{Z}\pi \otimes \Sigma^i \mathbf{Z}\pi/\Delta\pi \cong \mathbf{Z}\pi$ as a $\mathbf{Z}$-module. The associated right action of $\mathbf{Z}\pi$ is via conjugation $(g(a) = gag^{-1})$. When we apply the Steenrod construction we obtain the complex $C_{\mathbf{Z}/2} \otimes \Sigma^i \mathbf{Z}\pi \otimes \Sigma^i \mathbf{Z}\pi/\Delta\pi$, where $C_{\mathbf{Z}/2}$ is a free resolution of $\mathbf{Z}$ over $\mathbf{Z}(\mathbf{Z}/2)$. The action of the generator τ of $\mathbf{Z}/2$ on $\Sigma^i \mathbf{Z}\pi \otimes \Sigma^i \mathbf{Z}\pi/\Delta\pi$ is via the involution, $\tau(\sum n_i g_i) = \sum n_i g_i^{-1}$, and we easily calculate that

4.3 $$H_*(C \otimes_{\mathbf{Z}/2} (\Sigma^i \mathbf{Z}\pi \otimes \Sigma^i \mathbf{Z}\pi/\Delta\pi)) = \begin{cases} \mathbf{Z}\pi/\{a - \bar{a}\} & * = 2i \\ (\mathbf{Z}\pi)^\tau/\{a + \bar{a}\} & * > 2i \text{ odd} \\ 0 & * > 2i \text{ even} \\ 0 & * < 2i \end{cases}$$

Calculating explicitly we see that the general non-zero homology group is

4.4 $$\Lambda\pi = \mathbf{F}_2 < 1 > \oplus \coprod_{g \in \pi,\ g^2 = 1,\ g \neq 1} \mathbf{F}_2 < g >,$$

and if we replace $\mathbf{Z}\pi$ with $\mathbf{F}_2\pi$ in the above discussion then $\Lambda\pi$ will be the resulting homology group in all dimensions $> 2i$.

Lemma 4.5: *Let θ be a cocycle representing a cohomology class $\{\theta\} \in H^i_{cmpt}(\tilde{X}; \mathbf{F}_2)$, then $\theta \otimes \theta$ represents a well defined cohomology class*

$$\Gamma(\{\theta\}) \in H^{2i}(E_{\mathbf{Z}/2} \times_{\mathbf{Z}/2} \tilde{X} \times \tilde{X}/\Delta\pi; \Lambda\pi)$$

Proof: There is an equivalence between cochains with compact support in $C^*(\tilde{X}; \mathbf{F}_2)$ and $\mathbf{F}_2(\pi)$ equivariant cochains $C^*(\tilde{X}; \mathbf{F}_2(\pi))$ so we may assume $\theta \in C^*(\tilde{X}; \mathbf{F}_2(\pi))$. Moreover, since θ is a cocycle, it is equivalent to a chain map of $\mathbf{F}_2(\pi)$ complexes

$$Adj(\theta) : C_*(\tilde{X}) \longrightarrow \Sigma^i \mathbf{F}_2(\pi).$$

This induces

$$1 \otimes Adj(\theta) \otimes Adj(\theta): \; C_*(E_{\mathbf{Z}/2}) \otimes_{\mathbf{Z}/2} C_*(\tilde{X}) \otimes C_*(\tilde{X})/\Delta\pi \longrightarrow$$
$$C_*(E_{\mathbf{Z}/2}) \otimes_{\mathbf{Z}/2} \Sigma^i \mathbf{F}_2(\pi) \otimes \Sigma^i \mathbf{F}_2(\pi)/\Delta\pi$$

and, since $\Lambda\pi$ is the $2i^{th}$ stable homology group of this complex – where we stabilize over the periodicity inclusion $C_*(E_{\mathbf{Z}/2}) \rightarrow \Sigma^{-2} C_*(E_{\mathbf{Z}/2})$. The result follows.

Thus, we have

4.6 $S(\pi)^*(\Gamma(\theta)) = \sum e^j \otimes Sq_\pi^{i-j}(\theta) \in H^*(\mathbf{RP}^\infty \times X;\ \Lambda(\pi)) = H^*(\mathbf{RP}^\infty \times X;\ \mathbf{F}_2) \otimes \Lambda\pi$

and this defines the π-Steenrod operations $\{Sq_g^{i-j}(\theta) \mid g \in \pi, g^2 = 1\}$.

Let $tr:\ H^*(\tilde{X}) \longrightarrow H^*(X)$ be the transfer. This is a well known map on the level of chains and homology, but it is, in fact, induced by a stable map of spaces, and hence commutes with Steenrod squares.

Theorem 4.7: $Sq_g^{i-j}(\theta) \equiv 0$ *unless* $g = 1$.

Proof: Let $\{v\} \in H_i(X;\ \mathbf{F}_2)$ be represented by the cycle v, then

$$e_i \times v \xrightarrow{(\Delta^*)} \phi_i(v) + e_1 \otimes \phi_{i-1}(v) + \cdots + e_i \otimes \Delta(v)$$

is the ith Steenrod chain approximation to the geometric map $S(\pi)$, where each ϕ_i is, of course, constructed by the method of acyclic models. We have

$$g \times g(\phi_i(v)) = \phi_i(gv)$$

with the evident interpretation of (gv) in terms of a chosen lifting of v. There is a cochain λ defined over $\mathbf{Z}$, with compact support on $\tilde{X}$, and equivalent to θ so that we can write explicitly

$$<\theta \otimes \theta,\ \phi_i(v)> = <\sum_{g,h \in \pi} g\lambda \otimes h\lambda,\ \phi_i(v)> = <\sum \lambda \otimes g^{-1}h\lambda,\ \phi_i(v)>.$$

Thus, since the ϕ_i are tensored over the diagonal action $\Delta\pi$, we have that

$$g\lambda \otimes \lambda \ \ and\ \ \lambda \otimes g\lambda$$

both contribute a copy of $<\lambda \otimes g\lambda,\ \phi_i(v)>$ to the sum above. The result follows.

Actually, 4.7 does not depend on much. What we have actually shown above is that for any symmetric cochain $l \in C^*(\tilde{X}) \otimes C^*(\tilde{X})$, $<l, \phi_i(v)>$ has trivial coefficient on g for $G^2 = 1$, $g \neq 1$.

That was preliminary material to provide a geometric justification for the modification of A. Ranicki's definition of the symmetric L-groups, which we present next. In the rest of this section we will assume the reader is familiar with Ranicki's paper [R1].

If X is a Poincaré duality complex representing a class in $\Omega_n^{P.D}(B_\pi)$ then the Wu class $W(\theta)$ of $\theta \in H^*(\tilde{X};\ \mathbf{F}_2)$ is

$$[X^n] \cap Sq_\pi^{n-i}(\theta) \in \Lambda\pi.$$

and we have seen that this class reduces to the transfer of an ordinary Steenrod square. On the other hand Ranicki shows in [R1] that $W(\theta)$ is precisely the obstruction to doing algebraic surgery to "kill" θ.

Jim Davis recently pointed out to me that Ranicki's symmetric L-group $L_h^*(\mathbf{Z}\pi)$ does *not* detect R. Lee's semicharacteristic class [L], [D-M], [P], [D], which is so useful in showing the non-existence of free actions of many periodic groups on spheres. The failure traces back to the fact that the key to the detection of the obstruction was that the middle dimensional Wu classes behave in the restricted way above, and so any attempt to generalize Lee's techniques would seem to need a theory in which there were restrictions on the Wu classes. Thus we modify the definition of the symmetric L-groups to reflect this geometric condition.

Definition 4.8: *A Ronnie Lee (R.L for short) Poincaré structure on the finitely generated* $\mathbf{Z}\pi$*-free chain complex* $C = \{C_0, \cdots, C_n; \partial\}$ *is a sequence of maps*

$$\phi_i : C^* \longrightarrow C_*, \quad \phi_i : C^{n-j} \longrightarrow C_{j+i}$$

satisfying

(1) $0 = \phi_i\delta + (-1)^{i+1}\partial\phi_i + (-1)^{n-1}(\phi_{i-1} - (-1)^{(k^2+n^2+n-k)/2}\phi_{i-1}^*)$

(2) ϕ_0 *is a chain homotopy equivalence*

(3) The chain level Wu classes

$$W(\alpha) = \alpha(\phi_{2m-n}(\alpha)) \in \Lambda\pi$$

for $\alpha \in C^m$ *satisfy the condition that* $< W(\alpha), < g >> = 0$, $g^2 = 1$, *unless* $g = 1$.

Remark 4.9: An R.L Poincaré pair of dimension n, $(D, \partial D, f_*, \Phi_*, \phi_*)$ is defined similarly, with $(\partial D, \phi_*)$ an R.L-oriented Poincaré complex of dimension $n-1$ as above, and $f_* : \partial D \hookrightarrow D$ a chain map. 4.8.1 and 4.8.2 are replaced by

(1') The chain map $f : \partial D \hookrightarrow D$ *satisfies*

$$f_*\phi_k f^* = (-1)^{*+k-1}\{\Phi_k\delta + (-1)^{k+1}\partial\Phi_k + (-1)^{n-1}[\Phi_{k-1} - (-1)^{(k^2+n^2+n-k)/2}\Phi_{k-1}^*]\}$$

Here, the maps $\Phi_k : D^{n-i} \longrightarrow D_{i+k}$ are not chain maps, or chain homotopies. However, the map $((-1)^n\phi_0 f^*, \Phi_0) : D^{n-*} \longrightarrow MC(f_*)$ is a chain map, and it is required that

(2') $((-1)^n\phi_0 f^*, \Phi_0)$ *is a chain homotopy equivalence.*

Finally, 4.8.3 is unchanged, except that ϕ_i is replaced by Φ_i, and C^* by D^*.

Remark 4.10: An *R.L-symmetric structure* on D is a sequence of higher homotopies $\{\phi_i\}$ satisfying all the conditions of 4.8 above, except for 4.8.2. Since ϕ_0 is no longer a homotopy equivalence, we can consider the mapping cone $C = \Sigma^{-1}MC(\phi_0)$, which we denote as ∂D. It is defined by setting $C_j = D^{n-j} \oplus D_{j+1}$, so $C_{n-1-j} = D^{j+1} \oplus D_{n-j}$ and $\partial : C_j \to C_{j-1}$ is given by the formula $\partial(d^*, d) = (\delta d^*, (-1)^{n-j}\phi_0(d^*) + \partial d)$. As in [R1], we can define a Poincaré structure on C by defining

$$\mu_0 : C^j \longrightarrow C_{n-1-j}$$

by the formula

4.11 $$\mu_0(d, d^*) = (d^*, (-1)^{(n^2+n)/2}d + (-1)^{j+n}\phi_1(d^*))$$

while the higher μ's are given by

4.12 $$\mu_i(d, d^*) = (0, (-1)^{j+n}\phi_{i+1}(d^*)), \quad i > 0.$$

Note, here that $j = \dim(d^*)$, so $\dim(d) = n - j - 1$. Then the natural surjection $C \longrightarrow D^*$ – as in [R1] – gives rise to a Poincaré pair with the Φ_j all trivial. Since D was assumed to be R.L-oriented, it is direct to see that the resulting Poincaré pair is also R.L-oriented, and this gives rise to an equivalence between symmetric R.L-complexes and R.L-Poincaré pairs, associating to D the R.L-symmetric complex $D/\partial D$.

In a similar way to Ranicki's method of handling the ordinary symmetric L-groups we define the group

$$L^n_{R.L}(\mathbf{Z}\pi)$$

as the free abelian group with generators the set of all n-dimensional, finitely generated R.L-Poincaré complexes of free $\mathbf{Z}\pi$-modules, modulo those which are boundaries of Poincaré pairs. Note that more or less tautologically

4.13
$$L^*_h(\mathbf{Z}) = L^h_*(\mathbf{Z}[s]\pi) = L^*_{R.L}(\mathbf{Z}) = \begin{cases} \mathbf{Z} & * \equiv 0 \bmod 4 \\ \mathbf{Z}/2 & * \equiv 1 \bmod 4 \\ 0 & * \equiv 2,3 \bmod 4 \end{cases}$$

and there are pairings exactly as before between the R.L-L-groups for various groups π. Additionally, the forgetful functor $L^*_{R.L}(\mathbf{Z}\pi) \longrightarrow L^*_h(\mathbf{Z}\pi)$ has the usual naturality properties.

Of course, our previous remarks in this section show that Ranicki's orientation map $\Omega^{P.D}_*(B_\pi) \longrightarrow L^*_h(\mathbf{Z}\pi)$ factors as the composition of the forgetful map and the evident orientation

4.14
$$E_\pi : \Omega^{P.D}_*(B_\pi) \longrightarrow L^*_{R.L}(\mathbf{Z}\pi),$$

obtained by regarding the geometric complex together with its associated Steenrod maps as an R.L-oriented Poincaré complex.

Remark 4.15: These new symmetric L-groups seem very closely related to the "visible" L-groups of Michael Weiss [We]. I do not believe they are identical, since 4.8 and 4.9 require that the chain level Wu classes be identically 0, (a condition consistent with the geometry), but the definition in [We] only requires they that there be a specific sequence of chains with boundaries the chain level Wu classes. The visible L-groups in [We] are also stabilized so that they become 4-periodic, but this difference is not important.

The main application of these new groups is to the study of Poincaré duality bordism, and to provide invariants for detecting the non-triviality of surgery problems over Poincaré complexes. Specifically, there is the Levitt exact sequence [Lev], [B-M]

4.16
$$\cdots \longrightarrow \pi^s_{*+1}(B_{\pi+} \wedge MSG) \longrightarrow L^h_*(\mathbf{Z}\pi) \longrightarrow \Omega^{P.D}_*(B_\pi) \longrightarrow \pi^s_*(B_{\pi+} \wedge MSG) \longrightarrow \cdots$$

where MSG is the Thom spectrum of the universal fiber-homotopy-oriented sphere bundle. This space has been extensively studied and its homotopy type determined completely in [B-M-M] and [M-M]. When we take into account the orientations constructed above, and in

§2, we obtain the commutative diagram

4.17

$$\begin{array}{ccccccc}
\to \pi^s_{*+1}(B_{\pi+}\wedge MSG) \to & L^h_*(\mathbf{Z}\pi) & \to & \Omega^{P.D}_*(B_\pi) & \to \pi^s_*(B_{\pi+}\wedge MSG) \to \\
 & \Big\downarrow{=} & & \Big\downarrow{\pi} & \\
 & L^h_*(\mathbf{Z}\pi) & \xrightarrow{\sigma} & L^*_{R.L}(\mathbf{Z}\pi) & \\
 & & & \Big\downarrow{p} & \\
 & & & L^*_h(\mathbf{Z}\pi) & \\
 & & & \Big\downarrow{\cdot s} & \\
 & & & L^*_*(\mathbf{Z}[s]\pi) &
\end{array}$$

where σ is induced from the inclusion of even complexes into R.L-oriented complexes. The composition $\sigma \cdot p$ is the usual map, studied by Ranicki and others of the quadratic into the symmetric L-theory. Finally the map $\cdot s$ is induced by the operation $(C, \phi_*) \mapsto (B \otimes C, \phi_0 \cdot s)$ studied in 1.1, 1.2.

If $M^n \longrightarrow X^n$ is a degree one normal map, then it is not hard to show that, letting μ be the surgery obstruction, then

4.18
$$\sigma(\mu) = E_\pi(\{M^n\} - \{X^n\}).$$

This provides a method for detecting the surgery obstruction. The image of this class in $L^*_h(\mathbf{Z}\pi)$ tends to be both difficult to calculate and generally trivial, but we will show in the next sections that $L^*_{R.L}(\mathbf{Z}\pi)$ is very effective in detecting these obstructions.

§5. The structure of the R.L-L-groups

In this section we do two things. First we review and extend results of G. Carlsson on the structure of the suspension map for L^*-groups in 5.10, and apply them to the R.L-L-groups. Then we examine the structure of the $L^*_{R.L}(\mathbf{Z}\pi)$ when π is 2-hyperelementary. If the R.L-L-groups are a Mackey functor, then 5.13 and a direct generalization give structure results whenever π is finite. In any case, 5.13 applies to the space-form groups, and completely clarifies the arguments and results in [L], [M3], [H-Mad], on the non-trivial surgery obstructions to the existence of free actions on spheres in the critical cases.

Let A be an associative ring with identity 1, and involution τ. Let

$$C_* = \{C_0,\, C_1,\, \cdots, C_m,\, \partial\}$$

be a finitely generated chain complex of projective or free A-modules, we study the groups $Hom_A(C^p, C_q)$.

Definition 5.1: *A homogeneous chain of internal degree m, $\{\phi_{p,q}\}$, is a collection of maps*

$$\phi_{p,q} \in Hom_A(C^p,\, C_q),\ \textit{with}\ p + q = m,\ \textit{a constant.}$$

We denote by $N_m(C)$ the graded additive group of all homogeneous chains of internal degree m.

Remark 5.2: If $\{\phi_{p,q}\} \in N_m(C)$, then the sets $\{\partial\phi_{p,q}\}$ and $\{\phi_{p,q}\delta\}$ are both homogeneous chains of internal degree $m - 1$.

Definition 5.3: $Q_n(C) = \prod_{m \geq n} N_m(C)$. *A map*

$$\partial : Q_n(C) \longrightarrow Q_{n-1}(C)$$

is defined by

$$\partial\{\phi_s\}_s = (-1)^{s+s-1}\{\phi_s\delta + (-1)^{s+1}\partial\phi_s + (-1)^{n-1}[\phi_{s-1} - (-1)^{(s^2+n^2+n-s)/2}\phi^*_{s-1}]\}$$

where $\{\phi_s\}$ denotes the component of $\{\phi\} \in N_{n+s}(C)$.

Lemma 5.4: $(\partial)^2 : Q_n(C) \longrightarrow Q_{n-2}(C)$ *is identically zero, hence $\{Q_*(C), \partial\}$ is a chain complex.*

Let $f : C_* \longrightarrow D$ be a chain map (of degree 0), then, if $\{\phi_{p,q}\} \in Q_n(C)$, the family $\{f_*\phi_{p,q}f^*\} \in Q_n(D)$, and it is easy to verify that $\partial\{f_*\phi_{p,q}f^*\} = f_*\partial\{\phi_{p,q}\}f^*$, so that the correspondence

$$Q(f)(\{\phi_{p,q}\}) = \{f_*\phi_{p,q}f^*\}$$

is a functor on the appropriate category of chain complexes and chain maps over A.

Suppose that $\{\phi_{p,q}\} \in Q_n(C)$ is a cycle under ∂, then comparing the definition of ∂ in 5.3 with the condition in 4.10 for the existence of a *symmetric structure*, we see that the set of cycles in $Q_n(C)$ and the set of symmetric structures of dimension n on C are identical. Moreover, the set of Poincaré structures on C is exactly equivalent to the set of symmetric structures for which the *chain map* ϕ_0 is a chain homotopy equivalence.

There is a sort of suspension operation, $\sigma : Q_n(C, \tau) \longrightarrow Q_{n+2}(\Sigma C, -\tau)$, where $-\tau$ is the involution on A defined by $-\tau(a) = -(\tau(a))$, and ΣC is the suspension of C. This is defined by $\sigma(\{\phi_s\}) = \{\phi_s\}$, and the reader can quickly check that this is a chain map after we note that $(-1)^{(n^2+n)/2} = -(-1)^{((n+2)^2+(n+2))/2}$. With respect to the new involution $-\tau$, we obtain $\phi^{*(new)} = -\phi^{*(old)}$. The skew suspension map extends to Poincaré pairs as well, and consequently defines a homomorphism, also called the skew suspension

5.5 $$\sigma : L^*_d(A, \tau) \longrightarrow L^{*+2}_d(A, -\tau).$$

In [C] Gunnar Carlsson studied this skew-suspension map for the cases where $d = h$, or p. He showed that there was a construction, based essentially on the construction of a resolution of

5.6 $$H_\epsilon(\mathbf{Z}/2;\ A) = \{a \in A \mid a = \in \tau(a)\}/\{a + \in \tau(a)\}$$

and a group $W_*(A, \tau)$ which is given as a subgroup of a group of cycles modulo a subgroup of the set of boundaries in a certain associated resolution. This quotient detects the obstruction to desuspension in the sense that the following sequence is exact

$$L^*_d(A, \tau) \xrightarrow{\sigma} L^{*+2}_d(A, -\tau) \longrightarrow W_{*+2}(A, -\tau).$$

Note that the groups $\Lambda\pi$ discussed in 4.3, 4.4 are special cases of these groups, and just as the $\Lambda\pi$ are modules over $\mathbf{Z}\pi$, so too, the groups $H_\epsilon(\mathbf{Z}/2;\ A)$ are modules over A with action defined by $a(h) = ah\tau(a)$.

We now review this construction. Let

$$\mathcal{W} = \{\cdots K_n \longrightarrow K_{n-1} \longrightarrow \cdots K_1 \xrightarrow{\partial} K_1 \longrightarrow H_\epsilon(\mathbf{Z}/2;\ A) \longrightarrow 0\}$$

be a free resolution of $H_\epsilon(\mathbf{Z}/2;\ A)$ over A. We assume that all the K_i are finitely generated A-modules in what follows. Let $\mathcal{W}^n$ be the n-skeleton of $\mathcal{W}$, and consider the complexes

$$\cdots Q_*(\mathcal{W}^i) \subset Q_*(\mathcal{W}^{i+1}) \subset \cdots$$

Definition 5.7: *$\tilde{Z}_n \subset Q_n(K_n)$ is the subgroup of the set of cycles which satisfy the condition that the composite*

$$K^n \xrightarrow{\phi_0} K_0 \xrightarrow{\partial} H_\epsilon(\mathbf{Z}/2;\ A)$$

is precisely equal to the homomorphism $\kappa \longrightarrow \kappa(\phi_n(\kappa)) \in H_\epsilon(\mathbf{Z}/2;\ A)$.

This condition makes sense because we must have $\phi_{n+1} = 0$, so the condition

$$(-1)^n \partial\phi_{n+1} + \phi_{n+1}\delta + (-1)^{n-1}\{\phi_n - \epsilon\phi_n^*\} = 0$$

implies that ϕ_n is ϵ-symmetric. Similarly, we have

Definition 5.8: *$\tilde{B}_{n+1} \subset Q_{n+1}(K_{n+1})$ is the subgroup of chains having boundary contained in the image of $Q_n(K_n)$, and which satisfy the same condition as $\tilde{Z}_n$, with regard to the relation between Φ_{n+1} and $p \cdot \Phi_0$.*

It is direct to verify that $\partial(\tilde{B}_{n+1}) \subset \tilde{Z}_n$, and the resulting quotient group is defined to be $W_\epsilon(A,\ \tau)$. Carlsson shows that these groups do not depend on the particular resolution chosen. Indeed, the resolution does not even need to be free, merely projective.

Actually, these results can be extended in two directions. First, they extend to the R.L.-L-groups. The changes needed are to replace the resolution $\mathcal{W}$ above by the resolution

5.9 $$\mathcal{RW} = \{\cdots K_n \xrightarrow{\partial} K_{n-1} \longrightarrow \cdots K_1 \xrightarrow{\partial} K_0 \xrightarrow{\epsilon} \mathbf{Z}/2\}$$

where ϵ above is the usual augmentation. Then the groups $\tilde{Z}_n$, $\tilde{B}_{n+1}$ are defined exactly as before, except, of course, for the side condition that the Wu classes $W_T(a) \equiv 0$ whenever $T \neq 1$, and we define $W_n^{R.L}(\mathbf{Z}\pi,\ \tau) = \tilde{Z}_n/\partial(\tilde{B}_{n+1})$. Direct modification of the constructions in [C] leads to the exact sequence

$$L_{R.L}^{n-2}(\mathbf{Z}\pi,\ \tau) \xrightarrow{\sigma} L_{R.L}^n(\mathbf{Z}\pi,\ -\tau) \longrightarrow W_n^{R.L}(\mathbf{Z}\pi,\ -\tau).$$

Second, we can continue the exact sequence to the right. Indeed, we have

Theorem 5.10: *There are maps*

$$\partial : W_n(A,\ -\tau) \longrightarrow L_h^{n-3}(A,\ \tau), \quad \partial : W_n^{R.L}(\mathbf{Z}\pi,\ -\tau) \longrightarrow L_{R.L}^{n-3}(\mathbf{Z}\pi,\ \tau)$$

so the the resulting sequences

$$\cdots L_h^{n-2}(A,\tau) \xrightarrow{\sigma} L_h^n(A,\ -\tau) \longrightarrow W_n(A,-\tau) \xrightarrow{\partial} L_h^{n-3}(A,\tau) \cdots$$

$$\cdots L_{R.L}^{n-2}(\mathbf{Z}\pi,\tau) \xrightarrow{\sigma} L_{R.L}^n(\mathbf{Z}\pi,\ -\tau) \longrightarrow W_n^{R.L}(\mathbf{Z}\pi,-\tau) \xrightarrow{\partial} L_{R.L}^{n-3}(\mathbf{Z}\pi,\tau) \cdots$$

are long exact.

(The proof is tedius but quite direct. The key is to take any cycle in $\tilde{Z}_n(\mathbf{Z}\pi)$, say $(K^{(n)},\ \phi_*)$, and associate to it its boundary,

$$\Sigma^{-1}MC(\phi_0), \quad \phi_0 : K^{n-*} \longrightarrow K_*$$

as in 4.10. This boundary is not, to begin, an element in the image of $L^{n-3}()$. However, there is a more or less cannonical set of surgery data that we can associate to the quotient complex

$$\Sigma^{-1}MC(\phi_0)/\Sigma^{-1}MC(\phi_0)_{n-2},$$

as follows. Let

$$f: \Sigma^{-1}MC(\phi_0) \to \Sigma^{-1}MC(\phi_0)/\Sigma^{-1}MC(\phi_0)_{n-2} = \begin{cases} K^0 & \text{degree } n \\ K^1 \oplus K_n & \text{degree } n-1 \\ 0 & \text{otherwise} \end{cases}$$

be the surjection. Define Φ_n as a "characteristic element" map $\lambda: K_0 \longrightarrow K^0$. That is to say, λ is any A-map so that the correspondence

$$a \to a(\lambda(a)) \in H_\epsilon(\mathbf{Z}/2;\ \mathbf{Z}\pi)$$

is precisely the original augmentation map ϵ. Set $\Phi_n = \lambda$. Then Φ_{n-1} is defined to be $(0, -\partial\Phi_n)$, and it is direct to show that we can solve the equation

$$\phi_n = \pm\{(-1)^n \partial\Phi_{n-1} + \Phi_{n-1}\delta + (-1)^{n-1}[\Phi_{n-2} + \epsilon\Phi^*_{n-2}]\}$$

for Φ_{n-2}. The Poincaré complex which results from this surgery desuspends cannonically, and is defined to be the image of $(K_{(n)}, \phi_*)$ under ∂.

Of course, in order to show that ∂ is well defined we must check that the choices in the construction of Φ_n, Φ_{n-1}, Φ_{n-2} do not affect the result. The only map which actually matters is Φ_{n-2}, and any two choices differ by an element μ, with $\mu = -\epsilon\mu^*$. By enlarging the complex we can push this difference off, and we see directly that the two suspension Poincaré complexes obtained differ by the addition of a complex concentrated in degrees 1 and $n-2$ which is clearly a boundary.

With the definition in hand, the remaining steps in the proof of exactness are direct, though one needs the notion of a Poincaré triad at several points.)

Remark 5.11: Explicit elements can be constructed in the groups $W_n^{R.L}(\mathbf{Z}\pi, \tau)$ by taking manifold pairs $\{M^n, f\}$ representing elements in $\Omega_n(B_\pi)$ for which $f_*([M]) \neq 0$, and using the fact that

$$H^*(W_{\mathbf{Z}/2} \times \mathcal{R}W \times \mathcal{R}W/\{\mathbf{Z}/2 \times \pi\};\ \mathbf{Z}) = H^*(\mathbf{Z}/2 \times \pi;\ \mathbf{Z}/2)$$

to show that the resulting element in $W_n^{R.L}(\mathbf{Z}\pi, \tau) \neq 0$.

Remark 5.12: the R.L-L-groups $L^0_{R.L}(\mathbf{Z}\pi, \pm)$ and $L^1_{R.L}(\mathbf{Z}\pi, \pm)$ have very explicit descriptions:

1. $L^0_{R.L}(\mathbf{Z}\pi, \pm)$ is the Witt ring of non-singular R.L-$\pm$symmetric forms, and
2. $L^1_{R.L}(\mathbf{Z}\pi, \pm)$ represents equivalence classes of slightly modified types of formations as in [R1].

Thus the theorem above gives a reasonably constructive method of "building" the $L^*_{R.L}(\mathbf{Z}\pi)$ from $L^0_{R.L}()$ and $L^1_{R.L}()$.

Note (for 4.17) that the natural map $L^h_*(\mathbf{Z}\pi) \to L^*_{R.L}(\mathbf{Z}\pi)$ factors through the interated suspension map

$$L^h_*(\mathbf{Z}\pi) \xrightarrow{\sigma_0} L^\epsilon_{R.L}(\mathbf{Z}\pi, \pm) \xrightarrow{\sigma} \cdots \xrightarrow{\sigma} L^*_{R.L}(\mathbf{Z}\pi).$$

We now consider the situation when π is 2-hyperelementary. That is, π is a semi-direct product of a cyclic group $\mathbf{Z}/n$ of odd order and a finite 2-group G,

$$\pi = \mathbf{Z}/n \odot_\alpha G, \quad \alpha: G \longrightarrow Aut(\mathbf{Z}/n).$$

Since π is a semi-direct product there are natural splittings

$$L_*^h(R\pi) = L_*^d(RG) \oplus K_*^d, \quad L_{R.L}^*(R\pi) = L_{R.L}^*(RG) \oplus K_{R.L}^*$$

where K^* is the kernel of the projection map $L_{R.L}^*(R\pi) \longrightarrow L_{R.L}^*(RG)$.

Lemma 5.13: *For π a 2-hyperelementary group, we have that the natural map $K_*^h \longrightarrow K_{R.L}^*$ is an isomorphism provided that $H_1(\pi;\ \mathbf{Z}) \cong H_1(G;\ \mathbf{Z})$.*

Proof: Consider the localization sequence

$$\cdots \longrightarrow L_{p-R.L}^*(\mathbf{Z}\pi) \longrightarrow \coprod_{p|n} L_{R.L}^*(\hat{\mathbf{Z}}_p\pi) \oplus L_{R.L}^*(\hat{\mathbf{Z}}_2\pi) \oplus L_{R.L}^*(Q(1/2n)\pi) \longrightarrow$$

5.14

$$\coprod_{p|n} L_{R.L}^*(\hat{\mathbf{Q}}_p\pi) \oplus L_{R.L}^*(\hat{\mathbf{Q}}_2\pi) \xrightarrow{\partial} L_{p-R.L}^{*-1}(\mathbf{Z}\pi) \longrightarrow \cdots$$

The middle two terms are all just L_*-groups except for $L_{R.L}^*(\hat{\mathbf{Z}}_2\pi)$. So we study these groups next. We have

$$\hat{\mathbf{Z}}_2\pi = \hat{\mathbf{Z}}_2(\mathbf{Z}/n) \odot_\alpha G = \coprod_{m|n} \hat{\mathbf{Z}}_2 \otimes_{\mathbf{Z}} (\mathbf{Z}(\zeta_m) \odot_\alpha G) = \hat{\mathbf{Z}}_2 G \oplus W.$$

Clearly, for the Wu-classes involved with W we have $\mathrm{Hom}_{\hat{\mathbf{Z}}_2\pi}(W,\ \mathbf{Z}/2) = 0$, so it is easily verified that

$$L_{p-R.L}^*(W) \cong L_*^p(W).$$

Consider the commutative diagram

$$\begin{array}{ccccccc}
\cdots \to & L_*^h(\hat{\mathbf{Z}}_2\pi) & \to & L_*^p(W) \oplus L_*^h(\hat{\mathbf{Z}}_2 G) & \to & \hat{\mathrm{H}}_\epsilon(\mathbf{Z}/2;\ \tilde{K}_0(\hat{\mathbf{Z}}_2\pi)) & \to \cdots \\
 & \downarrow & & \downarrow & & \| & \\
\cdots \to & L_{R.L}^*(\hat{\mathbf{Z}}_2\pi) & \to & L_{R.L}^*(W) \oplus L_{R.L}^*(\hat{\mathbf{Z}}_2 G) & \to & \hat{\mathrm{H}}_\epsilon(\mathbf{Z}/2;\ \tilde{K}_0(\hat{\mathbf{Z}}_2\pi)) & \to \cdots
\end{array}$$

It is well known that $\tilde{K}_0(\hat{\mathbf{Z}}_2\pi) = \mathbf{Z}^{v-1}$ where v is the number of divisors of n. Here, the involution induced by τ is trivial. Consequently, the only spillover in the sequence between the two terms corresponds with the effect of reduction by the free modules. However, it is easily checked that there will be no spillover unless there is some odd torsion in the first homology group of π. As a result, when H_1 is strictly 2-torsion, the direct sum splitting of $L_{p-R.L}^*(\hat{\mathbf{Z}}_2\pi)$ results. But from this and the Meyer-Vietoris sequence 5.14, the splitting of $L_{p-R.L}^*(\mathbf{Z}\pi)$ follows. Finally, there is a natural splitting of $\tilde{K}_0(\mathbf{Z}\pi)$ into $\tilde{K}_0(\mathbf{Z}G) \oplus V$. Using this splitting in the Ranicki–Rothenberg exact sequence gives the result.

Remark 5.15: As an example, consider the compact space-form problem for the groups

$$Q(8a, b, c) = \mathbf{Z}/abc \odot_\alpha Q_8.$$

The first homology group of $Q(8a, b, c)$ satisfies the conditions of the Lemma. This group is periodic of period 4, and the indeterminacy of the surgery problems over it (the variations which can occur in the surgery obstruction when we vary the lifting of the Spivak normal bundle) are exactly in the image (in the Levitt sequence) of the ∂-map. It follows that the image of the resulting surgery problem in the R.L-L-groups completely detects the surgery obstruction. Indeed, a close reading of [M3], and the somewhat derivative [H-Mad], shows that this is exactly how the obstructions are being studied.

The semi-characteristic classes used by Lee in [L] to obtain the original non-existence results are also seen here. Indeed, they are already seen in $L^3_{R.L}(\hat{\mathbf{Z}}_2\pi)$. This follows directly from 5.13 and the work of J. Davis [D], and/or W. Pardon [P].

§6. The Clauwens transfer and the oozing conjecture

Now we return to the groups $L^h_*(\mathbf{Z}[s]\pi)$. The first thing we do is to construct maps

$$L^h_*(\mathbf{Z}[s]\pi) \longrightarrow L^h_{*+2}(\mathbf{Z}\pi)$$

which, when composed with the orientation map $\Omega^{TOP}_*(B_\pi) \to L^h_*(\mathbf{Z}[s]\pi)$ give the surgery obstruction to the surgery problem

$$M \times K^{4i+2} \xrightarrow{id \times \sigma} M \times S^{4i+2}$$

where $(K^{4i+2} \xrightarrow{\sigma} S^{4i+2})$ is the Kervaire problem. Based on this construction, the calculations reviewed in §2, and some extensions to $\mathbf{Z}/2$-coefficients, it is quite mechanical† to obtain complete results for surgery problems on closed manifolds.

We begin by giving the construction introduced origninally by F. Clauwens [Cl] in somewhat more generality than he originally did it. Lück-Ranicki [L-R] have recently generalized the construction still further.

Assumptions:

$\mathbf{A}$ is a unitary ring with involution τ, $\mathbf{L} \subset \mathbf{A}$ is a central unitary subring fixed under τ, $\mathbf{N}$ is a finitely generated free $\mathbf{L}$-module, $\mathbf{p}(s) \in \mathbf{L}[s]$ is a polynomial fixed under the extended involution on $\mathbf{L}[s]$, (i.e. $\mathbf{p}(s) = \mathbf{p}(1-s)$).

Let $\alpha \in \mathrm{Hom}_{\mathbf{L}}(\mathbf{N},\mathbf{N})$ satisfy $p(\alpha) = 0$, then

6.1) if $\beta \in \mathrm{Hom}_{\mathbf{L}}(\mathbf{N}^*, \mathbf{N}^*)$ is $1-\alpha^*\{\mathbf{N}^* = \mathrm{Hom}_{\mathbf{L}}(\mathbf{N}, \mathbf{L})\}$, then $\mathbf{p}(\beta) = 0$.

6.2) α defines an action of $\mathbf{L}[s]$ on $\mathbf{N}$ by $(\Sigma b_i s^i)(n) = (\Sigma b_i \alpha^i)(n), b \in \mathbf{L}$ and the action factors through the quotient $\mathbf{L}_p = \mathbf{L}[s]/(\mathbf{p}(s))$

6.3) Similarly, β defines an action of $\mathbf{L}_p$ on $\mathbf{N}^*$, and we will assume these actions in the remainder of this section.

We use the notation $\mathbf{A}_p = \mathbf{A} \bigotimes_{\mathbf{L}} \mathbf{L}_p$, and define functors

$$\otimes_\alpha : {}_{\mathbf{A}_p}\mathcal{C} \longrightarrow {}_{\mathbf{A}}\mathcal{C}, \quad \otimes_\beta : {}_{\mathbf{A}_p}\mathcal{C} \longrightarrow {}_{\mathbf{A}}\mathcal{C}$$

† The referee objects here that the author is being too liberal in his use of the word mechanical. The image of product surgery and twisted product surgery obstructions can be completely determined in $L^h_*(\mathbf{Z}[s]\pi)$, $L^h_*(\mathbf{Z}[s], \mathbf{Z}/2)$, and the actual obstructions are the images of these classes under the transfer above. However, the special class $\tilde{\kappa}_4$ for the $L^h_*(\mathbf{Z}[s])$ is non-zero, but no examples are known where its image in a surgery L-group is non-zero. In fact, it has been conjectured that its' image is always zero

by $\;_{\alpha}(\mathbf{M}) = \mathbf{M} \quad _{\mathbf{L}_p N}$, and $\;_{\beta}(\mathbf{M}) = \mathbf{M} \quad _{\mathbf{L}_p N^*}$. (This makes sense since $\mathbf{L}_p$ is central in $\mathbf{A}_p$.) Also, define a map $e'_c : \mathbf{M}^* \times \mathbf{N}^* \longrightarrow \mathrm{Map}(\mathbf{M} \times \mathbf{N}, \mathbf{A})$ by the rule

6.4 $$[e'_c(\phi \times \rho)](m \times n) = \rho(\phi(m) \otimes_{\mathbf{L}_p} n) = \phi(m)\rho(n).$$

Lemma 6.5:

a) *e'_c factors through a map $e_c : \mathbf{M}^* \bigotimes_{\mathbf{L}_p} N^* \to (\mathbf{M} \bigotimes_{\mathbf{L}_p} N)^*$*

b) *e_c is an isomorphism if $\mathbf{M}$ is a projective $\mathbf{A}_p$ module.*

c) *Suppose $\lambda \in Aut_{\mathbf{L}}(\mathbf{N}, \mathbf{N}^*)$ and $\beta = \lambda\alpha\lambda^{-1}$, then $\phi : \mathbf{M} \longrightarrow \mathbf{M}^*$ given implies that $(\phi \times \lambda) : \mathbf{M} \times \mathbf{M} \longrightarrow \mathbf{M}^* \times \mathbf{N}^*$ factors through a well defined map (denoted $\phi \bigotimes \lambda$ by a mild abuse of notation since the two tensor products are over* **Different** *actions) $\phi \bigotimes \lambda : \mathbf{M} \bigotimes_{\mathbf{L}_p} N \to \mathbf{M}^* \bigotimes_{\mathbf{L}_p} N^*$ and $(\phi \bigotimes \lambda)^* = \phi^* \bigotimes \lambda^*$ when $\mathbf{M}$ is projective.*

Remark 6.6: e_c induces a natural transformation (again called e_c) $e_c : \bigotimes_\alpha \longrightarrow \bigotimes_\beta$ and on the subcategory of projectives $(e_c)^2 = id$.

Example 6.7: Suppose $\lambda\mathbf{N} \longrightarrow \mathbf{N}^*$ is given as $\theta \pm \theta^*$ (so θ represents a non-singular $\pm$symmetric quadratic form on $\mathbf{N}$) then if we set $\alpha = \lambda^{-1}\theta$ we have $\beta = 1 - \alpha^* = 1 \mp \theta^*\lambda^{-1}$ but $\theta^* = \pm(\lambda - \theta)$ so $\beta = \theta\lambda^{-1} = \lambda\alpha\lambda^{-1}$, and non-singular quadratic forms over $\mathbf{L}$ provide non-trivial examples of the functors e_c, $\bigotimes_\alpha$, $\bigotimes_\beta$ as discussed above.

Given θ as above we can define a mapping τ_θ on quadratic complexes (C_*, ϕ_*) over $\mathbf{A}_p$ where the C_i are projective $\mathbf{A}_p$ modules by the correspondence

$$\tau_\theta((C_\cdot, \phi_\cdot)) = (\otimes_\alpha(C_\cdot), 1 \otimes \lambda e_c(\phi_0)).$$

By naturality we obtain

Theorem 6.8: *τ_θ induces maps of L-groups,*

$$\tau_\theta : L_*^{s,h,p}(\mathbf{A}_p) \longrightarrow L_{*+1\mp1)}^{s,h,p}(\mathbf{A}).$$

Now, consider the composite

6.9 $$\tau_\theta\mu : L^*(A) \longrightarrow L_{\cdot+(1\mp1)}(A \otimes B)$$

In the case when $A = \mathbf{Z}$, the map μ is the composite of the map $\cdot s : L^*(\mathbf{Z}\pi) \to L^*(\mathbf{Z}[s]\pi)$, and the map $p : L^*(\mathbf{Z}[s]\pi) \to L^*((\mathbf{Z}[s]/p(s))\pi)$ induced by projection onto the quotient ring.

Theorem 6.10: *The following diagram commutes*

$$\begin{array}{ccc} L^k(A) & \xrightarrow{\mu} & L_k(A_p) \\ \Big\downarrow {\scriptstyle 1\otimes <\theta>} & & \Big\downarrow {\scriptstyle \tau_\theta} \\ L^k(A) \otimes L_{1\mp1}(L) & \xrightarrow{(\times)} & L_{k+(1\pm1)}(A) \end{array}$$

where $(\times)$ is the Ranicki product pairing.

Proof: Explicitly, if (C, ϕ) represents $\alpha \in L^*(A)$ then $\mu(\alpha)$ is represented by

$$(C \otimes_A A[s], \phi s).$$

Consequently, $\tau_\theta\mu(C, \phi)$ is represented by $(\bigotimes_\alpha(C), e_c(\phi s \bigotimes \lambda))$, but

$$e_c(\phi s \bigotimes \lambda) = \{\phi_0 \bigotimes \lambda\alpha\} = \{\phi_\cdot \bigotimes \theta\},$$

and the result follows since this is the definition of the Ranicki product $(\times)$.

Here is how the above theory works for the Kervaire-Arf form

6.11
$$K = \begin{pmatrix} 1 & 1 \\ 0 & 1 \end{pmatrix}$$

Note that
$$\lambda^{-1}K = \begin{pmatrix} 0 & -1 \\ 1 & 0 \end{pmatrix}\begin{pmatrix} 1 & 1 \\ 0 & 1 \end{pmatrix} = \begin{pmatrix} 0 & -1 \\ 1 & 1 \end{pmatrix}$$

and this latter matrix satisfies the characteristic equation $p(x) = x^2 - x + 1$. But this is the irreducible equation of the 6th root of unity $-\zeta_3$. Consequently $\mathbf{Z}[s]/(p(x)) = \mathbf{Z}(\zeta_3)$, and from the naturality of the map τ_K in 6.9 we have

Theorem 6.12: *For the Kervaire form in 6.11 we have the factorization of the Ranicki product*

$$\begin{array}{ccccccc} \Omega_*^{TOP}(B_\pi) & \xrightarrow{E_\pi} & L^k_{R.L}(\mathbf{Z}\pi) & \xrightarrow{\cdot *} & L^*(\mathbf{Z}[s]\pi) & \xrightarrow{p} & L_k(\mathbf{Z}(\zeta_3)(\pi)) \\ & & \Big\downarrow 1\otimes <K> & & & & \Big\downarrow \mu_\theta \\ & & L^k_{R.L}(\mathbf{Z}(\pi)\otimes <K>) & & \xrightarrow{(\times)} & & L_k(\mathbf{Z}(\pi)) \end{array}$$

where $<K>$ represents the Kervaire form, E_π is defined in 4.14, the composite gives the value of the surgery obstruction for the surgery problem $M \times K^{4i+2} \xrightarrow{id\times\sigma} M \times S^{4i+2}$, and $\mathbf{Z}(\zeta_3)\pi$ is given the involution $(\Sigma\alpha_i g_i)^* = \Sigma\bar{\alpha}_i(g_i)^{-1}$.

Remark 6.13: There is nothing magic about $\mathbf{Z}(\zeta_3)$ in 6.12. Indeed, if we choose alternate representatives for the Kervaire form we factor through other quadratic extensions of $\mathbf{Z}$. For example if we set
$$\theta = \begin{pmatrix} 1 & 1 \\ 0 & n \end{pmatrix}$$

then this also represents $<K>$ when n is odd, but in this case the associated extension is $\mathbf{Z}\frac{(1+\sqrt{1-4n}}{2})$, and since n is odd, $1-4n$ is congruent to 5 (mod 8). Thus, there will be a factorization for each quadratic extension $\mathbf{Z}(\frac{1+\sqrt{v}}{2})$ where $v \equiv 5(\text{mod}8)$. Indeed, $\zeta_3 = \frac{-1\pm\sqrt{-3}}{2}$ is only singled out for convenience.

Remark 6.14: It is also possible to extend the entire sequence of ideas to $\mathbf{Z}/2$-coefficients and to factor the resulting surgery diagram as

$$\begin{array}{ccccccc} \Omega_*^{TOP}(B_\pi;\ \mathbf{Z}/2) & \longrightarrow & L^*_{R.L}(\mathbf{Z}\pi;\ \mathbf{Z}/2) & \longrightarrow & L^h_*(\mathbf{Z}[s]\pi;\ \mathbf{Z}/2) & \longrightarrow & L^h_*(\mathbf{Z}(w_v);\ \mathbf{Z}/2) \\ & & \Big\downarrow & & & & \Big\downarrow \mu_\tau \\ & & L^*_{R.L}(\mathbf{Z}\pi;\ \mathbf{Z}/2) & & \xrightarrow{\mu_0} & & L^h_{*-1}(\mathbf{Z}\pi;\ \mathbf{Z}/2) \end{array}$$

Here, we put $w_v = \frac{1+\sqrt{v}}{2}$ with $\begin{cases} v \text{ prime} \\ v > 0 \\ v \equiv 5(\text{mod}8) \end{cases}$, and where τ is the "twisted Kervaire probem". The point is that such rings have a unit ϵ with $N(\epsilon) = -1$. Since the involution is the Galois automorphism, this implies that multiplication by 2 anhilates $L^h_*(\mathbf{Z}(w)\pi)$, and provides the data for building the transfer μ_r. The diagrams above and of 6.12 fit together to give the composite

6.15
$$\begin{array}{ccccc} \Omega_*^{TOP}(B_\pi) \longrightarrow & L_*(\mathbf{Z}[s]\pi) & \longrightarrow L_*(\mathbf{Z}[s]\pi;\ \mathbf{Z}/2) \xrightarrow{\mu_r} & L_{*-1}(\mathbf{Z}\pi) \\ & \downarrow p & & \downarrow \partial \\ & L_*(\mathbf{Z}(\zeta_3)\pi) & \xrightarrow{\mu_0} & L_{*-2}(\mathbf{Z}\pi) \end{array}$$

where ∂ is the boundary map in the exact sequence

$$\cdots L_*^h(\mathbf{Z}\pi) \xrightarrow{\times 2} L_*^h(\mathbf{Z}\pi) \longrightarrow L_*^h(\mathbf{Z}\pi;\ \mathbf{Z}/2) \xrightarrow{\partial} L_{*-1}^h(\mathbf{Z}\pi) \xrightarrow{\times 2} \cdots.$$

In any case, we see that producting with the twisted Kervaire problem factors through the orientation $\Omega_*^{TOP}(B_\pi;\ \mathbf{Z}/2) \to L_*^h(\mathbf{Z}[s]\pi;\ \mathbf{Z}/2)$. We can easily check that the few classes which do not depend only on the rational representation ring of π in $L_*^h(\mathbf{Z}[s]\pi;\ \mathbf{Z}/2)$ go to 0 under μ_r. Consequently, the possible surgery obstructions on closed manifolds producted with the Kervaire or twisted Kervaire problem are determined by what happens for the Fontaine model groups given in 3.1. But this is direct, depending mostly on the calculation for the case $\pi = \mathbf{Z}/2$ given after 3.12. In any case, except for the groups $\mathbf{Z}/2^n$, $n > 1$, the complete result is given in [MII], [M-O], and [D-M]. But for $\mathbf{Z}/2^n$, the maps $\mathbf{Z}/2 \to \mathbf{Z}/2^n$, and $\mathbf{Z} \to \mathbf{Z}/2^n$ are detecting for $\mathbf{Z}/2$-bordism, so it is not difficult to extend the calculation.

The other basic product type surgery problem has the form

$$N^m \times M_8^{4i} \xrightarrow{id\times\sigma} N^m \times S^{4i}$$

where $\sigma :\ M_8^{4i} \to S^{4i}$ is the index 8 Milnor problem. For this product problem, using the theory above with the E_8-matrix, [M4] shows that there is a factorization

6.16
$$\Omega_*^{TOP}(B_\pi) \longrightarrow L_*^h(\mathbf{Z}[s]\pi) \longrightarrow L_*^h(\mathbf{Z}(\zeta_{15})\pi) \xrightarrow{tr} L_*^h(\mathbf{Z}\pi)$$

with image the surgery obstruction. An easy calculation then shows that the resulting map is exactly equal to the composition

6.17
$$\Omega_*^{TOP}(B_\pi) \longrightarrow \Omega_*^{TOP}(pt) \xrightarrow{tr} L_*^h(\mathbf{Z}) \longrightarrow L_*^h(\mathbf{Z}\pi).$$

Thus, there is no twisting of the resulting surgery formula by elements in $H^*(B_\pi;\ \mathbf{Z})$. (This result was first proved by L. Taylor and B. Williams in [T-W] using other methods.)

On the other hand every non-trivial surgery problem occuring in the bordism group $\Omega_*(B_\pi \times G/TOP,\ \mathbf{Z}/2)$ can be shown to trduce to a sum of a trivial problem, problems obtained by producting with the Kervaire problem, products with twisted Kervaire problems, or products with the index 8 Milnor problem. Thus, the above factorizations give complete control of the surgery obstructions!

Here is how we can describe the result. We note that the $L_*^h(\mathbf{Z}[s]\pi)$ are the homotopy groups of a certain spectrum $\mathcal{L}S_\pi$ (which turns out to consist entirely of Eilenberg-MacLane spaces at the prime 2). Moreover, the orientation $\cdot s :\ \Omega_*^{TOP}(B_\pi) \to L_*^h(\mathbf{Z}[s]\pi)$ is the map in homotopy associated to a map of spectra $\mu_\pi :\ \Sigma^0 B_\pi \to \mathcal{L}S_\pi$ which is given as the composite

$$\mu_\pi :\ S^0 \times B_\pi \xrightarrow{U\wedge id} MSTOP \wedge \Sigma^0 B_{\pi+} \xrightarrow{\sigma} \mathcal{L}S_\pi$$

where σ is the map of spectra associated to $\cdot s$, and U is the Thom class map. The classes in $H^*(B_\pi;\ \mathbf{Z}/2)$ which appear in the formulae like those after 3.12 or 3.16 are exactly the classes in image(μ_π^*) . These classes do not form an ideal or subring, but are closed under the action of the Steenrod algebra $\mathcal{A}_2$, and we have

Theorem 6.18: *As a module over $\mathcal{A}_2$, image(μ_π^*) is generated by classes in dimensions 1, 2, and 3 only. The classes in dimension 1 come from the image in $H^1(B_\pi;\ \mathbf{Z}/2)$ of $H^1(B_{\pi_{AB}};\ \mathbf{Z}/2)$. The generating classes in dimension 2 have similar descriptions in terms of stable transfer and projection onto subquotients of the form $\mathbf{Z}/2\ \times\ \mathbf{Z}/2,\ \mathbf{Z}/2^n,\ n\ >\ 1$, and dihedral groups. The generating classes in dimension 3 are associated to quaternion subquotients.*

(See e.g. [H-M-T-W] for details, though it appears to me that the proof outlined here is much more direct.)

Remark 6.19: The original examples showing the classes in dimension 1 are non-trivial are due to C.T.C. Wall. The classes in degree three were first seen by S. Cappell-J. Shaneson [C-S], while the first classes in dimension 2 were observed by J. Morgan-W. Pardon . However, their work was never published, and it is not entirely certain that they had a proof. L. Taylor and B. Williams took up these questions later and certainly knew how to prove the non-triviality of the Morgan-Pardon example, but the first complete published proof that I know of appears in [M-R].

Remark 6.20: 6.18 implies, in particular that the boundary maps in the surgery sequences

$$\cdots \xrightarrow{\partial} L^h_{n+1}(\mathbf{Z}\pi) \longrightarrow \mathcal{HT}(M^n) \longrightarrow [M^n,\ G/TOP] \xrightarrow{\partial} L^h_n(\mathbf{Z}\pi)$$

are completely understood for π finite. Thus the sets $\mathcal{HT}(M^n)$ are effectively evaluated in these cases. (Indeed, these boundary maps are determined as the homotopy maps in the composite

6.21
$$\Sigma^k S^n \xrightarrow{\text{eval}} \Sigma^k MSTOP \wedge \Sigma^0 B_{\pi+} \wedge \Sigma^{-k} G/TOP \longrightarrow$$
$$MSTOP \wedge \Sigma^0 B_{\pi+} \wedge G/TOP \longrightarrow \mathcal{LS}_\pi \wedge \mathcal{L}_0 \longrightarrow \mathcal{L}_\pi$$

where, given $h:\ M^n\ \longrightarrow\ \Omega^k G/TOP$, we have that eval$(h)$ is the Pontragin map associated to $(g,\ h) : M \longrightarrow B_\pi \times \Omega^k G/TOP$, with $g:\ M\ \longrightarrow\ B_\pi$ classifying the π-cover of M^n. Also, we have denoted the spectrum with homotopy groups the surgery L-groups $L^h_*(\mathbf{Z}\pi)$ as $\mathcal{L}_\pi$. Then standard formulae determine the map eval$_*$ on homotopy, thus completing the determination of the boundary map in the surgery exact sequence.)

Remark 6.22: When π is finite the boundary map in the Levitt exact sequence 4.16 is determined from the case $\pi\ =\ 1$ and the product formulae

$$\partial_{Levitt}\{M\ \times\ \alpha\}\ =\ \sigma(M\ \times\ K^{4i+2} \longrightarrow M\ \times\ S^{4i+2})$$

$$\partial_{Levitt}\{N\ \times\ \beta\}\ =\ \sigma(N\ \times\ \tau)$$

In the first case we assume M is oriented, and α has boundary the Kervaire class in the simply connected Levitt sequence. In the second case τ represents the twisted Kervaire problem, and we assume $\partial(\beta)\ =\ \tau$ in the simply connected Levitt sequence with $\mathbf{Z}/2$-coeffieients, while N is a $\mathbf{Z}/2-\pi$-manifold. From this it follows that

$$\text{im}(L^h_*(\mathbf{Z}\pi) \longrightarrow \Omega^{P.L}_*(B_\pi)\ =\ L^h_*(\mathbf{Z}\pi)/\{\textit{image of surgery on closed manifolds}\}.$$

Thus 6.18 implies that we have an effective determination of this map and the kernel of the "G-characteristic class" map

$$\Omega_*^{P.L}(B_\pi) \longrightarrow \pi_*^s(B_{\pi+} \wedge MSG).$$

§7 Extensions: Some results of R. Oliver

The final results in §5, §6 show that our orientation techniques give essentially complete techniques for analyzing surgery problems and classification questions when π is a finite group – at least for homotopy equivalence and $\mathcal{H}$-level classification.

However the surgery exact sequences

$$\cdots \longrightarrow L_{n+1}^s(\mathbf{Z}\pi) \longrightarrow \mathcal{ST}(M^n) \longrightarrow [M^n, G/TOP] \longrightarrow L_n^s(\mathbf{Z}\pi)$$

give an evaluation of the sets $\mathcal{ST}(M^n)$ which actually refines the homeomorphism classification of manifolds homotopy equivalent to M^n. Thus, it is natural to try to extend 6.18 and its consequences to the surgery obstruction map

$$\Omega_*^{TOP}(B_\pi \times G/TOP) \longrightarrow L_*^s(\mathbf{Z}\pi) \longrightarrow L_*^h(\mathbf{Z}\pi).$$

Once more, the factorization of product with the Kervaire and twisted Kervaire problems through $L_*^s(\mathbf{Z}[s]\pi)$ and its various quotients $L_*^s(\mathbf{Z}(w_v)\pi)$ goes through without change, as does the factorization, in 6.16, of product with the index 8 Milnor problem through $L_*^s(\mathbf{Z}(\zeta_{15})\pi)$.

At this point there is some advantage in working with the various quotients above since information on the structure of $L_*^s(\mathbf{Z}[s]\pi)$ seems hard to obtain, but, thanks to a long series of papers by R. Oliver, it is possible to say quite a lot about $L_*^s(\mathbf{Z}(w)\pi)$. In the remarks which follow we will summarize what is currently known about these questions.

Recall that the Whitehead group $Wh_1(\pi) = K_1(\mathbf{Z}\pi)/\{\pm\pi^{AB}\}$. Also, if we denote by $V \subset Wh_1(\pi)$ is any involution-invariant subgroup, there are groups $L_*^V(\mathbf{Z}\pi)$ and Ranicki-Rothenberg exact sequences

7.1
$$\cdots \longrightarrow \hat{H}_*(\mathbf{Z}/2;\, V) \longrightarrow L_*^s(\mathbf{Z}\pi) \longrightarrow L_*^V(\mathbf{Z}\pi) \longrightarrow \hat{H}_{*-1}(\mathbf{Z}/2;\, V) \longrightarrow \cdots$$
$$\cdots \longrightarrow \hat{H}_*(\mathbf{Z}/2;\, Wh_1(\pi)/V) \longrightarrow L_*^V(\mathbf{Z}\pi) \longrightarrow L_*^h(\mathbf{Z}\pi) \longrightarrow \hat{H}_{*-1}(\mathbf{Z}/2;\, Wh_1(\pi)/V) \longrightarrow \cdots$$

so it is natural to break the calculation into steps.

First, recall that $SW_1(\pi) \cong SK_1(\mathbf{Z}\pi) \subset Wh_1(\pi)$ is the torsion subgroup, and it is customary to denote $L_*^{SW_1(\pi)}(\mathbf{Z}\pi) = L_*'(\mathbf{Z}\pi)$. The calculations of [MI], [MII], [H-M-T-W] are actually carried out for the L_*'-groups, and the results of §6 completely reducing things to Fontaine model groups remain true at this level. Hence, 6.18 continues to give the full story for $L_*'(\mathbf{Z}\pi)$.

At this point Oliver's results show the existence of an exact sequence

7.2
$$0 \longrightarrow CL_1(\pi) \longrightarrow SW_1(\pi) \longrightarrow H_2^{AB}(B_\pi;\, \mathbf{Z}) \longrightarrow 0$$

where

$$H_2^{AB}(B_\pi;\, \mathbf{Z}) = H_2(B_\pi;\, \mathbf{Z})/ < \textit{subgroup generated by} \coprod_{\substack{K \subset \pi \\ K\ \textit{abelian}}} \{imH_2(B_H;\, \mathbf{Z})\} >$$

He also gives a set of generators (related to symbols in certain K_2-groups) for $Cl_1(\pi)$, [O], [O1]. What is lacking is a complete description of the relations among these generators, and also, the extension data for $SW_1(\pi)$ is currently unclear.

In any case Oliver and I decided to try to understand something of the surgery obstructions in $L_*^{Cl_1(\pi)}(\mathbf{Z}\pi)$ which might not be seen in $L_*^h(\mathbf{Z}\pi)$. This was partially motivated by the arguments of [M4] which actually showed that the factorization of 6.17 continues to hold in the $Cl_1(\pi)$ category.

The main result of this collaboration was a marvelous calculation by R. Oliver which shows

Theorem 7.3: *The surgery obstruction map*

$$\Omega_*^{TOP}(B_\pi) \longrightarrow L_0^{CL_1(\pi)}(\mathbf{Z}\pi)$$

defined by $\{M^2 \to \sigma(M^2 \times K^{4i+2} \to M^2 \times S^{4i+2})\}$ *hits every element in the kernel of the map*

$$L_0^{Cl_1(\pi)}(\mathbf{Z}\pi) \longrightarrow L_0^h(\mathbf{Z}\pi).$$

(The main idea is to observe that $\Omega_2(B_\pi) = H_2(B_\pi;\ \mathbf{Z})$, and if we represent an element in this homology group by a map of an oriented 2-manifold $M_g^2 \to B_\pi$, then this is exactly the same as specifying a conjugacy class of homomorphisms $\Gamma_g \longrightarrow \pi$, with Γ_g defined in 1.12.

But then the element in $L_s^2(\mathbf{Z}[s]\pi)$ represented by this map is the image of the orientation matrix described in 1.15 under this homomorphism! In particular, Oliver analyzed this image for the smallest 2-group π for which $H_2^{AB}(B_\pi;\ \mathbf{Z}) \neq 0$. In this case the surface was the 2-hole torus, and a computer calculation showed the image was non-trivial. Then a lifting argument completed the proof.)

At this point, nothing is yet known about the new elements in $L_*^s(\mathbf{Z}\pi)$ which are in the kernel of the map to $L_*^{Cl_1(\pi)}(\mathbf{Z}\pi)$, but based on their form, and the general behavior of product formulae I conjecture that at least some of the new elements occuring in $L_1^s(\mathbf{Z}\pi)$ are in the image of surgery on the product of some 3-dimensional manifolds with a Kervaire problem.

References

[B-M] G. W. Brumfiel, J. W. Morgan, Homotopy theoretic consequences of N. Levitt's obstruction theory to transversality for spherical fibrations, *Pacific J. Math*, 67(1976), 1-100.

M-M] ———————, Ib Madsen, R. J. Milgram, PL characteristic classes and cobordism, *Ann. of Math.*, 97(1973), 82-159.

[C-S] S. Cappell, J. Shaneson, A counterexample on the oozing conjecture, *Algebraic Topology, Aarhus 1978*, Lecture Notes in Math., #763, Springer-Verlag (1979), 627-634.

[C] G. Carlsson, Desuspension in the symmetric L-groups, *Algebraic Topology, Aarhus 1978*, Lecture Notes in Mathematics #763, Springer-Verlag, (1979), 175-197.

[Cl] F. Clauwens, The K-theory of almost symmetric forms, *Topological structures II*, 1979 Mathematical Centre Tracts, Proceedings of the symposium in Amsterdam, October 31-November 2, 1978 115(1979), 41-49.

[D] J. F. Davis, The surgery semicharacteristic, *Proc. London Math. Soc.* (3) 47 (1983), 411-428.

[D-M] J. Davis, R. J. Milgram, Semicharacteristics, bordism and free group actions, *(Trans. A.M.S., to appear)*.

[F] J. M. Fontaine, Sur la decomposition des algèbres de groupes, *Ann. Sce. École Norm. Sup.* 4(1971), 121-180.

-Mad] I. Hambleton, Ib Madsen, Actions of finite groups on $\mathbf{R}^{n+k}$ with fixed set $\mathbf{R}^k$, *(preprint)*.

-T-W] I. Hambleton, R. J. Milgram, L. Taylor, and B. Williams, Surgery with finite fundamental group, *(Proc. London Math. Soc., to appear)*.

-T-W] I. Hambleton, L. Taylor, and B. Williams, Detection theorems in K and L-theory, *preprint*.

[L] R. Lee, Semicharacteristic classes, *Topology* 12 (1973), 189-199.

[Lev] N. Levitt, Poincaré duality cobordism, *Ann. of Math.* 96(1972), 211-244.

[L-R] , W. Lück, A. Ranicki, (To appear)

[M-M] Ib Madsen, R. J. Milgram, *The Classifying Spaces for Surgery and Cobordism of Manifolds*, Ann. of Math. Studies #92, (Princeton U. Press, 1979).

MI] R. J. Milgram, Surgery with finite fundamental group I: the obstructions, *(preprint – (1985) Stanford University)*.

[MII] ———————, Surgery with finite fundamental group II: the oozing conjecture, *(preprint – (1985) Stanford University)*.

[M3] ———————, Patching techniques in surgery and the solution of the compact space form problem, *(Preprint, Stanford University 1981)*.

[M4] ——————— The Taylor-Williams theorem, *(Preprint, Stanford University 1985)*.

[M-O] ———————, R. Oliver, P-adic logarithms and the algebraic K, L groups for certain transcendence degree 1 extension of $\mathbf{Z}\pi$.

[M-R] ———————, A. Ranicki, Some product formulae for nonsimply connected surgery problems, *Trans. A.M.S.*, 297(1986), 383-413.

[O] R. Oliver, SK_1 for finite group rings II, *Math. Scand.* 47(1980), 195-231.

[O1] ———————, *A survey of Whitehead groups of finite groups*, Cambridge U. Press, to appear.

[P] W. Pardon, The exact sequence of a localization for Witt groups II. Numerical invariants for odd-dimensional surgery obstructions. *Pacific J. Math.* 102(1982), 123-169.

[R1] A. Ranicki, The algebraic theory of surgery I. Foundations *Proc. London Math Soc.* 40(1980), 87-192.

[S] R. E. Stong, Semi-characteristics and free group actions, *Compositio Math.* 29 (1974), 223-248.

[St] N. E. Steenrod, Homology groups of symmetric groups and reduced power operations, *Proc. Nat. Acad. Sci. USA.*, 29(1953), 195-218.

[S-E] ———————, D. E. Epstein, *Cohomology Operations*, Annals of Math Studies #50, Princeton U. Press (1962).

[T-W] L. Taylor, B. Williams, Surgery on closed manifolds, *(preprint)*.

[We] M. Weiss, On the definition of the symmetric *L*-groups, *(preprint)*

A Double Coset Formula for Levi Subgroups and Splitting BGL_n

by

Stephen A. Mitchell[1] and Stewart B. Priddy[2]

In **[Sn]** Snaith showed that the natural filtration on BGL_nK stably splits if $K = \mathbb{C}$ or $\mathbb{H}$. He also obtained coarser splittings when $K = \mathbb{R}$ or (after suitable localization) $K = \mathbb{F}_q$. For example, he showed that $BO(2n) \cong \bigvee_{1\leq k\leq n} BO(2k)/BO(2k-2)$. His method involved using the transfer associated to the normalizer of a maximal torus. However, there is another method: the idea is to use instead the transfers associated to the natural maps $BGL_i \times BGL_j \to BGL_{i+j}$. It is well-known that one can split $B\Sigma_n$ this way, and in fact on the homology level this idea goes back to Dold **[D1]**.

We show:

Theorem 4.1. *Let $K = \mathbb{R}, \mathbb{C}, \mathbb{H}$, or $\mathbb{F}_q$. If $K = \mathbb{F}_q$, assume q has been inverted. Then the filtration of BGL_nK by BGL_iK, $i \leq n$, stably splits: $BGL_nK \cong \bigvee_{1\leq i\leq k} BGL_iK/BGL_{i-1}K$.*

When $K = \mathbb{R}$ or $\mathbb{F}_q$ this theorem refines the splittings obtained by Snaith. However the method of proof is of interest in all cases; for example, it can easily be "restricted" to produce stable splittings of $\Omega SU(n)$, $\Omega(SU(2n)/Sp(n))$, and $\Omega(SU(n)/SO(n))$(**[M-R]** **[C-M]**). For $K = \mathbb{R}$ the result is that $BO(n) \cong \bigvee_{1\leq i\leq n} MO(i)$. This refinement of Snaith's splitting seems to have been noticed only recently by (independently) M. Crabb, B. Richter, and the authors (and perhaps others). The splitting of $BGL_n\mathbb{F}_q$ has been established independently, using a different method, by M. Crabb.

The key ingredient of the proof is the double coset formula associated to the pullback diagram

$$\begin{array}{ccc} E & \longrightarrow & BG_i \times BG_j \\ \downarrow & & \downarrow \\ BG_r \times BG_s & \longrightarrow & BG_k \end{array}$$

where $r + s = k = i + j$, and $G_i = GL_iK$. Now if $G_i = \Sigma_i$, the classical double coset formula applies and leads quickly to a proof of the splitting. Furthermore, if we replace $G_i \times G_j$ by the parabolic subgroup P_{ij}—the stabilizer of an i-dimensional subspace in K^n—the $P_{rs} - P_{ij}$ double cosets of GL_n are indexed by the $(\Sigma_r \times \Sigma_s) - (\Sigma_i \times \Sigma_j)$ double cosets of Σ_n, and one obtains essentially the same formula. But the kernel of the projection $P_{ij} \to G_i \times G_j$ is contractible in the continuous case and is a p-group in the finite case, so $BP_{ij} \to BG_i \times BG_j$ is an equivalence (after inverting p in the finite case). Thus it is not surprising that the double coset formula for the above diagram is essentially the same in all cases—but there will be coefficients arising from local fixed point indices. In fact this idea works for arbitrary Levi factors of parabolic subgroups of reductive Lie groups over $\mathbb{R}$ or $\mathbb{F}_q$, as we show in §2 and §3. In the case of real Lie groups various special cases of this formula have been known for some time—for example, Brumfiel and Madsen **[B-M]** showed that the composite $BT \xrightarrow{i} BK \xrightarrow{t} BT$, where T is a maximal torus of the

[1] Partly supported by NSF grant no. DMS-8601524.
[2] Partly supported by NSF grant no. DMS-8641403.

compact group K and t is the transfer, equals $\sum_{w \in W} w$, where W is the Weyl group. (This is the special case of Theorem (2.4) below obtained by taking $G_{\mathbb{R}} = K_{\mathbb{C}}$, $I = J =$ empty set). As transfers of this sort arise very frequently, it seems desirable to have a general formula.

In the case of real Lie groups, the "double coset formula" refers, of course, to the formula of Feshbach [**F**]. With modern technology—i.e., $G - CW$-complexes—the proof of Feshbach's theorem becomes extremely simple (G. Lewis; see [**L-M-S**], p. 203 ff.). On the other hand, the formula itself is not so easy to apply, as it often involves many extraneous terms. However, the statement and proof can easily be generalized so as to allow decompositions more general than a $G - CW$-decomposition. In §1 we give such a generalization, in which "G-cells" are replaced by "G- equivariant vector bundles". This formula (Theorems 1.1 and 1.5) is much more convenient for many purposes and is just as easy to prove. The key here, as in [**L-M-S**], is the Mayer-Vietoris property of the fixed point transfer: there is no need for the traditional equivariant vector fields.

The reader interested only in splitting BGL_n need not wade through the Lie theory of sections §2 and §3. It is not hard to work out the details for GL_n directly (using Theorem 1.5 in the continuous case).

The case $K = \mathbb{C}$ of Theorem 4.1 was first proved using this new approach by Bill Richter, who we would like to thank for a number of helpful conversations. We would also like to thank Albrecht Dold for some helpful discussions of the fixed point transfer.

§1. Some remarks on the double coset formula.

We begin by recalling some important properties of the fixed point transfer, as defined by Dold [**D2**]. Let B be a finite complex and let $\pi : E \to B$ be a fibre bundle with fibre a finite complex F. (More generally, E could be a Euclidean neighborhood retract over B.) Let $f : E \to E$ be a fibre-preserving self map. The associated *transfer* τ_f is then a stable map $B^+ \to E^+$ (See [**D2**] for the definition, which is remarkably simple). The transfer is a fibre-homotopy invariant, and is also *retraction invariant* in the following sense: suppose E is a retract over B of a larger bundle E_1. Define $f_1 : E_1 \to E_1$ by $f_1 = ifr$, where r is the retraction and i is the inclusion. Then the diagram

$$\begin{array}{ccc} E^+ & \xrightarrow{\ i\ } & E_1^+ \\ {\scriptstyle \tau_f} \nwarrow & & \nearrow {\scriptstyle \tau_{f_1}} \\ & B^+ & \end{array}$$

commutes. Now if the fixed point set $Fix\ f$ is a neighborhood deformation retract in E, it is immediate from the definition that τ_f factors as a stable map through $(Fix\ f)^+$. Thus if $Fix\ f$ has path-components $E_1, \ldots E_k$, we have $\tau_f = \eta_1 + \ldots + \eta_k$, where η_i is a stable map $B^+ \to E_i^+$ (and we have omitted the inclusion maps from the notation). Of course, E_i need not be a sub-bundle, and even if it is, η_i need not be the corresponding transfer (even when B is a point). But if E_i is an NDR over B, *and* the neighborhood can be taken to be invariant under f, it follows at once from the homotopy and retraction invariance that η_i is the transfer.

Next, we will need the Mayer-Vietoris property of the transfer, taking $f =$ identity for convenience. Suppose $E = E_1 \cup E_2$, where E_1, E_2, and hence $E_1 \cap E_2$ are sub-bundles. Let $\tau_1, \tau_2, \tau_{12}$ be the corresponding transfers (associated to the identity maps). Then $\tau = \tau_1 + \tau_2 - \tau_{12}$.

The proof of this powerful formula is so simple that we cannot resist giving it here (this proof was shown to us by Dold). One can find a neighborhood U of E_{12} and a deformation retraction $f_t(0 \le t \le 1)$ of U (over B) onto E_{12}. This can be arranged so that for $0 \le t < 1$, f_t extends to a global deformation of 1_E over B_1 and $Fix(f_{1/2}) = E_{12} \coprod E_1' \coprod E_2'$ (disjoint union), where $E_i' = E_i - U \cap E_i$. Let $f = f_{1/2}$. Then $\tau = \tau_f$, and hence $\tau = \eta_1 + \eta_2 + \tau_{1/2}$, where $\eta_i : B^+ \to (E_i')^+$ is as above. Similarly, $\tau_i = \eta_i + \tau_{12}$, and substituting $\tau_i - \tau_{12}$ for η_i yields the desired formula.

We now turn to Feshbach's double coset formula [F]. This formula is a special case of a general decomposition formula for bundles $P \times_G F \to B$, where G is a compact Lie group acting of F, and P is a principal G-bundle. If F is a finite $G-CW$-complex, the decomposition formula has a very simple proof [**L-M-S**]: one proceeds by induction on the number of G-cells, using the Mayer-Vietoris property at the inductive step. However, it is equally easy, and often more convenient, to replace G-cells by equivariant vector bundles in this argument: First of all, by a *finite bundle decomposition* of a space F, we will mean the obvious analogue of a finite CW-decomposition, in which cells are replaced by vector bundles. Explicitly, $F = \coprod_{1 \le i \le n} X_i$ (disjoint union), where each stratum X_i is homeomorphic to an open disc bundle $\mathring{D}_i$ of dimension N_i over a finite complex F_i, subject to the following conditions: (1) the homeomorphism $\mathring{D}_i \to X_i$ extends to a continuous map of the closed disc bundle D_i into F; (2) $\overline{X}_i - X_i$ lies in a union of strata X_j of lower fibre dimension. If F is a G-space, the decomposition will be called G-equivariant if the bundles are G-vector bundles and the attaching maps are G-equivariant. Now suppose given such a G-space F, with equivariant bundle decomposition $F = \coprod_{1 \le i \le n} \mathring{D}_i$, and let P be a principal G-bundle over B as above. Let S_i be the sphere bundle of D_i, and let χ_i be the universal Euler class of the sphere bundle $P \times_G S_i$ over $P \times_G F_i$. Here we recall that the Euler class of a fibre bundle $E \xrightarrow{\pi} B$ is the element $\chi \in \pi_S^0 B^+$ given by the composite $B^+ \xrightarrow{\tau} E^+ \xrightarrow{\varepsilon} S^0$, where τ is the transfer and ε is the augmentation. We have $\pi\tau = \chi \cdot 1_B$ (regarding $\{B^+, B^+\}$ as a module over $\pi_S^* B^+$). Here and in the sequel, "transfer" always means the transfer associated to the identity map.

(1.1) Theorem. *Let τ (resp. τ_i) denote the transfer associated to the bundle $P \times_G F \longrightarrow B$ (resp. $P \times_G F_i \longrightarrow B$).*

Then

$$\tau = \sum_{i=1}^{n} (1 - \chi_i)\tau_i$$

Remark. The notation $(1 - \chi_i)\tau_i$ stands for the composite

$$B^+ \xrightarrow{\tau_i} (P \times_G F_i)^+ \xrightarrow{\phi} (P \times_G F_i)^+ \xrightarrow{j} (P \times_G F)^+,$$

where ϕ is $(1 - \chi_i) \cdot$ (identity).

Proof. We use induction on n (the number of strata). For $n = 1$ the theorem is trivial. (Of course, we have ordered the strata so that the fibre dimension of D_K is less than or equal to the fibre dimension of D_{K+1}. In particular, D_1 is closed and has fibre-dimension zero, and $\chi_1 = 0$). At the inductive step we have $F = \mathring{D}_n \coprod X$, where $X = \coprod_{1 \le i \le n-1} \mathring{D}_i$, and X is closed. Hence we also have $F = D_n \cup X'$, where D_n is a closed G-disc bundle in $\mathring{D}_n$, $X' = \overline{F - D_n}$, and $D_n \cap X' = S_n$. Now for any G-subspace Y of F, let τ_Y denote the transfer for $P \times_G Y \to B$. Then by the Mayer-Vietoris property of the transfer, $\tau = \tau_{D_n} + \tau_{X'} - \tau_{S_n}$. Since F_n (resp. X) is an equivariant deformation retract of D_n (resp. X'), we obtain $\tau = \tau_n + \tau_X - \tau_{S_n}$. By induction, $\tau_X = \sum_{i=1}^{n-1}(1-\chi_i)\tau_i$, so it only remains to identify τ_{S_n}. Here, as usual, τ_{S_n} actually stands for the composite $B^+ \xrightarrow{\tau_{S_n}} (P \times_G S_n)^+ \longrightarrow (P \times_G F)^+$. But the transfer is transitive, so that $\tau_{S_n} = \tau_0 \tau_n$, where τ_0 is the transfer for $P \times_G S_n \to P \times_G F_n$. Furthermore, the inclusion of $P \times_G S_n$ in $P \times_G F$ factors up to homotopy through projection on $P \times_G F_n$. Hence $\tau_{S_n} = \chi_n \tau_n$, which completes the proof. □

(1.2) Corollary. *On ordinary homology,* $\tau_* = \sum_{i=1}^{n}(-1)^{n_i}(\tau_i)_*$. □

(1.3) Corollary. *Suppose the sphere bundles* $P \times_G S_n \longrightarrow P \times_G F_n$ *are all trivial. Then* $\tau = \sum_{i=1}^{n}(-1)^{n_i}\tau_i$. □

In particular, corollary (1.3) holds whenever the strata of F are product bundles over homogeneous spaces G/H—i.e., the decomposition is just the cell decomposition of a $G-CW$-complex. (This is almost the usual version of Feshbach's decomposition formula, as explained in [**L-M-S**]. Feshbach's version, which involves the "internal Euler characteristic" of the orbit- type components, is an immediate consequence—see [**L-M-S**], p. 185).

(1.4) Corollary. *Suppose each sphere bundle* $S_i \longrightarrow F_i$ *admits a G-equivariant self-map, equivariantly fibre-homotopic to the identity, which is fixed point free. Then* $\tau = \sum_{i=1}^{n}\tau_i$. *In particular this holds if F is decomposed into complex G-vector bundles.*

Proof. It follows that the transfers $(P \times_G F_i)^+ \longrightarrow (P \times_G S_i)^+$ are null, and hence $\chi_i = 0$. □

The double coset formula is obtained by applying (1.3) to the bundle $EK \times_K G/H \longrightarrow BK$, where K, H are subgroups of G. Of course, this involves extending the definition and properties of the transfer to bundles over infinite complexes, see [**L-M-S**], p. 187. Using (1.1) we get a more general formula. Suppose given a K-equivariant bundle decomposition of G/H in which each bundle has the form $K \times_{K_w} V_w$. Here K_w is the isotropy group of wH, V_w is a representation of K_w, and the zero section K/K_w is identified with the orbit of wH. Of course, w ranges over some finite set $w_1, \ldots w_n$. Note $K_w = K \cap wHw^{-1}$, and let $c_w : BK_w \longrightarrow BH$ be the map induced by $k \mapsto w^{-1}kw$. Let $i_{H,G}$ denote the inclusion $BH^+ \longrightarrow BG^+$, and let $tr_{G,H}$ denote the transfer $BG^+ \longrightarrow BH^+$. Then, retaining the notations of (1.3), we have:

(1.5) Theorem. $tr_{G,H} \circ i_{K,G} = \sum_w c_w(1-\chi_w)tr_{K,K_w}$ □

Of course the analogues of corollaries 2.2, 2.3, and 2.4 hold as well.

(1.6) Remark. Theorem (1.1) contradicts Corollary (2.11) of [B-M]. Consider for example $\mathbb{Z}/2$ acting on S^1 by conjugation. Clearly there is a $\mathbb{Z}/2$-equivariant vector field on S^1 with nondegenerate zeros at ± 1, so this example satisfies the hypothese there. We can take the index of the vector field to be 1 at 1 and - 1 at -1. Thus, according to [B-M], the transfer τ associated to $E\mathbb{Z}/2 \times_{\mathbb{Z}/2} S^1 \xrightarrow{\tau} B\mathbb{Z}/2$ satisfies $\tau = i_1^+ - i_{-1}^+$, where i_1, i_{-1} are the inclusions at the fixed points. On the other hand S^1 has an obvious bundle decomposition consisting of 1 and its complement. Applying (1.1) to this, we obtain $\tau = i_1^+ + (1-\chi)i_{-1}^+$, where $\chi : B(\mathbb{Z}/2)^+ \longrightarrow S^0$ is the Euler class of the transfer τ_0 for $E\mathbb{Z}/2 \longrightarrow B\mathbb{Z}/2$. Now observe that these two results cannot both be right (although they agree on homology). For if so, then after applying π^+ we conclude that $(2-\chi)Id = 0$ as a map $B\mathbb{Z}/2^+ \longrightarrow B\mathbb{Z}/2^+$. In other words, twice the identity factors through S^0, which is known to be false. As a check on the correctness of (1.1), we apply the usual version of the decomposition formula ([L-M-S], Theorem 6.1, p. 203). For this we need a $\mathbb{Z}/2-CW$-decomposition of S^1, and we take the obvious one: the two fixed points as zero-cells and a single free one-cell. The cited formula then yields $\tau = i_1^+ + i_{-1}^+ - f$, where f is the composite of τ_0 and the inclusion of a free orbit. But clearly f is homotopic to $\chi \cdot i_{-1}^+$.

§2 Levi subgroups over R.

In this section we assume the reader is familiar with the basic structure theory of semisimple real Lie groups (an excellent survey of the relevant results can be found in [Ma]). The groups we consider will, in fact, be real forms of connected reductive algebraic groups over $\mathbb{C}$.

Let G be a connected compact Lie group with complexification $G_{\mathbb{C}}$. Let σ be an anti-complex involution on $G_{\mathbb{C}}$. We assume that σ commutes with the compact involution defining G, and that σ preserves a fixed maximal torus of $G_{\mathbb{C}}$. Let $G_{\mathbb{R}} = (G_{\mathbb{C}})^{\sigma}, K = G^{\sigma}$. Then $G_{\mathbb{R}}$ is a real reductive Lie group with maximal compact subgroup K. There is a Tits system $(G_{\mathbb{R}}, B_{\mathbb{R}}, N_{\mathbb{R}}, S_{\mathbb{R}})$ associated with $G_{\mathbb{R}}$ as follows: Let $\mathfrak{g}_{\mathbb{R}} = \mathfrak{k} + \mathfrak{p}$ be a Cartan decomposition of the Lie algebra of $\mathfrak{g}_{\mathbb{R}}$, and let $\mathfrak{a}_p \subseteq \mathfrak{p}$ be a maximal abelian subspace. Then $N_{\mathbb{R}} = N_K \mathfrak{a}_p$ and $S_{\mathbb{R}}$ is a set of involutions generating the Weyl group $W_{\mathbb{R}} \equiv N_K \mathfrak{a}_p / C_K \mathfrak{a}_p$. To define $B_{\mathbb{R}}$, let Q be the parabolic subgroups of $G_{\mathbb{C}}$ associated to the black nodes of the Satake diagram of $(G_{\mathbb{C}}, \sigma)$. (In other words, Q is generated by the Borel subgroup of $G_{\mathbb{C}}$ together with the simple reflections s in the Weyl group of $G_{\mathbb{C}}$ which fix $\mathfrak{a}_p$). Then $\sigma Q = Q$ and we define $B_{\mathbb{R}} = Q^{\sigma}$. The standard parabolic subgroups $P_I, I \subseteq S_{\mathbb{R}}$, are then defined by $P_I = \langle B_{\mathbb{R}}, I \rangle$. There is a semidirect product decomposition $P_I = U_I L_I$, where U_I is the unipotent radical of P_I and L_I is a Levi factor. The exponential map $\mathfrak{u}_I \longrightarrow U_I$ is an L_I-equivariant diffeomorphism. Let H_I be a maximal compact subgroup of L_I. We will call H_I a subgroup of K of Levi type. Now consider a generalized flag manifold $G_{\mathbb{R}}/P_J, J \subseteq S$. Recall here that by the Iwasawa decomposition, $G_{\mathbb{R}}/P_J = K/H_J$. From the axioms for a Tit system one obtains the Bruhat decomposition:

(2.1) Proposition. $G_{\mathbb{R}}/P_J = \coprod_{w \in W_I \backslash W / W_J} P_I w P_J / P_J.$ □

In fact the orbits are vector bundles over homogeneous spaces of H_I: Let w be the minimal length representative of $W_I w W_J$, and let $U_w = \{u \in U_I : w^{-1} u w \in U^-\}$ (here U^- is the unipotent subgroup corresponding to the negative roots of the root system of $G_{\mathbb{R}}$). Then H_w normalizes U_w, and we may form the balanced product $H_I \times_{H_w} U_w$; it is a vector bundle over H_I/H_w.

(2.2) Proposition. *(a) The map $H_I \times_{H_w} U_w \longrightarrow G_{\mathbb{R}}/P_J$ given by $hu \mapsto huwP_J$ is a diffeomorphism onto the orbit $P_I w P_J/P_J$.* □

(b) The P_I-orbit decomposition of $G_{\mathbb{R}}/P_J$ is an H_I- equivariant bundle decomposition.

Proof. Part (a) is easily proved as in [Mi], Theorem 8.4. For part (b), recall (see e.g., [Mi]) that H_I/H_w has its own Bruhat decomposition, and in fact the Bruhat cells in $H_I \times_{H_w} U_w$ are just the inverse images of the cells of H_I/H_w. From this we easily obtain the required attaching maps. As for the closure condition, suppose the orbit of w lies in the closure of the orbit of w', where w, w' have minimal length. Then $w \leq w'$ (Bruhat order), and hence either $w = w'$ or $l(w) < l(w')$. □

Remark. This decomposition can also be obtained from a Morse-Bott function on $G_{\mathbb{R}}/P_J$.

From Theorem (1.5) we have at once:

(2.3) Theorem. $tr_{K,H_J} \circ i_{H_I,K} = \sum_{w \in W_I \backslash W/W_J} c_w(1-\chi_w) tr_{H_I,H_w}$. □

Of course, we may also apply the corollaries of (1.5). For example, on integral homology $1-\chi_w$ can be replaced by $(-1)^{n(w)}$, where $n(w) \equiv \dim U_w$ is determined by the root system of $G_{\mathbb{R}}$ in a standard way. In particular, if $G_{\mathbb{R}}$ is the split real form of $G_{\mathbb{C}}$ (e.g., $SL(n,\mathbb{R})$), $(1-\chi_w)$ can be replaced by $sgn\ w$ (on homology). If $G_{\mathbb{R}} = K_{\mathbb{C}}$, so that $G_{\mathbb{R}}$ is a complex algebraic group, then $\chi_w = 0$ (1.4). But in general, on the spectrum level, of course, it is not true that $1-\chi_w = (-1)^{n(w)}$. For example, take $K = 0(n), G_{\mathbb{R}} = GL(n,\mathbb{R})$, and $I = J = \emptyset$. Then $H_I = H_J = (\mathbb{Z}/2)^n$, and on homology $tr \circ i = \sum_{w \in \Sigma_n} (sgn w) c_w : B(\mathbb{Z}/2)^n \to B(\mathbb{Z}/2)^n$. But it is easy to see that this is false as maps of spectra. Indeed when $n = 2$, the action of $\mathbb{Z}/2^2$ on $\mathbb{R}P^1$ factors through a $\mathbb{Z}/2$-action, and the discrepancy arises exactly as in Remark (1.6). Thus Theorem (3.11) of [B-M] is true only on homology, not as stable maps.

§3 Levi subgroups over $\mathbb{F}_q$

Fix a prime p and let G be a finite Chevalley group over $\mathbb{F}_q$, where $q = p^m$. There is a Tits system (G, B, N, S) where B is a Borel subgroup. Thus for each $I \subseteq S$ we obtain a parabolic subgroup P_I, which fits into a split extension $U_I \to P_I \to L_I$. Here L_I is the Levi factor and U_I is the unipotent radical. In particular U_I is a p-group. Now fix $I, J \subseteq S$ and let $q^j = |U_J|$. Let $W = N/B \cap N$ denote the Weyl group. As in §2 we define $L_w = L_I \cap wL_J w^{-1}$ and let $c_w : BL_w \to BL_J$ denote the map induced by $h \mapsto w^{-1}hw$. Let W^{IJ} denote the set of minimal length representatives for the double cosets $W_I \backslash W/W_J$.

(3.1) Theorem. *On $H_*(-,\mathbb{Z}[\frac{1}{p}])$, $tr_{G,L_J} \circ i_{L_I,G} = q^j \sum_{w \in W^{IJ}} q^{l(w)} c_w tr_{L_I,L_w}$.*

Proof. We will use the following simple observation: suppose $U \to P \underset{s}{\overset{\pi}{\rightleftarrows}} L$ is a split extension of finite groups, and U is a p-group. Then on $H_*(-,\mathbb{Z}[\frac{1}{p}])$, π_* is an isomorphism with inverse s, and furthermore $\pi_* = |U|^{-1}(tr_{P,L})_*$. Here the first assertion is obvious from the Hochshild-Serre spectral sequence, and then the second follows from $(c_{L,P})_* \circ (tr_{P,L})_* =$ multiplication by $|U|(i_{L,P} = s)$.

For the remainder of the proof we will omit asterisks to simplify the notation: all maps are to be read as the induced maps on $H_*(-;\mathbb{Z}[\frac{1}{p}])$. It is enough to show that the two sides of the equation become equal after precomposition with $\pi : BP_I \to BL_I$ and postcomposition with i_{L_J,P_J}. First consider the left-hand side. Clearly $i_{L_I,G} \circ \pi = i_{P_I,G}$, and $i_{L_J,P_J} \circ tr_{G,L_J} = i_{L_J,P_J} \circ tr_{P_J,L_J} \circ tr_{G,P_J} = q^j tr_{G,P_J}$. Hence the left-hand side becomes

$$q^j tr_{G,P_J} \circ i_{P_I,G} = q^j \sum_{w \in W^{IJ}} c'_w tr_{P_I,P_w}. \tag{3.3}$$

Here $P_w = P_I \cap w P_J w^{-1}$, $c'_w : x \mapsto w^{-1}xw$, and we have used the Bruhat decomposition $G = \coprod_{w \in W^{IJ}} P_I w P_J$. Comparing this with the right-hand side of (3.1), we see that it is enough to show

$$c'_w tr_{P_I,P_w} = q^{l(w)} i_{L_J,P_J} c_w tr_{L_I,L_w} \pi. \tag{3.3}$$

Now P_w fits into a split extension $U'_w \to P_w \xrightarrow{\pi_w} H_w$, where $U'_w U_w = U_I$ and $U'_w \cap U_w = \{1\}$. It follows that the following diagram commutes:

$$\begin{array}{ccc} BP_w & \xrightarrow{\pi_w} & BL_w \\ tr\uparrow & & \uparrow q^{l(w)} tr \\ BP_I & \xrightarrow{\pi} & BL_I \end{array}$$

For $\pi = |U_I|^{-1} tr$, $\pi_w = |U'_w|^{-1} tr$, and $|U_I| = |U_w| \cdot |U'_w| = q^{l(w)} |U'_w|$. Thus $q^{l(w)} tr_{L_I,L_w} \pi = \pi_w tr_{P_I,P_w}$. Finally, we have $i_{L_J,P_J} c_w \pi_w = c'_w i_{L_w,P_w} \pi_w = c'_w$. This implies (3.3), and completes the proof of the theorem. □

We will need a slightly refined version of (3.1). The groups L_w are parabolic subgroups of L_I, and we want our formulas to involve only reductive groups. But this is easy, since the identity map of $H_* BL_w$ is $q^{-r(w)} i_{H_w,L_w} tr_{L_w,H_w}$, where H_w is the Levi factor of L_w and $q^{r(w)}$ is the order of the unipotent radical. Hence we obtain:

(3.2) Theorem. *On* $H_*(-;\mathbb{Z}[\frac{1}{p}])$, $tr_{G,L_J} \circ i_{L_I,G} = \sum_{w \in W^{IJ}} q^{s(w)} c_w tr_{L_I,H_w}$ *for certain integers* $s(w)$.

□

§4 Splitting BGL_n.

(4.1) Theorem. *Let* $K = \mathbb{R}, \mathbb{C}, \mathbb{H}$, *or* $\mathbb{F}_q$. *If* $K = \mathbb{F}_q$, *assume* p *has been inverted. Then the filtration of* $BGL_n K$ *by* $BGL_i K$, $1 \le i \le n$, *stably splits:*

$$BGL_n K \cong \bigvee_{1 \le i \le n} BGL_i K / BGL_{i-1} K.$$

Remarks. (1) As noted in the introduction, this is a stronger version of theorems of Snaith [**S**]. (2) It follows immediately that $BGL(K)$ splits, too; the obvious map $\bigvee_{1\le i\le\infty} BGL_iK/BGL_{i-1}K \to BGL(K$ is an equivalence. (3) When $K = \mathbb{F}_q$, of course, the theorem is false if p is not inverted: indeed by a theorem of Quillen [Q], $BGL(\mathbb{F}_q)$ is acyclic at p. If the filtration splits integrally, we conclude that $BGL_n\mathbb{F}_q$ is also p- acyclic, which is absurd.

It is well-known that theorem (4.1) holds with GL_n replaced by the symmetric group Σ_n. In fact all of these splittings can be obtained in the same way. (Here we recall that the maximal compact subgroup of $GL(n,K)$ is $0(n)$ for $K = \mathbb{R}, U(n)$ for $K = \mathbb{C}$, and $Sp(n)$ for $K\mathbb{H}$. In particular the inclusion $G_n \to GL(n,K)$ is a homotopy equivalence, so we may replace $GL(n,K)$ by G_n in Theorem 4.1.) Let G_n denote any of the groups Σ_n, $0(n)$, $U(n)$, $Sp(n)$, or $GL(n,\mathbb{F}_q)$. Let $G_{ij} = G_i \times G_j$ and let $m_{ij} : G_{ij} \to G_{i+j}$ be the usual inclusion. We use the same notation for the induced map $BG_{ij} = BG_i \times BG_j \to BG_{i+j}$. Adding a disjoint basepoint, we get $m_{ij}^+ : BG_i^+ \wedge BG_j^+ \to BG_{i+j}^+$. Let $\overline{BG}_j = BG_j/BG_{j-1}$. We obtain maps $\overline{M}_{ij} : \overline{BG}_i \wedge \overline{BG}_j \to \overline{BG}_{i+j}$; note $BG_0^+ = \overline{BG}_0 = S^0$. Finally, let $\tau_{ij} : BG_{i+j}^+ \to BG_{ij}^+$ denote the transfer associated to m_{ij}, and let f_{ij} denote the composite

$$BG_{i+j}^+ \xrightarrow{\tau_{ij}} BG_{ij}^+ = BG_i^+ \wedge BG_j^+ \xrightarrow{\pi_{ij}} \overline{BG}_j$$

where π_{ij} is the obvious projection (note $f_{n,o} : BG_n^+ \to S^0$ is the augmentation ε and $f_{o,n} : BG_n^+ \to \overline{BG}_n$ is the projection). Then Theorem (4.1) follows from:

(4.2) Theorem. *Let $G_n = \Sigma_n, 0(n), U(n), Sp(n)$, or $GL_n\mathbb{F}_q$. If $G = GL(n,\mathbb{F}_q)$, assume p has been inverted. Then the map*

$$\bigvee_{o\le j\le n} f_{n-j,j} : BG_n^+ \longrightarrow \bigvee_{o\le j\le n} \overline{BG}_j$$

is an equivalence.

(4.3) Lemma. *Suppose $i+j = r+s$. Then on integral homology (if $G_n \neq GL_n\mathbb{F}_q$) or on $\mathbb{Z}[\frac{1}{p}]$-homology (if $G_n = GL_n\mathbb{F}_q$)*

$$f_{ij} \circ m_{rs}^+ = \sum_{\substack{a+b=i\\ a\le r,b\le s}} \alpha_{ab}\overline{m}_{r-a,s-b} \circ (f_{a,r-a} \wedge f_{b,s-b}).$$

where the coefficients α_{ab} are integers, $\alpha_{o,j} = 1$ if $G_n \neq GL_n\mathbb{F}_q$, and $\alpha_{o,j} = p^k$ for some k if $G_n = GL_n\mathbb{F}_q$.

Remark. If $G_n = \Sigma_n, U(n)$, or $Sp(n)$, the proof will show that (4.3) holds as maps of spectra, with $\alpha_{a,b} = 1$ for all a, b. Hence we obtain a "Cartan formula" or "exponential property" for the splitting maps themselves. But in any case the main point is that the right-hand side has the right "leading term".

Proof of Theorem (4.2). It is enough to show that the composites $g_j : BG_j^+ \longrightarrow BG_n^+ \xrightarrow{f_{ij}} \overline{BG}_j$ $(i+j=n)$ coincide with the projections—at least on homology, or up to an automorphism of $\overline{BG}_j$. For then the map in (4.2) is filtration preserving (at least on homology), and is an equivalence on the associated graded objects. To see this, take $r = i, s = j$ in the lemma and precompose with $S^o \wedge BG_j^+ \to BG_i^+ \wedge BG_j^+$. On the left we obtain g_j. On the right, ignoring the coefficients α_{ab} for the moment, the terms of the sum have the form

$$S^o \wedge BG_j^+ \longrightarrow BG_i^+ \wedge BG_j^+ \xrightarrow{\tau_{a,i-a}\wedge\tau_{b,j-b}} (BG_a^+ \wedge BG_{i-a}^+) \wedge (BG_b^+ \wedge BG_{j-b}^+) \xrightarrow{\pi_{i-a}\wedge\pi_{j-b}} \overline{BG}_{i-a} \wedge \overline{BG}_{j-b}$$

$$\xrightarrow{\overline{m}_{i-a,j-b}} \overline{BG}_j.$$

But $\pi_o \overline{BG}_a = o$ if $a \neq o$, so the terms are zero except for $a = o, b = j$. Hence on the right-hand side we obtain α_{oj} times the composite

$$S^o \wedge BG_j^+ \longrightarrow BG_i^+ \wedge BG_j^+ \xrightarrow{\epsilon\wedge\pi} S^o \wedge \overline{BG}_j = \overline{BG}_j.$$

This last composite is indeed the projection. In view of our condition on α_{oj}, this complete the proof. □

Before proving Lemma (4.3), we note how the groups G_n fit into the context of the previous sections. First consider $G_n = \Sigma_n$. The minimal length representatives for Σ_n/Σ_{ij} are precisely the shuffles of type ij. The minimal length representatives for the $\Sigma_{rs} - \Sigma_{ij}$ double cosets are the shuffles w_{ab} defined by $w(k) = k$ if $k \leq a, w(k) = k + r - a$ if $a + 1 \leq k \leq i$. Here $a + b = i$, $a \leq r$, $b \leq s$. Note $l(w_{ab}) = b(r - a)$. Now, of course, the Weyl group for the usual Tits system on GL_nK is Σ_n, with $S = \{s_1, \ldots, s_{n-1}\}$ the usual set of Coxeter generators. If $I = S - \{s_i\}$, then $\Sigma_{ij} = \Sigma_I$. The corresponding parabolic P_{ij} is the stabilizer of an i-plane in K^n, and its Levi factor is $GL_i \times GL_j$. Now let $H_{ab} = G_{rs} \cap wG_{ij}w^{-1}$. Then it is clear that $H_{ab} = G_{a,r-a} \times G_{b,s-b}$ (except when $K = \mathbb{F}_q$; then this is only the Levi factor of H_{ab}; cf. §3).

Proof of (4.3). We apply the double coset formula to $\tau_{ij} \circ m_{rs}^+$. We obtain (on homology)

$$\tau_{ij} \circ m_{rs}^+ = \sum_{\substack{a+b=i \\ a\leq r, b\leq s}} \alpha_{ab} c_{w_{ab}} \circ (\tau_{a,r-a} \wedge \tau_{b,s-b})$$

with $\alpha_{o,j}$ as in the lemma. (If $G_n = \Sigma_n, U(n)$, or $Sp(n)$, this holds on the spectrum level with all $\alpha_{ab} = 1$). But

$$c_{w_{ab}} : BG_a \times BG_{r-a} \times BG_b \times BG_{s-b} \longrightarrow BG_a \times BG_b \times BG_{r-a} \times BG_{s-b} \longrightarrow BG_i \times BG_j$$

is just the obvious map switching the two inner factors. The lemma is now obvious on inspection.
□

Remark. In the case $K = \mathbb{R}, \mathbb{C}$, or $\mathbb{H}$, the G_{rs} orbit of $w_{ab}G_{ij}$ is a product of Grassmanians $Gr_{r,a} \times Gr_{s,b}$ (here $Gr_{r,a}$ denotes a-planes in r-space). The corresponding P_{rs} orbit is then a bundle of dimension $b(r - a)$ over this space. In fact, as one can easily check, it is just $\mathrm{Hom}(\gamma_{s,b}, \gamma_{r,a}^{\perp})$, where γ denotes the canonical bundle.

Remark. One can obtain Snaith's splitting of $BSO(2n+1)$ by the same method: if 2 is inverted, $BSO(2n+1) \cong \bigvee_{1 \le i \le n} BSO(2i+1)/BSO(2i-1)$. The splitting maps here are defined as in the proof of (4.1) using $BSO(2i) \times BSO(2j+1) \to BSO(2n+1)$ $(i+j=n)$. One cannot apply Theorem 2.4 directly, however, as $SO(2i) \times SO(2j+1)$ is not a subgroup of Levi type in $SO(2n+1)$ (thought of as the maximal compact subgroup of $SL(n, \mathbb{R})$). But it does have index 2 in $S(O(2i) \times O(2j+1))$, which is of Levi type, so this causes no difficulty. (Alternatively, one can simply use the fact that $BSO(2n+1) \to BO(2n+1)$ is a $\mathbb{Z}[\frac{1}{2}]$-equivalence).

References

[B-M] G. Brumfiel and I. Madsen. Evaluation of the transfer and the universal surgery classes, *Inventiones Math.* **32** (1976), 133–169.

[D1] A. Dold, Decomposition theorems for $S(n)$-complexes, *Annals of Math.* **75** (1962), 8–16.

[D2] A. Dold, The fixed point transfer of fibre-preserving maps, *Math. Z.* **148** (1976), 215–244.

[F] M. Feshbach, The transfer and compact Lie groups, *Trans. Amer. Math. Soc.* **251** (1979), 139–169.

[C-M] M. Crabb and S. Mitchell, The loops on $SU(n)/SO(n)$ and $SU(2n)/Sp(n)$, in preparation.

[L-M-S] L.G. Lewis, Jr., J.P. May, and M. Steinberger, *Equivariant Stable Homotopy Theory*, Springer Lecture Notes in Mathematics vol. **1213**, 1986.

[Ma] I.G. MacDonald, Algebraic structure of Lie groups, in *Representation Theory of Lie Groups* (G. Luke, ed.), Cambridge University Press, 1979, 91–150.

[Mi] S. Mitchell, Quillen's theorem on buildings and the loops on a symmetric space, preprint.

[M-R] S. Mitchell and B. Richter, A stable splitting of the loops on $SU(n)$, in preparation.

[Q] D. Quillen, On the cohomology and K-theory of the general linear groups over a finite field, *Annals of Math.* **96** (1972), 552–586.

[S] V. Snaith, Algebraic cobordism and K-theory, *Memoirs Amer. Math. Soc.* No. **221** (1979.

Stephen A. Mitchell
Department of Mathematics
University of Washington
Seattle, Washington 98195

Stewart Priddy
Mathematics Department
Northwestern University
Evanston, Illinois 60201

Browder-Fröhlich Symbols

Jack Morava
The Johns Hopkins University

Abstract: We construct a Galois-equivalent symbol, mapping the pro-p-component of the second K-group of maximal totally ramified abelian extension of a local number field to its Tate module; and we propose as a topic of further research, the problem of constructing the associated periodic localization for this K-functor, along the line of [1] and [13].

§1. Background from local number theory

1.0 We follow Fröhlich's notes [5], or Serre [11] or Lang [8]. L is a finite extension of $\mathbb{Q}_p$, with residue field of cardinality q; $\underline{\underline{O}}$ is its (maximal compact) valuation ring of integers, with maximal ideal $\underline{\underline{m}} = (\pi)$.

The logarithm

$$\ell_\pi(T) = \sum_{i\geq 0} \pi^{-i} T^{q^i}$$

associated to π defines the formal group law

$$F_\pi(X,Y) = \ell_\pi^{-1}(\ell_\pi(X)+\ell_\pi(Y)) \in \underline{\underline{O}}[[X,Y]]$$

cf. [2]. The group $LT(\bar{\underline{\underline{O}}})$ of its $\bar{\underline{\underline{O}}}$-valued points, where $\underline{\underline{O}}$ is the valuation ring of integers in $\bar{\mathbb{Q}}_p$, is the maximal ideal $\bar{\underline{\underline{m}}}$ of $\bar{\underline{\underline{O}}}$, given the composition-law

$$x[+]y = F_\pi(x,y)$$

and the $\underline{\underline{O}}$-module structure

$$a,x \longmapsto [a](x) = \ell_\pi^{-1}(a\ell_\pi(x)):$$

as such it is divisible [5, p.107] with torsion subgroup isomorphic to $L/\underline{\underline{O}}$.

1.1 An element of kernel of the self-map $[\pi^{n+1}]$ of $LT(\bar{\underline{\underline{O}}})$, $n \geq 0$, is called a

division point of the Lubin-Tate group; we write L_n for the subfield $\mathbb{Q}_p$ generated over L by the elements of $\mathrm{Ker}[\pi^{n+1}]$. The union L_∞ of the L_n is the maximal totally ramified abelian extension of L, and there is a canonical isomorphism of the units $\underline{\underline{O}}^\times$ of $\underline{\underline{O}}$ with the Galois group $\mathrm{Gal}(L_\infty/L)$.
In particular when L is $\mathbb{Q}_p$, L_∞ is its p-cyclotomic closure, obtained by adjoining the pth power roots of unity.

1.2 The classical pairing of Kummer generalizes to a homomorphism

$$L_n^\times \times LT(\underline{\underline{O}}_n) \longrightarrow \pi^{-n}\underline{\underline{O}}/\underline{\underline{O}} \subset L/\underline{\underline{O}}$$

defined by the formula

$$\alpha,\beta \mapsto (\alpha,\beta)_n = ([\pi^{-n}]\beta)^\alpha[-]([\pi^{-n}]\beta)$$

with exponentation used to denote the effect of the canonical isomorphism alluded to in 1.1, and where $[\pi^{-n}]\beta = \lambda$ is some element of $LT(\bar{\underline{\underline{O}}})$ (not necessarily abelian over L_n) such that $[\pi^n]\lambda = \beta$.

1.3 More generally, if $\alpha \in L_m^\times$ with $m \geq n$, then

$$[\pi^{m-n}](\alpha,\beta)_n = (\alpha, N_n^m\beta)_n,$$

the map $N_n^m\colon L_m^\times \rightarrow L_n^\times$ being the norm; alternately, the diagram

$$\begin{array}{ccc} LT(\underline{\underline{O}}_n) & \longrightarrow & \mathrm{Hom}(L_n^\times, L/\underline{\underline{O}}) \\ \big\downarrow \text{inclusion} & & \big\downarrow \text{dual to norm} \\ LT(\underline{\underline{O}}_m) & \longrightarrow & \mathrm{Hom}(L_m^\times, L/\underline{\underline{O}}) \end{array}$$

commutes. In the limit we obtain a morphism

$$LT(\underline{\underline{O}}_\infty) \longrightarrow \varinjlim Hom(L_n^\times, L/\underline{\underline{O}}).$$

§2. Proalgebraic K-theory, mod p^ν

2.0 A spectrum with finite homotopy groups defines, by Milnor's axioms, a cohomology theory taking values in profinite abelian groups: a good example is $BGl^+(A)$, where the ring A is finite; in particular, a quotient

$$\underline{\underline{O}}_{,k} = \underline{\underline{O}}/\pi^k\underline{\underline{O}},\ k \geq 1,$$

with $\underline{\underline{O}}$ the integers of a local field as above.
More generally, an indexed system of spectra with finite homotopy groups defines an indexed system of such cohomology theories with values in profinite abelian groups; the latter category having exact limits, a projective system of such spectra has as limit a cohomology theory taking profinite values.
On the other hand the projective system itself can be considered as a functor from finite complexes to the pro-category of finite abelian groups; this pro-category having exact limits, the functor extends to a cohomology theory taking values in pro(finite abelian) mapping to the theory just constructed, under

$$\varprojlim: \text{Pro(finite abelian groups)} \longrightarrow \text{(profinite abelian groups)}.$$

2.1 For example, the projective system

$$j \geq k,\ p_k^j: \underline{\underline{O}}_{,j} = \underline{\underline{O}}/\pi^j\underline{\underline{O}} \longrightarrow \underline{\underline{O}}/\pi^k\underline{\underline{O}} = \underline{\underline{O}}_{,k}$$

of rings leads to the projective system

$$\{BGl^+(\underline{\underline{O}}_{,k}), p_{k*}^j\} = BGl^+_{pro}(\underline{\underline{O}})$$

of spectra; nowadays [6], [12] one knows that the limit of this system is naturally isomorphic to the algebraic K-theory of $\underline{\underline{O}}$, modulo rational vectorspaces; but we do not need that here.

2.2 We denote by $i_*: K_*A \longrightarrow K_*B$ the ring-homomorphism induced by the ring-homomorphism $i: A \longrightarrow B$; if B becomes (by means of i) a projective A-module, then by $i^*: K_*B \longrightarrow K_*A$ we denote the associate norm or transfer, satisfying

$$i_*(i^* a \otimes b) = a \otimes i_* b.$$

2.3 With this notation we can extend the construction in §2.1 to a general algebraic extension of $\mathbb{Q}_p$, not necessarily finite: if $L \subset L_\infty \subset \bar{\mathbb{Q}}_p$ is some such field, let I_∞ be the category of its locally compact subfields, and write

$$i_n^m: \underline{\underline{O}}_m \longrightarrow \underline{\underline{O}}_n \quad (n \geq m \text{ in } I_\infty)$$

for the inclusions of its map of integers.

From homotopy commutativity of the diagram (cf. [10, IV §2.11])

$$\begin{array}{ccc} BGl^+(\underline{\underline{O}}_{m,k}) & \xrightarrow{i^m_{n_*}} & BGl^+(\underline{\underline{O}}_{n,k}) \\ \downarrow p_\ell^{k_*} & & \downarrow p_\ell^{k_*} \\ BGl^+(\underline{\underline{O}}_{m,\ell}) & \xrightarrow{i^m_{n_*}} & BGl^+(\underline{\underline{O}}_{n,\ell}) \end{array} \qquad (m \geq n,\ k \geq \ell)$$

we obtain the projective system

$$\{BGl^+(\underline{\underline{O}}_{n,k}),\ i^m_{n_*} p_\ell^{k*}\} = BGl^+_{pro}(\underline{\underline{O}}_\infty)$$

of spectra with finite homotopy groups.

A field extension $L_\infty \subset \tilde{L}_\infty \subset \bar{\mathbb{Q}}_p$ has an associated transfer

$$BGl^+_{pro}(\tilde{\underline{\underline{O}}}_\infty) \longrightarrow BGl^+_{pro}(\underline{\underline{O}}_\infty)$$

of systems.

2.4 We write $K^{*}_{pro}(\underline{\underline{O}}_{\infty})$ for the homotopy of this system, and define

$$K^{i}_{pro}(X,\underline{\underline{O}}_{\infty}) = [S^{-i}X, BGl^{+}_{pro}(\underline{\underline{O}}_{\infty})];$$

in fact this is the same as the more elaborate system

$$\{[S^{-i}X, \mathbb{Z}\times BGl^{+}(\underline{\underline{O}}_{n,k})],\ q^{m-n}\times i^{m}_{n_*} p^{k*}_{\ell}\}.$$

Reduction modulo some prime power p^{ν} leads to exact sequences

$$0 \longrightarrow K^{*}_{pro}(\underline{\underline{O}}_{\infty})\otimes\mathbb{Z}/p^{\nu}\mathbb{Z} \longrightarrow K^{*}_{pro}(\underline{\underline{O}}_{\infty},\mathbb{Z}/p^{\nu}\mathbb{Z}) \longrightarrow {}_{p^{\nu}}K^{*+1}_{pro}(\underline{\underline{O}}_{\infty}) \longrightarrow 0$$

(in which ${}_{n}A$ denotes the kernel of n-multiplication on A); in low dimensions we obtain the sequence

$$0 \longrightarrow K^{-2}_{pro}(\underline{\underline{O}}_{\infty}) \otimes \mathbb{Z}/p^{\nu}\mathbb{Z} \longrightarrow K^{-2}_{pro}(\underline{\underline{O}}_{\infty},\mathbb{Z}/p^{\nu}\mathbb{Z}) \longrightarrow {}_{p^{\nu}}K^{-1}_{pro}(\underline{\underline{O}}_{\infty}) \longrightarrow 0,$$

where ${}_{p^{\nu}}K^{-1}_{pro}(\underline{\underline{O}}_{\infty})$ is the system $\{{}_{p^{\nu}}\underline{\underline{O}}^{x}_{n,k}\}$.

2.5 We now assume that L_{∞} is as in §1: the paring

$$L^{x}_{n} \times LT(\underline{\underline{O}}_{n}) \longrightarrow LT(\underline{\underline{O}}_{n})_{tors} \subset L/\underline{\underline{O}}$$

restricted to the subgroup of torsion points in $LT(\underline{\underline{O}}_{n})$ defines what might be called the explicit symbol

$$L^{x}_{n} \longrightarrow \mathrm{End}_{\underline{\underline{O}}}(LT(\underline{\underline{O}}_{n})_{tors}) = \underline{\underline{O}}/\pi^{n}\underline{\underline{O}} \subset L/\underline{\underline{O}};$$

restricting to the proper units, we obtain a homomorphism

$$K^{-2}(\underline{\underline{O}}_{n,k},\mathbb{Z}/p^{\nu}\mathbb{Z}) \longrightarrow {}_{p^{\nu}}(\underline{\underline{O}}_{n,k})^{\times} \longrightarrow {}_{p^{\nu}}(L/\underline{\underline{O}})$$

when k is large enough so that $1+\pi^{k}\underline{\underline{O}}_{n} \subset \underline{\underline{O}}_{n}^{p^{\nu}}$; indeed the Fröhlich symbol can be reformulated as a pairing

$$K^{-2}_{pro}(\underline{\underline{O}}_{n},\mathbb{Z}/p^{\nu}\mathbb{Z}) \times LT(\underline{\underline{O}}_{n}) \longrightarrow L/\underline{\underline{O}}$$

along lines suggested by the work of Browder [1].

2.6 The homomorphism

$$K^{-2}_{pro}(\underline{\underline{O}}_{n},\mathbb{Z}/p^{\nu}\mathbb{Z}) \longrightarrow {}_{p^{\nu}}\mathrm{Hom}(LT(\underline{\underline{O}}_{n}),\ L/\underline{\underline{O}})$$

so obtained maps into the submodule of $\underline{\underline{O}}$-module homomorphisms, which can be naturally identified with the dual $LT(\underline{\underline{O}}_{n})^{\check{}}$ of the compact abelian group $LT(\underline{\underline{O}}_{n})$. This defines a morphism

$$K^{-2}_{pro}(\underline{\underline{O}}_{n},\mathbb{Z}/p^{\nu}\mathbb{Z}) \longrightarrow \{{}_{p^{\nu}}LT(\underline{\underline{O}}_{n})^{\check{}}\}.$$

2.7 By the p-component of the proalgebraic K-theory of L_{∞}, we mean the cohomology functor

$$K^{*}_{pro-p}(X,L_{\infty}) = K^{*}_{pro}(X \wedge M(\mathbb{Q}_{p}/\mathbb{Z}_{p}),\underline{\underline{O}}_{\infty}),$$

M signifying the Moore spectrum. The construction of the preceding paragraph defines a homomorphism of $K^{-2}_{pro-p}(L_{\infty})$ to what might be called the Tate module $\varprojlim_{\nu} \varprojlim_{n} {}_{p^{\nu}}LT(\underline{\underline{O}}_{n})^{\check{}}$; restricting to the torsion subgroup defines a homomorphism

$$b\colon K^{-2}_{pro-p}(L_{\infty}) \longrightarrow \varprojlim \mathrm{End}_{\underline{\underline{O}}}(\pi^{-n}\underline{\underline{O}}/\underline{\underline{O}}) = \underline{\underline{O}}.$$

§3. The explicit formula

3.0 Restricting to the torsion points makes it relatively easy to calculate the symbol b, as in [14, §2]: the homomorphism

$$K^{-2}_{\text{pro}}(\underline{\underline{O}},\mathbb{Z}/p^{\nu}\mathbb{Z}) \longrightarrow {}_{p^{\nu}}(L/\underline{\underline{O}})$$

sends β_n in $\underline{\underline{O}}_n^{\times}$, congruent to 1 modulo the maximal ideal $\underline{\underline{m}}_n$ of $\underline{\underline{O}}_n$, to the class of $\pi^{-(n+1)}(N\beta_n^{-1}-1)$ in $\underline{\underline{O}}$ mod $\pi^{n+1}\underline{\underline{O}}$; where $N: L_n^{\times} \longrightarrow L^{\times}$ is the absolute norm. In the limit, a sequence of elements β_n as above, such that $N_n^m\beta_m = \beta_n$, is sent to $\lim\limits_{n\to\infty} \pi^{-(n+1)}(N\beta_n^{-1}-1) \in \underline{\underline{O}}$. When L is unramified, this simplifies to the very explicit formula of Iwasawa [7]

$$b(\beta_n) = -\lim_{n\to\infty} p^{-(n+1)}\operatorname{Tr}\log\beta_n,$$

where $\operatorname{Tr}: L_n \longrightarrow L$ is the absolute norm, and $\log(x) = -\sum\limits_{n\geq 1} n^{-1}(1-x)^n$.

[This formula has been generalized by Wiles [14], and the case $L = \mathbb{Q}_p$ has been studied in detail by Coates and Wiles [3]; while Coleman [4, cf. Cor. 18] has profoundly increased our understanding of these projective systems. The symbol b is $\operatorname{Gal}(L_{\infty}/L) \cong \underline{\underline{O}}^{\times}$-equivalent, and in view of [9] perhaps also $\operatorname{Gal}(L_{\infty}/\mathbb{Q}_p)$-equivalent.]

3.1 It seems very reasonable to conjecture that <u>there exists a factorization of</u> $b\otimes\mathbb{Q}$ <u>through the singularity cobordism theory defined by a Lubin-Tate group law for</u> L, as in the following diagram:

$$\begin{array}{ccl} K^0_{pro\text{-}p}(S^2,L_\infty) & \dashrightarrow & K^0_L(S^2) \\ & \searrow & \downarrow \mathrm{ch} \\ & & K^0_L(S^2)\otimes\mathbb{Q} = H^0(S^2)\otimes(K^0_L\otimes\mathbb{Q}) \end{array}$$

$$K^0_L(S^2)\otimes\mathbb{Q} \xrightarrow{\ b\otimes\mathbb{Q}\ } H^2(S^2)\otimes(K_L^{-2}\otimes\mathbb{Q}) = L.$$

<u>ALternately</u>: b <u>lifts</u> <u>to</u> <u>a</u> <u>natural</u> <u>transformation</u>

$$K^*_{pro\text{-}p}(-,L_\infty) \longrightarrow K^*_L(-)$$

where K^*_L is the extraordinary K-theory of L.

3.2 There are some plausibility arguments in favor of this conjecture: it is very <u>natural</u>, i.e. Galois-equivariant; and a homomorphism from $K^*_{pro\text{-}p}$ to $H^2(-,K_L^{-2}\otimes\mathbb{Q}) \cong H^2(-,L)$ is the rationalization of a morphism

$$K^{-2}_{pro}(\underline{\underline{O}},\mathbb{Z}/p^\nu\mathbb{Z}) \longrightarrow H^2(S^2M(\mathbb{Z}/p^\nu\mathbb{Z}),\underline{\underline{O}}/\pi^n\underline{\underline{O}}),$$

i.e. a system

$$\{[-,BGl^+(\underline{\underline{O}}_{n,k})]\} \longrightarrow \{H^2(-,\underline{\underline{O}}/\pi^i\underline{\underline{O}})\}$$

represented by a class in some group cohomology

$$\varinjlim_{n,k}\ \varprojlim_{i}\ H^2(Gl(\underline{\underline{O}}_{n,k}),\underline{\underline{O}}/\pi^i\underline{\underline{O}}).$$

Restricting this hypothetical Chern class to a line bundle corresponds to mapping i· to

$$\varinjlim_{n,k}\ \varprojlim_{i}\ H^2(\underline{\underline{O}}^{\times}_{n,k},\ \underline{\underline{O}}/\pi^i\underline{\underline{O}}) = H^2(\{\underline{\underline{O}}^{\times}_{n,k}\},\{\underline{\underline{O}}/\pi^i\underline{\underline{O}}\});$$

at this point it seems to be worth remarking that the homomorphism which sends β in $\underline{\underline{O}}_n^{\times}$ to the endomorphism

$$\nu \longmapsto (\nu,\beta)_n \in \mathrm{End}(\underline{\underline{O}}/\pi^n\underline{\underline{O}})$$

defines a class in the continuous cochain cohomology

$$\varinjlim_n H_c^1(\underline{\underline{O}}_n^{\times}, L/\underline{\underline{O}});$$

the boundary homomorphism in the sequence $0 \longrightarrow \underline{\underline{O}} \longrightarrow L \longrightarrow L/\underline{\underline{O}} \longrightarrow 0$ sends this to a continuous cohomology class which defines a natural candidate for this hypothetical Chern class of a line bundle.

3.3 But perhaps it would be most interesting of all if the logarithmic derivative of Kummer [cf. 4] could be understood as a Chern class for some kind of line bundle.

References

1. W. Browder, Algebraic K-theory with coefficients Z/pZ, in Geometric Applications of Homotopy Theory I, Springer Lecture Notes 657.
2. P. Cartier, Groupes de Lubin-Tate generalisées, Inventiones Math. 35(1976) 273-284.
3. J. Coates, A. Wiles, Explicit reciprocity laws, Asterisque 41(1977) 7-17.
4. R.F. Coleman, Division values in local fields, Inventiones Math. 53(1979) 91-116.
5. A. Fröhlich, Formal Groups, Springer Lecture Notes 74.
6. H.A. Gillet, R.W. Thomason, The K-theory of strict Hensel local rings, J. Pure and Applied Algebra 34(1984) 241-255.
7. K.Iwasawa, On the explicit formulas for the norm residue symbol, J. Math. Soc. Japan 20(1968) 151-164.
8. S. Lang, Cyclotomic Fields, Vol.I, Addison-Wesley.
9. J. Morava, The Weil group as automorphisms of the Lubin-Tate group, Asterisque 63(1979) 169-178.
10. D. Quillen, Higher Algebraic K-theory I, Springer Lecture Notes 341.
11. J.P. Serre, Local class field theory, in J.W.S. Cassels, A. Fröhlich (eds) Algebraic Number Theory (conf. at Brighton).
12. A.A. Suslin, On the K-theory of local fields, J. Pure and Applied Algebra 34(1984) 310-319.
13. R.W. Thomason, Algebraic K-theory and etale cohomology, Ann. Sci. de l'Ecole Normale Supérieure 13(1980) 437-552.
14. A. Wiles, Explicit reciprocity laws, Ann. of Math. 107(1978) 235-254.

Further Notes:

1) The referee observes that the conjecture made in 3 is not shown to be true in even one example. In particular if $L = \mathbb{Q}_p$ then L_{n-1} is the field of $(p^n)^{th}$ roots of unity over $\mathbb{Q}_p$, and the map β (§3.0) is very explicit; but the algebraic K-theory of local 'cyclotomic' class-field towers is just beginning to come under investigation.

2) Since the Arcata meeting, it has become clear that explicit reciprocity laws in local fields have something to do with cyclic and Hochschild (co)homology [3; these functions have been studied (perhaps too) extensively on $\mathbb{Q}$-algebras.] It is easy to verify that the rings $\underline{\underline{O}}_n$ have Hochschild homological dimension one, and that

$$HH^1(\underline{\underline{O}}_n) = \text{the Kähler differentials of } \underline{\underline{O}}_n$$

is cyclic over $\underline{\underline{O}}_n$, with $\ell'(\underline{\underline{v}}_n)d\underline{\underline{v}}_n$ as generator (where $\underline{\underline{v}}_n$ is a generator of the π^{n+1}-torsion points on the Lubin-Tate group). Hence

$$HH^1(\underline{\underline{O}}_\infty) = \underline{\underline{O}}_\infty \otimes \mathbb{Q}/\mathbb{Z}$$

is some sort of dualizing module for

$$HH^0(\underline{\underline{O}}_\infty) = \underline{\underline{O}}_\infty,$$

cf. [1, Theorem 1.2].

The Chern class defined for Hodge cohomology by Grothendieck [in terms of Kummer's logarithmic derivative and Kähler differentials, cf. [1, VIII §6.7] seems to be closely related to this topic, and might lead to a proof of the conjecture when $L = \mathbb{Q}_p$.

3) I would like to thank Marcel Bökstedt and Friedhelm Waldhausen for intensive tutorials on cyclic and Hochschild (co)homology, and in particular for telling me

about the different. Finally, I should cite [4] for some related material.

References

1. G. Faltings, p-adic Hodge theory, Journal. of the A.M.S. 1(1988) 255-299.
2. A. Grothendieck, Dix exposes sur la cohomologie etale des schemas, North Holland.
3. J.L. Loday, D. Quillen, Cyclic homology and the Lie algebra homology of matrices, Comment. Math. Helvetici, 49(1984) 565-591.
4. J. Morava, Some Weil group representations motivated by algebraic topology, Proc. Conf. on Elliptic Cohomology, ed. P. Landweber, SLN 1326.

K–theory homology of spaces

by

Erik K. Pedersen* and Charles. A. Weibel**

Matematisk Institut, Odense Universitet and

Department of Mathematics, Rutgers University

Abstract. Let KR be a nonconnective spectrum whose homotopy groups give the algebraic K–theory of the ring R. We give a description of the associated homology theory $KR_*(X)$ associated to KR. We also show that the various constructions of KR in the literature are homotopy equivalent, and so give the same homology theory.

AMS/MOS(1980) Subject Classification. Primary 55N15, Secondary 19D55 ,55N20

Keywords and phrases. Generalized Homology theory, algebraic K–theory, nonconnective spectrum.

§0. Introduction.

There is a generalized homology theory E_* associated to every spectrum E, namely

$$E_n(X) = \lim_{i\to\infty} \pi_{n+i}(E_i\ X).$$

In particular this is true if E is the nonconnective K–theory spectrum KR of a ring R. In this paper, we give a geometric interpretation of $KR_n(X)$
or $n \leq 0$ (and a new interpretation for $n > 0$).

Let X be a subcomplex of S^n, and form the open cone O(X) on X inside $\mathbb{R}^{n+1}$ (which is the open cone on S^n). There is a category $\mathscr{C} = \mathscr{C}_{O(X)}(R)$, whose objects are based free R–modules parameterized in a locally finite way by O(X), and whose morphisms are linear maps moving the bases a bounded amount. (Compare with Quinn's geometric R–modules in [Qu3]). The group $K_1(\mathscr{C})$ is generated by the automorphisms in $\mathscr{C}$, with well–known relations [B]; our main theorem yields the formula:

$$KR_o(X) \cong K_1(\mathscr{C}).$$

When $R = \mathbb{Z}\pi$, these groups appear as obstruction groups of bounded (or thin) h–cobordisms parameterized by O(X) with constant fundamental group π. (See [P2] for a relatively elementary proof of this. This, of course, is in accordance with the basic results of Chapman [C1,C2] and Quinn [Qu1,Qu2]).

*The first author wants to thank the Sonderforschungsbereich für Geometrie und Analysis in Göttingen for a most agreeable year.

**Partially supportes by NSF grant MCS–8301686.

The negative KR–homology groups of X can be obtained from the formula $KR_{-n}(X) = KR_o(S^nX)$. The positive KR–homology groups are Quillen's higher K–groups of $\mathscr{C}$:

$$KR_n(X) \cong K_{n+1}(\mathscr{C}) \text{ for } n \geq 0.$$

To make the proofs easier, it turns out to be better to generalize the above discussion, replacing the category $\mathscr{F}_R$ of finitely generated based free R–modules by any additive category $\mathscr{A}$. We impose the semisimple exact structure of [B], i.e., declaring every short exact sequence split, in order to compute the K–theory of $\mathscr{A}$; this makes Bass' groups $K_1(\mathscr{A})$ the same as Quillen's by [W].

This being said, we can generalize the spectrum KR (for $\mathscr{A} = \mathscr{F}_R$) to the nonconnective spectrum $K\mathscr{A}$ constructed in [PW] and ask about $K\mathscr{A}$–homology. There is a category $\mathscr{C}_{O(X)}(\mathscr{A})$ generalizing $\mathscr{C}_{O(X)}(R)$ and described in §1 below. In §2 we make this construction functorial in X. Our Main Theorem is carefully stated in §3 and proved in §4:

Main Theorem. The $K\mathscr{A}$–homology of X is naturally isomorphic to the algebraic K–theory of the idempotent completion $\mathscr{C}^\wedge$ of $\mathscr{C} = \mathscr{C}_{O(X)}(\mathscr{A})$, with a degree shift:

$$K\mathscr{A}_{*-1}(X) \cong K_*(\mathscr{C}_{O(X)}(\mathscr{A})^\wedge).$$

Note that $K_*(\mathscr{C}^\wedge) = K_*(\mathscr{C})$ for $* \geq 1$ but that $K_o(\mathscr{C}^\wedge) \neq K_o(\mathscr{C})$ in general. For example if $\mathscr{C} = \mathscr{F}_R$ then $K_o(\mathscr{C})$ is $\mathbb{Z}$ (or a quotient of $\mathbb{Z}$) but $K_o(\mathscr{C}^\wedge) = K_o(R)$. If $* < 0$ and $\mathscr{C} \neq \mathscr{C}^\wedge$, the groups $K_*(\mathscr{C})$ are not even defined. As in [PW], the key technical step is an application of Thomason's double mapping cylinder construction from [T].

The knowledgeable reader will wonder about the relationship between the spectrum $K\mathscr{A}$ and other nonconnective spectra in the literature. We pin down this loose end in §6. When $\mathscr{A} = \mathscr{F}_R$ we show that our spectrum $K\mathscr{F}_R$ is homotopy equivalent to the Gersten–Wagoner spectrum of [G][Wag]. or general $\mathscr{A}$, we prove that our spectrum $K\mathscr{A}$ is homotopy equivalent to Karoubi's spectrum [K136]. Actually, the discussion in [K136] only mentions K_o and K_1, and not spectra, since it was written before higher K–theory emerged. We devote §5 to showing that Karoubi's prescription in [K136] actually gives an infinite loop spectrum. The authors want to thank W. Vogell for pointing out an error in an earlier draft of this paper.

§1. The functor $\mathscr{C}$.

In this section we generalize the functor $\mathscr{C}_i(\mathscr{A})$ considered in [PW] to a functor in two variables. We have already described how $\mathscr{C}_i(-)$ is an endofunctor of the category of filtered additive categories [PW]. We remind the reader that a filtered additive category is an additive category $\mathscr{A}$, that comes with a filtration of the homsets Hom(A,B),

$$O \subseteq F_0\mathrm{Hom}(A,B) \subseteq F_1\mathrm{Hom}(A,B) \subseteq \ldots \subset \mathrm{Hom}(A,B)$$

such that

a) $F_i\mathrm{Hom}(A,B)$ is a subgroup and $\mathrm{Hom}(A,B) = \cup F_i\mathrm{Hom}(A,B)$.

b) $F_0\mathrm{Hom}$ contains: O_A and 1_A for each A, all coherence isomorphisms of $\mathscr{A}$, all projections $A \oplus B \to A$ and all inclusions $A \to A \oplus B$.

c) if $f \in F_i\mathrm{Hom}(A,B)$ and $g \in F_j\mathrm{Hom}(B,C)$ then $g \circ f \in F_{i+j}\mathrm{Hom}(A,C)$.

Note that any additive category may be endowed with the "discrete filtration", in which $F_0\mathrm{Hom}(A,B) = \mathrm{Hom}(A,B)$ for every A,B.

Thinking of the lower index i as the metric space $\mathbb{Z}^i$ or $\mathbb{R}^i$, we shall now turn $\mathscr{C}$ into a functor of that variable. (See [PW], Remark 1,2,3.)

Definition 1.1. Let X be a metric space and $\mathscr{A}$ a filtered additive category. We then define the filtered additive category $\mathscr{C}_X(\mathscr{A})$ as follows:

1) An object A of $\mathscr{C}_X(\mathscr{A})$ is a collection of objects A(x) of $\mathscr{A}$, one for each $x \in X$, satisfying the condition that for each ball $B \subset X$, $A(x) \neq 0$ for only finitely many $x \in B$.
2) A morphism $\varphi : A \to B$ is a collection of morphisms

$\varphi^x_y : A(x) \to B(y)$

in $\mathscr{A}$ such that there exists r depending only on φ so that

a) $\varphi^x_y = 0$ for $d(x,y) > r$

b) all φ^x_y are in $F_r\mathrm{Hom}$

(We then say φ has filtration degree $\leq r$).

Composition of $\varphi : A \to B$ with $\psi : B \to C$ is given by $(\psi\varphi)^x_z = \sum_{y\in X} \psi^y_z\, \varphi^x_y$. Notice that the sum makes sense because the category is additive and because the sum will always be finite. The category $\mathscr{C}_{\mathbb{Z}^i}(\mathscr{A})$ is the $\mathscr{C}_i(\mathscr{A})$ of [PW].

We now introduce the category $\mathscr{M}$ of metric spaces and proper, eventually Lipschitz maps.

Definition 1.2. The category $\mathscr{M}$ has objects metric spaces X. A morphism $f{:}X\to Y$ must be both proper and eventually Lipschitz. We remind the reader that a map $f{:}X\to Y$ between metric spaces is Lipschitz if there exists a number $k \in \mathbb{R}_+$ such that

$$d(f(x),f(y)) \leq kd(x,y).$$

We say that f is eventually Lipschitz if

there exist r and k, only depending on f, so that

$\forall x,y \in X, \forall s \in \mathbb{R}_+$: if $s > r$ and $d(x,y) < s$ then $d(f(x),f(y)) < k\cdot s$.

Finally, we call f proper if the inverse image of a bounded set is bounded.

Example 1.3. One should note that maps in $\mathscr{M}$ are not necessarily continuous, but any jumps allowed must be universally bounded. For example the map $\mathbb{R} \to \mathbb{Z}$ sending a real number x to the greatest integer smaller than x is a map in $\mathscr{M}$.

Given a proper eventually Lipschitz map $f{:}X \to Y$ we obtain a functor $f_*{:}\mathscr{C}_X(\mathscr{A}) \to \mathscr{C}_Y(\mathscr{A})$ by defining $(f_*(A))_y = \bigoplus_{z\in f^{-1}(y)} A(z)$ for objects A in $\mathscr{C}_X(\mathscr{A})$. Since the inverse of a bounded set is bounded, there are only finitely many nonzero modules in a ball in Y, and $f_*(A)$ is well-defined. On morphisms f_* is induced by the identity. The eventually Lipschitz condition on f ensures that we indeed do get morphisms in the category $\mathcal{C}_Y(\mathcal{A})$. Hence $\mathcal{C}_{\_}(\mathcal{A})$ is a functor from $\mathcal{M}$ to (semisimple filtered) additive categories.

Lemma 1.4. Let X and Y be metric spaces and give X×Y the max metric,

$$d_{X\times Y}((x_1,y_1),(x_2,y_2)) = \max(d_X(x_1,x_2),d_Y(y_1,y_2)),$$

then $\mathcal{C}_{X\times Y}(\mathcal{A}) = \mathcal{C}_X(\mathcal{C}_Y(\mathcal{A}))$.

Proof. This is where we use that $\mathcal{C}$ takes values in filtered categories. The internal filtration degree will control distances in the Y component, while the external filtration degree will control distances in the X component and thus the max distance will be controlled.

Remark. Note that the isomorphism class in $\mathcal{M}$ is not affected if we change the metric to a proper Lipschitz equivalent metric, i.e. a metric so that the identity is a proper Lipschitz equivalence both ways. Therefore Lemma 1.4 remains true up to natural equivalence if the max metric is replaced by the usual product metric

$$d_{X\times Y}((x_1,y_1),(x_2,y_2)) = \sqrt{d_X(x_1,x_2)^2 + d_Y(y_1,y_2)^2}$$

The concept of homotopy is introduced in the category $\mathcal{M}$ in a standard fashion using the inclusions $X = X\times\{0\} \to X\times I$, $X = X\times\{1\} \to X\times I$ and the projections $X\times I \to X$. Since X×I has the max metric, these are maps in the category $\mathcal{M}$. Note that the inclusion $\mathbb{Z} \to \mathbb{R}$ is a homotopy equivalence in $\mathcal{M}$, with homotopy inverse the greatest integer function of example 1.3.

Lemma 1.5. A compact metric space X is homotopy equivalent to a point in $\mathcal{M}$ and hence $\mathcal{C}_X(\mathcal{A})$ is equivalent to the category $\mathcal{A}$.

Proof. Since maps are not required to be continuous and a compact metric space is globally bounded, any map $X\times I \to X$ which is constant on the top and the identity on the bottom will be a contracting homotopy. The second assertion follows from the evident fact that $\mathcal{C}_X(\mathcal{A}) = \mathcal{A}$ when X is a point.

Proposition 1.6. The functor $\mathcal{C}_{\_}(\mathcal{A})$ is homotopy invariant. That is, for each metric space X, the inclusion $X \subset X\times I$ as $X\times\{0\}$ and the projection $X\times I \to X$ induce an equivalence of categories.

$$\mathcal{C}_X(\mathcal{A}) \to \mathcal{C}_{X\times I}(\mathcal{A})$$

Proof. By 1.4 and 1.5, $\mathcal{C}_{X\times I}(\mathcal{A}) = \mathcal{C}_I(\mathcal{C}_X(\mathcal{A}))$ is homotopic to $\mathcal{C}_X(\mathcal{A})$.

§2. Open Cones

In this section we construct an open cone functor O(X) from finite PL complexes to $\mathcal{M}$, so that $\mathcal{C}_{O(X)}(\mathcal{A})$ depends functorially on X in a homotopy invariant way.

To fix notation let $S^o = \{-1,1\} \subset \mathbb{R}$. Then the n–sphere is the join $S^n = S^o * S^o * ... * S^o$ as a sub PL–complex. We shall be considering the category of finite sub PL–complexes of S^∞ (PL complexes that are subcomplexes of some S^n) and PL morphisms. We denote this category $\mathcal{PL}$. This is essentially the category of finite PL complexes, since any such may be embedded in some S^n, but we need the way the complex sits in S^n as part of the structure. We think of $\mathbb{R}^{n+1}$ as a metric space using the max metric

$$d(\underline{x},\underline{y}) = \max |x_i - y_i|.$$

This induces a metric on the n–sphere and hence on any subcomplex.

We now construct a functor O from $\mathcal{PL}$ to $\mathcal{M}$.

Definition 2.1. O sends an object $X \subset S^n$ to $O(X) = \{t\cdot x \in \mathbb{R}^{n+1} \mid t \in [0,\infty),\ x \in X\}$ with metric induced from $\mathbb{R}^{n+1}$. For example $O(S^n) = \mathbb{R}^{n+1}$. On morphisms $f: X \to Y$ we extend to $O(f): O(X) \to O(Y)$ by linearity: $f(t\cdot x) = t\cdot f(x)$. One checks easily that O(f) is proper and Lipschitz (and therefore eventually Lipschitz) using the following well–known

Lemma 2.2. A PL–map $f: X \to Y$ between finite complexes is Lipschitz.

Proof. Triangulate so that f is linear on each simplex. Since there are only finitely many simplices, f will be Lipschitz.

Remark 2.3. For a PL complex X, O(X) does not really depend on the embedding $X \subset S^n$, since a PL homeomorphism $X_1 \subset S^n$ to $X_2 \subset S^m$ will induce a proper Lipschitz homeomorphism from $O(X_1) \longrightarrow O(X_2)$.

Lemma 2.4. Let X and Y be two PL complexes and let * denote the join. Then

$$O(X*Y) \cong O(X) \times O(Y).$$

In particular, $O(\Sigma X) = O(X*S^o) \cong O(X) \times \mathbb{R}$ and $O(CX) \cong O(X) \times [0,\infty)$.

Proof. Embed $X \subset S^n$ and $Y \subset S^m$ so $X*Y \subset S^n * S^m = S^{n+m+1}$. A point in O(X*Y) is $s\cdot(t\cdot x + (1-t)\cdot y) = stx + s(1-t)y$ and stx lies in $O(X) \subset \mathbb{R}^{n+1} \times 0$, whereas $s(1-t)y$ lies in $O(Y) \subset 0 \times \mathbb{R}^{m+1}$. The last sentence follows from the identities $\Sigma X = X*S^o$, $CX = X*(\text{point})$, $O(S^o) = \mathbb{R}$ and $O(\text{point}) = [0,\infty)$.

Remark 2.5. The functor O is not homotopy invariant. We only obtain homotopy invariance after passing to K–theory.

§3 The Main Theorem

So far we have constructed homotopy functors

$\mathcal{O}:\mathcal{PL} \to \mathcal{M}$ and

$\mathcal{C}\_(\mathcal{A}):\mathcal{M} \to$ (semisimple) additive categories.

For our main theorem, we need a functor K_* from additive categories to graded abelian groups. For $n \geq 1$, there is no problem: given an additive category $\mathcal{A}$ we take $K_n\mathcal{A} = \pi_n(\Omega BQ\mathcal{A})$ as in [Q], using the semisimple exact structure (in which all exact sequences split). However, unless $\mathcal{A}$ is idempotent complete, the groups $K_o\mathcal{A}$ may be wrong for our purposes, and $\mathcal{A}$'s negative K–groups will not even be defined.

For example consider the category $\mathcal{F}_R$ of finitely generated based free R–modules. When R is a group ring, we know that the geometrically interesting group is not $K_o(\mathcal{F}_R) = \mathbb{Z}$, but rather $K_o(R)$, which measures projective modules.

To handle this problem, we pass to the idempotent completion $\mathcal{A}$^ of $\mathcal{A}$. This provides the correct group $K_o(\mathcal{A}$^$)$ and does not change the higher groups, since $K_n(\mathcal{A}) = K_n(\mathcal{A}$^$)$ for $n \geq 1$. For example, $\mathcal{F}_R$^ is equivalent to the category of finitely generated projective R–modules.

Scholium. $\mathcal{A}$^ inherits the structure of a filtered additive category from $\mathcal{A}$. The objects of $\mathcal{A}$^ are pairs (A,p), where $p:A \to A$ is idempotent. A morphism φ from (A_1,p_1) to (A_2,p_2) is an $\mathcal{A}$–morphism $\varphi:A_1 \to A_2$ with $\varphi = p_2\varphi p_1$. The filtration degree of φ is the smallest d such that $\varphi = p_2 f p_1$ for some $f \in F_d\mathrm{Hom}(A_1,A_2)$ satisfying $fp_1 = p_2 f$. This filtration should have been stated explicitly in [PW,1.4].

If $\mathcal{A}$ is idempotent complete, the negative K–groups of $\mathcal{A}$ were defined by Karoubi in [K136], and agree with the definition in [PW] (as we shall see in § 6 below). If $\mathcal{A} = \mathcal{F}_R$, they agree with Bass' negative K–groups $K_n(R)$.

The construction in [PW] actually gives us slightly more information. If $\mathcal{A}$ is idempotent complete, it yields a nonconnective infinite loop spectrum $K\mathcal{A}$, and the homotopy groups of $K\mathcal{A}$ are the groups $K_*\mathcal{A}$ above.

Now associated to any spectrum such as $K\mathcal{A}$ is a reduced homology theory $K\mathcal{A}_*$. It is defined by

$$K\mathcal{A}_n(X) = \lim \pi_{n+i}((K\mathcal{A})_n\, X).$$

The coefficients of this homology theory are the groups

$$K\mathcal{A}_n(S^o) = \pi_n(K\mathcal{A}) = K_n\mathcal{A}.$$

We can now state our main theorem.

Main theorem 3.1. If $\mathcal{A}$ is an idempotent complete additive category, the functor from $\mathcal{PL}$ to graded abelian groups sending X to $K_*(\mathcal{C}_{O(X)}(\mathcal{A})$^$)$ is the $K\mathcal{A}_*$–homology theory of X, with a degree shift:

$$K_*(\mathcal{C}_{O(X)}(\mathcal{A})\hat{\ }) \cong K\mathcal{A}_{*-1}(X).$$

Remark 3.2. This theorem had previously been known for spheres. For $X = S^i$, $O(S^i) = \mathbb{R}^{i+1}$, which is homotopy equivalent to $\mathbb{Z}^{i+1}$ in $\mathcal{M}$, so therefore by 1.4 and 2.5

$$K_*(\mathcal{C}_{\mathbb{R}^i}(\mathcal{A})^\wedge) = K_*(\mathcal{C}_{\mathbb{Z}^i}(\mathcal{A})^\wedge),$$

which was shown in [PW] to equal $K_{*-i-1}(\mathcal{A}^\wedge)$. The category $\mathcal{C}_{\mathbb{Z}^{i+1}}(\mathcal{A})$ was first studied in [P1], where $\mathcal{A}$ was the category of finitely generated free R–modules. There it was shown that

$$K_1(\mathcal{C}_{\mathbb{Z}^{i+1}}(R)) = K_{-i}(R)$$

which is equal to $KR_i(S^o) = KR_o(S^i)$, thus agreeing with our main theorem.

§4. Proof of main theorem.

Define the functor f as the composition

$$\mathcal{PL} \xrightarrow{O} \mathcal{M} \xrightarrow{C\_(\mathcal{A})} \text{filtered add. categ.} \xrightarrow{\square} \text{add.cat.} \xrightarrow{\wedge}$$

$$\text{idempotent complete add.cat.} \xrightarrow{\Omega^\infty K} \text{Top. spaces.}$$

Here $\square$ is the forgetful functor, ^ is idempotent completion and $\Omega^\infty K$ is the zero'th space of the infinite loop spectrum K giving the K–theory of the category, either ΩBQ or the result of applying an infinite loop machine to the symmetric monoidal category of isomorphisms (see Thomason for a very functorial construction [T]). That is:

$$f(X) = \Omega^\infty K(\mathcal{C}_{O(X)}(\mathcal{A})^\wedge).$$

By 2.5 and 1.4, the space f(X) is a homotopy invariant of the finite PL complex X.

Lemma 4.1. If X is a cone, X = CX, then f(X) is contractible.

Proof. $\mathcal{C}_{O(CK)}(\mathcal{A}) = \mathcal{C}_{[0,\infty)\times O(K)}(\mathcal{A}) = \mathcal{C}_+\mathcal{C}_{O(K)}(\mathcal{A})$ in the notation of [PW]. . It was proven in [PW,(3.1)] that $\Omega^\infty K\mathcal{C}_+(\mathcal{A})$ is contractible for an arbitrary category $\mathcal{A}$, and the argument given there applies verbatim to show that

$$f(X) = \Omega^\infty K((\mathcal{C}_+(\mathcal{C}_{O(K)}\mathcal{A})^\wedge)$$

is also contractible.

Proposition 4.2. For each X, $\Omega f(\Sigma X)$ is homotopy equivalent to f(X). In particular f(X) is an infinite loopspace.

Proof. By 1.6 and 2.4 we have

$$\mathcal{C}_{O(\Sigma X)} = \mathcal{C}_{O(X)\times\mathbb{R}} = \mathcal{C}_{\mathbb{R}}(\mathcal{C}_{O(X)}(\mathcal{A})) = \mathcal{C}_1(\mathcal{C}_{O(X)}(\mathcal{A})).$$

By [PW, Theorem (3.2)], applied to the filtered additive category $\mathcal{C}_{O(X)}(\mathcal{A})^\wedge$, we know that the loop space of $\Omega^\infty K\mathcal{C}_{O(\Sigma X)}(\mathcal{A})$ is homotopy equivalent to f(X). But by the cofinality theorem, the spaces

$\Omega^\infty K(\mathcal{C})$ and $\Omega^\infty K(\mathcal{C}\hat{\ })$ have homotopy equivalent connected components, and hence homotopy equivalent loop spaces, for any $\mathcal{C}$.

Theorem 4.3. The functor f sends cofibrations to fibrations.

Before proving this result, let us draw a quick consequence. It follows from Lemma 4.1 that f is homotopy invariant, since $X \to X \times I \to CX$ is a cofibration. The homotopy groups of the spectrum $\{f(X), f(\Sigma X), f(\Sigma^2 X), \ldots\}$ coincide with the groups $K_*(\mathcal{C}_{O(X)}(\mathcal{A})\hat{\ })$. These groups are homotopy invariants of X which vanish when X is contactible by 4.1. From 4.3 we immediately obtain

Corollary 4.4. The functor from $\mathcal{PL}$ to graded abelian groups, sending X to $K_*(\mathcal{C}_{O(X)}(\mathcal{A})\hat{\ })$ is a reduced homology theory.

Proof of Theorem 4.3. The special case $X \subset CX \to \Sigma X$ follows from 4.1 and 4.2. To do the general case, we proceed in a manner very much like the proof of Theorem 3.4 in [PW]. Consider a cofibration $A \subset X \to X \cup CA$. Then we get

$$O(A) \subset O(X) \subset O(X) \cup_{O(A)} O(CA)$$

and a diagram

$$\begin{array}{ccc} \mathcal{C}_{O(A)}(\mathcal{A}) & \longrightarrow & \mathcal{C}_{O(CA)}(\mathcal{A}) \\ \downarrow & & \downarrow \\ \mathcal{C}_{O(X)}(\mathcal{A}) & \longrightarrow & \mathcal{C}_{O(X \cup CA)}(\mathcal{A}) \end{array}$$

By 1.6 and 2.4, we have $O(CA) = O(A) \times [0,\infty)$ and

$$\mathcal{C}_{O(CA)}(\mathcal{A}) = \mathcal{C}_{[0,\infty)}(\mathcal{C}_{O(A)}(\mathcal{A})) = \mathcal{C}_{+}(\mathcal{C}_{O(A)}(\mathcal{A})).$$

To simplify notation, let us write $\underline{\underline{C}}_X$ for the category of isomorphisms in $\mathcal{C}_{O(X)}(\mathcal{A})\hat{\ }$. Using the "double mapping cylinder" pushout construction $\underline{\underline{P}}$ of Thomason [T,(5.1)], we get

$$\begin{array}{cccc} \underline{\underline{C}}_A & \longrightarrow & \underline{\underline{C}}_{CA} & \\ \downarrow & & \downarrow & \\ \underline{\underline{C}}_X & \longrightarrow & \underline{\underline{P}} & \\ & & & \overset{\Sigma}{\searrow} \\ & & & \underline{\underline{C}}_{X\cup CA} \end{array}$$

We wish to show that Σ induces a homotopy equivalence on certain components, i.e. that Σ induces a π_0-monomorphism and π_i-isomorphism. That π_0 behaves correctly is then obtained by applying the argument to the suspension of this cofibration. It is therefore enough to show that for every object Y of $\mathcal{C}_{O(X\cup CA)}(\mathcal{A})$, considered as an object of $\underline{\underline{C}}_{X\cup CA}$, that the category $Y{\downarrow}\Sigma$ is a contractible category. At this point we follow the proof of [PW,Theorem (3.4)] very closely. We use the bound d to filter

$Y\downarrow\Sigma$ as the increasing union of subcategories fil_d and show each of these has an initial object $*_d$. Fil_d is the full subcategory of all iso's $\alpha: Y \to \Sigma(B^A, B^X, A^+)$, where B^X is an object of $\underline{\underline{C}}_X$, B^A an object of $\underline{\underline{C}}_A$ and A^+ an object of $\underline{\underline{C}}_{CA}$. We define $Y_d{}^A$, $Y_d{}^X$ and $Y_d{}^+$ in $\underline{\underline{C}}_A$, $\underline{\underline{C}}_X$ and $\underline{\underline{C}}_{CA}$ as follows: Let

$$N_d = \{x \in O(X \cup CA) \mid \exists y \in O(A) : d(x,y) \leq d\}.$$

Since $A \subseteq X$ is a cofibration it is easy to see that N_d is proper Lipschitz homotopy equivalent to $O(A)$. Choose some proper Lipschitz homotopy equivalence $h: N_d \to O(A)$ and proceed as follows:

$$Y_d^X(x) = \begin{cases} Y(x) \text{ for } X \in O(X) - N_d \\ 0 \quad \text{otherwise} \end{cases}$$

$$Y_d^+(x) = \begin{cases} Y(x) \text{ for } X \in O(A) \times (d,\infty) = O(CA) - N_d \\ 0 \quad \text{otherwise} \end{cases}$$

and

$$Y_d{}^A = h_*(Y_d{}^A) \text{ where}$$

$$Y_d{}^A = \begin{cases} Y(x) \text{ for } x \in N_d \\ 0 \quad \text{otherwise.} \end{cases}$$

There is now an obvious isomorphism

$$\sigma : Y \cong Y_d{}^X \oplus Y_d{}^A \oplus Y_d{}^+ = \Sigma(Y_d^A, Y_d^X, Y_d^+)$$

bounded by d (essentially the identity) and we may proceed to prove this is an initial object in Fil_d exactly as in the proof of [PW, Theorem 3.4].

We finish off the proof of the main theorem as follows:

Proof of main theorem (3.1).

We need to identify the homology theory of 4.4 as the homology theory associated to the spec trum $K(\mathcal{A})$. It was proven by Thomas Gunnarson [Gunn] that when a homotopy functor f sends cofibrations to fibrations and contractible spaces to contractible spaces, then the homology theory obtained by applying homotopy groups has as its representing spectrum $\{f(S^o), f(S^2), f(S^2), \ldots\}$. This is a well-known fact when the homology theory is connective – see e.g. [Woolf, theorem 1.14], [Segal] or [May] – but in the general case we need to use [Gunn]. Now we have by Lemma 2.4 that

$$f(S^i) = \Omega^\infty K(\mathscr{C}_{i+1}(\mathcal{A})^\wedge).$$

The space $f(S^i)$ is therefore the $(i+1)^{st}$–space of the spectrum $K(\mathcal{A})$ constructed in Theorem B of [PW]. The representing spectrum for the homology theory of 4.4 is therefore $\Omega^{-1}K(\mathcal{A})$, and the main theorem follows.

We are finished, except we need to show that the delooping given in [PW] agrees with the other delooping in the literature. This is the subject of the final 2 sections.

§5 Karoubi's nonconnective spectra.

In this section, we follow Karoubi's ideas in [K136] and construct nonconnective K–theory spectra, whose negative homotopy groups are the negative K–groups defined by Karoubi in op.cit. We then show that these agree with the spectra constructed in [PW] (and in special cases, in [G], [Wagl]). More explicitly, but also more technically, we show that $\mathcal{C}_1(\mathcal{A}) \longrightarrow \mathcal{C}_+(\mathcal{A})/\mathcal{A}$ induces an isomorphism on Quillen K–theory. We are indebted to M. Karoubi, who suggested this possibility to us in 1983.

We shall use Karoubi's delooping construction from [K136], so we begin by recalling that construction. Let $\mathcal{A}$ be a full subcategory of an additive (hence semisimple exact) category $\mathcal{U}$. We shall use the notation that letters A–F (resp. U–Z) denote objects of $\mathcal{A}$ (resp. $\mathcal{U}$), and that $U = E_\alpha \oplus U_\alpha$ means an internal direct sum decomposition of U with $E_\alpha \varepsilon \mathcal{A}$. We say that $\mathcal{U}$ is $\mathcal{A}$–filtered if every object U has a family of decompositions $\{U = E_\alpha \oplus U_\alpha\}$ (called a filtration of U) satisfying the following axioms (cf. [K136, pp. 114 ff.]):

(F1) For each U, the decompositions form a filtered poset under the partial order that $E_\alpha \oplus U_\alpha \leq E_\beta \oplus U_\beta$ whenever $U_\beta \subseteq U_\alpha$ and $E_\alpha \subseteq E_\beta$.

(F2) Every map $A \to U$ factors $A \to E_\alpha \to E_\alpha \oplus U_\alpha = U$ for some α.

(F3) Every map $U \to A$ factors $U = E_\alpha \oplus U \to E_\alpha \to A$ for some α.

(F4) For each U,V the filtration on $U \oplus V$ is equivalent to the sum of the filtrations $\{U = E_\alpha \oplus U_\alpha\}$ and $\{V = F_\beta \oplus V_\beta\}$, i.e., to $\{U \oplus V = (E_\alpha \oplus F_\beta) \oplus (U_\alpha \oplus V_\beta)\}$.

We shall assume each filtration is saturated in the sense that if $U = E_\alpha \oplus U_\alpha$ is in the filtration and $E_\alpha = A \oplus B$ in $\mathcal{A}$, then $U = A \oplus (B \oplus U_\alpha)$ is also in the filtration. Finally, we say that $\mathcal{U}$ is flasque if there is a functor $\infty : \mathcal{U} \to \mathcal{U}$ and a natural isomorphism $U^\infty \cong U \oplus U^\infty$ ([K136, p. 147]).

Our favorite selection of $\mathcal{U}$ is the following:

Example 5.1. The category $\mathcal{C}_+(\mathcal{A})$ of [PW,(1.2.4)] is flasque and $\mathcal{A}$–filtered. Objects of $\mathcal{C}_+(\mathcal{A})$ are sequences $(\mathcal{A}_o, \mathcal{A}_1, \ldots)$ of objects in $\mathcal{A}$, and morphisms are given by "bounded" matrices. $\mathcal{A}$ is the full subcategory of objects $(\mathcal{A}_o, 0, 0, \ldots)$. The $\mathcal{A}$–filtration on an object $U = (A_o, A_1, \ldots)$ contains the decompositions

$$U \cong \mathrm{Fil}_n(U) \oplus (0, \ldots, 0, A_{n+1}, \ldots)$$
$$\mathrm{Fil}_n(U) = A_o \oplus \ldots \oplus A_n \text{ in } \mathcal{A}.$$

We proved that $\mathcal{C}_+(\mathcal{A})$ was flasque in [PW,(1.3)], using the translation $t(A_o, A_1, \ldots) = (0, A_o, A_1, \ldots)$ on $\mathcal{C}_+(\mathcal{A})$.

We now suppose given an $\mathcal{A}$–filtered category $\mathcal{U}$. Call a map $U \to V$ completely continuous (cc) if it factors through an object of $\mathcal{A}$. Karoubi defines $\mathcal{U}/\mathcal{A}$ to be the category with the same objects as $\mathcal{U}$, but with $\mathrm{Hom}_{\mathcal{U}/\mathcal{A}}(U,V) = \mathrm{Hom}_{\mathcal{U}}(U,V)/\{\text{cc maps}\}$.

Lemma 5.2. Suppose $\mathcal{A}$ is idempotent complete and that $\bar{\varphi}: U \to V$ is an isomorphism in $\mathcal{U}/\mathcal{A}$. Then there are decompositions $U = E_\alpha \oplus U_\alpha$ and $V = F_\alpha \oplus V_\alpha$ in the (saturated) filtrations and a $\mathcal{U}$-isomorphism $U_\alpha \cong V_\alpha$ such that $\bar{\varphi}$ is represented by $U \to U_\alpha \cong V_\alpha \to V$.

Proof. Choose representatives φ, ψ for $\bar{\varphi}, \overline{\varphi}^{-1}$. Since $1_U - \psi\varphi$ is c.c., there is a decomposition $U = E_\beta \oplus U_\beta$ with $1 = \psi\varphi$ on U_β. Replace U by U_β to assume $1_U = \psi\varphi$. Similarly, write $V = F \oplus W$ with $\varphi\psi = 1_W$. Write $\varphi = (\varepsilon, i) : U \to F \oplus W$ and $\psi = (\delta, j) : F \oplus W \to U$. Observe that $ij = 1_W$ and that

(*) $\delta\varepsilon + ji = 1_U$.

Multiplying (*) by i,j and by (*) yields the equations $i\delta\varepsilon = 0, \delta\varepsilon j = 0$ and $(\delta\varepsilon)^2 = \delta\varepsilon$. Replacing ε by $\varepsilon\delta\varepsilon$ makes $(\varepsilon\delta)^2 = \varepsilon\delta$ without affecting $\bar{\varphi}$ or (*). Set $F_\alpha = \ker(\varepsilon\delta)$ and $V_\alpha = (\varepsilon\delta F) \oplus W$; $F_\alpha \in \mathcal{A}$ because $\mathcal{A}$ is idempotent complete. The rest of the proof is straightforward, and left to the reader.

Remark. The failure of this lemma when $\mathcal{A}$ is not idempotent complete is the Bass–Heller–Swan phenomenon. See [P2,(1.16)].

Theorem 5.3. Let $\mathcal{A}$ be semisimple and idempotent complete. Then for every A–filtered category $\mathcal{U}$ the sequence

$$K^Q(\mathcal{A}) \to K^Q(\mathcal{U}) \to K^Q(\mathcal{U}/\mathcal{A})$$

is a homotopy fibration, where $K^Q(\mathcal{A})$ denotes the space whose homotopy groups give the algebraic K–theory of $\mathcal{A}$.

Before proving this theorem, we draw our main conclusion. Suppose that in addition $\mathcal{U}$ is flasque; from additivity and the equation $\infty \cong 1 + \infty$ we conclude that $K^Q(\mathcal{U}) \simeq *$. Since $\mathcal{U}/\mathcal{A}$ shares the same objects as $\mathcal{U}$, we conclude that $K_0(\mathcal{U}/\mathcal{A}) = K_0(\mathcal{U}) = 0$. Finally, applying 5.3 to the diagram $\mathcal{U} \to \mathcal{C}_+(\mathcal{U}) \leftarrow \mathcal{C}_+(\mathcal{A})$ shows that $K^Q(\mathcal{U}/\mathcal{A}) \simeq K^Q(\mathcal{C}_+(\mathcal{A})/\mathcal{A})$. This proves:

Theorem/Definition 5.4. Let $\mathcal{A}$ be semisimple and idempotent complete. Choose a flasque, $\mathcal{A}$–filtered category $\mathcal{U}$ and define $S\mathcal{A}$ to be $\mathcal{U}/\mathcal{A}$. Then $K^Q(S\mathcal{A})$ is a connected space with $\Omega K^Q(S\mathcal{A}) \simeq K^Q(\mathcal{A})$, and the homotopy type of $K^Q(S\mathcal{A})$ is independent of the choice of $\mathcal{U}$.

Our proof of Theorem 5.3 follows the proof of [PW,(3.4)]. Let $\underline{\underline{A}}$, $\underline{\underline{U}}$ and $\underline{\underline{S}}$ denote the categories of isomorphisms of $\mathcal{A}, \mathcal{U}$ and $\mathcal{U}/\mathcal{A}$. It is well–known that $K^Q(\mathcal{A})$ is the group completion of $B\underline{\underline{A}}$, and that $K^Q(\mathcal{A})$, $B\underline{\underline{A}}^{-1}\underline{\underline{A}}$ and $Spt_o(\underline{\underline{A}})$ are homotopy equivalent. Let $\underline{\underline{Q}}$ denote the double mapping cylinder of $(0 \leftarrow \underline{\underline{A}} \to \underline{\underline{U}})$ given by Thomason in [T,(5.1)]. Thus objects of $\underline{\underline{Q}}$ are pairs (A,U), and a $\underline{\underline{Q}}$–map from (A,U) to (B,V) is an equivalence class of data $(E,F,A \cong E \oplus B \oplus F, F \oplus U \cong V)$. Thomason proves in [T,(5.2) and (5.5)] that $Spt_o(\underline{\underline{A}}) \to Spt_o(\underline{\underline{U}}) \to Spt_o(\underline{\underline{Q}})$ is a homotopy fibration. Since $\underline{\underline{U}} \to \underline{\underline{S}}$ factors through $\underline{\underline{Q}}$, we see that 5.3 follows from [T,(2.3)] and the following result:

Proposition 5.5. The functor $\Sigma:\underline{Q} \to \underline{\underline{S}}$ given by $\Sigma(A,U) = A \oplus U$ is a homotopy equivalence when $\mathcal{A}$ is idempotent complete.

Proof. Fix an object S of $\underline{\underline{S}}$; we will show that $S\downarrow\Sigma$ is a contractible category. The desired result will then follow from Quillen's Theorem A [Q]. In order to do this, we need to thicken $S\downarrow\Sigma$ up a bit. Let $\mathcal{S}$ denote the category whose objects are tuples

$$\alpha = (A,U,S \cong D_\alpha \oplus S_\alpha, U \cong E_\alpha \oplus U_\alpha, f_\alpha: S_\alpha \cong U_\alpha)$$

where $A \in \mathcal{A}$, $U \in \mathcal{U}$, f_α is a $\mathcal{U}$-isomorphism, and the direct sum decompositions belong to the $\mathcal{A}$-filtrations of S and U. A map from α to

$$\beta = (B,V,S \cong D_\beta \oplus S_\beta, V \cong F_\beta \oplus V_\beta, f_\beta: S_\beta \cong V_\beta)$$

is just a map in $\underline{Q}$ from (A,U) to (B,V), say given by

$$e = (E,F,A \cong E \oplus B \oplus F, F \oplus U \cong V)$$

such that $D_\alpha \oplus S_\alpha \geq D_\beta \oplus S_\beta, (F \oplus E_\alpha) \oplus U_\alpha \geq F_\beta \oplus V_\beta$ in the filtrations of S and V, and such that there is a commutative square in $\mathcal{U}$:

$$\begin{array}{ccc} S_\beta & \cong & V_\beta \\ \uparrow & & \uparrow \\ S_\alpha & \cong & U_\alpha. \end{array}$$

There is a natural functor $\mathrm{pr}\,\mathcal{S} \to (S\downarrow\Sigma)$ sending α to the object $\mathrm{pr}_\alpha: S \to S_\alpha \cong U_\alpha \to U \to A\oplus U = \Sigma(A,U)$ and e to itself. By Lemma 5.2, pr is onto. In fact, pr is cofibered. We assert that pr is a homotopy equivalence, which follows from [Q, p.93] and the following:

Sublemma 5.6. Given an isomorphism $\bar{\varphi}: S \to \Sigma(A,U)$ in $\mathcal{U}/\mathcal{A}$, the fiber category $\mathrm{pr}^{-1}(\bar{\varphi})$ is a cofiltered poset, and hence is contractible.

Proof. Set $\phi = \mathrm{pr}^{-1}(\bar{\varphi})$; it is clear from the definition of $\mathcal{S}$ that ϕ is a poset. We need only show that for each α,β in ϕ there is a diagram $\alpha \leftarrow \gamma \rightarrow \beta$ in ϕ. To do this, we introduce some notation. Write $\alpha = (A,U,S \cong D_\alpha \oplus S_\alpha, U \cong E_\alpha \oplus U_\alpha, f_\alpha)$ and $\beta = (A,U,S \cong D_\beta \oplus S_\beta, U \cong E_\beta \oplus U_\beta, f_\beta)$. If $D_\alpha \oplus S_\alpha \leq D_\gamma \oplus S_\gamma$ we set $D_{\gamma\alpha} = D_\gamma \cap S_\alpha$, $E_{\gamma\alpha} = f_\alpha(D_{\gamma\alpha})$ and $\alpha^1 = (A,U,S \cong D_\gamma \oplus S_\gamma, U \cong (E_\alpha \oplus E_{\gamma\alpha}) \oplus f_\alpha(S_\gamma), f_\alpha|S_\gamma: S_\gamma \cong f_\alpha(S_\gamma))$. We shall refer to α^1 as "α cut down to $D_\gamma \oplus S_\gamma$"; note that there is a map $\alpha^1 \to \alpha$ in ϕ.

By axiom (F1), there is a decomposition $S \cong D_o \oplus S_o$ larger than both $D_\alpha \oplus S_\alpha$ and $D_\beta \oplus S_\beta$. Cutting α and β down, we can assume that $D_\alpha = D_\beta = D_o$ and $S_\alpha = S_\beta = S_o$. Now $f_\alpha - f_\beta: S_o \to D$ is completely continuous, so after cutting α and β down further we can assume that $f_\alpha = f_\beta$. Note that we still may have $E_\alpha \neq E_\beta$. Consider the maps

$$E_\alpha \to U \to U_\beta \cong S_o \text{ and } E_\beta \to U \to U_\alpha \cong S_o.$$

By axiom (F2) there is a decomposition $S \cong D_\gamma \oplus S_\gamma$ for which these two maps factor through D_γ. Let γ be α cut down to $D_\gamma \oplus S_\gamma$; evidently γ is also β cut down to $D_\gamma \oplus S_\gamma$. The resulting maps $\alpha \leftarrow \gamma \rightarrow \beta$ in ϕ were what we needed to prove Sublemma 5.6, so we are done.

Resuming the proof of (5.5) we let $e: \alpha \to \beta$ be the map described at the proof's outset. Let $D_{\alpha\beta} = D_\alpha \cap S_\beta$ and $E_{\alpha\beta} = f_\beta(D_{\alpha\beta})$, so that $D_{\alpha\beta} \oplus D_\beta = D_\alpha$ and $D_{\alpha\beta} \oplus S_\alpha = S_\beta$. We first observe that from the definition of the map e there is a natural identification of subobjects of V:

$$(*)\ F_\beta \oplus E_{\alpha\beta} = F \oplus E_\alpha.$$

Using this, there is a natural isomorphism in $\mathcal{A}$:

$$s_e : D_\alpha \oplus A \oplus E_\alpha \cong (D_{\alpha\beta} \oplus D_\beta) \oplus (E \oplus B \oplus F) \oplus E_\alpha \cong (D_{\alpha\beta} \oplus E) \oplus (D_\beta \oplus B \oplus F_\beta) \oplus E_{\alpha\beta}.$$

Now define q_α to be the object $S \to S_\alpha \to \Sigma(D_\alpha \oplus A \oplus E_\alpha, S_\alpha)$ of $S \downarrow \Sigma$ defined naturally by α, and let $q_e : q_\alpha \to q_\beta$ be the map

$$(D_{\alpha\beta} \oplus E, E_{\alpha\beta}, s_e, E_{\alpha\beta} \oplus S_\alpha \cong S_\beta).$$

It is easy to see that q is a functor from $\mathcal{S}$ to $S \downarrow \Sigma$.

Now let z denote the object $1 : S \to \Sigma(0,S)$ of $S \downarrow \Sigma$. There is a map $\theta_\alpha : q_\alpha \to z$ given by the data

$$(A \oplus E_\alpha, D_\alpha, D_\alpha \oplus A \oplus E_\alpha \cong (A \oplus E_\alpha) \oplus 0 \oplus D_\alpha, D_\alpha \oplus S_\alpha \cong S)$$

and a map $\eta_\alpha : q_\alpha \to pr_\alpha$ given by the data

$$(D_\alpha, E_\alpha, D_\alpha \oplus A \oplus E_\alpha = D_\alpha \oplus A \oplus E_\alpha, E_\alpha \oplus S_\alpha \cong E_\alpha \oplus U_\alpha \cong U).$$

Using (*), it is a straightforward matter to check that θ and and η are natural transformations. This proves that the maps z, q and pr from $B\mathcal{S}$ to $B(S \downarrow \Sigma)$ are homotopic. Since pr is a homotopy equivalence, this shows that $B(S \downarrow \Sigma)$ is contractible. This finishes the proof of Proposition 5.5, and hence of Theorem 5.3.

Next we show how to remove the hypothesis that $\mathcal{A}$ is idempotent complete from Theorem 5.4. Let $\mathcal{U}$ be a flasque $\mathcal{A}$-filtered category. Let $\mathcal{A}^\wedge$ denote the idempotent completion of $\mathcal{A}$, and let $\mathcal{U}^\wedge$ be the full subcategory of the idempotent completion of $\mathcal{U}$ on objects $P \oplus U$, P in $\mathcal{A}^\wedge$ and U in $\mathcal{U}$. Then $\mathcal{U}^\wedge$ is $\mathcal{A}^\wedge$-filtered but not flasque. However, it is easy to see that $K^Q(\mathcal{U}^\wedge)$ is contractible, and Theorem 5.3 applies to show that $\Omega K^Q(\mathcal{U}^\wedge/\mathcal{A}^\wedge) \cong K^Q(\mathcal{A}^\wedge)$. On the other hand, $\mathcal{U}^\wedge/\mathcal{A}^\wedge \cong \mathcal{U}/\mathcal{A}$, so we have proven:

<u>Corollary 5.7</u>. If $\mathcal{A}^\wedge$ denotes the idempotent completion of $\mathcal{A}$, and $\mathcal{U}$ is a flasque $\mathcal{A}$-filtered category, then

(i) $K^Q(\mathcal{U}/\mathcal{A}) \simeq K^Q(\mathcal{U}^\wedge/\mathcal{A}^\wedge)$

(ii) $\Omega K^Q(\mathcal{U}/\mathcal{A}) \simeq K^Q(\mathcal{A}^\wedge)$

(iii) $K^Q(\mathcal{A}) \to K^Q(\mathcal{U}) \to K^Q(\mathcal{U}/\mathcal{A})$ is a homotopy fibration if and only if $K_o(\mathcal{A}) \cong K_o(\mathcal{A}^\wedge)$.

<u>Definition 5.8</u> (Karoubi [K136]). Given a semisimple exact category $\mathcal{A}$, we define $K_{-n}(\mathcal{A})$ to be $K_1(S^{n+1}\mathcal{A})$. Note that $K_{-o}(\mathcal{A}) = K_o(\mathcal{A}^\wedge)$ by 5.7 (ii). This is well-defined because by 5.4 the homotopy type of $S^{n+1}\mathcal{A}$ is independent of the choice of the flasque category $\mathcal{U}$ used to construct $S\mathcal{A} = \mathcal{U}/\mathcal{A}$. Note that $K_o(S^n\mathcal{A}) = 0$ for $n \geq 1$ because $S^n\mathcal{A}$ need not be idempotent complete. In fact $K_o((S^n\mathcal{A})^\wedge) = K_{-n}(\mathcal{A})$. (Cf. [K136, p. 151].)

<u>Definition 5.9.</u> By 5.4, 5.7 and 5.8, there is an Ω-spectrum

$$\{K_{-n}(\mathcal{A}) \times K^Q(S^n\mathcal{A})\} \simeq \{K^Q((S^n\mathcal{A})^\wedge)\}$$

We shall call it Karoubi's non-connective K-theory spectrum for $\mathcal{A}$, since Karoubi gave the prescription for this spectrum in [K136].

§6. Agreement of spectra.

Our task is now to show that Karoubi's spectrum agrees with the other spectra in the literature. We first recall Wagoner's construction in [Wag]. Given a ring R, let ℓR denote the ring of locally finite $\mathbb{N}$-indexed matrices over R, i.e., matrices (r_{ij}) with $1 \le i,j < \infty$ such that each row and each column has only finitely many nonzero entries. The finite matrices form an ideal mR of ℓR, and we set $\mu R = \ell R/mR$. Wagoner's spectrum is

$$\{K_o(\mu^n R) \times BGL^+(\mu^n R)\}.$$

We shall show that this spectrum is the same as Karoubis spectrum for the category $\mathcal{F}(R)$ of (based) finitely generated free R-modules. Note that $\mathcal{F}$^(R) is equivalent to the category of fin. gen. projective R-modules, so K($\mathcal{F}$^(R)) is the usual space $K_o^Q(R) \times BGL^+(R)$. The following argument was shown to us by H.J. Munkholm and A. Ranicki.

Proposition 6.1. Let $\mathcal{U}$ denote the category of countably (but not necessarily infinitely) based free R-modules and locally finite matrices over R. Then $\mathcal{U}/\mathcal{F}(R)$ is equivalent to the category $\mathcal{F}(\mu R)$. Consequently, Wagoner's spectrum for R is homotopy equivalent to Karoubi's K-theory spectrum for $\mathcal{F}(R)$.

Proof. (Munkholm–Ranicki). Choose an infinitely based R-module R^∞ in $\mathcal{U}$, and observe that $End_{\mathcal{U}}(R^\infty) = \ell R$. Now $\mathcal{U}$ is $\mathcal{F}(R)$-filtered, and the completely continuous endomorphisms of R^∞ form the ideal mR. Thus $End_{\mathcal{U}/\mathcal{F}}(R^\infty) \cong \mu R$. The additive functor $\mathcal{F}(\mu R) \to \mathcal{U}/\mathcal{F}(R)$ which sends ${}^1\mu R$ to ${}^1R^\infty$ is therefore full and faithful. But every object of $\mathcal{U}/\mathcal{F}(R)$ is either isomorphic to 0 or to R^∞, so this functor is also an equivalence. Done.

Gersten has also constructed a nonconnective spectrum for the K-theory of a ring in [G]. Since Wagoner showed in [Wag] that it agreed with Wagoner's spectrum, Gersten's spectrum is also homotopy equivalent to Karoubi's spectrum."

Finally, we must compare Karoubi's K-theory spectrum with the spectrum $\{f(S^n)\}$ of section 4 above, constructed in [PW] using the categories $\mathcal{C}_n(\mathcal{A})$. We assume that $\mathcal{A}$ is filtered in the sense of [PW,(1.1)],or §2 above, so that in the notation of op cit. we have $\mathcal{C}_{n+1}(\mathcal{A}) = \mathcal{C}_1(\mathcal{C}_n(\mathcal{A}))$. The choice of the filtration affects the morphisms allowed in $\mathcal{C}_n(\mathcal{A})$ and $C_+(\mathcal{A})$, but not the homotopy type of $K^Q(\mathcal{C}_n(\mathcal{A}))$, as [PW,(3.2)] shows. The real point of the filtration on $\mathcal{A}$ is to reduce our discussion to the case n = 1.

Recall that objects of $\mathcal{C}_1(\mathcal{A}) = \mathcal{C}_{\mathbb{Z}}(\mathcal{A})$ are $\mathbb{Z}$–indexed sequences $A_\cdot = (\ldots, A_{-1}, A_0, A_1, \ldots)$ in $\mathcal{A}$. A map $\varphi: A \to B$ is a matrix of maps $\varphi_{ij}: A_j \to B_i$ such that for some bound $b = b(\varphi)$ we have $\varphi_{ij} = 0$ whenever $|i-j| > b$. Composition is given by matrix multiplication. Define $\mathrm{trunc}(A_\cdot)$ to be the object $(A_0, A_1, \ldots)$ of $\mathcal{C}_+(\mathcal{A})$ and $\mathrm{trunc}(\varphi)$ to be the submatrix of φ_{ij} with $i,j \geq 0$. If $\psi: B_\cdot \to C_\cdot$, then $\mathrm{trunc}(\psi\varphi) - \mathrm{trunc}(\psi)\mathrm{trunc}(\varphi)$ is completely continuous, being bounded by $b(\psi) + b(\varphi)$. Hence trunc defines a functor from $\mathcal{C}_1(\mathcal{A})$ to $C_+(\mathcal{A})/\mathcal{A}$.

Theorem 6.2. The functor trunc induces a homotopy equivalence

$$K^Q(\mathcal{C}_1(\mathcal{A})) \xrightarrow{\sim} K^Q(\mathcal{C}_+(\mathcal{A})/\mathcal{A}) \simeq K^Q(S\mathcal{A})$$

Assuming this result, it follows inductively from the above remarks that we have $K^Q(\mathcal{C}_{n+1}(\mathcal{A})) \xrightarrow{\sim} K^Q(\mathcal{C}_1(\mathcal{C}_n(\mathcal{A}))) \simeq K^Q(S\mathcal{C}_n(\mathcal{A})) \simeq K^Q(S^{n+1}\mathcal{A})$. Hence the spaces $\hat{B}_n$ of [PW] are $K_{-n}(\mathcal{A}) \times K^Q(\mathcal{C}_n(\mathcal{A}))$, and we have

Corollary 6.3. The nonconnective K–theory spectra of [PW] agree with Karoubi's. In fact, trunc induces a homotopy equivalence of spectra:

$$\{K_{-n}(\mathcal{A}) \times K^Q(\mathcal{C}_n\mathcal{A})\} \xrightarrow{\sim} \{K_{-n}(\mathcal{A}) \times K^Q(S^n\mathcal{A})\}.$$

Proof of Theorem 6.2. We shall use the notation of [PW], only remarking that $\underline{\underline{C}}_\varepsilon$ is the category of isomorphisms of $\mathcal{C}_\varepsilon(\mathcal{A})$, and that $\mathrm{Spt}_0(\underline{\underline{C}}_\varepsilon) \simeq K^Q(\mathcal{C}_\varepsilon(\mathcal{A}))$. By (5.7) and [PW,§3], we can assume that $\mathcal{A}$ is idempotent complete. There is a map of squares

$$\begin{array}{ccccc} \underline{\underline{A}} \to \underline{\underline{C}}_+ & & \underline{\underline{A}} \to \underline{\underline{C}}_+ \\ \downarrow \quad \downarrow & \to & \downarrow \quad \downarrow \\ \underline{\underline{C}}_- \to \underline{\underline{C}}_1 & & 0 \to \underline{\underline{S}}. \end{array}$$

By Theorem 5.3 and [PW,§3], applying Spt_0 yields homotopy cartesian squares with $\mathrm{Spt}_0(\underline{\underline{C}}_1) \simeq K^Q(\mathcal{C}_1(\mathcal{A}))$ and $\mathrm{Spt}_0(\underline{\underline{S}}) \simeq K^Q(S\mathcal{A})$ connected. Since $\mathrm{Spt}_0(\underline{\underline{C}}_-)$ and $\mathrm{Spt}_0(\underline{\underline{C}}_+)$ are contractible, it follows that $K^Q(\mathcal{C}_1(\mathcal{A})) \to K^Q(S\mathcal{A})$ is a weak homotopy equivalence, hence a homotopy equivalence.

References

[B] Bass,H., Algebraic K–theory, Benjamin, 1968.

[C1] Chapman,T.A., Controlled boundary and h-cobordism theorems, Trans. A.M.S. 280, 1983, 73–95.

[C2] Chapman,T.A., Controlled simple homotopy theory and applications, Springer Lecture Notes 1009, 1983.

[G] Gersten,S., On the spectrum of algebraic K–theory, Bull. A.M.S. 78, 1972, 216–219.

[Gunn] Gunnarson T. Algebraic K–theory of spaces as K–theory of monads. Aarhus preprint series No 21, 1981/1982.

[K136] Karoubi,M., Foncteurs derivées et K–theorie, Springer Lecture Notes 136, 1970.

[May] May,J.P., E^{∞}–spaces, group completions and permutative categories, New developments in topology, LMS Lecture Notes 11, 1974, 61–94.

[PW] Pedersen,E.K. and Weibel,C., A nonconnective delooping of algebraic K–theory, Proc. 1983 Rutgers Topology Conference, Springer Lecture Notes 1126, 1985, 166–181.

[P1] Pedersen,E.K., K_{-i} invariants of chain complexes, Proc. 1982 Leningrad Topology Conference, Springer Lecture Notes 1060, 1984, 174–186.

[P2] Pedersen,E.K., On the K_{-i} functors, J. of Algebra 90, 1984, 461–475.

[Q] Quillen,D., Higher algebraic K–theory I, Proc. 1972 Battelle Seattle Conf. on Algebraic K–theory, Springer Lecture Notes 341, 1973, 85–147.

[Qu1] Quinn,F., Ends of Maps II, Inv. Math. 68, 1982, 353–424.

[Qu2] Quinn,F., Ends of maps I, Ann. of Maths. 110, 1979, 275–331.

[Qu3] Quinn,F., Proc. of Geometric Algebra conference, Springer Lecture Notes of Mathematics 126, 1985, 182–198.

[Segal] Segal,E., Categories and cohomology theories, Topology 13, 1974, 293–312.

[T] Thomason,R., First quadrant spectral sequences in algebraic K–theory via homotopy colimits, Comm. in Alg. 10, 1982, 1589–1668.

[Wag] Wagoner, J., Delooping classifying spaces in algebraic K–theory, Topology 11, 1972, 349–370.

[W] Weibel, C., K–theory of Azumaya algebras, Proc. AMS 81, 1981, 1–7.

[Woolf] Woolfson, R., Hyper–Γ–spaces and hyperspectra, Quart. J. Math. 30 1979, 229–255

STIRLING AND BERNOULLI NUMBERS FOR COMPLEX ORIENTED HOMOLOGY THEORY

Nigel Ray
Mathematics Department, The University,
Manchester, M13 9PL.

Introduction

The Roman-Rota umbral calculus of Δ-operators over a field has recently achieved a wider audience with the appearance of Roman's book [13]. We have suggested in [9] that this calculus may naturally be considered over a graded ring, and in [10] that it then provides an attractive setting for complex oriented homology theories. Such a description is related to the more usual one of formal groups, e.g see [8], but differs in perspective and emphasis. For example, several old and venerable combinatorial formulae may be immediately translated into the homological context, where they become of significant interest to homotopy theorists.

Our aim here is to describe some of these formulae, and simultaneously to simplify and extend certain results of [9] and [10]. There are points, such as in Proposition (3.6), at which the parallel between the new and the classical proofs is particularly close and elegant.

In fact, it is possible to expurgate any mention of umbral calculus from this work; however, without its suggestive guidance we would not easily have been led to the formulae described below.

In view of recent developments the case of Morava K-theory appears to have a special role, so we have emphasised this example wherever possible.

It is a pleasure to acknowledge a host of helpful and enjoyable disussions with Andrew Baker and Francis Clarke. Baker's initial computations [2] were our inspiration for investigating the relevance of Rota's work, whilst Clarke first proved Theorem (3.7) by a more intricate method, obtaining deeper results [5] and pointing out the importance of Carlitz's work [3]. We were in turn all influenced by Haynes Miller's article [7], which pioneered the direct application of such number theory to the chromatic method of calculation in stable homotopy.

1. Complex oriented homology theories

Throughout, we assume that E is a complex orientable spectrum, for discussion of which see Adams [1], whose coefficient ring E_* we suppose for convenience to be free of additive torsion. The universal example is the Thom spectrum MU, and we shall find it convenient to identify the polynomial subalgebra $CP_* \subset MU_*$ generated by the bordism classes $CP_1, CP_2, \ldots$ of the complex projective spaces. Thus $CP\mathbf{Q}_* = MU\mathbf{Q}_*$.

If $MU_*(X)$ is free over MU_* on preferred basis elements $b_1, b_2, \ldots$, we write $CP_*(X)$ for the free CP_* module on the same basis.

As expounded in [10], we regard $E_*(CP^\infty)$ and $E^*(CP^\infty)$ as determined by their respective images under the Hurewicz and Boardman maps

$$E_*(CP^\infty_+) \longrightarrow E\mathbb{Q}_*[x] \quad \text{and} \quad E^*(CP^\infty_+) \longrightarrow E^*((D)).$$

Here $x \in H_2(CP^\infty)$ is carried by the standard inclusion $S^2 \longrightarrow CP^\infty$, its dual $D \in H^2(CP^\infty)$ is the first Chern class of the Hopf bundle η, and $E^*((D))$ denotes the ring of formal divided power series over E^* on D.

The Kronecker pairing

$$E^m(CP^\infty_+) \otimes E_n(CP^\infty_+) \xrightarrow{\langle\ |\ \rangle} E_{n-m}$$

is then described by

$$\langle D^k \mid p(x) \rangle = \frac{d^k}{dx^k} p(x)|_{x=0}$$

for any $p(x) \in E_*[x]$; in other words, $D = \frac{d}{dx}$.

In this parlance, an orientation for E corresponds to a <u>Δ-operator</u>

$$\Delta^E = D + \theta_1 \frac{D^2}{2!} + \ldots + \theta_{i-1} \frac{D^i}{i!} + \ldots$$

in $E^2((D))$. This may be expessed umbrally as

$$\Delta^E = e^{\theta D} - 1, \qquad \theta^i \equiv \theta_{i-1}$$

with $\theta_i \in E_{2i}$. Such an element is a Thom class for η, and so is represented by a suitable stable map

$$t^E : CP^\infty \longrightarrow S^2E.$$

Every Δ^E has a <u>conjugate</u>, or compositional inverse

$$\bar{\Delta}^E = e^{\bar{\theta} D} - 1, \qquad \bar{\theta}^i \equiv \bar{\theta}_{i-1}$$

in $E^2((D))$, specified by the <u>Lagrange inversion principle</u>. Note that

$$D = e^{\bar{\theta}\Delta^E} - 1, \qquad \bar{\theta}^i \equiv \bar{\theta}_{i-1}. \tag{1.1}$$

<u>(1.2) Examples.</u>

(i) The <u>universal</u> case is $E = MU$, and we write

$$\Delta^U = e^{\Phi D} - 1$$

for the universal Δ-operator Δ^{MU}, represented by the inclusion

$$t^U : CP^\infty \simeq MU(1) \longrightarrow S^2MU.$$

Then $\Phi_i \in MU_{2i}$ is the bordism class with Hurewicz image $(i+1)!b_i$ in $H_{2i}(MU)$. We write $\Phi_* \subset MU_*$ for the polynomial subalgebra generated by $\Phi_1, \Phi_2, \ldots$.

The Lagrange inversion principle becomes Misčenko's theorem, that $\bar{\Phi}_i = i!CP_i$. Using (1.1) we can express this as

$$D = \sum_{i=1}^{\infty} CP_{i-1} \frac{(\Delta^U)^i}{i} \tag{1.3}$$

in $MU^2((D))$, whence $\Phi_* \subset CP_*$.

Using this example, a general t^E always extends via

$$\begin{array}{ccc} CP^\infty & \xrightarrow{\;t^U\;} & S^2MU \\ & {\scriptstyle t^E}\searrow \quad \swarrow{\scriptstyle \gamma^E} & \\ & S^2E & \end{array} \tag{1.4}$$

to a map γ^E of ring spectra, which satisfies

$$\gamma^E_* \Phi_i = \theta_i \,, \quad \gamma^E_* CP_i = \bar{\theta}_i / i!.$$

(ii) The universal <u>p-typical</u> case is $E = BP$, where for each prime p

$$BP_* = \mathbb{Z}_{(p)}[v_1, v_2, \ldots]$$

and Δ^{BP} is defined by choosing γ^{BP} in (1.4) to be the Quillen idempotent $MU_{(p)} \longrightarrow BP$. Thus in (1.3)

$$D = \sum_{j=1}^{\infty} CP_{p^j-1} \frac{(\Delta^{BP})^{p^j}}{p^j} \,,$$

and the CP_{p^j-1}'s are related to the v_j's by a standard recursive formula.

(iii) Complex K-theory is the case $E = K$, where

$$K_* = \mathbb{Z}[u, u^{-1}]$$

and

$$\Delta^K = u^{-1}(e^{uD} - 1),$$

with γ^K of (1.4) being the Todd genus. Thus $\gamma^K_* \Phi_i = u^i$.

Lagrange inversion is trivial, giving

$$D = u^{-1}\log(1 + u\Delta^K)$$

and $\gamma_*^K CP_i = (-1)^i u^i$.

(iv) For each prime p, then <u>m-th Morava K-theory</u> is the case $E = K(m)$, where

$$K(m)_* = \mathbb{Z}_{(p)}[u^{p^m-1}, u^{1-p^m}] \subset K_{(p)*}$$

and $\Delta^{K(m)}$ is defined by choosing $\gamma^{K(m)}$ in (1.4) to be the p-typical Todd genus

$$MU_{(p)} \longrightarrow BP \longrightarrow K(m).$$

Thus

$$\gamma_*^{K(m)} CP_i = \begin{cases} p^{j(m-1)} u^{p^{jm}-1} & \text{if} \quad i = p^{jm}-1 \\ 0 & \text{otherwise}. \end{cases} \qquad \square$$

Usually, $K(m)$ is defined by reducing the above theory mod p, which enhances certain of its algebraic properties. For spaces such as CP^∞, however, it is clearly legitimate to work with the <u>integral</u> theory of (1.2 (iv)), and reduce mod p at the end if so desired.

With this formulation, we have

$$E^*(CP_+^\infty) \cong E^*[[\Delta^E]].$$

Dualising, a basis for $E_*(CP_+^\infty)$ is given by a sequence of polynomials

$$1,\ b_1^E(x),\ b_2^E(x), \ldots \tag{1.5}$$

where $b_n^E(x) \in EQ_{2n}[x]$. This is the normalised <u>associated sequence</u> of Δ^E, mapped one to the next by Δ^E and having the divided power property (e.g. see [9]).

A host of examples of associated sequences is discussed in [13]. Those related to (1.2) are a major ingredient of the next section.

2. Stirling numbers

(2.1) Definition. If we write

$$b_n^E(x) = \frac{1}{n!} \sum_{k=1}^{n} s^E(n,k) x^k,$$

then we call the elements $s^E(n,k) \in E_{2(n-k)}$ the E-theory <u>Stirling numbers</u> of the <u>first</u> kind.

Similarly, if we write

$$x^n = \sum_{k=1}^{n} k! S^E(n,k) b_k^E(x),$$

then we call the elements $S^E(n,k) \in E_{2(n-k)}$ the E-theory <u>Stirling numbers</u>

of the second kind.

Thus the matrices s^E and S^E over $E*$ are mutually inverse.

Note that from (1.5),

$$< \frac{(\Delta^E)^k}{k!} \mid x^n > = S^E(n,k).$$

Therefore

$$\frac{(\Delta^E)^k}{k!} = e^{S^E(\ ,k)D} \qquad (S^E(\ ,K))^i = S^E(i,k). \tag{2.2}$$

(2.3) Example. When $E = K$, we have

$$b_n^K(x) = \frac{1}{n!} x(x-u)(x-2u)\dots(x-(n-1)u),$$

so

$$s^K(n,k) = u^{n-k} s(n,k),$$

where $s(n,k) \in \mathbb{Z}$ is the classical Stirling number (e.g. see [6]). By the same token,

$$S^K(n,k) = u^{n-k} S(n,k).$$ □

This example motivates the nomenclature of (2.1).

The classical Stirling numbers are given by traditional combinatorial formulae, usefully expressed in terms of Riordan's partition polynomial notation [12]. Thus $\pi(n)$ denotes a partition of n into k_1 1's, k_2 2's,... . Hence

$$k_1 + 2k_2 + \dots + nk_n = n.$$

The partition has

$$k = k_1 + k_2 + \dots + k_n$$

parts, and if k is fixed we write $\pi(n;k)$.

Then

$$s(n,k) = \sum_{\pi(n;k)} \frac{(-1)^{n-k} n!}{k_1! 1^{k_1} \dots k_n! n^{k_n}} \tag{2.4}$$

$$= (-1)^{n-k} \frac{n!}{k!} \sum_{\substack{i_1+\dots+i_k = n \\ i_j > 0}} \frac{1}{i_1 \dots i_k},$$

and

$$S(n,k) = \sum_{\pi(n;k)} \frac{n!}{k_1! 1!^{k_1} \dots k_n! n!^{k_n}} \tag{2.5}$$

$$= \frac{1}{k!} \sum_{\substack{i_1+\dots+i_k = n \\ i_j > 0}} \binom{n}{i_1, \dots, i_k},$$

e.g. see [13]. We now generalise these.

It clearly suffices to consider the universal Stirling numbers $s^U(n,k) \in MU_{2(n-k)}$ and $S^U(n,k) \in MU_{2(n-k)}$, since by (1.4)

$$\begin{aligned}\gamma_*^E s^U(n,k) &= s^E(n,k)\\ \gamma_*^E S^U(n,k) &= S^E(n,k).\end{aligned} \tag{2.6}$$

(2.7) Proposition. In $MU_{2(n-k)}$, we have

$$\begin{aligned} s^U(n,k) &= n! \sum_{\pi(n;k)} \frac{CP_1^{k_2}\dots CP_{n-1}^{k_n}}{k_1!1^{k_1}\dots k_n!n^{k_n}} \\ &= \frac{n!}{k!} \sum_{\substack{i_1+\dots+i_k=n\\ i_j>0}} \frac{CP_{i_1-1}\dots CP_{i_k-1}}{i_1\dots i_k}\end{aligned} \tag{2.8}$$

and

$$S^U(n,k) = \frac{1}{(k-1)!} \sum_{\pi(n-k)} (-1)^j (n+j-1)! \frac{CP_1^{j_1}\dots CP_{n-k}^{j_{n-k}}}{j_1!2^{j_1}\dots j_{n-k}!(n-k+1)^{j_{n-k}}}. \tag{2.9}$$

Proof The formulae of [13 p.82] can be immediately adapted to our setting, following [9], and yield (2.8).

By the same procedure, a formula of [4, p.161] can be rearranged to give (2.9). This is tantamount to applying Lagrange inversion to (2.2), or equivalently (see Comtet [6]), to inverting the matrix s^U. □

(2.10) Corollary The universal Stirling numbers lie in CP_*.

Proof The rationals $\dfrac{n!}{k_1!1^{k_1}\dots k_n!n^{k_n}}$ and $\dfrac{(n+j_1+\dots+j_{n-k}-1)!}{(k-1)!j_1!2^{j_1}\dots j_{n-k}!(n-k+1)^{j_{n-k}}}$ are both integers; in fact they enumerate permutations of a specific cycle class, as in Riordan [11]. □

As we have shown in [10], (2.8) and (2.9) admit a combinatorial interpretation in terms of paths in posets. Many classical congruences can be generalised to (2.8) and (2.9), and we give a small sample, selected for their usefulness in §3.

(2.11) Lemma For each n,

$$x^n CP_*(CP^\infty) \subset CP_*(CP^\infty).$$

<u>Proof</u> A basis

$$1,\ cp_1(x),\ cp_2(x),\ldots$$

for $CP_*(CP^\infty)$ can be defined by

$$cp_k(x) = \sum_{j=0}^{k} CP_{k-j}\, b_j^U(x).$$

Then from (2.1) and (2.8),

$$xcp_k(x) = (k+1)b_{k+1}^U(x). \qquad \square$$

<u>(2.12) Lemma</u> For each prime p,

$$x^{d(p-1)+e} \equiv CP_{p-1}^{d-1}\, x^e \mod pCP_*(CP^\infty)$$

for all d, and $0 < e \leqslant p-1$.

<u>Proof</u> Remark from (2.8) that

$$s^U(p,p) = 1,\quad s^U(p,1) = (p-1)!CP_{p-1},$$

$$s^U(p,k) \equiv 0 \mod pCP_* \quad \text{for} \quad 1 < k < p.$$

Thus in (2.1),

$$p!b_p^U(x) \equiv x^p + (p-1)!CP_{p-1}x \mod pCP_*(CP^\infty),$$

whence

$$x^p \equiv CP_{p-1}x \mod pCP_*(CP^\infty).$$

Now iterate, using (2.11). $\square$

<u>(2.13) Lemma</u> Modulo $2CP_*[x]$ and $12CP_*(CP^\infty)$, we have that

$$x^{2d} \equiv CP_1^{2(d-1)}x^2,$$

$$x^{2d+1} \equiv CP_1^{2(d-1)}x^3.$$

<u>Proof</u> From (2.1) and (2.8),

$$24b_4^U(x) = x^4 + 6CP_1x^3 + (8CP_2+3CP_1^2)x^2 + 6CP_3x.$$

Thus

$$x^4 \equiv CP_1^2x^2,$$

which is iterated as in (2.12). $\square$

We can now deduce

<u>(2.14) Proposition.</u>

(i) If $n \not\equiv 0(p-1)$, then

$S^U(n,p-1) \in pCP_*$.

Otherwise

$$S^U(d(p-1),p-1) \equiv CP_{p-1}^{d-1} \mod pCP_*.$$

(ii) If n is odd, then

$$S^U(n,3) \equiv CP_1^{n-3} \mod 2CP_*.$$

Otherwise

$$S^U(2d,3) \in 2CP_*.$$

Proof. Using (2.2) and (2.12)

$$(p-1)!S^U(n,p-1) \equiv \langle (\Delta^U)^{p-1} \mid CP_{p-1}^{d-1}x^e \rangle$$

$$\equiv CP_{p-1}^{d-1} \langle (\Delta^U)^{p-1} \mid x^e \rangle \mod pCP_*.$$

This establishes (i).

For (ii), (2.2) and (2.13) imply that

$$S^U(2d+1,3) \equiv \langle \frac{(\Delta^U)^3}{3!} \mid CP_1^{2(d-1)}x^3 \rangle$$

and

$$S^U(2d,3) \equiv \langle \frac{(\Delta^U)^3}{3!} \mid CP_1^{2(d-1)}x^2 \rangle$$

mod $2CP_*$, as required. □

We can now obtain information about each of the examples in (1.2) by applying γ_*^E to our universal formulae.

(2.15) Examples

(i) The case $E = BP$ yields

$$S^{BP}(n,k) = n! \sum_{\pi_p(n;k)} \frac{CP_{p-1}^{k_1} \dots CP_{p^n-1}^{k_n}}{k_0! \dots k_n! p^{k_1+\dots nk_n}}$$

$$= \frac{n!}{k!} \sum_{p^{i_1}+\dots+p^{i_k}=n} \frac{CP_{p^{i_1}-1} \dots CP_{p^{i_k}-1}}{p^{i_1+\dots+i_k}},$$

where $\pi_p(n)$ denotes p-typical partitions, with k_0 1's, k_1 p's,...,k_n p^n's.

Similarly

$$S^{BP}(n,k) = \frac{1}{(k-1)!} \sum_{\pi_p(n-k+j)} (-1)^j \frac{(n-j+1)!CP_{p-1}^{j_1} \dots CP_{p^n-1}^{j_n}}{j_1! \dots j_n! p^{j_1+\dots+nj_n}}.$$

Also

$$S^{BP}(n,p-1) = 0 \quad \text{if} \quad n \not\equiv 0 \mod (p-1)$$

$$S^{BP}(d(p-1),p-1) \equiv CP^{d-1}_{p-1} = v_1^{d-1} \mod pCP_*.$$

(ii) The case $E = K$ reduces to (2.4) and (2.5), plus some standard congruences for $S(n,k)$, e.g. see [11].

(iii) The case $E = K(m)$ yields

$$s^{K(m)}(n,k) = \frac{1}{n!} \sum_{\pi_{p^m}(n;k)} \frac{u^{n-k}}{k_0!\dots k_n! p^{k_1+\dots+nk_n}}$$

$$= \frac{n!}{k!} \sum_{p^{i_1 m}+\dots+p^{i_k m} = n} \frac{u^{n-k}}{p^{i_1+\dots+i_k}},$$

and

$$s^{K(m)}(n,k) = \frac{1}{(k-1)!} \sum_{\pi_{p^m}(n-k+j)} (-1)^j \frac{(n-j+1)!u^{n-k}}{j_1!\dots j_n! p^{j_1+\dots nj_n}}.$$

The congruences of (2.14) simply tell us that

$$s^{K(m)}(d(p^m-1)+(p-1),p-1) \equiv 0 \mod p^{\delta(m-1)}u^{d(p^m-1)}$$

in $K(m)_*$, where the expansion

$$d = d_0 + d_1p^m + \dots + d_jp^{jm}$$

is such that $d_0 \geq d_1 \geq \dots \geq d_j$ and $\delta = d_0 + d_1 + \dots + d_j$ is a mimimum. □

3. Bernoulli numbers

(3.1) Definition If we write

$$\frac{D}{\Delta^E} = e^{B^E D}, \qquad (B^E)^i \equiv B_i^E$$

then we call the elements $B_i^E \in EQ_{2i}$ the E-theory Bernoulli numbers.

(3.2) Example When $E = K$, we have

$$\frac{D}{\Delta^K} = \frac{uD}{e^{uD}-1}.$$

So $B_n^K = u^n B_n$, where B_n is the classical Bernoulli number (e.g. see [11]). □
This example explains the nomenclature of (3.1).

The classical Bernoulli and Stirling numbers are related by a beautiful formula [11]

$$B_n = \sum_{k=0}^{n} \frac{(-1)^k}{k+1} k!S(n,k). \tag{3.3}$$

This in turn leads to a famous result of von Staudt (whose work in this area has an intriguing history, see [4]); for n even,

$$B_n \equiv - \Sigma \frac{1}{p} \mod \mathbb{Z} \tag{3.4}$$

summed over all primes p such that $(p-1)|n$. We now generalise these two formulae.

It clearly suffices to consider the universal Bernoulli numbers $B_n^U \in MUQ_{2n}$, since by (1.4),

$$\gamma_*^E B_n^U = B_n^E. \tag{3.5}$$

(3.6) Proposition In MUQ_{2n},

$$B_n^U = \sum_{k=0}^{n} \frac{CP_k}{k+1} k! S^U(n,k).$$

Proof By (3.1),

$$B_n^U = \langle e^{B^U D} \mid x^n \rangle$$

$$= \langle \frac{D}{\Delta^U} \mid x^n \rangle$$

$$= \langle \sum_{k \geq 0} CP_k \frac{(\Delta^U)^k}{k+1} \mid x^n \rangle \text{ by (1.3).}$$

So

$$B_n^U = \sum_{k \geq 0} \frac{CP_k}{k+1} k! \langle \frac{(\Delta^U)^k}{k!} \mid x^n \rangle,$$

which yields the result by utilising (2.2). □

This leads to the universal von Staudt theorem.

(3.7) Theorem If n is even then

$$B_n^U \equiv - \sum_{n=d(p-1)} \frac{CP_{p-1}^d}{p} \mod CP_*,$$

whilst if $n \geq 3$ is odd, then

$$B_n^U \equiv \tfrac{1}{2}(CP_1^n + CP_1^{n-3} CP^3) \mod CP_* .$$

Proof Note first that $k!/k+1$ is integral unless $k = p-1$, p prime, or $k = 3$. Thus from (2.10) and (3.6),

$$B_n^U \equiv \tfrac{3}{2} CP_3 S^U(n,3) + \sum_p \frac{CP_{p-1}}{p}(p-1)! S^U(n,p-1) \mod CP_*,$$

and the result follows from (2.14). □

Of course, $\frac{1}{2}(CP_1^3 + CP_3) \in MU_6$, so (3.7) tells us that $B_n^U \in MU_{2n}$ for odd $n \geq 3$. This fact is due to Miller.

Finally, we may combine (2.9) and (3.6) to obtain.

<u>**(3.8) Proposition**</u> In MU_{2n},

$$B_n^U = n \sum_{\pi(n)} (-1)^{j-1}(n+j-2)! \frac{CP_1^{j_1} \dots CP_n^{j_n}}{j_1! 2^{j_1} \dots j_n! (n+1)^{j_n}} . \qquad \square$$

We can now obtain information about each of the examples in (1.2) by applying γ_*^E to our universal formulae.

<u>**(3.9) Examples**</u>

(i) The case $E = BP$ (and p odd), yields

$$B_n^{BP} = 0 \quad \text{if} \quad n \not\equiv 0 \mod (p-1)$$

and

$$B_{d(p-1)}^{BP} \equiv -\frac{v_1^d}{p} \mod CP_*.$$

(ii) The case $E = K$ reduces to (3.3) and (3.4).

(iii) The case $E = K(m)$ is more subtle. In particular, from (2.15) and (3.3) we deduce that

$$B_n^{K(m)} = 0 \quad \text{if} \quad n \not\equiv 0 \mod (p^m-1)$$

and

$$B_{d(p^m-1)}^{K(m)} \in K(m)_{2d(p^m-1)}.$$

For more complete information we employ (3.8), and find

$$B_{d(p^m-1)}^{K(m)} = d(p^m-1) \sum_{\pi_{p^m}(d(p^m-1)+j)} (-1)^{j-1} \frac{(d(p^m-1)+j-2)!}{j_1! \dots j_n!} \left[\frac{u^{p^m-1}}{p}\right]^{j_1} \dots \left[\frac{u^{p^{nm}-1}}{p^n}\right]^{j_n}$$

$$= \frac{d(p^m-1)u^{d(p^m-1)}}{(d(p^m-1)+j)(d(p^m-1)+j-1)} \sum (-1)^{j-1} \binom{d(p^m-1)+j}{j_1 p^m, \dots, j_n p^{nm}} \frac{(j_1 p^m)!}{j_1! p^{j_1}} \dots \frac{(j_n p^{nm})!}{j_n! p^{n j_n}} .$$

It is now straightforward to analyse the p exponent ν_p of each of these terms $(m > 1)$. As in [10] we use

$$p^n! = \zeta p^{p^{n-1}+\dots+p+1} \quad \text{with} \quad \zeta \equiv (-1)^n \mod p ,$$

and hence

$$\nu_p\left(\frac{(j_i p^{im})}{j_i! p^{ij_i}}\right) = j_i(p^{im-1}+\ldots+p+1-i),$$

to show that the smallest power of p occurs in the term for which

$$j_1 = d, \quad j_2 = \ldots = j_n = 0.$$

Thus we reprove

$$B^{K(m)}_{d(p^m-1)}/p^{d(p^{m-1}+\ldots+p)+1-m} \equiv \frac{\epsilon u^{d(p^m-1)}}{p} \bmod K(m)_*,$$

where $\epsilon \equiv (-1)^{d(m-1)+1} \bmod p$.

Such computations greatly simplify (and extend) the results of [9, §10].

References

[1] J.F. Adams, 'Stable homotopy and generalised homology', Chicago Univ. Press (1972)

[2] A. Baker, 'Combinatorial and arithmetic identities based on formal group laws', preprint, Manchester Univ. (1984).

[3] L. Carlitz, 'The coefficients of the reciprocal of a series', Duke Math. J. 8, 689-700 (1941).

[4] F. Clarke, 'The proofs of von Staudt's theorems on the Bernoulli numbers', preprint, Univ. Coll. Swansea, (1985).

[5] F. Clarke, 'The universal von Staudt theorems', preprint, Univ. Coll. Swansea, (1986).

[6] L. Comtet, 'Advanced combinatorics', Reidel (1974).

[7] H. Miller, 'Universal Bernoulli numbers and the S^1-transfer', Current trends in algebraic topology 2, pt. 2, 437-449, CMS-AMS (1982).

[8] D. Ravenel, 'Complex cobordism and stable homotopy groups of spheres', Academic Press (1986).

[9] N. Ray, 'Extensions of umbral calculus: penumbral coalgebras and generalised Bernoulli numbers', Adv. in Math. 61, 49-100 (1986).

[10] N. Ray, 'Symbolic calculus : a 19th century approach to MU and BP', to appear in Proceedings of Durham symposium on homotopy theory, LMS Lecture Note Ser..

[11] J. Riordan, 'An introduction to combinatorial analysis', Wiley (1958).

[12] J. Riordan, 'Combinatorial identities', Wiley (1968).

[13] S. Roman, 'The umbral calculus', Academic Press (1984).

[14] S. Roman & G-C Rota, 'The umbral calculus', Adv. in Math. 27, 95-188 (1978).

COMPOSITION PRODUCTS IN RHom, AND RING SPECTRA OF DERIVED ENDOMORPHISMS

Alan Robinson

It is a well-known idea in algebra that useful rings arise as endomorphism rings of modules. In this paper we develop a related construction in stable homotopy theory, which gives rise to some new ring spectra.

The chief difference in the homotopy-theoretical case is that the construction needs to be made homotopy invariant. We use A_∞ (or E_∞) ring and module spectra throughout. By reformulating our definitions in terms of Boardman and Vogt's valuable theory of general homotopy-invariant structures, we obtain composition products among the spectra $RHom_X(Y,Z)$ of derived module homomorphisms which we introduced in [15], and among the associated spectral sequences. The composition product in RHom satisfies the A_∞ associativity condition. Taking the source and target to be the same module spectrum, we obtain an A_∞ ring spectrum of derived endomorphisms.

As an application, we construct some new A_∞ ring spectra which are closely related to the Baas-Sullivan theories of complex bordism with singularities.

1 Spectra, operads and the tree construction

We work in a model category for stable homotopy theory, the objects of which we call spectra. We assume this category has smash products which are associative and commutative up to coherent natural isomorphism. This assumption is realistic in view of recent work of Elmendorf [6] which can be applied to the category of [8] or that of [16]. We also assume that the model category has function spectra which are adjoint to the smash products at the level of maps.

For the definition of operad we refer the reader to [8]. Our operads are generally composed of unbased spaces, which will be cofibrant whenever this is necessary. We use both operads with permutations and operads without.

An A_∞ <u>operad</u> is an operad B (without permutations) in which each space B_n is contractible. An E_∞ <u>operad</u> is an operad B (with permutations) in which each space B_n is contractible and Σ_n-free.

We refer the reader to [2] or [3] for the important <u>bar construction</u> or <u>tree construction</u> which assigns to each operad B a new operad WB. (In the references cited, this construction is introduced for "topological-algebraic theories", which are more general than operads. However, the construction specializes to operads as it preserves "split theories".) If B is A_∞ or E_∞, then WB has the same property. There are natural maps ε: $WB \longrightarrow B$ and δ: $WB \longrightarrow WWB$ making W into a cotriple.

The virtue of WB is that it possesses certain cofibrancy or freeness properties which B generally does not. These properties are essential in any obstruction theory of operad actions, as in the Boardman-Vogt lifting theorem ([2] 7.1; [3] III 3.17). It follows from this theorem that an action of the operad WA can be replaced by an action (uniquely defined up to homotopy) of the operad WB, if A and B are both A_∞ or both E_∞.

We require some Boardman-Vogt theories which are slightly more general than operads. In particular, we use the theories $W(B \otimes L_n)$ to handle morphisms of WB-structures. Let us recall the significance of these. The theory $B \otimes L_1$, where B is an operad, acts on an ordered pair X, Y of spaces or spectra: such an action encodes separate B-actions on X and Y together with a map $X \longrightarrow Y$ which strictly preserves the B-action. The theory $W(B \otimes L_1)$ is the homotopy-invariant analogue: an action of it on (X,Y) corresponds to separate WB-actions on X and Y together with a map $X \longrightarrow Y$ which preserves the WB-action up to all necessary higher homotopies. The theory $W(B \otimes L_n)$ acts on $(n + 1)$-tuples of spaces or spectra: an action of it corresponds to a composable n-tuple of maps which preserve WB-actions up to coherent higher homotopies.

2 Ring and module spectra

Every spectrum X in our model category has an endomorphism operad $End(X)$ in which the n'th space is the space of maps from the n-fold smash product $X^{(n)}$ to X.

An A_∞ ring spectrum is a spectrum X together with a map of operads $\mu: WB \longrightarrow End(X)$, where B is an A_∞ operad.

Given such a ring spectrum (X,μ), we define an A_∞ right X-module to be a spectrum Y together with a family ν of maps $WB_n \longrightarrow Map(Y \wedge X^{(n-1)}, Y)$ such that μ and ν carry the composition in the operad WB

$$WB_n \times (WB_{i_1} \times WB_{i_2} \times \ldots \times WB_{i_n}) \longrightarrow WB_{i_1+i_2+\ldots+i_n}$$

into composition of maps

$$Map(Y \wedge X^{(n-1)}, Y) \times \{Map(Y \wedge X^{(i_1-1)}, Y) \times Map(X^{(i_2)}, X) \times \ldots \times Map(X^{(i_n)}, X)\}$$
$$\longrightarrow Map(Y \wedge X^{(i_1+i_2+\ldots+i_n-1)}, Y) .$$

Left modules and bimodules are defined analogously.

There are corresponding definitions of E_∞ ring and module spectra, in which B is required to be an E_∞ operad, the ring structure maps $\mu_n: WB_n \longrightarrow (End\,X)_n$ must preserve the action of the symmetric groups Σ_n, and the module structure maps $\nu_n: WB_n \longrightarrow Map(Y \wedge X^{(n-1)}, Y)$ must preserve the action of $1 \times \Sigma_{n-1}$. In this case, right and left modules are equivalent.

It does not matter which A_∞ or E_∞ operad arises in the above definitions, because the lifting theorem implies that an action of one can be replaced by an action of any other.

In the A_∞ case, we ought to reconcile the above definitions with those in [14], which use Stasheff cells. In the above, we may assume without loss of generality that B is the theory of monoids, i.e. that B_n is a point for all n. Then the Stasheff cell K_n is isomorphic with the subspace of WB_n generated by B_r, $r \geqslant 2$ ([3],p.21). It follows that the only difference between the definitions of [14] and those above (in the A_∞ case) is that we are here imposing a stronger unit condition. In particular, the results of [15] continue to be valid under our present conventions.

3 Spaces and spectra of derived homomorphisms

Let X be an A_∞ ring spectrum, and let Y and Z be A_∞ right X-modules. We can then consider the spectrum $\mathrm{RHom}_X(Y,Z)$ of derived X-module homomorphisms from Y to Z, which was introduced in [15]. We shall outline a proof that three different definitions of this spectrum are equivalent up to homotopy type. The first is a simple definition using mapping cylinders; the second is the definition used in [15]; and the third is a Boardman-Vogt style definition which we use in later sections. The equivalence of these is of interest in its own right, and it enables us to use the results of [15] in the present paper. However, the latter aim could be achieved more easily by re-working the paper [15] in terms of the Boardman-Vogt definition. Thus the reader who dislikes the technicalities of 3.3 has an alternative route to the next section.

3.1 <u>Definitions</u> A <u>derived</u> X-<u>homomorphism</u> from Y to Z can be defined in three different ways, as follows.

(i) It is a map $f: Y \longrightarrow Z$ together with a right X-module structure on the mapping cylinder M_f which restricts to the given X-module structures on $Y, Z \subset M_f$ (cf.[14], §3).

(ii) It is a collection of maps $f_n: WB_{n+2} \ltimes (Y \wedge X^{(n)}) \longrightarrow Z \quad (n \geqslant 0)$ such that the following diagram commutes whenever $n+2 = i_0 + \dots + i_{r+1}$:

$$\begin{array}{ccc}
WB_{r+2} \ltimes (WB_{i_0} \ltimes (Y \wedge X^{(i_0-1)}) \wedge (WB_{i_1} \ltimes X^{(i_1)}) \wedge \dots \wedge (WB_{i_r} \ltimes X^{(i_r)}) \wedge (WB_{i_{r+1}} \ltimes X^{(i_{r+1}-1)})) & & \\
\Big\downarrow {\scriptstyle 1 \ltimes \nu \wedge \mu^{(r)} \wedge 1} & & \\
WB_{r+2} \ltimes (Y \wedge X^{(r)}) \wedge (WB_{i_{r+1}} \ltimes X^{(i_{r+1}-1)}) & & \Big\downarrow {\scriptstyle \theta} \\
\Big\downarrow {\scriptstyle f_r \wedge 1} & & \\
WB_{i_{r+1}} \ltimes (Z \wedge X^{(i_{r+1}-1)}) & & \\
\Big\downarrow {\scriptstyle \sigma} & & \\
Z & \xleftarrow{\quad f_n \quad} & WB_{n+2} \ltimes (Y \wedge X^{(n)})
\end{array}$$

(Here WB is the A_∞ operad used in the ring and module structures μ, ν, σ, and θ is induced by composition in the operad.) (Cf. [15], §1.)

(iii) It is a $W(B \otimes L_1)$-structure on the pair Y, Z which extends the given WB-structures on Y and Z. (In explicit terms, such a structure consists of maps $\rho_k : W(B \otimes L_1)_k \ltimes (Y \wedge X^{(k-1)}) \longrightarrow Z$, $k \geqslant 1$; the spaces $W(B \otimes L_1)_k$ and the necessary identities are described in [2], [3].)

3.2 <u>Definitions</u> The <u>space</u> of derived X-homomorphisms from Y to Z is defined by introducing a parameter into the chosen form of the previous definition. In terms of simplicial sets, for instance, the k-simplices of this space are the derived X-homomorphisms from $\Delta[k] \ltimes Y$ to Z.

The usual naturality of mapping spaces with respect to suspension of the domain extends at once to derived homomorphisms, and shows that the spaces of derived homomorphisms from $S^n Y$ to Z fit together to form a <u>spectrum</u> $RHom_X(Y,Z)$ <u>of derived</u> X-<u>homomorphisms</u> from Y to Z.

3.3 <u>Proposition</u> The homotopy type of the spectrum $RHom_X(Y,Z)$ is independent of which of the three versions of 3.1 is used in the definition.

<u>Proof</u> It is clearly sufficient to show that the space of derived X-module homomorphisms from Y to Z is independent of the choice of definition, where Y and Z are arbitrary X-module spectra. We prove this by using the theories of Boardman and Vogt.

We first compare definitions (ii) and (iii). Definition (iii) specifies a $W(B \otimes L_1)$-structure on the pair Y, Z. This incorporates contractible spaces of maps $Y \wedge X^{(k-1)} \longrightarrow Z$ for all $k \geqslant 1$ as well as the ring and module structures on X, Y and Z. It is an instance of a split Boardman-Vogt theory. Definition (ii) is an instance of another theory D of the same kind, because it comprises families of maps $Y \wedge X^{(k-1)} \longrightarrow Z$ parametrized by contractible spaces in addition to the given ring and module structures. The theory D is augmented over the theory $B \otimes L_1$ of strict module homomorphisms, because a strict homomorphism has the structure prescribed in definition (ii). One can use acyclicity and cofibrations just as in the proof of the lifting theorem to obtain functors as indicated by dotted arrows

$$\begin{array}{ccc} W(B\otimes L_1) & \dashrightarrow & D \\ & \searrow \quad \swarrow & \\ & B\otimes L_1 & \end{array} \qquad\qquad \begin{array}{ccc} D & \dashrightarrow & W(B\otimes L_1) \\ & \searrow \quad \swarrow & \\ & B\otimes L_1 & \end{array}$$

together with homotopies of functors making the triangles homotopy commutative. By the same reasoning, the composite either way round of these two morphisms is homotopic to the identity functor. Therefore definitions (ii) and (iii) are equivalent up to homotopy.

Definitions (i) and (iii) are compared by means of a modification of the same method. We shall describe only the main differences from the previous case. First, the map $f:Y \to Z$ in (i) corresponds to the restriction of $\rho_1: W(B\otimes L_1)_1 \ltimes Y \to Z$ to a point in $W(B\otimes L_1)_1$. Then, we decompose the mapping cylinder M_f into $I \ltimes Y$ and Z, so that the module structure maps for M_f in definition (i) give maps $(I \times WB_k) \ltimes Y \wedge X^{(k-1)} \to M_f$. By using acyclicity and a generalized cube decomposition for $I \times WB_k$, we can then compare this structure with the theory in definition (iii) involving maps $W(B\otimes L_1)_k \ltimes Y \wedge X^{(k-1)} \to Z$, in view of the standard homotopy equivalence $M_f \simeq Z$. By verifying all the details, one can thus prove that definitions (i) and (iii) are equivalent up to homotopy. □

4 Composition products in RHom

Let M be an A_∞ ring spectrum, and let X,Y,Z be A_∞ right M-modules.

4.1 <u>Theorem</u> (i) There is a composition product

$$RHom_M(Y,Z) \wedge RHom_M(X,Y) \longrightarrow RHom_M(X,Z) \quad ,$$

well-defined up to homotopy, which is A_∞ associative in the sense that an A_∞ operad D acts by natural maps

$$D_n \ltimes (RHom_M(X_{n-1},X_n) \wedge RHom_M(X_{n-2},X_{n-1}) \wedge \ldots \wedge RHom_M(X_0,X_1)) \longrightarrow RHom_M(X_0,X_n)$$

in such a way that the composition product is given by the action of a point in D_2.

(ii) When M is the sphere spectrum S, this is the usual composition in $RHom_S(X,Y) \simeq Map(X,Y)$.

(iii) If M is the Eilenberg-MacLane spectrum HA associated with a ring A and X,Y,Z arise in the same way from right A-modules P,Q,R, then the above composition induces the Yoneda products in the homotopy groups $\pi_r RHom_{HA}(HP,HQ) \approx Ext_A^{-r}(P,Q)$.

<u>Proof</u> (i) We apply the theory of ([2], §10) to the definition of $RHom_M(X,Y)$ in 3.1(iii). Since $RHom_M(Y,Z)$ and $RHom_M(X,Y)$ are the spectra of $W(B \otimes L_1)$-actions on the pairs (Y,Z) and (X,Y), there is a canonical product

$$RHom_M(Y,Z) \wedge RHom_M(X,Y) \longrightarrow C_M(X,Y,Z)$$

where $C_M(X,Y,Z)$ is the spectrum of actions on the triple (X,Y,Z) of the subtheory of $W(B \otimes L_2)$ generated by the faces $d_0W(B \otimes L_2)$ and $d_2W(B \otimes L_2)$. It is proved in ([2],10.2) that there is a retraction functor $W(B \otimes L_2) \longrightarrow C$, and this is unique up to homotopy. The composite functor

$$W(B \otimes L_1) \approx d_1W(B \otimes L_2) \subset W(B \otimes L_2) \longrightarrow C$$

induces a map of spectra $C_M(X,Y,Z) \longrightarrow RHom_M(X,Z)$, which is also well-defined up to homotopy. By composing this with the canonical product described above, we obtain the promised pairing

$$RHom_M(Y,Z) \wedge RHom_M(X,Y) \longrightarrow RHom_M(X,Z) . \tag{4.2}$$

Let $C(n)$ be the subcategory of $W(B \otimes L_n)$ generated by the n principal edges $d_0d_1..d_{i-1}d_{i+2}..d_nW(B \otimes L_n)$ of $W(B \otimes L_n)$. Each of these edges is isomorphic to $W(B \otimes L_1)$. Let D_n be the space of retraction functors $W(B \otimes L_n) \longrightarrow C(n)$. As in the case $n = 2$ above, every point of D_n determines an n-fold composition product, by restriction to the edge $d_1d_2...d_{n-1}W(B \otimes L_n)$. There are evident structural maps which make the spaces D_n into an operad D which acts on the spectra $RHom_M(X,Y)$ through compositions.

We still have to show that D is an A_∞ operad: that is, D_n is contractible for every n. The proof is like that of ([2],10.2). The inclusion $C(n) \subset W(B \otimes L_n)$ is a homotopy equivalence, so a natural extension of the uniqueness part of the lifting theorem shows that the space D_n of retraction functors is contractible.

(ii) When M is the sphere spectrum S, we have an equivalence $RHom_S(X,Y) \longrightarrow Map(X,Y)$ given by restriction to unary module operations. At the level of unary operations, all our compositions are simply composition of maps of spectra.

(iii) When the module spectrum Z runs over an exact triangle, the product (4.2) is a morphism of induced exact triangles. In particular, it preserves the homotopy boundary. For Eilenberg-Mac Lane spectra, this means that the induced product on the homotopy groups $\pi_r RHom_{HA}(HP,HQ) \approx Ext_A^{-r}(P,Q)$ satisfies $\partial(ab) = (\partial a)b$, where ∂ is the connecting homomorphism for Ext arising from a short exact sequence $0 \longrightarrow Q' \longrightarrow Q \longrightarrow Q'' \longrightarrow 0$. Similarly $\partial(ab) = (-1)^{|a|}a(\partial b)$ when ∂ arises from a short exact sequence in the variable P. These properties, together with the fact that the composition in $Ext_A^0(P,Q)$ is the usual composition in $Hom(P,Q)$, characterize the Yoneda product [21]. □

4.2 <u>Corollary</u> There is a graded additive category $\mathcal{D}(M)$ in which the objects are the A_∞ right M-modules, and the morphisms from X to Y form the graded abelian group $\pi_* RHom_M(X,Y)$. This is the <u>extraordinary derived category</u> associated with the A_∞ ring spectrum M (see [17]).

We recall from [15] that there is a spectral sequence $\{E^r(X,Y)\}$ for each pair of right M-module spectra X,Y, such that

$$E^2_{s,t}(X,Y) \approx Ext^{-s,-t}_{\pi_* M}(\pi_* X, \pi_* Y) \Longrightarrow \pi_{s+t} RHom_M(X,Y) \quad .$$

Pursuing the analogy with the derived category, we call this the <u>hypercohomology spectral sequence</u>.

4.3 <u>Theorem</u> The above spectral sequence admits associative composition products

$$E^r_{s,t}(Y,Z) \otimes E^r_{s',t'}(X,Y) \longrightarrow E^r_{s+s',t+t'}(X,Z)$$

such that the product in E^2 is the Yoneda product in Ext, and the product in E^∞ is induced by the composition products in RHom.

<u>Proof</u> In the construction of the hypercohomology spectral sequence, $RHom_M(X,Y)$ is filtered according to the degree by which representative maps increase Cartan filtration. This degree is preserved by the composition in

4.1(i): the degree of a composite is at least the sum of the degrees of the components. Therefore there is an induced pairing of spectral sequences in which each differential d^r is a derivation, the pairing on E^r induces that on E^{r+1}, and the pairing on E^∞ is compatible with the composition pairing of homotopy groups induced by the products in RHom .

By construction of the spectral sequence, the pairing in the E^2 term is induced by the composition of the graded spectra associated with the filtration of RHom . According to ([15],§4) the associated graded spectrum of $RHom_M(X,Y)$ is $RHom_{Gr\,M}(Gr\,X, Gr\,Y)$, where $Gr\,M$, $Gr\,X$ and $Gr\,Y$ are the graded spectra associated with the Cartan filtrations on M, X and Y . The identification of the E^2 term with Ext arises from the fact that $Gr\,M$, $Gr\,X$ and $Gr\,Y$ are generalized Eilenberg-MacLane spectra. The induced composition at the level of graded spectra is the pairing of 4.1 for generalized Eilenberg-MacLane spectra. On homotopy groups (that is, on the E^2 term of our spectral sequence) this induces the Yoneda product, by 4.1(iii). □

When M is an E_∞ ring spectrum, there is an M-module structure on $RHom_M(X,Y)$ arising from the bimodule structure on X (or Y). The hypercohomology spectral sequence is a spectral sequence of modules over the commutative ring $\pi_* M$, and the pairings of 4.3 are $\pi_* M$-bilinear.

5 Ring spectra of derived endomorphisms

If X is an A_∞ right (or left) module over the A_∞ ring spectrum M , we define the <u>spectrum of derived</u> M-<u>endomorphisms</u> of X to be

$$REnd_M(X) = RHom_M(X,X) \quad .$$

By 4.1, this is another A_∞ ring spectrum.

We consider the case where M is MU . This is known to be an E_∞ ring spectrum [8]. (It is conjectured that this is also true of BP.)

Let S be a family of closed U-manifolds. There is an associated Sullivan-Baas homology theory $MU(S)_*(-)$ defined in terms of bordism of U-manifolds with singularities from S . Its representing spectrum $MU(S)$

is an MU-module. Suppose from now on that the bordism classes of the elements of S form a regular sequence in MU_* . Then the coefficient group $MU(S)_*$ is isomorphic to the factor ring of MU_* by the ideal I generated by the set S . We shall denote the spectrum

$$REnd_{MU}(MU(S)) = RHom_{MU}(MU(S),MU(S))$$

by R(S) , for the sake of brevity.

5.1 <u>Proposition</u> The homotopy ring of R(S) is an exterior algebra

$$\pi_* R(S) \approx \Lambda_{MU(S)_*}(\tau_1,\tau_2,\ldots,\tau_n,\ldots)$$

over the ring $MU(S)_* \approx MU_*/I$. This algebra has one generator $\tau_i \in \pi_{-2a_i-1}(R(S))$ for every class $[S_i] \in \pi_{2a_i} MU$ in the regular sequence S.

<u>Proof</u> We consider the hypercohomology spectral sequence

$$E^2_{s,t} \approx Ext^{-s,-t}_{\pi_* MU}(\pi_* MU(S),\pi_* MU(S)) \Longrightarrow \pi_{s+t} R(S) .$$

Since the sequence S is regular, the Koszul resolution shows that the E^2 term is an exterior algebra on generators $u_i \in E^2_{-1,-2a_i}$, one for each S_i in S , with $2a_i = \dim S_i$.

For each S_i we have a corresponding Sullivan-Bockstein operation

$$\partial_i : MU(S)_r(-) \longrightarrow MU(S)_{r-2a_i-1}(-)$$

(see [1]), which represents an element of $\pi_{-2a_i-1}Map(MU(S),MU(S))$. The geometric description of ∂_i shows that it is linear at the chain level. It follows that ∂_i lifts in a standard way to an element $\tau_i \in \pi_{-2a_i-1}R(S)$, with $\tau_i\tau_j = -\tau_j\tau_i$ and $\tau_i^2 = 0$. Chasing the definitions shows that τ_i is represented in the spectral sequence by u_i . Therefore u_i survives to E^∞. The MU_*-algebra structure now shows that the spectral sequence collapses.

To complete the proof, we check the convergence of the spectral sequence. When S is finite, the filtration is finite and convergence is automatic. Suppose S is an infinite regular sequence $\{S_1,S_2,\ldots,S_n,\ldots\}$. Let S(n) be the truncated sequence $\{S_1,\ldots,S_n\}$. Then MU(S) is the cofibred limit of the spectra MU(S(n)) . The hypercohomology spectral sequences for $\pi_* RHom_{MU}(MU(S(n)),MU(S))$ converge automatically, and these homotopy groups map epimorphically as n is decreased. Hence

$$\pi_*R(S) \approx \lim_n \pi_* RHom_{MU}(MU(S(n)), MU(S)) \quad ,$$

the Rlim term being zero. It follows that the spectral sequence for $\pi_*R(S)$ is also convergent. □

It has been shown by a number of workers that the spectra $MU(S)$ admit product maps [9,10,18,19]. There are limits to what can be achieved in this direction. For example, $P(n)$ can not be an E_∞ ring spectrum because of Nishida's theorem ([11], 3.4; see also [5], III 4.1) ; it is not even homotopy-commutative for the prime 2 [7,12,20] . It is not known whether $MU(S)$ is an A_∞ ring spectrum.

We propose $R(S)$ as a substitute for $MU(S)$ in this respect. Its multiplication is A_∞ , but it has the disadvantage of a slightly larger coefficient ring.

5.2 <u>Proposition</u> Let S be a regular sequence in π_*MU . Then the spectra $MU(S)$ and $R(S)$ belong to the same Bousfield class.

<u>Proof</u> First, $MU(S)$ is a module spectrum over $R(S)$, by evaluation. Therefore every $R(S)$-acyclic space is $MU(S)$-acyclic: that is, in the notation of [4], $\langle MU(S)\rangle \leqslant \langle R(S)\rangle$ ([13], 1.24).

We need to prove the converse. Let X be any space. There is a natural map

$$R(S)\wedge X = RHom_{MU}(MU(S),MU(S))\wedge X \longrightarrow RHom_{MU}(MU(S),MU(S)\wedge X)$$

which is an equivalence when X is a sphere. Since both source and target are exact functors of X, it is an equivalence for all X .

Suppose X is $MU(S)$-acyclic: that is, $\pi_*(MU(S)\wedge X) \approx 0$. By the hypercohomology spectral sequence, $\pi_*RHom_{MU}(MU(S),MU(S)\wedge X) \approx 0$. The equivalence above now shows that $\pi_*(R(S)\wedge X) \approx 0$, i.e. X is $R(S)$-acyclic. Hence $\langle R(S)\rangle \leqslant \langle MU(S)\rangle$. □

REFERENCES

1. N.A. Baas, On bordism theory of manifolds with singularities, Math. Scand. 33 (1973) 279–302.

2. J.M. Boardman, Homotopy structures and the language of trees, AMS Proc. Symp. Pure Math. 22 (1971) 37–58.

3. J.M. Boardman and R.M. Vogt, Homotopy invariant algebraic structures on topological spaces, Springer Lecture Notes in Mathematics 347 (1973).

4. A.K.Bousfield, The Boolean algebra of spectra, Comment. Math. Helv. 54 (1979) 368–377.

5. R.R. Bruner, J.P. May, J.E. McClure and R. Steinberger, H_∞ ring spectra and their applications, Springer Lecture Notes in Mathematics 1176 (1986).

6. A. Elmendorf, The grassmannian geometry of spectra (to appear).

7. D.C. Johnson and W.S. Wilson, BP operations and Morava's extraordinary K-theories, Math. Z. 144 (1975) 55–75.

8. L.G. Lewis, J.P. May, J.E. McClure and M. Steinberger, Equivariant stable homotopy theory, Springer Lecture Notes in Mathematics 1213 (1986).

9. O.K. Mironov, Multiplications in cobordism theories with singularities, and Steenrod – tom Dieck operations, Izv. Akad. Nauk SSSR Ser. Mat. 42 (1978) 789–806 = Math. of USSR - Izvestiya 13 (1979) 89–106.

10. J. Morava, A product for the odd-primary bordism of manifolds with singularities, Topology 18 (1979) 177–186.

11. G. Nishida, The nilpotency of elements of the stable homotopy groups of spheres, Manifolds – Tokyo 1973, 285–289.

12. A.V. Pazhitnov, Yu.B. Rudyak, On commutative ring spectra of characteristic 2, Mat. Sbornik 124(166) 4(8) (1984) 486–494 = Math. of USSR – Sbornik 52 (2) (1985) 471–479.

13. D.C. Ravenel, Localization with respect to certain periodic homology theories, Amer. J. Math. 106 (1984) 351–414.

14. C.A. Robinson, Derived tensor products in stable homotopy theory, Topology 22 (1983) 1–18.

15. C.A. Robinson, Spectra of derived module homomorphisms, Math. Proc. Camb. Phil. Soc. (to appear).

16. C.A. Robinson, Spectral sheaves: a model category for stable homotopy theory, J. Pure and Applied Algebra (to appear).

17. C.A. Robinson, The extraordinary derived category (to appear).

18. N. Shimada and N. Yagita, Multiplications in the complex bordism theory with singularities, Publ. Res. Inst. Math. Sci. Kyoto Univ. 12 (1976) 259–293.

19. U. Würgler, On products in a family of cohomology theories associated to the invariant prime ideals of $\pi_* BP$, Comment. Math. Helv. 52 (1977) 457–481.

20. U. Würgler, Commutative ring-spectra of characteristic 2, Comment. Math. Helv. 61 (1986) 33–45.

21. N. Yoneda, On the homology theory of modules, J. Fac. Sci. Univ. Tokyo Sect. I, 7 (1954) 193–227.

Mathematics Institute, Univ. of Warwick.

CONVEXITY AND ROOT CLOSURE IN NEGATIVELY CURVED MANIFOLDS

Christopher W. Stark

I. A root-closure theorem

A subgroup H of a group G is said to be r-th root closed in G if and only if every nontrivial element g of G such that g^r is an element of H must itself belong to H. Splitting problems in manifold topology lead to root-closure questions, after the seminal analysis of the surgery of splitting problems by Sylvain Cappell [4,5,6]. Recall that the setting for this discussion has as data a pair (M, N) and a homotopy equivalence $f: X \to M$; here M is a closed, connected n-manifold, X is another closed n-manifold, and N is a connected, codimension-one, two-sided submanifold of M. The homotopy equivalence f is said to split along N if f can be deformed to another homotopy equivalence g which is transverse to N (in the sense appropriate to the category of manifold structures under consideration) and such that g is a homotopy equivalence of pairs $(X, g^{-1}(N)) \to (M, N)$. A suitably relativized version is considered if M has nonempty boundary. The analysis of such a splitting problem is more tractable if we make the conventional assumption that the inclusion of N into M induces a monomorphism on the fundamental groups, so the fundamental group of M carries a graph of groups structure with the fundamental group of N as amalgamating subgroup. Cappell's splitting theorems show that the L-theoretic contribution to the splitting obstructions will vanish in this setting if the fundamental group of N is a square-root closed subgroup of the fundamental group of M. We presently have no other convenient vanishing criterion for these obstructions. This paper explores the root-closure problem for certain subgroups of fundamental groups of negatively-curved manifolds.

Let X^n be a complete, simply connected Riemannian manifold of sectional curvature $K \leq -a < 0$. The isometries of X are classified as elliptic, parabolic, or hyperbolic, in analogy with the familiar classification of isometries of hyperbolic h-space, according to the properties of the displacement function $D: G \times X \to \mathbf{R}$, $D(g, x) := \mathrm{dist}(x, gx)$. If $D(g,—)$ realizes a minimum of zero then g is elliptic; if

D(g,—) realizes a positive minimum then g is hyperbolic; and if D(g,—) has an unrealized infimum of zero then g is parabolic. In general, as in the familiar setting of $K = -1$, if a torsion-free group of isometries acts properly discontinuously on X, then it contains no elliptic elements. Such a manifold X compactifies by adding a sphere $S(\infty)$ at infinity whose points are asymptotic classes of geodesic rays [7,2], and isometries of X extend to homeomorphisms of $X \cup S(\infty)$, which is topologized to be homeomorphic to the closed unit n-disk. The classification of isometries via translation length is equivalent to a classification in terms of fixed point behavior at infinity. In particular, the hyperbolic isometries are those with exactly two fixed points in $S(\infty)$, and hyperbolics preserve the unique geodesic (the "axis" of the isometry) joining these two fixed points.

In any closed manifold N a variational argument shows that each free homotopy class of loops contains at least one smooth closed geodesic (i.e. a geodesic $c: [0,1] \to N$ which is smooth everywhere and has $c(0) = c(1)$ and $c'(0) = c'(1)$). In manifolds of strictly negative curvature (not assumed to be compact), the visibility property gives a better result — let $p: X \to N$ be the universal cover and lift arbitrarily long iterates of a closed loop based at $z \in M$ to X, basing the lifts at $x \in p^{-1}(z)$, and take the geodesic joining the two points at infinity thus determined we can prove versions of familiar facts from hyperbolic geometry.

Fact 1 Let M^n be a connected, complete Riemannian manifold of sectional curvature $K \leq -a < 0$. If $g \in \pi_1(M)$ is nontrivial and nonparabolic (when viewed as an isometry of the universal cover of M) then there is a unique smooth closed geodesic in the free homotopy class of loops determined by g.

Fact 2 Let X be a complete, simply connected, Riemannian n-manifold of sectional curvature $K \leq -a < 0$ and let G be a discrete group of isometries of X containing no elliptic elements. If G has nontrivial center then the elements of $G - \{1\}$ have common fixed point set in $S(\infty)$, i.e. they are either parabolic with a common fixed point or hyperbolic with a common axis. In the hyperbolic case G must be infinite cyclic.

We will be concentrating on hyperbolic elements of G below. This restriction is motivated partly by the nice geometry available in the hyperbolic case from the two facts above, but is also necessary for decent root closure results in general. Recall that any Bieberbach group (a fundamental group of a complete, Riemannian flat manifold) can appear as a discrete, parabolic group of isometries in an appropriate hyperbolic space; well-known examples of submanifolds of flat manifolds with non-square root closed fundamental groups [8] lead to parabolic groups of isometries of hyperbolic spaces which have the same bad features.

If M is a complete, connected Riemannian manifold of sectional curvature $K < 0$, let $p: X \to M$ be the universal cover of M, let $S(\infty)$ be the sphere at infinity of X, and let $X' := X \cup S(\infty)$ be the Eberlein-O'Neill compactification of X.

A submanifold N (perhaps with boundary) of a Riemannian manifold M is convex if N is connected and if any path in N can be deformed within N to a geodesic by a homotopy fixing endpoints. (We do not assume that the boundary of N is smooth — in particular, these remarks apply to the convex hulls made familiar by Thurston's notes on three-manifolds [11].) For example, any totally geodesic submanifold of a negatively curved M is convex in this sense. If M is a complete, connected Riemannian manifold of sectional curvature $K < 0$ and if N is a convex submanifold of M then each component of $p^{-1}(N)$ in X is a convex subset of X, implying that the inclusion of N into M induces a monomorphism $\pi_1(N) \to \pi_1(M)$. Pick a component Y of $p^{-1}(N)$ and consider its closure Y' in X': this is a convex subset of X'.

Lemma 3 Let M be a complete, connected Riemannian manifold of sectional curvature $K \leq -a < 0$ and suppose that N is a convex submanifold of M which is a closed subset of M. If τ is a loop in N, based at z, which represents a nonparabolic element of $\pi_1(M,z)$ then the smooth closed geodesic in M which is freely homotopic to τ also lies in N. (The homotopy from the loop τ to the smooth closed geodesic may be taken in N as well.)

The lemma is proved by considering the proof of Fact 1. The endpoints of the lift to X of our infinitely iterated loop lie in the intersection of the closure $cl_{X'}(Y)$ with $S(\infty)$, so the infinitely iterated lift of our geodesic also lies in Y and the smooth closed geodesic lies in N. The lemma suggests the following construction. Let Scg(M) denote the set of constant-speed, smooth closed geodesics in M, where we agree to parametrize geodesics over $I = [0,1]$ and we agree to identify a smooth closed geodesic with its time-translates, so if α is a constant then we identify $t \to c(t)$ with $t \to c(t + \alpha)$. We distinguish between a smooth closed geodesic and its iterates in both positive and negative directions and we write c^m for the m-fold iterate of the geodesic c. In particular, c^{-1} denotes the time-reversed version of c.

Lemma 4 (a) If M is a connected, complete Riemannian manifold of sectional curvature $K \leq -a < 0$ then free homotopy defines a unique function

$$F: \text{hyperbolics}(\pi_1(M, z)) \to \text{Scg}(M)$$

so that a loop c based at z is freely homotopic to F([c]).

(b) If N is a convex, connected submanifold of M (perhaps with boundary) which is a closed subset of M and if $i: N \to M$ is the inclusion map then for each $[c] \in i_\#(\pi_1(N, z))$, $F([c]) \in \text{Scg}(N)$.

Scg(M) is not a group and the function F above does not preserve products, but in some ways F behaves well.

Lemma 5 Let M be a connected, complete Riemannian manifold of sectional curvature $K \leq -a < 0$ and let N be a convex submanifold of M which is a closed subset of M. Let $F: \text{hyperbolics}(\pi_1(N, z)) \to \text{Scg}(N)$ be the function described above. Then F(g) is a k-th iterate in Scg(N) if and only if g is a k-th power in $\pi_1(N, z)$.

Proof If $g = h^k$ in the fundamental group of N then let $J = F(h) \in \text{Scg}(N)$ and let τ be a loop based at z representing h, so τ is freely homotopic in N to J, implying that $\tau * \tau * \cdots * \tau$ (k times) is freely homotopic in N to J^k. Since $g = h^k = [\tau * \tau * \cdots * \tau]$ (k times), this means that $J^k = F(h^k)$.

Conversely, if $F(g) = L = J^k$, then L is freely homotopic in N to a loop τ based at z that represents g, implying that there is a path θ in N from z to a basepoint y on L so that τ is homotopic (rel z) to the product $\theta * L * \theta^{-1}$. But $\theta * L * \theta^{-1}$ is homotopic (rel z) in N to

$$(\theta * J * \theta^{-1}) * (\theta * J * \theta^{-1}) * \cdots * (\theta * J * \theta^{-1})$$

(k times), so

$$g = [\theta^* L^* \theta^{-1}] = [\theta^* J^* \theta^{-1}]^k$$

and g is seen to be a k-th power.

(Take N = M in the Lemma to obtain the corresponding statement for M.)

Theorem 6 Let $M^n = X/G$ have as its universal cover the complete, simply connected Riemannian manifold X^n of sectional curvature $K \leq -a < 0$, so $G = \pi_1(M)$ acts on X as a discrete and torsion-free group of isometries. If N is a convex manifold of M which is closed as a subset of M then $H := \pi_1(N)$ is root-closed in G, except possibly for parabolic elements.

Proof Given an element g of G such that $g^k \varepsilon H$ for some non-zero integer k, we first show that there is an element h of H such that $g^k = h^k$. Since N is convex and closed in M the free homotopy function F: hyperbolics(G) → Scg(M) restricts to H and N, producing a commutative diagram

$$\begin{array}{ccc} \text{hyperbolics}(H) & \rightarrow & \text{Scg}(N) \\ \downarrow & & \downarrow \\ \text{hyperbolics}(G) & \rightarrow & \text{Scg}(M) \end{array}$$

Apply Lemma 5 to conclude that the preimage

$$(F|H)^{-1}(F(g^k))$$

consists entirely of k-th powers in H. In particular, since g^k lies in this preimage, there is an element h of H such that $g^k = h^k$.

Now consider the subgroup K of G generated by the elements g and h. K has non-trivial center, and if g is not parabolic then Fact 2 shows that K is infinite cyclic. In an infinite cyclic group our relation implies that g = h, and thus g lies in the subgroup H.

The theorem implies that any Fuchsian subgroup H of a Kleinian group G for which the quotient by H of the totally geodesic plane spanning the limit set of H is imbedded in $\mathbf{R}\, H^3/G$, the quotient of hyperbolic 3-space by G, is root-closed in G. Millson's proof that many members of the Borel family of arithmetically defined closed hyperbolic manifolds have positive first Betti number gives higher-dimensional analogs to this observation on Fuchsian subgroups of Kleinian groups, since the codimension-one submanifolds produced by Millson are totally geodesic [10,3]. The recent work of Farrell and Jones [9] on K-theoretic vanishing theorems for negatively curved manifolds then shows that in both these categories of examples all of the K- and L-theoretic obstructions to splitting homotopy equivalences along these totally geodesic submanifolds will vanish.

II. Remarks on quasi-Fuchsian and related groups

H is said to be a quasi-Fuchsian group of isometries of X^n if the limit set L(H) of H is an imbedded (n-2)-sphere in $S(\infty)$. Such groups are most familiar in the classical dimension, n = 3, but have been considered in higher dimensions as well [1]. The observations on Fuchsian subgroups of Kleinian groups ending the first section lead one to ask if these considerations might give root-closure information for quasi-Fuchsian subgroups H of a torsion-free, discrete group G of isometries of a negatively curved manifold X^n as above. More generally, one might consider a subgroup H of G such that L(H) is merely a proper subset of L(G). Consider the convex hull Y = Hull(L(H)) in X of L(H), following [11]. If the hull quotient Hull(L(H))/H imbeds in X/G then we are in the situation of Section I, since the subgroup H of G is carried by a convex submanifold of X/G, so the root closure problem is evidently related to the self-intersection properties of the map Hull(L(H))/H $\to$ X/G (which is the restriction of the quotient map X/H $\to$ X/G). In particular, if a hyperbolic element h of H has an r-th root g in G - H then the axes of g and h agree and this axis lies in Y, covering the smooth closed geodesic F(g), which in turn lies within the image of Y in M = X/G. Consider the effects of g and its powers on Y: we know that $g^r Y = hY = Y$, but gY might or might not coincide with Y, although Y and gY meet at least in Axis(g). Observe also that gY = Y if and only if gL(H) = L(H).

Lemma 7 Let X and G be as usual and suppose that H is a quasi-Fuchsian subgroup of G which is the fundamental group of a two-sided submanifold N^{n-1} of M^n = X/G such that

(a) a copy of the universal cover N' of N imbedded in X over N has the property that its closure in $X \cup S(\infty)$ meets $S(\infty)$ only in L(H), and

(b) N' separates the two components of $S(\infty) - L(H)$.

[Recall that $L(H)$ is an $(n - 2)$-sphere imbedded in the $(n - 1)$-sphere at infinity, so the "domain of discontinuity" $S(\infty) - L(H)$ has exactly two components.]

Then $H = H' := \{g \in G: gL(H) = L(H)$ and g preserves the components of $S(\infty) - L(H)\}$.

Note that hypothesis (a) is satisfied if N is compact and that both assumptions are tameness conditions on the imbedding of N' in X. If H is a geometrically finite group of isometries without parabolics then the quotient $(X \cup (S(\infty) - L(H)))/H$ is compact and hypothesis (b) is satisfied. The lemma is proved by an intersection theory argument: take an arc A running from one boundary component of $(X \cup (S(\infty) - L(H)))/H'$ to the other and meeting N transversely in a single point, and transfer this intersection picture via the covering $f: (X \cup (S(\infty) - L(H)))/H \to (X \cup (S(\infty) - L(H)))/H'$ to get $1 = \#(N \cap A) = \#(f^{-1}(N) \cap A')$, where A' is any lift to $(X \cup (S(\infty) - L(H)))/H$ of the arc A. Since the latter intersection set is in bijection with H'/H we conclude that $\#(H'/H) = 1$ and $H' = H$.

Let $H'' := \{g \in G: gL(H) = L(H)\}$, so $H < H' < H''$, and H''/H' has order one or two. The lemma implies that if H and G are as hypothesized then any root g which lies in H'' must either belong to H or be a square root which acts on $X \cup S(\infty)$ preserving $L(H)$ but interchanging the two components of $S(\infty) - L(H)$. Now consider all the ways in which $q: Y/H \to X/G$ might fail to be an imbedding:

(a) The overlap set $S := \{p \in q(Y/H) : \#(q^{-1}(p)) > 1\}$ might contain no smooth closed geodesics;

(b) A hyperbolic element g of G might have Axis(g) contained in Y even though no power of g is an element of H;

(c) A hyperbolic element h of H might have Axis(h) contained in $Y \cap gY$ for some $g \in G$ (with $g \in G - H$) even though h has no roots in $G - H$; or

(d) A hyperbolic element h of H might have an r-th root g in $G - H$ (the case considered above).

Observe first that (c) is a red herring, since the action of G on X permutes the set of axes of hyperbolic elements of G, so in case (c) we know that if Axis(h) is contained in $Y \cap gY$ then there is a hyperbolic element a of G such that gAxis(a) = Axis(h), i.e. Axis(h) = Axis(gag^{-1}), so we are in either case (b) or case (d).

Proposition 8 Let M = X^n/G be a complete Riemannian manifold of sectional curvature K ≤ —a < 0 and let H be a subgroup of G such that

(1) the limit sets L(G) and L(H) are unequal,

(2) H = H'', i.e. H is the maximal subgroup of G preserving L(H),

(3) Hull(L(H))/H is compact, and

(4) dim Hull(H) = n or Hull(H) is a subset of a totally geodesic k-plane in X with nonempty interior in the subspace topology for that k-plane.

Alternative (b) above is then impossible.

The proposition is proved by considering a compact fundamental domain W for the hull quotient Hull(L(H))/H and a maximal set of geometrically independent points w_0, ..., w_k within W, where by "geometrically independent" we mean that the geodesic simplex spanned by these points is nondegenerate. Pick a basepoint v on Axis(g) and take a ball centered at v whose diameter is twice the diameter of W. Any translate g^rY of Y includes a translate of W containing v, and the proper discontinuity of the action of G on X implies that only a finite number of these translates can meet our ball centered at v. By specifying a fundamental domain plus boundary identifications for an action (zHz^{-1} , zY) we completely specify the translated hull zY and the action of zHz^{-1} on that translated hull, so the infinite cyclic subgroup of G generated by g produces a finite set of hull translates $\{g^rY : r \in \mathbf{Z}\}$. This implies there is an integer s such that $g^sY = Y$, so $g^s \in H$.

The proposition gives a geometric correlate for the failure of root closure, in the sense that the overlap set for the convex hull quotient contains a smooth closed geodesic if and only if the conjugacy class corresponding to that loop gives root closure trouble.

References

[1] M. T. Anderson, Complete minimal hypersurfaces in hyperbolic n-manifolds, Comment. Math. Helvetici 58 (1983), 264-290.

[2] W. Ballmann, M. Gromov, V. Schroeder, Manifolds of Nonpositive Curvature, Birkhauser (1985).

[3] A. Borel, Compact Clifford-Klein forms of symmetric spaces, Topology 2 (1963), 111-122.

[4] S. E. Cappell, Unitary nilpotent groups and Hermitian K-theory. I, Bull. Amer. Math. Soc. 80 (1974), 1117-1122.

[5] S. E. Cappell, Manifolds with fundamental group a generalized free product. I, Bull. Amer. Math. Soc. 80 (1974), 1193-1198.

[6] S. E. Cappell, A splitting theorem for manifolds, Invent. Math. 33 (1976), 69-170.

[7] P. Eberlein and B. O'Neill, Visibility manifolds, Pacific J. Math. 46 (1973), 45-109.

[8] F. T. Farrell and W. C. Hsiang, The topological-Euclidean space form problem, Invent. Math. 45 (1978), 181-192.

[9] F. T. Farrell and L. E. Jones, Algebraic K-theory of hyperbolic manifolds, Bull. Amer. Math. Soc. (N. S.) 14 (1986), 115-119.

[10] J. J. Millson, On the first Betti number of a constant negatively curved manifold, Ann. of Math. 104 (1976), 235-247.

[11] W. P. Thurston, The geometry and topology of three-manifolds. Princeton University lecture notes (1978).

University of Florida
Department of Mathematics
Gainesville, Florida 32611

COHOMOLOGY OF FINITE GROUPS AND BROWN-PETERSON COHOMOLOGY

M. Tezuka and N. Yagita

§1. Introduction.

Let G be a finite group and $H^*(G)_{(p)}$ be the cohomology of G with the coefficient $Z_{(p)}$ for a prime number p. Since the restriction map to a Sylow p-group P of G is injective, it is important to know the cohomology of p-groups. However it seems a very difficult problem to compute $H^*(P)$ when P is a non abelian group. The smallest non abelian p-groups P have the order p^3, which have two types E and M (, for $p=2$ D and Q ; the dihedral and the quaternion groups).

The cohomology $H^*(P)$, $P=Q,D,M,E$ are determined by Atiyah, Evens, Lewis, Lewis respectively [A], [E], [L]. These cohomology rings are complicated, and there appear no relations among them. For example, $H^{even}(P)$ is generated as a ring by $p+1$ generators, however $H^{odd}(P)$ is generated as an $H^{even}(P)$-module by 0 (resp. 1, 1, 2) generators for $P=Q$ (resp. $=D,M,E$).

In this paper, we explain the above facts by using the Brown-Peterson cohomology. Recall that $BP^*(-)$ is the Brown-Peterson cohomology with the coefficient $BP^*=Z_{(p)}[v_1,v_2,\ldots]$, $v_i=-2(p^i-1)$. Let us write $BP^*(BG^+)=BP^*(G)$ simply. Suppose P is a abelian p-group of $|P|=p^3$, i.e., $P=Q,D,M$ or E.
We show that $BP^*(P)$ is more accessible than $H^*(P)$, and $H^*(P)$ is deduced from $BP^*(P)$,

(1.1) $BP^*(P)\otimes_{BP^*}Z_{(p)} \cong H^{even}(P)$

(1.2) $(\mathrm{Ker}\cdot v_n \text{ in } BP^*(P)\otimes_{BP^*}Z[v_n]) \cong S^{2p^n-1}H^{odd}(P)$

where S^{2p^n-1} is the suspension map of degree $2p^n-1$.

Moreover the Morava K-theory is given by

(1.3) $BP^*(P)\otimes_{BP^*}K(n)^* \cong K(n)^*(P)$

where $K(n)^*(-)$ is the cohomology theory with the coefficient $K(n)^* = Z/p[v_n,v_n^{-1}]$. These are the first examples for which $BP^*(G)$ or $K(n)^*(G)$ are explicitly known for non abelian p-groups G.

The outline of arguments are following. For each case of P, there is a central extension

$$1 \longrightarrow Z/p \longrightarrow P \longrightarrow Z/p \oplus Z/p \longrightarrow 1$$

which induces the Atiyah-Hirzebruch type spectral sequence

$$(1.4) \qquad E_2^{*,*} = H^*(Z/p \oplus Z/p; BP^*(Z/p)) \Longrightarrow BP^*(P).$$

The fact that $BP^*(Z/p)$ is even dimensional and p-torsion free, makes the computation easier than that of the spectral sequence convergencing $H^*(G)$. The only non zero differentials are d_3 and d_{2p-1}. Using this, we show

$$(1.5) \qquad BP^*(P) \cong BP[\![y_1, y_2, c_1, \ldots, c_p]\!]/(\text{relations})$$

where y_i is the first Chern class of some 1-dimensional representation and c_i are i-th Chern classes of the other representation of P. After changing adequate generators, the relations in (1.5) contain $c_1 + v_1 c_p = 0$ for $P = Q, D, E$ ($c_1 + v_1 c_p = y_1$ for $P = M$), which explains the number of generators in $H^{even}(P)$. The fact (Ker v_1 in $BP^*(Q) \otimes Z_{(p)}[v_1]$) $= 0$, induces $H^{odd}(Q) = 0$ and $v_1 c_p y_i = 0$ (resp. $v_1 c_p (y_2^p - y_1^{p-1} y_2) = 0$, $v_1 c_2 y_1 = 0$) mod $(v_2, v_3, \ldots)$ in $BP^*(E)$ (resp. $BP^*(M)$, $BP^*(D)$) induces the odd dimensional parts of the ordinary cohomology.

From (1.3) and (1.5) it is immediate $K(n)^*(P) = K(n)^{even}(P)$. Moreover all elements in $K(n)^*(P)$ are also deduced from Chern classes of some representations.

The p-th product $[p](y_i) = py_i + v_1 y_i^p + \ldots$ of the formal group law of BP^*-theory is zero in $BP^*(P)$ and this implies $py_i = 0$ in $H^*(P)$, which suggests the relation between $[p]$ and the exponent of $H^*(P)$. In §5, we give examples for which the exponent of G is small but that of $H^*(G)$ is large for large p by using BP^*-theory.

At last we note that (1.1) does not hold for all p-groups, for example $P = Z/p \oplus Z/p \oplus Z/p$. However, for all known examples, we have

$$(1.6) \qquad BP^*(G) \otimes_{BP^*} Z_{(p)} \cong Ch(G)^*_{(p)}$$

where $Ch(G)^*$ is the subring of $H^*(G)$ generated by Chern classes of representations of G (Thomas [T]). For $p \geq 3$, we give some examples such that

$$(1.7) \qquad BP^*(G) \otimes_{BP^*} Z_{(p)} \longrightarrow H^*(G)_{(p)}/\sqrt{0}$$

are epic where $\sqrt{0}$ is the ideal of nilpotent elements in $H^*(G)$ in §6.

The authors thank to D. Johnson, S. Wilson, J. Morava and

D. Ravenel for many conversations and suggestions. The authors also thank to the referee who point out errors in the first version.

§2. Hochschild-Serre spectral sequence.

Let $k^*(-)$ be a BP-oriented cohomology theory with coefficients the BP^*-module $k^*=Z_{(p)}[v_{j_1},v_{j_2},\ldots]$, $j_1=n<j_2<j_3 \ldots$. Write $k^*(BG_+)=k^*(G)$. Let P be a non abelian group of $|P|=p^3$. Then P is one of the following groups for $p\geq 3$;

$$E = \langle a,b,c \mid a^p=b^p=c^p=1,\ [a,b]=c,\ [a,c]=[b,c]=1 \rangle,$$

$$M = \langle a,b \mid a^{p^2}=b^p=1,\ [a,b]=a^p \rangle.$$

When $p=2$, $E=M$ and we denote it by D, and there is an another group Q;

$$Q = \langle a,b \mid a^4=b^4=1,\ [a,b]=a^2=b^2 \rangle.$$

For each group P, there is a central extension

$$(2.1)\qquad 1 \longrightarrow \langle c\rangle \xrightarrow{\ i\ } P \xrightarrow{\ p\ } \langle a_1,a_2\rangle \longrightarrow 1$$

where $i(c)=a^p$ for $P=Q,D,M$ and $p(a)=a_1$, $p(b)=a_2$. This induces the spectral sequence for $k^*(P)$

$$(2.2)\qquad E_2^{*,*} = H^*(\langle a_1,a_2\rangle;k^*(\langle c\rangle))$$

$$\cong \tilde{Z}/p[\![y_1,y_2]\!] \otimes \Lambda(\alpha) \otimes k^*[\![u]\!]/[p](u) \Longrightarrow k^*(P)$$

where $|y_i|=|u|=2$ and $|\alpha|=3$, $\tilde{Z}/p[a]=Z[a]/(pa)$, and $i(\alpha)=x_1y_2-x_2y_1$ for $i: H^*(\langle a_1,a_2\rangle;Z) \longrightarrow H^*(\langle a_1,a_2\rangle;Z/p) \cong Z/p[y_1,y_2] \otimes \Lambda(x_1,x_2)$, (when $p=2$, $x_i^2=y_i$), and $[p](u)=pu+v_{j_1}u^{p^{j_1}}+\ldots$ is the p-th product of the formal group law in $k^*(CP^\infty) \cong k^*[\![u]\!]$ (See Ravenel [R]).

It is known that the first differentials of the spectral sequence for $H^*(P;Z/p)$ are ([Q2],[T-Y-2])

$$d_2z = x_1x_2+\varepsilon_1y_1+\varepsilon_2y_2,\quad d_3u = \beta(x_1x_2) \quad (\text{when } p=2,\ Sq^1(x_1x_2))$$

here $H^*(\langle c\rangle;Z/p) \cong Z/p[u] \otimes \Lambda(z)$ and $\varepsilon_i=0$ or 1. By the naturality of the spectral sequence, we chose

$$(2.3)\qquad d_3u=\tilde{\alpha} \text{ with } \alpha=\lambda\tilde{\alpha} \bmod (v_{j_1},\ldots),\ \lambda\neq 0 \bmod p.$$

For $p\geq 3$ we can easily compute

$$(2.4)\qquad E_4^{*,*} \cong (k^*\{pu,\ldots,pu^{p-1}\} \oplus \tilde{k}^*/p[[y_1,y_2]]\{1,u^{p-1}\tilde{\alpha}\}) \otimes_{k^*} k^*[[u^p]]/([p](u))$$

where $A\{a_i\}$ means a free A-module generated by a_i.

Here we give the computation for $x \in E_2^{2n,*}$, $2n > 0$. The other cases are given by similar computations. Let K be the k^*-module generated by $y_1{}^i y_2{}^j u^{ps}$ in $E_2^{2n,*}$. We need to prove $K = \operatorname{Ker} d_2$. If $0 \neq x \notin K$, then we can take $b \in K$ such that

$$x - b = (a_1 u^s + a_2 u^{s+1} + \ldots) y_1{}^i y_2{}^j + c y_1{}^{i+1} y_2{}^{j-1} + \ldots$$

with $0 \neq a_1 \in k^*/p$ and $s \neq 0 \bmod p$, and moreover if $a_1 \in \text{Ideal } v_{j_1}$ then $s < p^{j_1}$ since $[P](u) = v_{j_1} u^{p^{j_1}} + \ldots = 0$ in $E_2^{2n,*}$. Then

$$dx = d(x-b) = ((s a_1 u^{s-1} + \ldots) y_1{}^i y_2{}^j + \ldots)\alpha \neq 0 \quad \text{in } E_2^{*,*},$$

since $0 \neq s a_1 \in k^*/p$ and $s-1 < p^{j_1}$ if $a_1 \in \text{Ideal } v_{j_1}$.

Consider restriction maps i^*, j^* and the transfer map $i_!$

$$k^*(\langle c \rangle) \xleftarrow{\;j^*\;} k^*(\langle a,c \rangle) \underset{i_!}{\overset{i^*}{\rightleftarrows}} k^*(P).$$

Applying the double coset formula, we get $j^* i^* i_!(u^i) = p u^i$. Hence $p u^i$ are permanent cycles. In the spectral sequence for $H^*(P;Z/p)$. the Kudo's transgression theorem says

$$d_{2p-1}(u^{p-1}(x_1 y_2 - x_2 y_1)) = \beta \mathcal{P}(x_1 y_2 - x_2 y_1) = y_1{}^p y_2 - y_1 y_2{}^p$$
$$(\text{when } p=2,\ \alpha^2 = (x_1 y_2 - x_2 y_1)^2)$$

and hence by the naturality

$$(2.5) \qquad d_{2p-1}(u^{p-1}\tilde{\alpha}) = y_1{}^p y_2 - y_1 y_2{}^p.$$

Therefore we get for all primes p

$$(2.6) \qquad E_{2p}{}^{*,*} \cong k^* \otimes (Z\{pu, \ldots, pu^{p-1}\} \oplus \tilde{Z}/p[\![y_1, y_2]\!]/(y_1{}^p y_2 - y_1 y_2{}^p)) \otimes Z[\![u^p]\!]/[p](u).$$

Since all non zero elements are even dimensional, we have ;

Theorem 2.7. The E_∞-term of the spectral sequence from (2.1) is

$$E_\infty^{*,*} \cong E_{2p}^{*,*} \cong (2.6).$$

From the above theorem, the natural map $\rho : BP^*(P) \longrightarrow k^*(P)$ is epic and $k^*(P) \cong BP^*(P) \otimes_{BP^*} k^*$. In particular, let $\tilde{k}(n)^* = Z_{(p)}[v_n]$. Then by the Sullivan-Bockstein exact sequence

$$\begin{array}{ccc} \tilde{k}(n)^*(P) & \xrightarrow{\;v_n\;} & \tilde{k}(n)^*(P) \\ {\scriptstyle\delta}\nwarrow & & \swarrow{\scriptstyle\rho} \\ & H^*(P) & \end{array}$$

we get the following theorem ;

Theorem 2.8.

(1) $H^{\text{even}}(P) \cong \tilde{k}(n)^*(P)/(v_n) \cong BP^*(P) \otimes_{BP^*} Z_{(p)}$

(2) $S^{2p^{n-1}} H^{odd}(P) \cong (v_n\text{-torsion in } BP^*(P) \otimes_{BP^*} Z_{(p)}[v_n])$.

§3. BP^*-module structure.

Consider the representation $\phi : A = \langle b,c\rangle \longrightarrow \mathbb{C}^\times$ defined by $\phi(c)=\exp(2\pi i/p)$ and $\phi(b)=1$ (, for $P=Q$, $\phi: A=\langle a\rangle \longrightarrow \mathbb{C}^\times$ defined by $\phi(a)=\exp(2\pi i/p^2)$). Let us write the induction $\mathrm{Ind}_A{}^G\phi : P \longrightarrow GL_p(C)$. Recall $BP^*(BGL_p(\mathbb{C})) \cong BP^*[[c_1,\ldots,c_p]]$ where c_i is the BP*-Chern class. Then

(3.1) $\bar{c}_i = (\mathrm{Ind}_A{}^G\phi)^*(c_i) \in BP^*(P)$

represents pu^i for $1\le i\le p-1$ and u^p for $i=p$ in $E_\infty^{*,*}$ (2.6) , since

$$j^*(\mathrm{Ind}_A{}^G\phi)^*(1+c_1+\ldots+c_p) = (1+u)^p \text{ for } j : \langle c\rangle \hookrightarrow P .$$

Hence from (2.6), $BP^*(P)$ is generated by $y_1,y_2,c_1,\ldots,c_p$.

First we note that the map $p^* : BP^*(\langle a_i\rangle) \longrightarrow BP^*(P)$ shows

(3.2) $[p](y_i) = py_i + v_1y_i^p + \ldots = 0$

also holds in $BP^*(P)$.

For $P=E,M,Q$, there is an automorphism of P such that

$$f_\lambda \begin{cases} a \longmapsto ab^\lambda \\ b \longmapsto b \\ c \longmapsto c \end{cases} \qquad f_\lambda^* \begin{cases} y_1 \longmapsto y_1 \\ y_2 \longmapsto y_2 +_{BP} [\lambda](y_1) \end{cases}$$

where $+_{BP}$ means the product of the formal group law of BP-theory and $[\lambda](y)$ is the λ-th product of the formal group law.

Recall the filtration F^s of $BP^*(P)$ with $F^s/F^{s+1} \cong E_\infty^{s,*}$. Since $y_1{}^p y_2 - y_1 y_2{}^p = 0$ in $E_\infty^{2p+2,\,0}$, there is $b \in F^{2p+4}$ such that $\Delta' = y_1{}^p y_2 - y_1 y_2{}^p - b$ is exactly zero in $BP^*(P)$. Note that $b=b'y_1$ since $\Delta'|\langle a,c\rangle=0$ and $b' \in \mathrm{Ideal}(v_{j_1},\ldots)$ by dimensional reason. The fact $f_\lambda^*\Delta'=0$ in $BP^*(P)$ implies

$$f_\lambda^*\Delta' \in \mathrm{Ideal}\Delta' \text{ in } BP^*/p[[y_1,y_2,u^p]]/([p](u)) .$$

The element Δ' contain y_1 as a product factor so contains $f_\lambda^*\Delta'$. Hence

$$\Delta' = (1+a')y_1 \prod_{0\le\lambda\le p-1} (y_2 +_{BP} [\lambda] y_1) \text{ for some } a' \in \mathrm{Ideal}(v_1,\ldots) .$$

We note that the topology of $BP^*(P)$ is given by positive valuation y_i, c_p . Hence $(1+a')$ is unit and we can take

(3.3) $\Delta = y_1\Pi(y_2 + {}_{BP}[\lambda]y_1)$

such that Δ is exactly zero in $BP^*(P)$.
For $P=D$, we consider the map f'_λ with $f'_\lambda(a)=a$, $f'_\lambda(b)=a^\lambda b$ and we get $\Delta=0$ also in $BP^*(D)$.

The element $c_1+v_1c_p$ is expressed in $E_\infty^{0,*}$

$$pu + v_1u^p = [p](u) \bmod J = (p^2, pv_1, v_1^2, v_2, \ldots)$$

which is zero in $E_\infty^{0,*}$. Hence there is $a_1 \in F^2$ such that $c_1+v_1c_p = a_1$ in $BP^*(P)$. Similarly in $E_\infty^{0,*}$ (2.6),

$$pu^i[p](u) = p^2u^{i+1}+pu^iv_1u^p$$

mod $J=(p^3, pv_1, v_1^2, v_2, \ldots)$. Hence in $BP^*(P)$

$$\bar{c}_1+v_1\bar{c}_p = a_1$$

$$p\bar{c}_2+v_1\bar{c}_1\bar{c}_p = a_2$$

(3.4) $$\vdots$$

$$p\bar{c}_{p-1}+v_1\bar{c}_{p-2}\bar{c}_p = a_{p-1}$$

$$p^2\bar{c}_p+v_1\bar{c}_{p-1}\bar{c}_p = a_p$$

where a_i depends on type of P and $a_i \in \mathrm{Ideal}(y_1,y_2,J)$.
Moreover $pu^iy_i, v_1u^py_i$, $(pu^i)(pu^i)-p^2u^{i+j}$ are zero in $E_\infty^{*,*}$, we get

(3.5) $\bar{c}_iy_j=d_{ij}$, $v_1\bar{c}_py_j=e_j$, $\bar{c}_i\bar{c}_j-p\bar{c}_{i+j}=f_{ij}$ for $i+j\neq p$

if $i+j=p$, $\bar{c}_i\bar{c}_j-p^2\bar{c}_p=f_{ij}$ here d_{ij}, e_j, $f_{ij} \in \mathrm{Ideal}(y_1^2, y_1y_2, y_2^2)$.

Theorem 3.6.

$$BP^*(P) \cong BP^*[\![y_1,y_2,c_1,\ldots,c_p]\!]/((3.2),(3.3),(3.4),(3.5)).$$

Now we compute a_i, e_i for each case of P and consider the ordinary cohomology $H^*(P)$, after adequate generators changing.

(i) $P=E$ case. For inclusion $i : \langle b,c\rangle \hookrightarrow E$, let us write the transfer $i_!(u^i)=\tilde{c}_i$, $1\le i\le p-1$. Hence $a_i|\langle a\rangle=a_i|\langle b\rangle=0$. Here $x|G$ means restriction to the group G, so $a_i \in \mathrm{Ideal}(y_1y_2)$. The element a_i is invariant under f_λ^*-action. By an argument similar to the proof of $\Delta=0$, we see $a_i \in \mathrm{Ideal}\,\Delta$, that is $a_i=0$ in $BP^*(E)$.

Corollary 3.8. (Lewis [L])

$$H^{even}(E) = (Z/p\{\tilde{c}_2,\ldots,\tilde{c}_{p-1}\} \otimes \tilde{Z}/p[y_1,y_2]) \otimes \tilde{Z}/p^2[\bar{c}_p]/(y_1^py_2-y_1y_2^p)$$

$$H^{odd}(E) = Z/p[y_1,y_2,\bar{c}_p]\{d_1,d_2\}/(y_1^py_2-y_1y_2^p), \quad |d_i|=3.$$

Proof. First note $\Delta=y_1^py_2-y_1y_2^p \bmod (v_1,v_2,\ldots)$. Since $a_i=0$, we see $\tilde{c}_1, p\tilde{c}_2, .., p\tilde{c}_{p-1}$ and $p^2\bar{c}_p$ are in the ideal$(v_1,...)$. This shows the even degree parts of the ordinary cohomology. Since $\tilde{c}_1+v_1\bar{c}_p=0$ and

$$\tilde{c}_1y_1 = i_!(ui^*(y_1)) = 0 ,$$

we have $v_1\bar{c}_py_1=0$. Similarly we have $v_1\bar{c}_py_2=0$. We can prove (Ker v_1 in $BP^*(E) \otimes_{BP^*} Z_{(p)}[v_1]$) is generated by $\bar{c}_py_1$ and $\bar{c}_py_2$. Hence we have the odd degree parts.

(ii) P=M cases. Consider the restrictions to the subgroups $\langle a\rangle$, $\langle ab^\lambda\rangle, \langle b\rangle$ and $y_1|\langle a\rangle=[p](u)$. We can easily see

$$\tilde{c}_1+v_1\bar{c}_p = y_1 \bmod J .$$

Since $\tilde{c}_1y_1=i_!(u[p](y_1))=0$, we get $v_1\bar{c}_py_1=y_1^2$. Moreover $v_1u^py_2=0$ in $E_\infty^{*,*}$ implies

$$v_1c_py_2 = \lambda y_1y_2+v_1y_1^py_2+... .$$

Thinking restriction to $\langle ab\rangle$ we have $\lambda\neq 0 \bmod P$. These facts mean that $\tilde{c}_1=y_1$, $y_1^2=0$, $y_1y_2=0$ in $H^*(M)$. On the other hand, (Ker v_1 in $BP^*(M) \otimes_{BP^*} Z_{(p)}[v_1]$) is generated by $\bar{c}_p(y_2^p-y_1^{p-1}y_2)$.

Corollary 3.9. (Lewis [L])

$$H^{even}(M) \cong (Z/p\{c_1=y_1,\tilde{c}_2,..,\tilde{c}_{p-1}\} \oplus \tilde{Z}/p[y_2]) \otimes \tilde{Z}/p^2[c_p]$$

$$H^{odd}(M) \cong Z/p[y_2,\bar{c}_p]\{e\} \qquad |e|=2p+1 .$$

(iii) P=D case. Considering restriction of y_1 , y_2 , c_1 to subgroups $\langle a\rangle$, $\langle b\rangle$ and $\langle ab\rangle$. Then we can prove

$$\tilde{c}_1+v_1\bar{c}_2 = 0 .$$

Since $\tilde{c}_1y_1=0$, we have $v_1\bar{c}_2y_1=0$. Since $v_1u^2y_2=0$ in E_∞ and by considering restriction to $\langle ab\rangle$, we get $v_1\bar{c}_2y_2=y_1y_2$. Moreover we have

$$4\bar{c}_2+v_1\bar{c}_2\tilde{c}_1 = y_1y_2 .$$

Corollary 3.10. (Evens [E])

$$H^{even}(D) \cong \tilde{Z}/2[y_1,y_2]/(y_1y_2) \otimes \tilde{Z}/4[\bar{c}_2]$$

$$H^{odd}(D) \cong Z/2[y_1,y_2,\bar{c}_2]/(y_1y_2)\{e\}, |e|=3 .$$

(iv) P=Q case. Let us write $\tilde{c}_1=i_!(u)$ for $j : \langle a\rangle \hookrightarrow Q$. Considering restrictions, we get $\tilde{c}_1+v_1\bar{c}_2=0$. Since $v_1u^2y_1=0$ in $E_\infty^{*,*}$, we get

$$v_1\bar{c}_2y_1 = y_1^2 \quad \text{and} \quad v_1\bar{c}_2y_2 = y_2^2 ,$$

considering restrictions to subgroups. We can also prove (Ker v_1 in $BP^*(Q)\otimes_{BP^*}Z_{(p)}[v_1])=0$ and

$$4\bar{c}_2+v_1\tilde{c}_1\bar{c}_2 = y_1^2+y_1y_2+y_2^2 .$$

These follow ;

Corollary 3.11. (Atiyah [A])

$$H^*(Q) \cong H^{even}(Q)$$
$$\cong \tilde{Z}/2[y_1,y_2]/(y_1^2,y_2^2)\otimes\tilde{Z}/8[\bar{c}_2]/(4c_2=y_1y_2) .$$

§4. Morava K-theory.

We recall the Morava K-theory. Let $k(n)^*=Z/p[v_n]$ and $K(n)^*=[v_n^{-1}]k(n)^*$. Let $P(n)^*=Z/p[v_n,v_{n+1},..]=BP^*/(p,v_1,..)$ and let $\tilde{k}(n)^*=Z_{(p)}[v_n]$ and $\tilde{P}(n)^*=Z_{(p)}[v_n,...]$. Ravenel [R] proved that $K(n)^*(G)$ is a finite $K(n)^*$-module for each finite group.

From Theorem 2.7, we have seen

$$\tilde{P}(n)^*(P) \cong \tilde{P}(n)^*\otimes_{BP^*}BP^*(P) .$$

Lemma 4.1. $P(n)^*(P)\cong P(n)^*\otimes_{BP^*}BP^*(P)$.

Proof. From the Sullivan-Bockstein exact sequence

$$\begin{array}{ccc} \tilde{P}(n)^*(P) & \xrightarrow{\ p\ } & \tilde{P}(n)^*(P) \\ {\scriptstyle\delta}\nwarrow & & \swarrow{\scriptstyle\rho} \\ & P(n)^*(P) & \end{array}$$

We only need to prove $\tilde{P}(n)^*(P)$ has no p-torsion. Let F_t be the filtration associated with the spectral sequence in §2. Each element $a\in F_0$ with $a\neq 0$ in $E_\infty^{0,*}$ is not p-torsion because $E_2^{0,*}$ in Theorem 2.7 has no p-torsion. Let $a\in F_t$, $t\geq 1$. Then $a=y_1a'$ or $=y_2a'$. Suppose $a\neq 0$, $a=y_1a'$ and $pa=0$. Then

$$a'[p](y_1) = py_1a'+v_ny_1^{p^n}a'+\ldots = 0 \text{ in } \tilde{P}(n)^*(P)$$

implies $v_ny^na'+\ldots=0$. The fact $a\notin$ Ideal $y_1\prod_\lambda(y_2+_{BP}[\lambda]y_1)=\Delta$ implies $y_1^{p^n}a'\notin\Delta$. Hence $v_ny_1^{p^n}a'\neq 0\in E^{t+2p^n-2,*}$ since $[p](u)=v_nc_p^{p^{n-1}}+\ldots+v_{n+1}c_p^{p^n}+\ldots$. This is a contradiction. q.e.d.

Corollary 4.2. $K(n)^*(P)\cong K(n)^*\otimes_{BP^*}BP^*(P)$.

Proof. From the Johnson-Wilson's theorem [J-W], $K(n)^*(X) \cong K(n)^* \otimes_{BP^*} P(n)^*(X)$, we have the corollary. q.e.d.

Theorem 4.2. There are $K(n)^*$-algebra isomorphisms

$$K(n)^*(P) \cong K(n)^*(Z/p[\![y_1,y_2]\!] \oplus Z/p\{\tilde{c}_2,\dots,\tilde{c}_{p-1}\}) \otimes Z/p[\![\bar{c}_p]\!]/$$

$$(y_1^p y_2 - y_1 y_2^p, y_1^{p^n}, y_2^{p^n}, v_n^2 \bar{c}_p^{2p^{n-1}} - \alpha_0, v_n \bar{c}_p y_1 - \alpha_1, v_n \bar{c}_p y_2 - \alpha_2)$$

where α_0 (resp. α_1, α_2) is modulo $(y_i y_j y_k)$

$$= \begin{cases} 0 \quad (\text{resp. } 0,\ 0) \quad \text{for } P=E \\ y_1^2 \quad (\text{resp. } y_1^2, y_1 y_2) \quad \text{for } P=M \\ y_1 y_2 \quad (\text{resp. } 0,\ y_1 y_2) \quad \text{for } P=D \\ y_1 y_2 + y_1^2 + y_2^2 \quad (\text{resp. } y_1^2, y_2^2) \quad \text{for } P=Q, \end{cases}$$

and $\tilde{c}_i \tilde{c}_j$, $y_j \tilde{c}_i$ $(1 \le i,j \le p-1)$ are in Ideal$(y_1 y_2)$.

Proof. First let $P=E$. Then $\tilde{c}_1 + v_n \bar{c}_p^{p^{n-1}} = 0$ in $\tilde{k}(n)^*(P)$ by the reason (i) in §3. Moreover $p\bar{c}_2 + v_n \tilde{c}_1 \bar{c}_p^{p^{n-1}} = 0$ in $\tilde{k}(n)^*(P)$. Therefore $(v_n \bar{c}_p^{p^{n-1}})^2 = -v_n \tilde{c}_1 \bar{c}_p^{p^{n-1}} = 0$ in $K(n)^*(P)$. Similarly we have $v_n \bar{c}_p y_i = 0$. From the results similar to (ii) - (iv) in §4, we can seek each α_i. q.e.d.

Remark 4.3. $k(n)^*(P) \not\cong k(n)^* \otimes_{BP^*} BP^*(P)$ for $P=E,M,D$, for example, there is $\alpha \in k(n)^{2^{n+1}+1}(D)$ such that $v_n \alpha = 0$ and $\rho(\alpha) = Q_n u = u^{2^n}(x_1+x_2) + \dots + u(x_1+x_2)^{2^{n+1}-1}$ in $H^*(D)$.

The additive structure of $K(n)^*(D)$ was first obtained by Ravenel using the Hochschild-Serre spectral sequence converging to $K(n)^*(D)$. He also conjectured that $K(n)^*(G) = K(n)^{even}(G)$. Hopkins et al. [H-K-R] have found way to compute the $K(n)^*$ Euler number $\chi(G) = \dim_{K(n)^*} K(n)^{even}(G) - \dim_{K(n)^*} K(n)^{odd}(G)$. Of course, in our cases $\chi(P) = \dim_{K(n)^*} K(n)^*(P)$.

§5. Exponent of $H^*(G)$.

Recall that the exponent $ex(G)$ of G is the maximum order of cyclic subgroups of G. It is important to study $ex(H^*(G))$ for G-action on spaces (see [B]). When A is an abelian group, it is immediate $ex(A_{(p)}) = ex(H^*(A)_{(p)})$. However for nonabelian p-groups, there seems no relation $ex(G)$ and $ex(H^*(G))$. Here we give examples such that $ex(G)$ is small but $ex(H^*(G))$ is large if p is large.

Let K_m be an extra special p-group of the order p^{2m+1}, that is, K_m is a group such that $\mathrm{center}(K_m)=Z/p$ and there is a central extension

$$(5.1) \qquad 0 \longrightarrow Z/p \longrightarrow K_m \longrightarrow \oplus^{2m} Z/p \longrightarrow 0 .$$

of course $ex(K_m) | p^2$.

Theorem 5.2. If $p \geq m$, then $p^{m+1} | ex(H^*(K_m))$.

Proof. Consider the maps of extensions

$$\begin{array}{ccccccccc} 1 & \longrightarrow & K_s & \longrightarrow & K_{s+1} & \longrightarrow & \oplus^2 Z/p & \longrightarrow & 1 \\ & & \uparrow j_1' & & \uparrow j_2' & & \| & & \\ 1 & \longrightarrow & Z/p & \longrightarrow & K_1 & \longrightarrow & \oplus^2 Z/p & \longrightarrow & 1 \end{array}$$

and the spectral sequences

$$EK(s+1)_2^{*,*} = H^*(\oplus^2 Z/p; \tilde{k}(1)^*(K_s)) \Longrightarrow \tilde{k}(1)^*(K_{s+1}) ,$$

$$E(s)_2^{*,*} = H^*(\oplus^{2m} Z/p; \tilde{k}(1)^*(Z/p)) \Longrightarrow \tilde{k}(1)^*(K_s) .$$

The element $p^s u^t \in E(s)^{*,*}$ is permanent, since it represents $\tilde{c}_t(s) = j_{1!}(u^t)$ in $k(1)^*(K_s)$ by the arguments similar to the case $K_1 = P$. The differential d_2 in $EK(s+1)^{*,*}$ is

$$d_2 \tilde{c}_t(s) = d_2 j_{1!}(u^t) = j_{2!}(d_2 u^t) = t j_{2!}(u^{t-1} \otimes \alpha)$$
$$= t j_{1!}(u^{t-1}) \otimes \alpha = t \tilde{c}_{t-1}(s-1) \otimes \alpha .$$

If $t\tilde{c}_{t-1}(s-1) \neq 0 \bmod p$ in $\tilde{k}(1)^*(K_{s-1})$, then $\tilde{c}_t(s) \in EK(s+1)_2^{0,*}$ is not permanent and so is $p^s u^t \in E(s+1)_2^{0,*}$, hence $\tilde{c}_t(s+1) \neq 0 \bmod p$ in $\tilde{k}(1)^*(K_{s+1})$. Therefore by induction on s we can see

$$(5.3) \qquad p^{s-1} u^{pk+s} \in E(s)_2^{0,*} , \quad 1 \leq s \leq p-1 \quad \text{is not permanent.}$$

The element $u^{p^m} \in E(m)_2^{0,*}$ is permanent, since it represents the Chern class $\bar{c}_{p^m}$ by arguments similar to the case $K=P$. We shall show $p^m \rho(\bar{c}_{p^m}) \neq 0$ in $H^*(K_m)$.

Suppose $p^m \rho(\bar{c}_{p^m}) = 0$ in $H^*(K_m)$. Then $p^m c_{p^m} \in \mathrm{Ideal} \cdot v_1$ in $\tilde{k}(1)^*(K_m)$, hence $p^m u^{p^m} \in \mathrm{Ideal} \cdot v_1$ in $E(m)_\infty^{0,*}$. Therefore $p^m u^{p^m} = v_1 a$, $a \in E(m)_\infty^{0,*}$. On the other hand, since,

$$[p](u) p^{m-1} u^{p^m - 1} = p^m u^{p^m} + p^{m-1} v_1 u^{p^m + p - 1} + \ldots = 0$$

and since $E(m)_\infty^{0,*} \subset E(m)_2^{0,*} \cong k(1)^*[\![u]\!]/[p](u)$ is v_1-torsion free,

$$-a = p^{m-1} u^{p^m + p - 1} \mod (v_1^i p^{n-i}) .$$

That $a \in E(m)_\infty^{0,*}$ contradicts to (5.3). q.e.d.

§6. General cases

In general, $BP^*(G) \otimes_{BP^*} Z_{(p)} \not\cong H^{even}(G)_{(p)}$, indeed, when $G=Z/p \oplus Z/p \oplus Z/p$ this does not hold. However when G is an abelian group or $|G|=p^3$,

(6.1) $\rho : BP^*(G) \otimes_{BP^*} Z_{(p)} \cong Ch(G)^*_{(p)}$

where $Ch(G)^*$ is the Chern subring which is a subring of $H^*(G)$ generated by Chern classes of representations of G (Theorem [T]). Moreover for $p\geq 3$, if G is an abelian or $|G|=p^3$, then

(6.2) $\rho : (BP^*(G) \otimes_{BP^*} Z_{(p)})/\sqrt{0} \cong H^*(G)/\sqrt{0}$

where $\sqrt{0}$ is the ideal generated by nilpotent elements.

Question 6.3. For what group G , do (6.1) or (6.2) hold?

Let P be a p-Sylow subgroup of G and $N_G(P)$ (resp. $C_G(P)$) is normalizer (resp. centralizer) or P in G . Let $W = N_G(P)/C_G(P)$ and $W'=N_G(P)/P$. The exact sequence

$$1 \longrightarrow P \longrightarrow N_G(P) \longrightarrow W' \longrightarrow 1$$

induces the Hochschild-Serre spectral sequence, and we get

$$BP^*(N_G(P)) \cong BP^*(P)^{W'} \cong BP^*(P)^W .$$

It is known (Nishida [N]) that $BN_G(P)$ and BG is stably p-equivalent if P is abelian.

Proposition 6.4. Let A be an abelian group and be a p-Sylow subgroup of G . Then $BP^*(A)^W \cong BP^*(G)$ and for $p\geq 3$ the map ρ in (6.2) is epic.

Proof. Given $a \in H^*(G)_{(p)}$, take $\tilde{a} \in BP^*(A)$ such that $\rho(\tilde{a})=a|<A>$. Then

$$\tilde{\tilde{a}} = \frac{1}{|W|} \Sigma_{\omega \in W} \omega^* \tilde{a}$$

is in $BP^*(A)^W=BP^*(G)$. Moreover $\rho(a)=\tilde{\tilde{a}}$ since $H^*(G)_{(p)} \hookrightarrow H^*(A)_{(p)}$.

q.e.d.

Let $GL_3(Z/p)$ be the general linear group of 3×3-matrices over Z/p and let B be the Borel subgroup generated by upper triangular matrices. Let U be the maximal unipotent subgroup of G generated by upper triangular matrices with the diagonals 1.

Then U is a p-Sylow subgroup of $GL_3(Z/p)$ and isomorphic to E (, for $p=2$ to D) .

By using the transfer map, we can see the Cartan-Eilenberg stable theorem (Proposition 9.3, 9.4 Chapter XII [C-E]), for BP*-theory. By the arguments similar to the ordinary cohomology case [T-Y 1], we can prove

$$(6.5)\quad BP^*(B) \cong BP^*(U)^T$$

$$(6.6)\quad BP^*(GL_3(Z/p)) \cong (i_1^{*-1}BP^*(A_1)^{(2,3)} \cap i_2^{*-1}BP(A_2)^{(1,2)}) \cap BP^*(B)$$

where T is the subgroup generated by diagonal matrices, $i_j : A_j \hookrightarrow U$ $(j=1,2)$ is the inclusion of 1-st line group and 3-rd column group, and (2,3), (1,2) are the induced action from the transpositions. (for details see [T-Y 1]).

Since the order of the action in T is prime to p, all invariants $a \in H^*(U)^T$ can be extended to $a' \in BP^*(U)^T$ with $\rho(a')=a$. The order of (2,3) and (1,2) is 2, we have ;

Proposition 6.7. For $p \geq 3$, the maps ρ

$$\rho : (BP^*(B) \otimes_{BP^*} Z_{(p)})/\sqrt{0} \longrightarrow H^*(B)/\sqrt{0} ,$$

$$\rho : (BP^*(GL_3(Z/p)) \otimes_{BP^*} Z_{(p)})/\sqrt{0} \longrightarrow H^*(GL_3(Z/p))/\sqrt{0} .$$

are epic.

References

[A] M.F.Atiyah, Character and cohomology of finite groups, Publ. I.H.E.S. 9 (1964), 23-64.

[B] W.Browder, Cohomology and group actions, Invt. Math. 71 (1983), 599-607.

[C-E] H.Cartan and S.Eilenberg, "Homological algebra" Princeton Univ. Press 1956.

[E] L.Evens, On the Chern class of representation of finite groups, Trans. Amer. Math. Soc. 115 (1965), 180-193.

[H-K-R] M.Hopkins, N.Kuhn and D.Ravenel.

[J-W] D.Johnson and S.Wilson, BP operations and Morava's extraordinary K-theories, Math. Zeit. 144 (1975), 55-75.

[L] G.Lewis, The integral cohomology ring of groups of order p^3, Trans. Amer. Math. Soc. 132 (1968), 501-529.

[N] G.Nishida, Stable homotopy types of classifying spaces of finite groups, "Algebraic and Topological theories" to memory of T.Miyata, 1986 Kinokuniya Comp. Ltd.

[Q1] D.Quillen, A cohomological criterion for p-nilpotency J. Pure and Appl. Algebra 4 (1971), 373-376.

[Q2] D.Quillen, The mod 2 cohomology ring of extra-special 2-groups and Spinor group, Math. Ann. 194 (1971), 197-223.

[R] D.Ravenel, Morava K-theory and finite groups, contemporary Math. 12 (1982), 289-292.

[T] C.B.Thomas, Chern classes of representations, Bull. London Math. Soc. 18 (1986), 225-240.

[T-Y-1] M.Tezuka and N.Yagita, The mod p cohomology ring of $GL_3(F_p)$, J. Algebra 81 (1983), 295-303.

[T-Y-2] M.Tezuka and N.Yagita, The varieties of the mod p cohomology rings of extra special p-groups for an odd prime p, Math. Proc. Camb. Phil. Soc. 94 (1983), 443-459.

M.Tezuka
Department of Mathematics
Tokyo Institute of Technology
Ohokayama Meguroku
Tokyo Japan

N.Yagita
Department of Mathematics
Musashi Institute of Technology
Tamazutumi Setagayaku
Tokyo Japan

Current Address
Department of Mathematics
Ryukyu University
Okinawa Japan

THE ARTIN-HASSE LOGARITHM FOR λ-RINGS

Tammo tom Dieck
Mathematisches Institut, Bunsenstraße 3-5
D-3400 Göttingen

We show that for p-adic λ-rings I in the sense of Atiyah-Tall [4] the map

$$(1 - \frac{\psi^p}{p})\log : 1 + I \longrightarrow I$$

is essentially an isomorphism. This can be used to produce explicit equivalences of infinite loop spaces

$$BSU_p^{\otimes} \longrightarrow BSU_p^{\oplus} \text{ and } BSO_p^{\otimes} \longrightarrow BSO_p^{\oplus}.$$

For BSO another such equivalence was constructed by Madsen-Snaith-Tornehave [8], using Atiyah-Segal [3]. The existence of an equivalence for BSU was shown by Adams and Priddy [2]. This logarithm is related to analogous logarithmic maps in algebraic K-theory, see e.g. Oliver [9].

Let I be a special p-adic γ-ring in the sense of Atiyah-Tall [4], III.2.1. We refer to [4] for generalities about λ- and γ-rings. The ring I has a filtration $I = I(1) \supset I(2) \supset \ldots$ by λ- and γ-ideals $I(k)$ and I is Hausdorff and complete with respect to the topology defined by the $I(k)$. Moreover I is an algebra over the p-adic integers $\mathbb{Z}_p$. The ideal $I(k)$ is the closure of the $\mathbb{Z}_p$- module I_k generated by elements of the form $\gamma^{n_1}(x_1)\cdot\ldots\cdot\gamma^{n_r}(x_r)$, $x_j \in I$, $\Sigma n_j \geq k$. In particular $I(a)I(b) \subset I(a+b)$. A polynomial in the operations λ^i resp. γ^i is called a λ-resp. γ-operation. The Adams operation ψ^p satisfies $x^p \equiv \psi^p(x) \bmod p$.

This leads to the definition of λ-operations θ^q for $q = p^t$, $t > 0$.

<u>Lemma 1.</u> There exists a unique λ-operation θ^q which satisfies

$$q\theta^q(x) = x^q - \psi^p(x^{q/p})$$

for $x \in I$.

<u>Proof.</u> This follows for $t = 1$ from $x^p \equiv \psi^p(x) \bmod p$. For $t > 0$ and $r = q/p$ one uses

$$\begin{aligned}\theta^q(x) &= q^{-1}(x^q - \psi^p(x^r))\\ &= \sum_{j=1}^{r} p^{j-t}\binom{r}{j}(\psi^p x)^{r-j}(\theta^p x)^j\end{aligned}$$

and $p^{j-1}\binom{r}{j} \in \mathbb{Z}$. □

Consider the infinite series

(1) $$L(1-x) = \sum_{(n,p)=1} \frac{1}{n}\left(\sum_{t=1}^{\infty} \theta^{p^t}(x^n)\right)$$

where the first sum is over integers $n > 0$ prime to p. This yields a continuous map

(2) $$L : 1 + I \longrightarrow I,\ 1 - x \longmapsto L(1-x).$$

If one computes in $I \otimes \mathbb{Q}$ then

(3) $$L(1-x) = \left(1 - \frac{\psi^p}{p}\right)\log(1-x).$$

One can check natural identities between λ-operations in torsion-free rings. The characteristic property of log and (3) therefore yields

<u>Lemma 2.</u> $$L((1-x)(1-y)) = L(1-x) + L(1-y).$$ □

Lemma 3.

(i) For $x \in I$, we have $x + \theta^p(x) \in I(2)$.

(ii) For $x \in I(k)$, $t > 1$, we have $\theta^{p^t}(x) \in I(2k)$.

(iii) $L(1+I(k)) \subset I(k)$.

(iv) $L(1+I) \subset I(2)$.

Proof. (iii) follows from (1) and (ii) from the proof of Lemma 1. (i) and (ii) imply (iv). In order to show (iii) one verifies that mod I^2 the following identity holds

$$x + \theta^p(x) = \sum_{j=2}^{p-1} (-1)^{j-1}\frac{1}{p}\binom{p}{j} j\gamma^j(x) + (p-1)!\gamma^p(x). \square$$

Proposition 4. $L : 1 + I(2) \longrightarrow I(2)$ is an isomorphism.

Proof. For $k \geq 2$ we have induced homomorphisms

$$L(k) : 1 + I(k)/1 + I(k+1) \longrightarrow I(k)/I(k+1)$$

by Lemma 3. As in [4], III.4 it suffices to show that the $L(k)$ are isomorphisms. Lemma 3 shows $L(k)(1-x) \equiv x + \theta^p(x) \bmod I(k+1)$. By [4], III.2.6 we have $\theta^p(x) \equiv -p^{k-1} \bmod I(k+1)$ for $x \in I(k)$ and therefore $L(k)(1-x) = (1-p^{k-1})x$. But $1 - p^{k-1}$ is for $k > 1$ a unit in $\mathbb{Z}_p$. $\square$

We now discuss the difference between $I(1)$ and $I(2)$, We assume that the special λ-ring A is finite-dimensional. We assume given an augmentation $\varepsilon : A \longrightarrow \mathbb{Z}$ which is a λ-map and satisfies $\varepsilon(x) = \dim x$ if x is finite-dimensional. We assume that $A(1) = \{x | \dim x = 1\}$ is a group with respect to multiplication. If $x - y \in A$, $m = \dim x$, $n = \dim y$, then

$$\dim \lambda^m x = \binom{m}{m} = 1 = \binom{n}{n} = \dim \lambda^n y$$

and we can define

(4) $$D(x-y) = (\lambda^m x)(\lambda^n y)^{-1} \in A(1).$$

We obtain a well-defined map $D : A \longrightarrow A(1)$ satisfying

(5) $$D(u+v) = D(u)D(v),\ D(xy) = D(x)^{\varepsilon(y)}D(y)^{\varepsilon(x)}.$$

If $I = \mathrm{Ker}\ \varepsilon$ and $1 + I = \{1+x \mid x \in I\}$ then, by (5),

(6) $$D(xy) = D(x)D(y) \text{ for } x,y \in 1 + I.$$

Lemma 5. The sequence $0 \longrightarrow I_2 \underset{\subset}{\longrightarrow} I_1 \underset{D}{\longrightarrow} A(1) \longrightarrow 1$ is exact.

Proof. Fulton-Lang [7], III.1.7. □

Suppose now that A is finitely generated as abelian group and that the γ-topology is finer than the p-adic topology. Then

$$0 \longrightarrow I \otimes \mathbb{Z}_p \longrightarrow A \otimes \mathbb{Z}_p \overset{\varepsilon}{\longrightarrow} \mathbb{Z}_p \longrightarrow 0$$

is exact and $J = I \quad \mathbb{Z}_p$ is, by definition, a p-adic γ-ring. The map D above induces a homomorphism $D : A_p —— A(1)_p$ and

$$0 \longrightarrow J(2) \longrightarrow J(1) \underset{D}{\longrightarrow} A(1)_p \longrightarrow 1$$

is also exact. If we combine Proposition 4 and Lemma 5 we obtain

Theorem 6. Under the hypotheses of Proposition 4 and Lemma 5 the following sequence is exact

$$0 \longrightarrow 1+J \underset{(L,D)}{\longrightarrow} J \oplus A(1)_p \underset{\binom{D}{0}}{\longrightarrow} A(1)_p \longrightarrow 1. \ \square$$

We describe a few applications. In [4], III.4 it is shown that, for $p \neq 2$, the unit group $\mathbb{Z}/p^*$ acts on the p-adic γ-ring I and yields a decomposition $I = \bigoplus I_i$, $i = \{0,\dots,p-2\}$. From Proposition 4 we obtain an isomorphism

(7) $$L : 1 + I_0 \overset{\cong}{\longrightarrow} I_0.$$

Compare [4], III.4.4, where an isomorphism $\rho : I_o \longrightarrow 1 + I_o$ is constructed.

Atiyah-Segal [3] define for $p \neq 2$ and a real p-adic γ-ring I an isomorphism $I \longrightarrow 1 + I$. This isomorphism has to be constructed separately on the summands $I(i)$. Since I is real (i.e. ψ^{-1} = identity) one has $I_1 = 0$ and $I(1) = I(2)$. Therefore, Proposition 4 yields an isomorphism $1 + I \longrightarrow I$ in this case too. For $p = 2$ it is assumed in [3] and [4] that I is orientable, and this implies ([4], p. 285) $I(1) = I(2)$, so that again Proposition 4 is applicable.

Let G be a finite group, X a connected finite G-complex and $KU_G(X)$ the Grothendieck ring of complex G-vector bundles over X. The bundle dimension yields an augmentation $KU_G(X) \longrightarrow \mathbb{Z}$ with kernel $\widetilde{KU}_G(X)$ and the p-adic completion $\widetilde{KU}_G(X)_p$ is a p-adic γ-ring (e.g. [6], p. 45). The kernel of the determinant map is denoted $\widetilde{KSU}_G(X)$, since elements in it are representable as difference E - F of G-vector bundles E and F which are associated to (G,SU(n))-bundle in the sense of [5]. In this case [3] is not applicable. Proposition 4, however, yields an isomorphism

(8) $$L : 1 + \widetilde{KSU}_G(X)_p \longrightarrow \widetilde{KSU}_G(X)_p.$$

This isomorphism is already interesting in the case $G = \{1\}$ because it can be considered as a homotopy equivalence of p-local spaces

(9) $$L : BSU^{\otimes}_{(p)} \longrightarrow BSU^{\oplus}_{(p)};$$

see Adams-Priddy [2], Adams [1], and Madsen-Snaith-Tornehave [8] for a context. It is interesting to remark that L is a map of infinite loop spaces. In order to see this it suffices, according to [8], that for cyclic groups the map

(10) $$L : 1 + I(\mathbb{Z}/p^r) \longrightarrow I(\mathbb{Z}/p^r)$$

is compatible with additive and multiplicative induction. (We have denoted by $I(\mathbb{Z}/p^r)$ the augmentation ideal of the p-adic completion $A(r) := R(\mathbb{Z}/p^r)_p$ of the complex representation ring of $\mathbb{Z}/p^r$.) The ordinary induced representation yields tr: $I(\mathbb{Z}/p^r) \longrightarrow I(\mathbb{Z}/p^{r+1})$. Let t_r be the onedimensional standard representation of $\mathbb{Z}/p^r$. Then $A(r)$ has $\mathbb{Z}_p$-basis $1, t_r, \ldots, t_r^k$, $k = p^r-1$. Let $N = \sum_{i=o}^{p-1} t_{r+1}^{ip^r}$. Given $x = \Sigma a_i t_r^i \in A(r)$ and let $\tilde{x} = \Sigma a_i t_{r+1}^i \in A(r+1)$. Then $tr(x) = N\tilde{x}$. We also need a description of the multiplicative induction $m : 1 + I(\mathbb{Z}/p^r) \longrightarrow 1 + I(\mathbb{Z}/p^{r+1})$. The composition of m with the restriction res: $1 + I(\mathbb{Z}/p^{r+1}) \longrightarrow 1 + I(\mathbb{Z}/p^r)$ satisfies $res \circ m(x) = x^p$. Elements in $R(\mathbb{Z}/p^{r+1})$ are detected by their restriction to $R(\mathbb{Z}/p^r)$ and $R(\mathbb{Z}/p^{r+1})/(N)$. By [8], Proposition 1.15, the multiplicative induction is additive modulo N and satisfies

$$m(\Sigma a_i t_r^i) = \Sigma a_i t_{r+1}^{ip} \bmod N. \tag{11}$$

Therefore m has the form

$$m(\sum_{i=o}^{k} a_i t_r^i) = \Sigma a_i t_{r+1}^{ip} + N(\sum_{i=o}^{k} b_i t_{r+1}^i).$$

The b_i can be determined by restriction. Note that res (N) = p. We conclude

$$(\Sigma a_i t_r^i)^p = \Sigma a_i t_r^{ip} + p(\Sigma b_i t_r^i).$$

Using the Adams operation ψ^p we can write the multiplicative induction in the form

$$\begin{aligned} m(x) &= \psi^p\tilde{x} + Np^{-1}(\tilde{x}^p - \psi^p\tilde{x}) \\ &= \psi^p\tilde{x} + N\theta^p(\tilde{x}). \end{aligned} \tag{12}$$

<u>Theorem 7.</u> The following diagram is commutative

$$\begin{array}{ccc} 1 + I(\mathbb{Z}/p^r) & \xrightarrow{L} & I(\mathbb{Z}/p^r) \\ \downarrow m & & \downarrow t_r \\ 1 + I(\mathbb{Z}/p^{r+1}) & \xrightarrow{L} & I(\mathbb{Z}/p^{r+1}). \end{array}$$

<u>Proof.</u> Since L is compatible with restriction, we see

$\mathrm{res} \circ \mathrm{tr} \circ L = \mathrm{res} \circ L \circ m$. Therefore it suffices to show that the diagram commutes modulo N. Since tr is zero mod N we have to show $Lm \equiv 0 \bmod N$. Since $I(\mathbb{Z}/p^{r+1})$ is torsion-free we can use (3)

$$Lm(1-x) = (1-\frac{\psi^p}{p})\log(\psi^p(1-\tilde{x}) + N\theta^p(1-\tilde{x}))$$

$$= (1-\frac{\psi^p}{p})\sum_{i=1}^{\infty}\frac{1}{i}(\psi^p\tilde{x} - \frac{N}{p}((1-\tilde{x})^p - \psi^p(1-\tilde{x})))^i.$$

We compute modulo N and use $N^2 = pN$ and $\psi^p N = p$ and obtain for the sum

$$\sum_{i=1}^{\infty}\frac{1}{i}(\psi^p\tilde{x})^i - \frac{1}{p}\sum_{i=1}^{\infty}\frac{1}{i}(1-\psi^p(1-\tilde{x})^p)^i$$

$$= \log(1-\psi^p\tilde{x}) - \frac{1}{p}\log\psi^p(1-\tilde{x})^p = 0. \quad \square$$

A similar computation can be applied to BSO. One obtains an equivalence $L : BSO_p^{\otimes} \longrightarrow BSO_p^{\oplus}$ of infinite loop spaces which differs from the one in [8].

References

1. Adams, J. F.: Infinite loop spaces. Princeton, Princeton Univ. Press 1978.

2. Adams, J. F., and S. B. Priddy: Uniqueness of BSO. Math. Proc. Camb. Phil. Soc. 80, 475 - 509 (1976).

3. Atiyah, M. F., and G. B. Segal: Exponential isomorphisms for λ-rings. Quart. J. Math. Oxford (2), 371 - 378 (1971).

4. Atiyah, M. F., and D. O. Tall: Group representations, λ-rings and the J-homomorphism. Topology 8, 253 - 297 (1969).

5. tom Dieck, T.: Faserbündel mit Gruppenoperation. Arch. Math. 20, 136 - 143 (1969).

6. tom Dieck, T.: Transformation groups and representation theory. Berlin-Heidelberg-New York, Springer 1979.

7. Fulton, W., and S. Lang: Riemann-Roch algebra. New York-Berlin-Heidelberg, Springer 1985.

8. Madsen, I., V. Snaith, and J. Tornehave: Infinite loop maps in geometric topology. Math. Proc. Camb. Phil. Soc. 81, 399 - 430 (1971).

9. Oliver, R.: SK_1 for finite group rings: II. Math. Scand. 47, 195-231 (1980).

HIGHER COHOMOLOGY OPERATIONS
THAT DETECT HOMOTOPY CLASSES

Zhou Xueguang
Nankai University,
Tianjin, China.

Let $p \geq 5$ be a prime, R.Cohen had proved in [3] that $b_k \otimes h_o$ is an infinite cycle and represents a nontrivial element ζ_k of order p of the stable homotopy groups of sphere. R.Cohen and P.Goerss proved in [4] that $h_k \otimes h_o$ is an infinite cycle and represents a nontrivial element η_k of order p for $k \geq 2$. Thus they obtained a kind of secondary cohomology operations and a kind of thirdary cohomology operations which detect homotopy classes. The main purpose of this paper is to obtain more kinds of higher cohomology operations that detect homotopy classes and then obtain more new infinite families of the stable homotopy groups of sphere. We prove that certain Toda's secondary product of $\eta_2, \eta_3, \ldots, \zeta_1, \zeta_2, \ldots$ and p, certain product of $\beta_1, \beta_2, \eta_2, \eta_3, \ldots, \zeta_1, \zeta_2, \ldots, \alpha_1$ are nonzero and we obtain their representation in the Adams spectral sequence, where α_t and β_t denote respectively α family and β family.

The main results of the present paper are the following:

Let $R = \{k_1, k_2, \ldots, k_s, \ell_1, \ell_2, \ldots, \ell_s, \varepsilon_1, \varepsilon_2, \ldots, \varepsilon_s, j\}$, $s \geq 0$ be a set of 3s+1 nonnegative integers satisfying the following conditions:

(1) $k_i > k_{i+1}+1$, $\quad s-1 \geq i \geq 1$ and $k_s \geq 2$.

(2) $\varepsilon_i \leq 1$, $\quad s \geq i \geq 1$.

(3) $\ell_i + \varepsilon_i \geq 1$, $\quad s \geq i \geq 1$.

(4) $\Sigma(2\ell_i+\varepsilon_i)+j+1 \leq p-1$.

(5) $j+1 \geq \Sigma(\ell_i+\varepsilon_i)$.

then we say that R is an H set and we put $\ell(R)=\Sigma(2\ell_i+\varepsilon_i)+j+1$, $r(R)=(\Sigma p^{k_i}(\ell_i+\varepsilon_i)+j+1)q+j$, where $q=2(p-1)$ and sometimes we write j as j(R). We use A_R to denote the expression

$$\underbrace{\tau_1 \otimes \tau_1 \otimes \cdots \otimes \tau_1}_{j \text{ copies}} \otimes h_{k_1}^{\varepsilon_1} \otimes b_{k_1-1}^{\ell_1} \otimes \cdots \otimes h_{k_s}^{\varepsilon_s} \otimes b_{k_s-1}^{\ell_s} \otimes h_o$$

where $(\ )^{\ell}$ means that ( ) appears ℓ times. We use also $(\bar{A}_R) \in \mathrm{ext}^{\ell(R),r(R)}(Z_p,Z_p)$ to denote the cohomology class containing $\bar{A}_R$, For the meaning of $\bar{A}_R$, see §1 and §4--6 in Part I, then we have

Theorem 1: Let $p\geq 5$ be a prime, R be an H set, then

(a) $(\bar{A}_R) \neq 0$.

(b) $(\bar{A}_R)$ is an infinite cycle in the Adams spectral sequence and represents a nontrivial element P(R) of order p of the stable homotopy groups of spheres. If $\ell(S)=\ell(R)$, $r(S)=r(R)$, all the $(\bar{A}_S)$ constitute a basis of $\mathrm{Ext}^{\ell(R),r(R)}(Z_p, Z_p)$ and all the P(S) are linear independent in $\pi_*(S^0)$.

Let $D=\{\ell, k, \varepsilon_1, \varepsilon_2, \varepsilon_3, \varepsilon_4\}$ be a set of six nonnegative integers satisfying

(1) $\varepsilon_3+2(\ell+\varepsilon_1+\varepsilon_2+\varepsilon_3)+\varepsilon_4\leq p$.

(2) $\varepsilon_i\leq 1$, $1\leq i\leq 4$.

(3) $\varepsilon_2+\varepsilon_3+\varepsilon_4=1$

(4) $k\geq 2$.

Then we say that D is a G set, we use B_D to denote the cohomology class $\overline{(\xi_2 \otimes h_1)^{\varepsilon 1} \otimes (h_k \otimes h_o)^{\varepsilon 2} \otimes (b_{k-1} \otimes h_o)^{\varepsilon 3} \otimes h_o^{\varepsilon 4} \otimes b_o^{\ell})}$ and we put

$\ell(D)=2(\ell+\varepsilon_1+\varepsilon_2+\varepsilon_3)+\varepsilon_4+\varepsilon_3$, $r(D)=q(p^k(\varepsilon_2+\varepsilon_3)+p(2\varepsilon_1+\ell)+(\varepsilon_1+\varepsilon_2+\varepsilon_3+\varepsilon_4)$, then we have

Theorem 2: Let D be a G set, then

(a) $B_D\neq 0$ in $\mathrm{Ext}^{\ell(D),r(D)}(Z_p,Z_p)$.

(b) $\beta_2^{\varepsilon 1}\, \beta_1^{\ell}\, \eta_k^{\varepsilon 2}\, \zeta_k^{\varepsilon 3}\, \alpha_1^{\varepsilon 4} \neq 0$ in $\pi_*(S^0)$.

In order to prove the main results, we need some new information of Ext groups of Steenrod algebra. In Part I, we use order cochain complex to reformulate Ext group of Hopf algebra. We prove that the E_2 term of May spectral sequence inherit a coboundary operation from the cobar construction and Ext is the homology group with respect to the coboundary operation. Thus we get a simple cochain complex which is homology equivalent to the cobar construction, For the Adams spectral sequence which relating the homology group of Hopf algebra G, H and $G/G\cdot\tilde{H}$, the cobar construction is also homology equivalent to $H^{**}(G/G\cdot\tilde{H}) \otimes H^{**}(H)$, our results give an effective step to calculate the higher coboundary operations of these spectral sequence.

In Part II, we give some properties of Massey product and Toda's secondary pro-uct from η_k, ... , ζ_j, ... , p. Using these results, we give the proof of theorem 1 and 2.

I would like to thank Professor R.Cohen for his helpful conversation during his visitation in Nankai University.

PART I.

§1. Ordered cochain complex.

Let p be a prime, C be an ordered set, we use $Z_p(C)$ to denote the vector space over Z_p having C as base. If there is a homomorphism δ: C→C such that $\delta\delta=0$, we say that $Z_p(C)$ is an ordered cochain complex. Let $\alpha \in C$, if $\delta(\alpha)$ is linearly independent on all $\delta(\beta)$, $\beta<\alpha$, we say that α is a ba-element and a linearly combination of ba-elements is said to be a ba-chain. If α is not a ba-element, then we say that α is a c-element and there exists a unique finite set

$$\{\alpha_i,\ \lambda_i,\ \alpha_i \in C,\ \alpha_i \text{ ba-element},\ \alpha_i<\alpha,\ \lambda_i \in Z_p, 1\leq i\leq m\}$$

such that $\delta(\alpha+\sum_{i=1}^{m}\lambda_i\alpha_i)=0$. We use $\bar{\alpha}$ to denote tha cocycle $\alpha+\sum_{i=1}^{m}\lambda_i\alpha_i$. We use also $(\bar{\alpha})$ to denote the cohomology class containing $\bar{\alpha}$. If $(\bar{\alpha})$ is not linearly dependent on all $(\bar{\alpha}')$, $\alpha'<\alpha$, α' a c-element, then we say that α is an h-element. If α is a c-element but isn't an h-element, then we say that α is a b0-element and there exists a unique set

$$\{\alpha_i,\ \lambda_i,\ \alpha_i \text{ h-element},\ \alpha_i<\alpha,\ \lambda_i \in Z_p,\ 1\leq i\leq m\}$$

such that $\bar{\alpha}+\Sigma\lambda_i\alpha_i \sim 0$. It can be easily seen that all the cohomology classes $(\bar{\alpha})$ of h-element form a basis of $H^*(Z_p(C))$ and we can easily obtain the following proposition.

Let $A=\{\alpha\}$ be a set of ba-elements, we use $\bar{A}$ to denote the subcochain complex $Z_p\{\alpha,\ \delta\alpha,\ \alpha \in A\}$.

<u>Proposition 1</u>: For any set $A=\{\alpha\}$ of ba-elements, the projection induces a homology equivalence between $Z_p(C)$ and the cochain complex $Z_p(C)/\bar{A}$.

<u>Corollary</u>: $H^*(Z(C))\cong Z_p(C)/\bar{B}$, where B is the set of all ba-elements of C.

§2. The cobar construction of Hopf algebra.

Let G be an anticommutative connected truncated polynomial algebra with generators $x_1, x_2, \ldots; x_n, \ldots$, let λ_i be the height of x_i, $i\leq\lambda_i\leq\infty$, i.e. $x_i^{\lambda}\neq 0$ for $\lambda<\lambda_i$ and $x_i^{\lambda}=0$ for $\lambda\geq\lambda_i$. We use also $\tilde{G}$ to denote the subalgebra of G of elements with grade > 0. It can be easily seen that the set $V(G)=\{\prod_{j=1}^{m} x_{i_j}^{\mu_j},\ 1\geq\mu_j\leq\lambda_{i_j}\}$ form a basis of G. We call $\prod_{j=1}^{m} x_{i_j}^{\mu_j}$ a letter in G. Suppose that there exists a diagonal algebra homomorphism $\psi : G\to G\otimes G$ such that G forms a Hopf algebra. Let $C(G)=Z_p+\sum_{i=1}^{m} \underbrace{\tilde{G}\otimes \ldots\ldots \otimes \tilde{G}}_{i \text{ copies}}$ be the cobar construction of G and $\delta:C(G)\to(G)$ be the boundary homomorphism. Then the set $W(G)=\{y_1\otimes \ldots \otimes y_n,\ y_i\in V(G)\}$ and $1\in Z_p$ constitute a basis of C(G), we call $y_1\otimes y_2\otimes \ldots \otimes y_n$ a word with length n in G. An order relation in W(G) is said to be commutative with the length of W(G) whenever $\ell(x)<\ell(y)$ implies $x<y$.

Let H be a subHopf algebra of G generated by some subset $(\bar{x}_1, \bar{x}_2,\ldots;.)$. Let $\{\bar{\bar{x}}_1, \bar{\bar{x}}_2, \ldots\}=\{x_1, x_2, \ldots\}-\{\bar{x}_1, \bar{x}_2, \ldots\}$, then $G/G\cdot\tilde{H}$ is also a Hopf algebra generated by $\bar{\bar{x}}_1, \bar{\bar{x}}_2, \ldots$. Suppose that there exists order relations on W(H) and $W(G/G\cdot\tilde{H})$ respectively. Hereafter, we shall extend these relations to an order relation in W(G).

It can be easily seen that any letter in W may be written in one of the following forms:

(a) $\bar{x}$, $\bar{x}\in V(H)$

(b) $(-1)^{(\ )}\bar{x}\bar{\bar{x}}$, $\bar{x}\in V(H)$, $\bar{\bar{x}}\in V(G/G\cdot\tilde{H})$.

(c) $\bar{\bar{x}}$, $\bar{\bar{x}}\in V(G/G\cdot\tilde{H})$.

In cases (a) —— (b), we say that $\bar{x}$ is the H factor of x.

In cases (b) —— (c), we say that $\bar{\bar{x}}$ is the $G/G\cdot\tilde{H}$ factor of x.

In case (b), we say that x is a mixed letter. Let $\alpha=x_1\otimes \ldots \otimes x_n$ be a word in G, let $\bar{x}_{i_1}, \bar{x}_{i_2}, \ldots, \bar{x}_{i_s}$ be the sequence of H factors of $x_1, x_2, \ldots, x_n$ and $\bar{\bar{x}}_{j_1}$, . ; . , $\bar{\bar{x}}_{j_t}$ be the sequence of $G/G\cdot\tilde{H}$ factors of $x_1, x_2, \ldots, x_n$.

We use $\bar{\alpha}$, $\bar{\bar{\alpha}}$ respectively to denote the word $\bar{x}_{i_1}\otimes \ldots \otimes \bar{x}_{i_s}$ and $\bar{\bar{x}}_{j_1}\otimes ..;\otimes\bar{\bar{x}}_{j_t}$. We also use $P_1(\alpha)$, $P_2(\alpha)$ to denote the sequence $(i_1, i_2, \ldots, i_s)$ and $(j_1, j_2, \ldots, j_t)$ of positive integers and call them the first position sequence and the second position sequence respectively. We also call s+t-n= the numbers of mixed letter in $x_1, \ldots, x_n$ as the genus of α. Let $a=(a_1, \ldots, a_m)$, $b=(b_1, \ldots, b_n)$ be two sequence of real numbers. We define the order of a and b as follows. If m>n, then a>b.

If $m=n$ and $a_1=b_1, \ldots, a_i=b_i$ but $a_{i+1}>b_{i+1}$, $1\leq i+1\leq n$, then $a>b$. We define an order relation in $W(G)$ as follows. Let x, y be two words in $W(G)$,

(a) If $\ell(x)<\ell(y)$, then $x<y$.

(b) If $\ell(x)=\ell(y)$, the genus of x < the genus of y, then $x>y$.

(c) If $\ell(x)=\ell(y)$, the genus of x = the genus of y, $\bar{\bar{x}}<\bar{\bar{y}}$ or $\bar{\bar{x}}=\bar{\bar{y}}$, $\bar{x}<\bar{y}$, then $\bar{\bar{x}}<\bar{\bar{y}}$.

(d) If $\ell(x)=\ell(y)$, the genus of x = the genus of y, $\bar{\bar{x}}=\bar{\bar{y}}$, $\bar{x}=\bar{y}$, $P_2(x)<P_2(y)$ or $P_2(x)=P_2(y)$, $P_1(x)<P_1(y)$, then $x<y$.

It can be easily seen that if $\ell(x)=\ell(y)$, the genus of x = the genus of y, $x=y$, $\bar{x}=\bar{y}$, $P_1(x)=P_1(y)$, $P_2(x)=P_2(y)$ then $x=y$. Therefore, for any two words x,y in $W(G)$, (a) —(d) define an order relation. If x, y are all in H or in $G/G\cdot\tilde{H}$, then their order coincides with those in H or in $G/G\cdot\tilde{H}$. We call the ordered relation defined above as the natural ordered relation with respect to the given ordered relation of $C(G/G\cdot\tilde{H})$ and $C(H)$.

We say that a wird $\alpha=\alpha_1\otimes\alpha_2\otimes\ldots\otimes\alpha_t$ is a proper word if there exists a $j\geq 0$ such that $\alpha_i\in V(G/G\cdot\tilde{H})$ for $i<j$ and $\alpha_i\in V(H)$ for $i\geq j$, then we have

<u>Proposition 2</u>: Let $x=x_1\otimes\ldots\otimes x_n$ be a mixed word in G and x_i be the first mixed letter, then a necessary and sufficient condition for x to be a ba-word is that $x_1\otimes\ldots\otimes x_{i-1}$ is a proper word.

Proof: Suppose that $x_1\otimes\ldots\otimes x_{i-1}$ is a proper word, let $x_i=k\otimes h$, $k\in V(G/G\cdot\tilde{H})$, $h\in V(H)$, the word $x_1\otimes\ldots\otimes x_{i-1}\otimes h\otimes k\otimes\ldots\otimes x_n$ appears in δx but doesn't appear in δy, $y<x$, so x is a ba-element. Suppose that $x_1\otimes\ldots\otimes x_{i-1}$ is not proper, then there exists a j, $1\leq j\leq i-1$, such that $x_1\otimes\ldots\otimes x_{j-1}\otimes x_j$ is a proper word, where $x_j\in V(H)$ and $x_{j+1}\in V(G/G\cdot\tilde{H})$, then it can be easily seen that x is the biggest word in $\delta(x_1\otimes\ldots x_{j-1}\otimes(x_jx_{j+1})\otimes x_{j+2}\otimes\ldots\otimes x_{i-1}\otimes x_i\ldots\otimes x_n)$ then x is a bo-word, so the condition is also necessary.

By calculating the coboundary of words of genus ≥ 2, $\delta(\ldots(\bar{x}_i\cdot\bar{\bar{x}}_i)\otimes\ldots\otimes(\bar{x}_j\cdot\bar{\bar{x}}_j)\otimes\ldots)$ and by induction, the following proposition follows easily.

<u>Proposition 3</u>: Let x be a mixed word in G, suppose that x isn't a ba-word, then there exists a set

$\{x_i, \lambda_i, x_i<x, x_i$ being a mixed ba-word, $\lambda_i\in Z_p, 1\leq i\leq m\}$ such that $x+\sum_{i=1}^{m}\lambda_i x_i$ is

the coboundary of a linearly combination of ba-mixed words.

Let M(G,H) be the set formed from all ba-mixed words, and it can be proved that

$$\bar{M}(G,H)=Z_p\{m,\ \delta m,\ m \text{ takes all mixed words}\}$$

It follows from proposition 2 and 3 that $H^*(M(G,H))=0$.

Let $\tilde{C}(G)=C(G)/\bar{M}(G,H)$, then we have

<u>Theorem 3</u>: $\tilde{\tilde{C}}(G)\cong C(G/G\cdot\tilde{H})\otimes C(H)$ and the projection is a homology equivalence.

The main reason for theorem 3 holds is that mixed words = 0 mod $\bar{M}(G,H)$ and $\delta(hk)=0 \bmod \bar{M}(G,H)$ implies that $k\otimes h=(-1)^{|k|\cdot|h|}h\otimes k+$chain of length $2<k\otimes h$ where $k\in W(G/G\cdot\tilde{H})$, $h\in W(H)$. The proof is straightforward and is omitted.

It should be noticed thay M(G,H) is a two sided ideal in C(G) with respect to the tensor product $\otimes$, so $\tilde{\tilde{C}}(G)$ inherits a tensor product from C(G). Since $\tilde{\tilde{C}}(G)$ may be considered as the subvector space $C(G/G\cdot\tilde{H})\otimes C(H)\subset C(G)$, so it also inherit an ordered relation defined above for C(G).

Let b be a ba-word in $G/G\cdot\tilde{H}$, then $b\otimes\alpha$ is a ba-word in C(G) and $\tilde{\tilde{C}}(G)$ for any word α in H, let y be an h-word in $G/G\cdot\tilde{H}$, then it can be proved by some straightforward calculation that $y\otimes a$ is a ba-word in C(G) and $\tilde{\tilde{C}}(G)$ for any ba-word a in H. Let $N(G,H)=\{b\otimes\alpha,\ y\otimes a$, b ba-word in $G/G\cdot\tilde{H}$, α word in H, y h-word in $G/G\cdot\tilde{H}$ and a ba-word in H$\}$. It is easily to prove that $C(G)/\bar{N}(G,H)\cong H^{**}(G/G\cdot\tilde{H})\otimes H^{**}(H)$. Then the following theorem follows from proposition 1 – 4 at once.

<u>Theorem 4</u>: (Adams) The natural projection from C(G) to $C(G)/\bar{N}(G,H)=H^{**}(G/G\cdot\tilde{H})\otimes H^{**}(H)$ is a homology equivalence.

<u>§3. The cobar construction of P(ξ)</u>.

Let ξ be an element of even grade, we use $P(\xi)$ to denote the Hopf algebra of polynomial algebra with generator ξ and diagonal homomorphism $\psi(\xi)=\xi\otimes 1+1\otimes\xi$. Let $\alpha=\sum_{i=0}^{n}\alpha_i p^i\ (0\le\alpha_i\le p-1)$ be the p-adic expansion of α. If there exists two integers i, j, $0\le i<j\le n$ such that $\alpha_i\ne 0$, $\alpha_j\ne 0$, we say that α is a mixed number and ξ^α is a mixed letter. Let $x=\xi^{i_1}\otimes\xi^{i_2}\otimes\ldots\otimes\xi^{i_n}$, if some i_j is a mixed number, we say that x is a mixed word. We define the genus of x as number of mixed number in $i_1, i_2, \ldots, i_n$.

Let $i_j=\Sigma\alpha_{j,t}p^t$, $0\le\alpha_{j,t}\le p-1$, be the p-adic expansion of i_j, we use $\alpha(x,t)$ to denote the finite subsequence

$\{\alpha_{j_1,t}, \alpha_{j_2,t}, \ldots, \alpha_{j_{\phi(t)},t}\}$ of $\alpha_{1,t}, \alpha_{2,t}, \ldots$ constituting from all $\alpha_{j,t}$

such that $\alpha_{j,t}>0$, we use also $P(x,t)$ to denote the t-th position sequence $\{j_1, j_2, \ldots, j_{\phi(t)}\}$ of x.

It should be noticed that $\phi(t)$ may be equal to zero. We establish an order relation commutative with length as follows.

(a) Let x and y be two words in $P(\xi)$, $\ell(x)=\ell(y)$, genus of x > genus of y then $x<y$.

(b) Let x and y be two words in $P(\xi)$, $\ell(x)=\ell(y)$, the genus of x = the genus of y. Suppose that there exists an integer s such that $\alpha(x,t)=\alpha(y,t)$ for $t>s$, but $\alpha(x,s)>\alpha(y,s)$, then $x>y$.

(c) Suppose that x and y be two words in $P(\xi)$, the genus of x = the genus of y, if $\alpha(x,s)=\alpha(y,s)$ for all $s\geq 0$ and there exists an integer $t\geq 0$ such that $P(x,s)=P(y,s)$ for all $s\geq t+1$ and $P(x,t)>P(y,t)$, then $x>y$.

The order relation defined above is called the natural order relation in $C(P(\xi))$.

In the following, we shall determine all the ba, bo, h-words of $P(\xi)$.

Let $x=\xi^{i_1 p^{\lambda_1}} \otimes \ldots \otimes \xi^{i_n p^{\lambda_n}}$, $0\leq i_t \leq p-1$, $1\leq t\leq n$, be a nonmixed word. If $\lambda_j \geq \lambda_{j+1}$ for all $j+1\leq n$, we say that x is a regular word in $P(\xi)$. By the argument as proposition 2 in the preceeding section, we have the following.

<u>Proposition 1</u>: Let $x=x_1 \otimes x_2 \otimes \ldots \otimes x_n$ be a mixed word in $P(\xi)$, $x_i (i\geq 1)$ be the first mixed letter, then a necessary and sufficient condition for x to be a ba-mixed word is that $i=1$ or $x_1 \otimes x_2 \otimes \ldots \otimes x_{i-1}$ is a regular word in $P(\xi)$.

By the same argument as proposition 3 in §1, we have

<u>Proposition 2</u>: Let x be a non ba-mixed word in $P(\xi)$, then there exists a set $\{x_i, \lambda_i, x_i<x, x_i \text{ being a mixed ba-word}, \lambda_i \in Z_p\}$ such that $x+\sum_{i=1}^{n}\lambda_i x_i$ is the coboundary of a linearly combination of ba-mixed word in $W(P(\xi))$.

Let $M(\xi)$ = the set of ba-mixed words in $P(\xi)$. It can be easily proved from proposition 2 that $\bar{M}(\xi)=Z_p\{x, \delta x, x \text{ being mixed word in } P(\xi)\}$ and $\bar{M}(\xi)$ is a two sided ideal of $C(\xi)$.

Since $\sum_{j=1}^{n-1} (C_j^p/p)\cdot\xi^{jp^i} \otimes \xi^{p^i(p-j)}$ is a cocycle which isn't homologous to zero in $C(\xi)$, so $\xi^{(p-1)p^i} \otimes \xi^{p^i}$ is an h-word in $P(\xi)$. Let $2\leq \ell \leq p-1$, $1\leq \ell_i \leq p-1$, $1\leq i\leq n$, it can be easily seen that

$$\xi^{\ell_1 p^i} \otimes \dots \otimes \xi^{\ell_n p^i} \otimes \xi^{(\ell-1)p^i} \otimes \; p^i \otimes \underbrace{\xi^{(p-1)p^i} \otimes \xi^{p^i} \otimes \dots \xi^{(p-1)p^i} \otimes \xi^{p^i}}_{t \text{ tuples}}$$

is the biggest word in

$$\delta(\xi^{\ell_1 p^i} \otimes \dots \otimes \xi^{\ell_n p^i} \otimes \xi^{\ell p^i} \otimes \underbrace{\xi^{(p-1)p^i} \otimes \xi^{p^i} \otimes \dots \otimes \xi^{(p-1)p^i} \otimes \xi^{p^i}}_{t \text{ tuples}}) .$$

In virtue of this and some straightforward calculation, we have

Proposition 3: Every non-mixed, nonregular word x is a bo-word and is homologous to a regular word < x.

Proposition 4: Let x be a regulat word, then a necessary and sufficient condition for x to be a h-word is that $\alpha(x,t)$ is one of the following sets for all $t \geq 0$,

(a) Empty set.

(b) $(1, \underbrace{p-1, 1, \dots, p-1, 1}_{k \text{ tuples},\ k \geq 0})$.

(c) $(\underbrace{p-1, 1, \dots, p-1, 1}_{k \text{ tuples},\ k > 0})$

Any one of (a) — (c) is said to be an h-set. Let $A = (\lambda_1, \dots, \lambda_n)$ be a finite sequence of positive integers, $k > 0$ be an integer such that $(\lambda_{k+1}, \dots, \lambda_n)$ is an h-set but $(\lambda_k, \dots, \lambda_n)$ isn't, we say that $(\lambda_{k+1}, \dots, \lambda_n)$ is the h part of A.

Proposition 5: Let x be a regular word in $P(\xi)$, let $i \geq 0$ be an integer such that $\alpha(x,j)$ is an h-set for all $j > i$ and $\alpha(x,i)$ isn't, then a necessary and sufficient condition for x to be a ba-word is that the h-part of $\alpha(x,i)$ is of type (c) in proposition 4. A necessary and sufficient condition for x to be a bo-word is that $\alpha(x,i)$ is of type (b).

We use $\xi(i,1)$ to denote the letter ξ^{p^i} and $\xi(i,2)$ to denote the h-word $\xi^{(p-1)p^i} \otimes \xi p^i$. It can be easily seen that

$$\overline{\xi}(i,2) = \sum_{j=1}^{p-1} (C^p_j / p)\, \xi^{(p-j)p^i} \otimes \xi^{jp^i}$$

we have

Proposition 6: $H^{**}(P(\xi)) = \prod_{i=1}^{\infty} (P(\xi(i,2)) \times E(\xi(i,1)))$.

A word x in $P(\xi)$ is said to be of second kind if ξ is a ba-word or a bo-word, a linearly combination of words of second kind is said to be a chain of second kind, then it follows from proposition 1 and 5 that we have the following

Proposition 7: Let α, β be two words in $P(\xi)$ and β be a word of the second kind, then $\alpha\otimes\beta$ is a chain of second kind.

§4. The cobar construction of $E(\tau)$.

In this section, it is assumed that $|\tau|$ = odd and $E(\tau)$ denote the exterior algebra of τ, $\psi(\tau)=\tau\otimes 1+1\otimes\tau$ be the diagonal homomorphism, thus $E(\tau)$ becomes a Hopf algebra. The only word of length ℓ in $C(E(\tau))$ is $\underbrace{\tau\otimes \dots \otimes\tau}_{\ell \text{ copies}}$, we have $\delta(\tau\otimes \dots \otimes\tau) = 0$, so all the word in $C(E(\tau))$ are h-words, so we have

Proposition 1: $H^{**}(E(\tau)) = P(\tau)$.

It should be noticed that there exists one and only one order relation in $C(E(\tau))$ commutative with length, we call it also tha natural order relation in $C(E(\tau))$.

§5. Cochain complex homology equivalent to the cobar construction.

In this section, it is assumed that

(a) The Hopf algebra $T(x_o, x_1, \dots , x_m, x_{m+1} , \dots)$ is the truncated polynomial algebra with generators $x_o, x_1, \dots$

(b) $|x_i|\leq|x_{i+1}|$

(c) If $|x_i| = 0 \pmod 2$, then x_i is of infinite order.

In the following, we use $T(\infty)$ to denote $T(x_o, x_1, \dots)$ and $T(m)$ to denote the subHopf algebra generated by $x_o, x_1, \dots , x_m$. Then $T(m)/T(m)\cdot\tilde{T}(m-1) = T(x_m)$, it can be easily seen that $T(x_m)$ is the exterior algebra of x_m if $|x_m|$ is odd and $T(x_m)$ is the polynomial algebra with generator x_m if $|x_m|$ is even. Therefore there exist a natural ordered relation inductively. Suppose that we have defined the natural order relation in $C(T(m))$, then there is a natural order relation in $C(T(m+1))$ with respect to the natural order relation of $C(T(m))$ and $C(T(x_{m+1}))$ and so on, so we can define an ordered relation in $T(x_o, x_1, \dots , x_m, \dots)$.
The ordered relation defined above is called then natural ordered relation in $C(T(x_o, x_1, \dots , x_m, \dots))$.

We define a set $B(m)$ of ba-words in $C(T(m))$ inductively as follows.

Let $B(0) = \{b,\ b \text{ ba-word in } C(T(x_o))\}$. Suppose that $B(m)$ had been defined, then we define B(m+1) as follows.

$$B(m+1) = B(m) \cup \{t,\ t \text{ ba-mixed word of } T(m+1) \text{ with respect to } T(m)\} \cup$$

$$\cup \{b \otimes \alpha,\ b \text{ ba-word in } C(T(x_{m+1})),\ \alpha \in W(T(m))\} \cup$$

$$\cup \{x \otimes a,\ x \text{ h-word in } C(T(x_{m+1})),\ a \text{ ba-word in } B(m)\}$$

then it can be proved by induction that

$$C(T(m))/\bar{B}(m) = \prod_{n=o}^{m} H^{**}(T(x_n))$$

If m=∞, we use the notation $\prod_{n=o}^{\infty} H^{**}(T(x_n))$ to denote $\bigcup_{1\leq m<\infty} \prod_{n=o}^{m} H^{**}(T(x_n))$.

Since all b, $b \in B(m)$, are ba-words, so $H^{**}(\bar{B}(m))=0$ and C(T(m)) and $C(T(m))/\bar{B}(m)$ are homology equivalent cochain complexes, therefore we have the following

Theorem 3: (May) Let $\bar{\delta}$ be the coboundary operation in $C(T(m)/\bar{B}(m)$ inherited from C(T(m)), then C(T(m)) and the cochain complex $(\prod_{n=o}^{m} H^{**}(T(x_n)), \bar{\delta})$ are homology equivalent for $0\leq m\leq\infty$.

As usual, we use $\xi_1, \xi_2, \ldots, \xi_n, \ldots, \tau_o, \tau_1, \ldots$ to denote the generators of dual algebra A_p^* of Steenrod algebra,

$$|\tau_i| = 2(p^i-1) + 1$$

$$|\xi_i| = 2(p^i-1)$$

$$\psi(\tau_i) = \sum_{k=o}^{i} \xi_{i-k}^{p^k} \otimes \tau_k + \tau_i \otimes 1$$

$$\psi(\xi_i) = \sum_{k=o}^{i} \xi_{i-k}^{p^k} \otimes \xi_k$$

In the following, we set $x_o=\tau_o$, $x_1=\xi_1$, ... $x_{2i-1}=\xi_i$, $x_{2i}=\tau_i$, ... , $h_k=\xi_1^{p^k}$, $b_k=\xi_1^{(p-1)p^k} \otimes \xi_1^{p^k}$, then it can be easily seen that $\bar{b}_k = \sum_{i=1}^{p-1} (C_j^p/p)\ \xi_1^{jp^k} \otimes \xi_1^{(p-j)p^k}$.

§6. Some properities of C(T(i)) for i≤2.

Proposition 1: For any word $\alpha \in C(T(1))$, we have $\alpha \otimes \bar{B}(1) \in \bar{B}(1)$ that is to say $\bar{B}(1)$ is a left side ideal of C(T(1)).

Proof: It should be pointed out that $\bar{M}(m) = Z_p\{u, \delta u$, u being mixed word with respect to C(T(m-1)) in C(T(m))} , so we have

$$\xi_1^{\ell} \otimes \tau_o = \psi(\xi_1^{\ell}\tau_o) - \tau_o \otimes \xi_1^{\ell} \quad \text{mod } \bar{M}(1)$$

that is to say

$$\xi_1^{\ell} \otimes \tau_o = (-1)\tau_o \otimes \xi_1^{\ell} \qquad \text{mod } \bar{M}(1)$$

therefore we have

$$\alpha_1 \otimes \dots \otimes \alpha_s \otimes \xi_1^{\ell} \otimes \tau_o \otimes \alpha_{s+1} \otimes \dots \otimes \alpha_r = - \alpha_1 \otimes \dots \otimes \alpha_s \otimes \tau_o \otimes \xi_1^{\ell} \otimes \alpha_{s+1} \otimes \dots \otimes \alpha_r \mod \bar{M}(1)$$

for $\alpha_1, \dots, \alpha_r \in V(T(1))$.

Now for any mixed word β in $C(T(1))$, we have

$$\alpha \otimes \beta \in \bar{M}(1) \subset \bar{B}(1)$$

$$\alpha \otimes \delta\beta \in \bar{M}(1) \subset \bar{B}(1)$$

Suppose that α is a mixed word, then $\alpha\otimes\beta \in \bar{M}(1) \subset \bar{B}(1)$, if α isn't mixed word, it may be assumed that α is a nonmixed letter in $C(T(1))$. Suppose that $\alpha=\tau_o$, then for ba-word $a \in C(T(\xi_1))$

$$\tau_o \otimes a \otimes \underbrace{\tau_o \otimes \dots \otimes \tau_o}_{j \text{ copies}} = -a \otimes \underbrace{\tau_o \otimes \tau_o \otimes \dots \otimes \tau_o}_{(j+1) \text{ copies}} \mod \bar{M}(1)$$

and

$$\tau_o \otimes \delta(a) \otimes \underbrace{\tau_o \otimes \dots \otimes \tau_o}_{j \text{ copies}} = \delta(a) \otimes \underbrace{\tau_o \otimes \dots \otimes \tau_o}_{(j+1) \text{ copies}}$$

$$= \delta(a \otimes \underbrace{\tau_o \otimes \dots \otimes \tau_o}_{(j+1) \text{ copies}}) \in \bar{B}(1)$$

If $\alpha=\xi_1^a$, then it follows from proposition 7 in §3 that $\alpha\otimes a$ is a chain of second kind in $C(T(\xi_1))$, so we have

$$\alpha \cdot \bar{B}(1) \subset \bar{B}(1)$$

so the proposition 1 is proved.

By definition, $B(2)=\{\underbrace{\tau_1 \otimes \dots \otimes \tau_1}_{\ell} \otimes b,\ w,\ \ell \geq 0$, b ba-word in $T(1)$, w ba-mixed in $T(1)$, w ba-mixed word in $T(2)$ with respect to $T(1)\}$. It follows from proposition 1 that we have

<u>Proposition 2</u>: $\{\bar{B}(2)= \underbrace{\tau_1 \otimes \dots \otimes \tau_1}_{\ell} \otimes \beta,\ u,\ \ell \geq 0,\ \beta \in \bar{B}(1),\ u \in \bar{M}(T(2),T(1)).\}$

<u>Proposition 3</u>: For $i \geq 0$, $j \geq 0$, $\alpha \in V(T(1))$, $\ell+\varepsilon>0$, $\varepsilon=0, 1$, then

(a) $\underbrace{\tau_1 \otimes \dots \otimes \tau_1}_{i \text{ copies}} \otimes \xi_1^{\ell} \otimes \tau_o^{\varepsilon} \otimes \underbrace{\tau_1 \otimes \dots \otimes \tau_1}_{j \text{ copies}} \otimes \alpha - (-1)^j \underbrace{\tau_1 \otimes \dots \otimes \tau_1}_{(i+j) \text{ copies}} \otimes \xi_1^{\ell} \otimes \tau_o^{\varepsilon} \otimes \alpha$

may be written into a linearly combination of words of the following forms

$$\underbrace{\tau_1 \otimes \cdots \otimes \tau_1}_{(i+k)\text{ copies}} \otimes \xi_1^{m_1}\tau_o^{\varepsilon_1} \otimes \xi_1^{m_2}\tau_o^{\varepsilon_2} \otimes \cdots \otimes \xi_1^{m_s}\tau_o^{\varepsilon_s} \otimes \alpha \quad \text{mod } \bar{M}(2)$$

where $\varepsilon_i=0,1$, $m_i+\varepsilon_i>0$, $1\leq i\leq s$, $m_1+m_2+\ldots+m_s=j-k+\ell$, $\varepsilon_1+\varepsilon_2+\ldots+\varepsilon_s=j-k+\varepsilon$, $s=j-k+2$ if $\ell>0$, $\varepsilon=1$ and $s=j-k+1$ if $\ell=0$ or $\varepsilon=0$.

Proof: Since

$$\psi(\xi_1^{\ell}\tau_o\tau_1) - [(\xi_1^{\ell}\tau_o)\otimes\tau_1+\tau_1\otimes\xi_1^{\ell}\tau_o]$$

may be written by a linearly combination of the following words

$$\xi_1^k\tau_o \otimes \xi_1^j\tau_o, \quad k+j=\ell+1, \quad \text{mod } \bar{M}(2)$$

$\psi(\xi_1^{\ell}\tau_1) - [(\xi_1^{\ell}\otimes\tau_1+\tau_1\otimes\xi_1^{\ell})]$ is a linearly combination of the following words

$$\xi_1^j \otimes \xi_1^k\tau_o \quad \text{mod } \bar{M}(2), \quad j+k=\ell+1$$

and $\tau_o\otimes\tau_1-\tau_1\otimes\tau_o=\psi(\tau_o\tau_1)-(\tau_o\xi_1)\otimes\tau_o$, using this and $\bar{M}(2)$ is a two sided ideal of $C(T(2))$, proposition 2 can be easily proved by induction with respect to j, ℓ, ε.

Corollary 1: Let $0\leq j\leq p-1$, then

$$\delta(\underbrace{\tau_1\otimes\cdots\otimes\tau_1}_{j\text{ copies}}) - (-1)^j j(\underbrace{\tau_1\otimes\cdots\otimes\tau_1}_{(j-1)\text{ copies}}\otimes\xi_1\otimes\tau_o)$$

may be written into a linearly combination of the following form

$$\underbrace{\tau_1\otimes\cdots\otimes\tau_1}_{i\text{ copies}}\otimes\xi_1^{\rho_1}\tau_o^{\theta_1}\otimes\cdots\otimes\xi_1^{\rho_{j-i+1}}\tau_o^{\theta_{j-i+1}} \quad \text{mod } \bar{M}(2)$$

where $1\leq i\leq j-2$, $0\leq\rho_k$, $0\leq\theta_k\leq1$, $0\leq k\leq j-i+1$, $\Sigma\theta_k=\Sigma\rho_k=j-i$.

Proof: $\delta(\underbrace{\tau_1\otimes\cdots\otimes\tau_1}_{j\text{ copies}})=(-1)^k(\underbrace{\tau_1\otimes\cdots\otimes\tau_1}_{(k-1)\text{ copies}}\otimes\xi_1\otimes\tau_o\otimes\underbrace{\tau_1\otimes\cdots\otimes\tau_1}_{(j-k)\text{ copies}})$

Applying proposition 3 to the word $\underbrace{\tau_1\otimes\cdots}_{(k-1)\text{ copies}}\otimes\xi_1\otimes\tau_o\otimes\underbrace{\tau_1\otimes\cdots\otimes\tau_1}_{(j-k)\text{ copies}}$,

we get the proof of this corollary.

The number t of τ_1 factors in the biggest word in a chain in $C(T(2))$ is said to be the τ_1 length of u and we say that u is of τ_1-length t.

Corollary 2: If α is a cocycle in $T(1)$, $1\leq j\leq p-2$, then

$$\delta(\underbrace{\tau_1\otimes\cdots\otimes\tau_1}_{j\text{ copies}}\otimes\alpha) - (-1)^j j(\underbrace{\tau_1\otimes\cdots\otimes\tau_1}_{(j-1)\text{ copies}}\otimes\xi_1\otimes\tau_o\otimes\alpha) \in \bar{B}(2)$$

and is a chain of τ_1-length at most j-2.

Proof: There are only two possibilities for the word

$$\tau_1 \otimes \ldots \otimes \tau_1 \otimes \xi_1^{\rho_1} \tau_o^{\theta_1} \otimes \ldots$$

(a) There exists a k such that $\rho_k \theta_k \neq 0$, then $\xi_1^{\rho_k} \tau_o^{\epsilon_k} = \xi_1^{\rho_k} \tau_o$ is a mixed letter.

(b) There exists a k such that $2 \leq \rho_k \leq p-2$, $\theta_k = 0$, then the word reduces to the form $\ldots \otimes \xi^{\rho_k} \otimes \ldots$ which is also contained in $\bar{B}(2)$, so

$$\delta(\underbrace{\tau_1 \otimes \ldots \otimes \tau_1}_{j \text{ copies}} \otimes \alpha - (-1)^j j(\underbrace{\tau_1 \otimes \ldots \otimes \tau_1}_{(j-1) \text{ copies}}) \otimes \xi_1 \otimes \tau_o \otimes \alpha) \in \bar{B}(2)$$

It is evidently that the τ_1 length of above cochain is at most j-2.

It should be noticed that the condition $j \leq p-2$ is necessary. If $j = p-1$, the corollary may be failed. For example, we have

$$\delta(\underbrace{\tau_1 \otimes \ldots \otimes \tau_1}_{(p-1) \text{ copies}} \otimes \xi_1) = b_o \mod \bar{B}(2)$$

<u>Corollary 3</u>: Let $k_i > 1$, $\ell_i \geq 0$, $1 \leq i \leq s$, $0 \leq \varepsilon_i \leq 1$, $\alpha = \prod_{i=1}^{s} (h_{k_i} \otimes \bar{b}_{k_i - 1}^{\ell i})$, then

$$\delta(\underbrace{\tau_1 \otimes \ldots \otimes \tau_1}_{p-1 \text{ copies}} \otimes \alpha + \underbrace{\tau_1 \otimes \ldots \otimes \tau_1}_{p-2 \text{ copies}} \otimes \xi_1 \otimes \tau_o \otimes \alpha) \in \bar{B}(2)$$

Proof: Although it may be happened that the word

$$\tau_1 \otimes \ldots \otimes \tau_1 \otimes \xi_1^{\rho_1} \tau_o^{\theta_o} \otimes \ldots \otimes \xi_1^{\rho_s} \tau_o^{\theta_i} \otimes \alpha$$

may be reduced to $\xi_1^{p-1} \otimes \tau_o \otimes \ldots \otimes \alpha$, but $\xi_1 = h_o$ doesn't appear in α, so

$$\delta(\underbrace{\tau_1 \otimes \ldots \otimes \tau_1}_{p-1 \text{ copies}} \otimes \alpha) + (\underbrace{\tau_1 \otimes \ldots \otimes \tau_1}_{p-2 \text{ copies}} \otimes \xi_1 \otimes \tau_o \otimes \alpha) \in \bar{B}(2)$$

<u>Proposition 4</u>: Let $\alpha \in \bar{B}(2)$ be a cocycle of τ_1 length m, then there exist ba-chain $b_o, b_1, \ldots, b_m \in B(1)$ and ba-mixed chain $w \in M(T(2), T(1))$ such that $\delta(\sum_{i=o}^{m} \underbrace{\tau_1 \otimes \ldots \otimes \tau_1}_{i \text{ copies}} \otimes b_i + w) = \alpha$.

Proof: Since $H^{**}(\bar{B}(2)) = 0$ and α is a cocycle, so there exist ba-chains $b_o, b_1, \ldots, b_n$ and ba-mixed chain $w \in M(T(2), T(1))$ such that

$$\delta(\sum_{i=1}^{n} \underbrace{\tau_1 \otimes \ldots \otimes \tau_1}_{i \text{ copies}} \otimes b_i + w) = \alpha$$

Let $\beta = \sum_{i=1}^{n} (\underbrace{\tau_1 \otimes \ldots \otimes \tau_1}_{i \text{ copies}} \otimes b_i + w)$. If $n > m$, then the biggest word in $\delta(\beta)$ is contained in $\underbrace{\tau_1 \otimes \ldots \otimes \tau_1}_{n \text{ copies}} \otimes \delta(b_n)$ and doesn't appear in α, so $\alpha \neq \delta\beta$.

If $n<m$, then the biggest word in α is of τ_1 length m and doesn't appear in $\delta\beta$, so $\alpha \neq \delta\beta$. It is impossible, so we have $n=m$.

Proposition 5: Let R be an H set, then

(a) A_R is an h-word.

(b) There exist ba-chains $b_o, b_1, \ldots, b_{j-1}$ in B(2) and ba-mixed chain w such that $\bar{A}_R = \underbrace{\tau_1 \otimes \ldots \otimes \tau_1}_{j \text{ copies}} \otimes h_{k_1}^{\varepsilon_1} \otimes \bar{b}_{k_1-1}^{\ell_1} \otimes \ldots \otimes h_{k_s}^{\varepsilon_s} \otimes \bar{b}_{k_s-1}^{\ell_s} \otimes h_o + \sum_{i=o}^{j-1} \underbrace{\tau_1 \otimes \ldots \otimes \tau_1}_{i \text{ copies}} \otimes b_i + w$

(c) Let S be an H set such that $\ell(S)=\ell(R)$, $r(R)=r(S)$, then all $(\bar{A}_S)$ form a basis of $\mathrm{Ext}^{\ell(R),r(R)}(Z_p,Z_p)$.

Proof: Let $A = \underbrace{\tau_1 \otimes \ldots \otimes \tau_1}_{j \text{ copies}} \otimes h_{k_1}^{\varepsilon_1} \otimes \bar{b}_{k_1-1} \otimes \ldots \otimes h_o$. It can be easily seen that A_R is the biggest word in A and $A-A_R$ is a basic chain. Let $\alpha = b_{k_1}^{\varepsilon_1} \otimes \bar{b}_{k_1-1}^{\ell_1} \otimes \ldots \otimes h_o$, then it follows from proposition 3 that

$$\delta A = (-1)^j\, j(\underbrace{\tau_1 \otimes \ldots \otimes \tau_1}_{j-1 \text{ copies}}) \otimes \xi_1 \otimes \tau_o \otimes \alpha \in \bar{B}(2)$$

Now $(\underbrace{\tau_1 \otimes \ldots \otimes \tau_1}_{j-1 \text{ copies}}) \otimes \xi_1 \otimes \tau_o \otimes \alpha = (\underbrace{\tau_1 \otimes \ldots \otimes \tau_1}_{j-1 \text{ copies}}) \otimes \xi_1 \otimes \tau_o \otimes \alpha$

$$= \frac{(-1)^{j+\varepsilon_1+\ldots+\varepsilon_s-1}}{2}\, \delta(\underbrace{\tau_1 \otimes \ldots \otimes \tau_1}_{j-1 \text{ copies}} \otimes h_{k_1}^{\varepsilon_1} \otimes \bar{b}_{k_1-1}^{\ell_1} \otimes \ldots \otimes \ldots \otimes \xi_1^2 \otimes \tau_o) \bmod \bar{B}(2).$$

Therefore we have

$$\delta A = \frac{(-1)^{\varepsilon_1+\ldots+\varepsilon_s}}{2}\, \delta(\underbrace{\tau_1 \otimes \ldots \otimes \tau_1}_{j-1 \text{ copies}} \otimes h_{k_1}^{\varepsilon} \otimes \bar{b}_{k_1-1}^{\ell} \otimes \ldots \otimes \xi_1^2 \otimes \tau_o) \in \bar{B}(2).$$

It is evidently that $\delta + \frac{1}{2}(-1)^{\varepsilon_1+\ldots+\varepsilon_s}\delta(\underbrace{\tau_1 \otimes \ldots \otimes \tau_1}_{j-1 \text{ copies}} \otimes h_{k_1}^{\varepsilon_1} \otimes \bar{b}_{k_1-1}^{\ell_1} \otimes \ldots \otimes \xi^2 \otimes \tau_o)$ is a cocycle and is of τ_1 length at most j-1. It should be noticed that $\tau_1 \otimes \ldots \otimes \tau_1 \otimes h_{k_1}^{\varepsilon_1} \otimes b_{k_1-1}^{\ell_1} \otimes \ldots \otimes \xi_1^2 \otimes \tau_o$ is a ba-word, so it follows from proposition 4 that there exist ba-chains $b_o, b_1, \ldots, b_{j-1}$ in B(2) and ba-mixed chain w in M(T(2),T(1)) such that

$$\delta(A + \sum_{i=o}^{j-1} \underbrace{\tau_1 \otimes \ldots \otimes \tau_1}_{i \text{ copies}} \otimes b_i + w) = 0$$

It is obvious that A_R is the biggest word in $A+\sum_{i=o}^{j-1}\underbrace{\tau_1\otimes\ldots\otimes\tau_1}_{i\ \text{copies}}\otimes b_i+w$, so A_R is a c-word and $\bar{A}_R = A+\sum_{i=o}^{j-1}\tau_1\otimes\ldots\otimes\tau_1\otimes b_i+w$.

We shall prove that A_R is an h-word. If A_R is not an h-word, then the image of $\bar{A}_R$ in $C(T(\infty)/\bar{B}(\infty))$ is A_R and is a boundary of $u \in C(T(\infty)=\bar{B}(\infty))$.

For a bigraded cochain complx V, $\alpha \in V^{s,t}$, we use $|\alpha|_\ell$ to denote the first grade s of α and $|\alpha|_r$ to denote the second grade t of α , then

$$|u|_\ell = |A_R|_\ell - 1 = 2(\ell_1+\ldots+\ell_s)+\varepsilon_1+\ldots\varepsilon_s+j \leq p-2$$

$$|u|_r = |A_R|_r = q(\Sigma p^{k_i}(\ell_i+\varepsilon_i)+j+1)+j$$

Since

$$|\tau_i| = q(p^{i-1}+p^{i-2}+\ldots+1)+1$$

$$|\xi_i(t,1)| = qp^t(p^{i-1}+\ldots+1)$$

$$|\xi_i(t,2)| = qp^{t+1}(p^{i-1}+p^{i-2}+\ldots)$$

It is evidently that the number of τ_i factors of u is j. Let $\Sigma\alpha_t p^t$ be the p-adic expansion of $\frac{|u|_r-j}{q}$, then $\alpha_o=j+1$, $\alpha_{k_s}=(\ell_s+\varepsilon_s)$, ..., $\alpha_{k_1}=(\ell_1+\varepsilon_1)$, otherwise $\alpha_m=0$. It follows that $\alpha_i\alpha_{i+1}=0$. Therefore all τ factors of u are τ_1 and all ξ factors in u are b_i or h_k and $\xi_1=h_o$ appears in u (otherwise $\alpha_o=j$). So we have $\delta u=0$. It is impossible that $\delta u=A_R$ in $C(T(\infty)/\bar{B}(\infty)$. Therefore A_R is an h-word.

It can be easily seen by the same method as above that all the image of A_S with $\ell(S)=\ell(R)$, $r(S)=r(R)$ form a basis of $(C(T(\infty)/B(\infty))^{\ell(R),r(R)}$, so all $(\bar{A}_S)$ are linearly independent, thus we get the proof of (c).

<u>Proposition 6</u>: Let R be an H set, $t\geq 2$, then

$$\mathrm{Ext}^{\ell(R)-t,r(R)-t+1}(Z_p,Z_p) = 0$$

Proof: Since $\mathrm{Ext}^{m,n}(Z_p,Z_p)=H^{m,n}(C(T(\infty)/B(\infty))$, it needs only to prove that $C(T(\infty)/B(\infty))^{\ell(R)-t,r(R)-t+1}=0$.

It may be assumed that $t\leq\ell(R)$. Let $u\neq 0 \in (C(T(\infty)/B(\infty))^{\ell(R)-t,r(R)-t+1}$ If $t\geq j+1$, then $|u|_r=|A_R|_r-t+1=q(\Sigma p^{k_i}(\ell_i+\varepsilon_i)+j+1)+j-t+1$, Since $t\geq j+1$, we have $|u|_r=q-(t-j-1)$ mod q. Now

$$0\leq t-j-1<t-1\leq p-2, \quad q-(t-j-1)=2p-2-(t-j-1)\geq p$$

Therefore u contains at least p τ factors which is impossible.

If $t<j+1$, then

$$\frac{|u|_{r}-j+t-1}{q} = \Sigma p^{k_i}(\ell_i+\varepsilon_i)+j+1$$

By the some argument as in the proof of proposition 5 that ξ_i, τ_j doesn't appear in u for $i\geq 2$, $j\geq 2$ so every word in u contains the factor $\tau_1\otimes\ldots\otimes\tau_1$ (j copies). Now $|u|_r=j+1-t \bmod q$ and $t\geq 2$, so $j+1-t\leq j-1<p$. It is also impossible that every word in u contain the factor $\tau_1\otimes\ldots\otimes\tau_1$ (j copies).

Therefore it is impossible that

$$(C(T(\infty)/B(\infty))^{\ell(R)-t,r(R)-t+1} \neq 0.$$

and so

$$\mathrm{Ext}^{\ell(R)-t,r(R)-t+1}(Z_p, Z_p) = 0$$

for all $t\geq 2$.

<u>Proposition 7</u>: Let R be an H set, then

$$h_o \otimes [\bar{A}_R] = \tau_o \otimes [\bar{A}_R] = 0$$

Proof: Since $h_o \otimes \bar{A}_R = h_o \otimes \tau_1 \otimes \ldots \otimes \tau_1 \ldots \otimes h_o + \Sigma h_o \otimes \tau_1 \otimes \ldots \otimes b_i + h_o \otimes w = 0$ mod $\bar{B}(2)$ and $\bar{A}_R \otimes \tau_o = \tau_1 \otimes \ldots \otimes \tau_1 \otimes \ldots \otimes h_o \otimes \tau_o + \sum_{i=o}^{j-1} \tau_1 \otimes \ldots \otimes \tau_1 \otimes b_i \otimes \tau_o =$ $=(-1)^{j+1+\Sigma\varepsilon_i}\delta(\underbrace{\tau_1 \otimes \ldots \otimes \tau_1}_{(j+1)\text{ copies}} \otimes h_{k_1}^{\varepsilon_1} \otimes b_{k_1-1}^{\ell_1} \otimes \ldots \otimes h_{k_s}^{\varepsilon_s} \otimes b_{k_s-1}^{\ell_s})$ mod $\bar{B}(2)$, so we have

$$h_o \otimes [\bar{A}_R] = \tau_o \otimes [\bar{A}_R] = 0$$

Let $\{R= k_1, \ldots, k_s, \varepsilon_1, \ldots, \varepsilon_s, \ell_1, \ldots, \ell_s, j\}$ be an H set, we use j also to denote the H set $\{j\}$ of one integer j, then we have

<u>Proposition 8</u>: Let R be an H set, then

$$(\bar{A}_R) = (-1)^{\varepsilon_1+\ldots+\varepsilon_s}(\bar{A}_j) \otimes h_{k_1}^{\varepsilon_1} \otimes \bar{b}_{k_1-1}^{\ell_1} \otimes \ldots \otimes h_{k_s}^{\varepsilon_s} \otimes \bar{b}_{k_s-1}^{\ell_s}$$

$$= (-1)^{j(\varepsilon_1+\ldots+\varepsilon_s)} h_{k_1}^{\varepsilon_1} \otimes \bar{b}_{k_1-1}^{\ell_1} \otimes \ldots \otimes h_{k_s}^{\varepsilon_s} \otimes \bar{b}_{k_s-1}^{\ell_s} \otimes (\bar{A}_j)$$

PART II.

<u>§1. Properties of Massey product of $(\bar{A}_R)$.</u>

Let R be an H set, since $[\bar{A}_R]\otimes\tau_o=\tau_o\otimes h_o=0$, so we can define the Massey product $\langle(\bar{A}_R), \tau_o, h_o\rangle$. Because $\mathrm{Ext}^{\ell(R),r(R)}(Z_p, Z_p)$ is generated by all $[\bar{A}_S]$ with $\ell(S)=\ell(R)$, $\mathrm{Ext}^{\ell(R),r(R)}(Z_p,Z_p) \otimes h_o = 0$ Since $\mathrm{Ext}^{1,q}(Z_p, Z_p)$ is generated by h_o, we have also $(\bar{A}_R)\otimes\mathrm{Ext}^{1,q}(Z_p, Z_p)=0$. It follows that the indeterminacy of $\langle(\bar{A}_R), \tau_o, h_o\rangle=0$.

Since $(\bar{A}_R) = (-1)^{j\Sigma\varepsilon_i} h_{k_1}^{\varepsilon_1} \otimes \bar{b}_{k_1-1}^{\ell_1} \otimes \ldots \otimes (\bar{A}_j)$, so we have

<u>Proposition 1</u>: $<(\bar{A}_R), \tau_o, h_o> = (-1)^{j\Sigma\varepsilon_i} h_{k_1}^{\varepsilon_1} \otimes \bar{b}_{k_1-1}^{\ell_1} \otimes \ldots \otimes (\bar{b}_{k_s-1})^{\ell_s} \otimes$

$$\otimes <(\bar{A}_j), \tau_o, h_o>.$$

<u>Proposition 2</u>: Let $0 \leq j \leq p-3$, then

$$<(\bar{A}_j), \tau_o, h_o> = (-1)^{j+1}\left(\frac{2+j}{1+j}\right)(A_{j+1})$$

Proof: Since $\bar{A}_j$ may be expressed in the following form

$$\underbrace{\tau_1 \otimes \ldots \otimes \tau_1}_{j \text{ copies}} \otimes \xi_1 + \sum_{i=o}^{j-1} \underbrace{\tau_1 \otimes \ldots \otimes \tau_1}_{i \text{ copies}} \otimes b_i + c$$

where $b_1, \ldots, b_{j-1}$ are ba-chains in $B(1)$ and c is a ba-mixed chain in $C(T_2)$ with respect to $C(T_1)$.

Since $\underbrace{\tau_1 \otimes \ldots \otimes \tau_1}_{j \text{ copies}} \otimes \xi_1 \otimes \tau_o - \frac{(-1)^{j+1}}{j+1} \delta(\underbrace{\tau_1 \otimes \ldots \otimes \tau_1}_{j+1 \text{ copies}}) \in \bar{B}(2)$

and $A_j \otimes \tau_o - \frac{(-1)^{j+1}}{j+1} \delta(\underbrace{\tau_1 \otimes \ldots \otimes \tau_1}_{j+1 \text{ copies}})$ is a cocycle and is contained in $\bar{B}(2)$, so

there exist ba-chains $b'_o, \ldots, b'_j$ in $B(1)$ and ba-mixed chain c' in $M(T(2),T(1))$ such that

$$A_j \otimes \tau_o = \frac{(-1)^{j+1}}{j+1} \delta(\underbrace{\tau_1 \otimes \ldots \otimes \tau_1}_{j+1 \text{ copies}}) + \delta\Big(\sum_{i=o}^{j} \underbrace{\tau_1 \otimes \ldots \otimes \tau_1}_{i \text{ copies}} \otimes b'_i + c'$$

Now $\tau_o \otimes \xi_1 = -\delta(\tau_1) + \delta(\tau_o \xi_1)$, so

$$<(\bar{A}_j), \tau_o, h_o> = \frac{(-1)^{j+1}}{j+1} [\underbrace{\tau_1 \otimes \ldots \otimes \tau_1}_{j+1 \text{ copies}} + \sum_{i=o}^{j} \underbrace{\tau_1 \otimes \ldots \otimes \tau_1}_{i \text{ copies}} \otimes b'_i + c'] \otimes h_o.$$

$$+ (-1)^{j+4} \bar{A}_j \otimes \tau_1 + (-1)^{j+3} \bar{A}_j \otimes (\tau_o \xi_1)$$

$$= \left(1 + \frac{1}{j+1}\right)(-1)^{j+1} (\underbrace{\tau_1 \otimes \ldots \otimes \tau_1}_{j+1 \text{ copies}} \otimes \xi_1) + \ldots$$

therefore, the image of $<A_j \ \tau_o, h_o>$ on $C(T(\infty)/\bar{B}(\infty))$ is

$$\frac{(2+j)(-1)^{j+1}}{1+j} \underbrace{\tau_1 \otimes \ldots \otimes \tau_1}_{j+1 \text{ copies}} \otimes h_o$$

so we have $<(\bar{A}_j), \tau_o, h_o> = \left(\frac{2+j}{1+j}\right)(-1)^{j+1} \ (\bar{A}_{j+1})$.

Let $R = \{k_1, \ldots, k_s, \varepsilon_1, \ldots, \varepsilon_s, \ell_1, \ldots, \ell_s, j\}$, $j>0$, be an H set, we define an H set R' as follows

$$R'=\begin{cases}\{k_1, \ldots, k_s, \varepsilon_1, \ldots, \varepsilon_{s-1}, 0, \ell_1, \ldots, \ell_s, j-1\} & \text{if } \varepsilon_s=1,\ \ell_s>0\\ \{k_1, \ldots, k_{s-1}, \varepsilon_1, \ldots, \varepsilon_{s-1}, \ell_1, \ldots, \ell_{s-1}, j-1\} & \text{if } \varepsilon_s=1,\ \ell_s=0\\ \{k_1, \ldots, k_s, \varepsilon_1, \ldots, 0, \ell_1, \ldots, \ell_s-1 & \text{if } \varepsilon_s=0,\ \ell_s>1\\ \{k_1, \ldots, k_{s-1}, \varepsilon_1, \ldots, \varepsilon_{s-1}, \ell_1, \ldots, \ell_{s-1}, j-1\} & \text{if } \varepsilon_s=0,\ \ell_s=1\\ \{j-1\} & \text{if } j>0,\ s=0\end{cases}$$

We define $R''=(R')'$ ... and $R^{(n+1)}=(R^{(n)})'$. If $j+\sum_{i=1}^{s}(\ell_i+\varepsilon_i)\geq n+1$.It can be easily seen that

$$R^{(\sum_{i=1}^{s}(\ell_i+\varepsilon_i))} = \{j-\sum_{i=1}^{s}(\ell_i+\varepsilon_i)\}$$

It follows from proposition 8 of the last section and proposition 2 of the present section that we have the following

<u>Proposition 3</u>: Let R be an H set, then

$$(\bar{A}_R)=\begin{cases}(-1)^{\ell(R)}(\frac{j}{1+j})\langle A_{R'}, \tau_o, h_{k_s}\otimes h_o\rangle & \text{if } \varepsilon_s=1,\ s>0.\\ (-1)^{\ell(R)}(\frac{j}{1+j})\langle A_{R'}, \tau_o, \bar{b}_{k_s-1}\otimes h_o\rangle & \text{if } \varepsilon_s=0,\ s>0.\\ (-1)^{\ell(R)}(\frac{j}{1+j})\langle A_{R'}, \tau_o, h_o\rangle & \text{if } s=0,\ j>0.\end{cases}$$

<u>§2. Toda's secondary product $\langle\alpha, p, \beta\rangle$, the definition of P(R) and the proof of theorem 1.</u>

Let $\alpha, \beta\in\Pi_*(S^o)$, $p\alpha=0$, $p\beta=0$, then we can define Toda's secondary product $\langle\alpha, p, \beta\rangle \in\Pi_*(S^o)$.

Proposition : $p\langle\alpha, p, \beta\rangle = 0$.

Proof: Let $\alpha : S^{m+n}\to S^n$, $\beta: S^n\to S^o$ be two maps, since $p\alpha=0$, α may be extended to a map $\alpha' : S^{m+n}\cup_p e^{m+1+n}\to S^n$. Since p is odd, it may be assumed that $p\alpha'=0$, so we can define $\langle\alpha', p, \beta\rangle$ and $\langle\alpha, p, \beta\rangle$ may be considered as the restriction of $\langle\alpha', p, \beta\rangle$ on S^{m+n+1}, so we have $p \langle\alpha, p, \beta\rangle = 0$.

For H set R, we define $P(R) \in \Pi_*(S^o)$ as follows.

For R = {0}, we define $P(R) = \alpha_1$. It is evident that $(\bar{A}_R)=(\bar{A}_o)=h_o$ represents P(R).

Suppose that P(R) had been defined for $j(R)<j_o$ such that $pP(R)=0$ and $(\bar{A}_R)$ represents P(R). If $j(R)=j_o$, then $j(R')=j_o-1$, so P(R') had been defined, we define

$$\begin{cases}(-1)^{\ell(R)}\frac{j_o}{1+j_o}\langle P_{R'}, p, \eta_{k_s}\rangle, & \text{if } \varepsilon_s=1,\ s>0\end{cases}$$

$$P(R)=\begin{cases} (-1)^{\ell(R)} \dfrac{j_o}{1+j_o} \langle P_{R'}, p, \zeta_{k_s -1}\rangle, & \text{if } \varepsilon_s=0,\ s>0 \\ (-1)^{\ell(R)} \dfrac{j_o}{1+j_o} \langle P_{R'}, p, \alpha_1\rangle, & \text{if } j>0,\ s=0. \end{cases}$$

It shoulds be pointed out that $j_o \neq 0 \pmod p$, $(1+j_o) \neq 0 \pmod p$ since $0 \leq j_o \leq p-2$.

It follows from proposition 1 in this section that $pP(R)=0$, since $(\bar{A}_{R'})$ represents $P_{R'}$, τ_o represents p, $h_k \otimes h_o(\bar{b}_{k-1} \otimes h_o)$ represents $\eta_k(\zeta_{k-1})$. It follows from theorem 8.1 in [5] and proposition 2 of the preceeding section that we have $(\bar{A}_R)$ represents P_R, thus we have

Proposition 2: Let R be an H set, then $(\bar{A}_R)$ represents P_R and therefore is an infinite cycle.

Since $\mathrm{Ext}^{\ell(R)-t,r(R)-t+1}(Z_p,Z_p) = 0$ for all $t \geq 2$, it follows that $d_t(\mathrm{Ext}^{\ell(R)-t,r(R)-t+1}(Z_p,Z_p)) = 0$, where d_t denotes the t-th differential of Adams spectral sequence, therefore we have

Proposition 3: Let R be an H set, then all P_s with $\ell(S)=\ell(R)$, $r(R)=r(S)$ are linearly independent in $\Pi_*(S^o)$.

The proof of theorem 1: (a) of theorem 1 follows directly from proposition 1 of §5 of Part I and proposition 2 in this section and (b) follows directly from proposition 1 in this section and proposition 5 of §6 of Part I.

§3. The proof of theorem 2.

Proposition 1: Let $0 \leq r \leq p$, $s=0 \pmod q$, $2 \leq j \leq r$, then

$$\mathrm{Ext}^{r-j,s-j+1}(Z_p,Z_p) = 0$$

Proof: We shall prove that $(C(T(\infty)/\bar{B}(\infty))^{r-j,s-j+1} = 0$.

Let $u \neq 0$ and $u \in (C(T(\infty)/B(\infty))^{r-j,s-j+1}$ and $\tau_{t_1} \otimes \dots \otimes \tau_{t_m}$ be the τ factor of u, then on the one hand $|u|_r = m \pmod q$, on the other hand, $|u|_r = s-j+1 = -j+1 \pmod q$, so we have $m = -j+1 \pmod q$. Therefore $m+j=q+1=2p-1$.

Now $m \leq r-j \leq p-2$, $j \leq p$ so $m+j \leq 2p-2 < 2p-1$ and $m+j=2p-1$ is impossible, so

$$(C(T(\infty)/\bar{B}(\infty))^{r-j,s-j+1} = 0$$

Therefore we have

$$\mathrm{Ext}^{r-j,s-j+1}(Z_p,Z_p) = H^{r-j,s-j+1}(C(T(\infty)/B(\infty)) = 0.$$

The proof of theorem 2: Let

$$w = (\overline{\xi_2 \otimes h_1})^{\varepsilon_1} \otimes (h_k \otimes h_o)^{\varepsilon_2} \otimes (\bar{b}_{k-1} \otimes h_o)^{\varepsilon_3} \otimes h_o^{\varepsilon_4} \otimes \bar{b}_o^{\ell}$$

Since all factors in w are all infinite cocycles in Adams spectral sequence. In order to prove 2, it is sufficient to prove that $(w) \neq 0$ in $\mathrm{Ext}^{**}(Z_p,Z_p)$ and $w \notin d_j(\mathrm{Ext}^{|w|_\ell - j, |w|_r - j+1}(Z_p,Z_p))$. for all $j \geq 2$.

If $(w)=0$, then there is a $v \in C(T(\infty)/\bar{B}(\infty)$ such that $\delta v = w$, so we have $|v|_\ell = |w|_\ell - 1 \leq p-1$ and $|v|_r = |w|_r = 0 \pmod q$. Therefore no τ_i appears in v. Let $\Sigma \lambda_n p^n$ be the p-adic expansion of $|w|_r/q = p^k(\varepsilon_2+\varepsilon_3) + p(\ell + 2\varepsilon_1) + (\varepsilon_1+\varepsilon_2+\varepsilon_3)$, then

$$\lambda_n = \begin{cases} \varepsilon_2 + \varepsilon_3, & n=k \\ \ell + 2\varepsilon_1, & n=1 \\ \varepsilon_1+\varepsilon_2+\varepsilon_3, & n=0 \\ 0 & \text{otherwise} \end{cases}$$

Since

$$|\xi_i(j,1)|_r = p^j(p^{j-1} + p^{i-2} + \dots + 1)$$

$$|\xi_i(j,2)|_r = p^{j+1}(p^{i-1} + p^{i-2} + \dots + 1)$$

therefore v may be expressed as $\lambda \xi_2 \otimes x$ or λx, $x \in H^{**}(P(\xi_1))$, $\lambda \in Z_p$, $\lambda \neq 0$. If $v = \lambda x$, then $\delta x = 0 \neq w$. It is impossible. If $v = \lambda(\xi_2 \otimes x)$, so $w = \delta(\lambda \xi_2 \otimes x) = \lambda h_1 \otimes h_o \otimes x$. If $\varepsilon_1 = 1$, it is impossible that $w = \overline{\xi_2 \otimes h_1} \otimes \dots$ may be written in the form $\lambda(h_1 \otimes h_o \otimes x)$ $x \in h^{**}(P(\xi_1))$. If $\varepsilon_4 = 1$, $\varepsilon_2 = \varepsilon_3 = \varepsilon_1 = 0$, it is also impossible that $w = h_o \otimes b_o^\ell$ may be written in the form $\lambda(h_1 \otimes b_o \otimes x)$. If $\varepsilon_1 = 0$, $\varepsilon_2 + \varepsilon_3 = 1$, $\varepsilon_4 = 0$ then $w = b_o^\ell \otimes b_{k-1} \otimes h_o$ or $b_o^\ell \otimes b_k \otimes h_o$, it is also impossible that w may be written in the form $h_1 \otimes h_o \otimes x$. Thus $(w) \neq 0$ in $\mathrm{Ext}^{\ell(w),r(w)}(Z_p,Z_p)$.

Since $|w|_r = 0 \pmod q$, we have $\mathrm{Ext}^{|w|_\ell - j, |w|_r - j+1}(Z_p,Z_p) = 0$ for $j \geq 2$, therefore

$$(w) \notin d_j(\mathrm{Ext}^{|w|_\ell - j, |w|_r - j+1}(Z_p,Z_p)) = 0$$

So w represents a nontrivial element of the stable homotopy groups of spheres, thus theorem 2 is proved.

REFERENCES

[1] Adams J.F. : On the structure and applications of the Steenrod algebra. Comm. Math. Helv, 32(1958) 180-274.

[2] Adams J.F. : On the nonexistence of elements of Hopf invariant one. Ann. of Math. 72(1960) 20-104.

[3] Cohen R.L. : Odd primary infinite families in stable homotopy theory. Memoirs of Amer. Math. Soc. 242(1981).

[4] Cohen R.L. and Goerss P. : Secondary cohomology operations that detects homotopy classes. Topology 23(1984) 177-194.

[5] Kochmann S.O. : A chain functor for bordism. Trans. Amer. Math Soc. 239(1978) 167-190.

[6] Kochmann S.O. : Uniqueness of Massey products on the stable homotopy groups of spheres. Can. J. Math. 32(1980) 576-589

[7] Liulevicius : The factorization of cycle reduced power by secondary cohomology operation. Memoirs of Amer. Math. Soc. no. 42(1962).

[8] Mahowald M.E. : A new infinite family in ${}_2\Pi^s_*$. Topology 16(1977).

[9] May J.P. : The cohomology of restricted Lie algebra and Hopf algebra. J. algebra 3(1966) 123-146.

[10] Ravenel D.C. : Complex cobordism and stable homotopy groups of spheres. Academic Press INC 1986.

[11] Toda H. : Composition methods in homotopy groups of spheres. Ann. of Math. Studies No. 49(1962).

PROBLEM SESSION

Problem Session for Homotopy Theory

J.F. Adams

1. (A. Adem) Let G be a finite p-group. If $H^n(G;Z)$ has an element of order p^r, for some value of n, does the same follow for an infinity of n?

2. (J. Berrick) Recall that a group P is perfect if $P=[P,P]$. Let PG be the maximal perfect subgroup of G. Consider fibrations $F \to E \to B$ of connected spaces, where $P\pi_1 E = 1$ and F is of finite type. If $H_*(F;Z) \to H_*(E;Z)$ is iso, does it follow that $\pi_1 B = 1$?

   This question arises from the consideration of "plus-constructive fibrations", i.e. those for which $F^+ \to E^+ \to B^+$ is also a fibration.

3. (F. Cohen) Let $[2]: S^n \to S^n$ be a map of degree 2. When is $\Omega^q[2]: \Omega^q S^n \to \Omega^q S^n$ homotopic to the H-space squaring map $2: \Omega^q S^n \to \Omega^q S^n$? If $q=1$, it is so iff $n = 1, 3, 7$. If $q = 2$, it can only be so if $n = 2^k - 1$; what happens in this case? What happens for $q = 3$, $n = 5$?

4. (M. Feshbach) Call a p-group P "necessary" if there is a non-zero class $x \in H^*(P; Z/(p))$ which restricts to zero on all proper subgroups of P. Can one give a useful alternative description of the necessary p-groups?

   Conjecture: for $p = 2$, P is necessary iff every element of order 2 is central, i.e. iff P contains a unique maximal elementary abelian 2-subgroup and this subgroup is contained in the centre of P.

   (The obvious generalisation of this conjecture to p odd is false.)

5. (B. Gray) For which values n, r is the fibre $W_{n,r}$ of the iterated suspension map $S^n \to \Omega^r S^{n+r}$, localised at $p = 2$, an H-space?

   For p odd one should assume n even and replace S^n by the subspace S^n_{p-1} of the James construction; one then asks if the fibre is a double loop space.

6. (B. Gray) Suppose $\Phi: \Omega^2 S^{2np+1} \to S^{2np-1}$ satisfies

   a) $\Phi \mid S^{2np-1}$ has degree p (>2)

   b) The composition

$$\Omega^2 S^{2n+1} \xrightarrow{\Omega H_p} \Omega^2 S^{2np+1} \xrightarrow{\Phi} S^{2np-1}$$

   is nullhomotopic.

Is it then true that the composition

$$\Omega^3 S^{2np+1} \xrightarrow{\Omega\Phi} \Omega S^{2np-1} \xrightarrow{\Sigma^3} \Omega^3 S^{2np+1}$$

is the triple loops on the degree p map?

7. (K. Ishiguro) Let π be a finite group and G a compact Lie group. Is $[B\pi, BG]$ a finite set?

Let π_p run over the Sylow p-subgroups of π. Is the map

$$[B\pi, BG] \to \prod_p [B\pi_p, BG]$$

injective?

8. (N. Kuhn) Find conditions on X and Y such that

$$X_{K(n)} \simeq Y_{K(n)} \ \forall n \geq 1 \ \text{ implies } \ X \simeq Y .$$

(Here $K(n)$ is the nth Morava K-theory.)

9. (N. Kuhn after D.C. Ravenel) Are all suspension spectra E harmonic (meaning that E is $\bigvee_{n \geq 0} K(n)$-local)?

10. (N. Kuhn) Describe the equivariant cobordism $MU^*_G(pt)$. Find an equivariant version of Landweber's "exact functor" theorem.

11. (N. Kuhn) Set up equivariant $K(n)^*$, $E(n)^*$ and prove a "completion theorem" for these theories.

12. (N. Kuhn) Find good models for the infinite loop spaces representing complex-oriented cohomology theories such as $K(n)^*$, $E(n)^*$.

13. (M.E. Mahowald) Work of Barratt, Mahowald and Jones (The Kervaire invariant and the Hopf invariant, to appear in the Proceedings of the emphasis year in Seattle) leaves the following conjecture open:

Conjecture. If $\alpha \in \pi_{2j-2}$ represents a framed manifold with Kervaire invariant one, then the composite

$$QS^0 \xrightarrow{s} QP \to \Omega^\infty P \wedge J$$

where s is the Snaith map, induces a nonzero map on α.

The conjecture is true if α has order 2. Also, there are other Kervaire invariant questions in the Proceedings of the Northwestern conference 1982, Contempory Mathematics AMS vol 19.

14. (M.E. Mahowald) In numerical analysis it is often very useful to consider differences in a sequence to gain understanding of how the sequence is put together. In topology, differences might be construed to be fiber of a map which is an isomorphism in the first non-zero dimension of both spaces. Thus, the EHP sequence tells us that, at the prime 2, the difference between S^n and ΩS^{n+1} is ΩS^{2n+1}. Let $W(n)$ be the difference between S^{2n-1} and $\Omega^2 S^{2n+1}$. Work in Mahowald, The double suspension homomorphism, Trans AMS 214(1975)169-178, shows that each of the spaces $W(n)$ has a resolution by $K(Z/2,n)$'s which is a good approximation to the Adams resolution for the stable $Z/2$ Moore space. Using the above language, the difference between $W(n)$ and $\Omega^4 W(n+1)$, which we call $X(n)$, is constructed and shown to have a resolution which has many properties of the stable resolution of a spectrum whose cohomology is free on one generator over the sub algebra, A_1, of the Steenrod algebra generated by Sq^1 and Sq^2. It would be interesting to construct a map form $X(n)$ to the omega spectrum having as its stable cohomology an appropriate A_1. To do this at the chain level in the Λ-algebra sense, as was done in the Trans. paper above, would already be very interesting. The conjecture is that the difference between the $X(n)$'s should approximate A_2 and the differences between these differences should approximate A_3, etc.

15. (M.E. Mahowald) Let C_n be the fiber of $\Omega^{2n+1}S^{2n+1} \to \Omega^\infty P^{2n} \wedge J$. Find a stable space B_n and a map $C_n \to QB_n$ which is a v_2 equivalence. B_n could be a subspace of the space $L(2)$ of Mitchell and Priddy.

16. (M.E. Mahowald) The Work of Devinatz, Hopkins and Smith suggests that the Freyd generating hypothesis is now approachable.

17. (J.P. May) Let G be a finite p-group, X a finite G-complex, SX its singular subspace. If G has order p^e, then

$$\Sigma (p^e - 1)^i \dim H^i(SX, Z_p) \leq \Sigma (p^e - 1)^i \dim H^i(X, Z_p).$$

Are there other such generalizations of Smith theory? The point is that the homotopical structure of finite G-complexes is much more restricted than Smith theory alone dictates.

18. (J.P. May) Give an algebraic analysis of the rationalized stable category of G-spectra. Rational G-spectra fail to split as products of Eilenberg-MacLane G-spectra for general compact Lie groups G, although this does hold for finite G.

19. (J.P. May) Let $H \subset J \subset K \subset G$, where G is a compact Lie group. Suppose $q(H, p) = q(K, p)$. Is $q(J, p) = q(H, p)$? This is one of many questions involving the complexity of the lattice of closed subgroups of G.

20. (J.P. May) Let $B_G\Pi$ be the classifying G-space for principal (G, Π)-bundles. For a G-cohomology theory h^*_G, $h^*_G(B_G\Pi)$ gives all h^*_G-characteristic classes for (G,Π)-bundles. Calculate these groups in interesting cases. When h^*_G is Borel cohomology, complete information is easily obtained and is quite unilluminating. When h^*_G is stable cohomotopy or K-theory, a complete theoretical answer has been obtained [Adams, Haeberly, Jackowski, May, 1985] via generalizations of the Segal conjecture and the Atiyah-Segal completion theorem. When Π is Abelian, $B_G\Pi \simeq B\Pi \times K(R, 0)$, $R(G/H) = \mathrm{Hom}(H, \Pi)$, and the Bredon cohomology of $B_G\Pi$ is computable. This casts doubt on methods based on reduction to a maximal torus.

21. (J.P. May) The Borel construction on (G, Π)-bundles corresponds to a G-map $\alpha: B_G\Pi \to \mathrm{Map}(EG, B\Pi)$. When G and Π are discrete or when G is compact Lie and Π is Abelian compact Lie, α is a G-homotopy equivalence. When G is a finite p-group and Π is a compact Lie group, Dwyer's results show, essentially, that α is a p-adic equivalence. What can be said when G is a general finite group?

22. (F. Quinn) Get information about the space of based maps $\mathrm{Map}_0(B_\pi, B_{G(X)})$, where π is a finite group and $G(X)$ is the monoid of self-equivalences of the finite complex X.

23. (F. Quinn) Get information about

$$\mathrm{Map}(B_\pi, E_{G(X)} \times X / G(X)).$$

(For background and discussion of (22), (23) see F. Quinn's paper "Spaces of group actions" in these proceedings.)

24. (D.C.Ravenel) **Formal A-modules.** Let A be the ring of integers in a finite extension of the p-adic numbers Q_p. A formal A-module F over an A-algebra R is a formal group law over R equipped with power series $[a](x)$ for each $a \in A$ satisfying

i. $[1](x) = x$

ii. $[a_1 + a_2](x) = F([a_1](x), [a_2](x))$

iii. $[a](x) \equiv ax \bmod x^2$.

There is a well developed theory of such objects beginning with Lubin-Tate's work on local class field theory. In particular there is a universal p-typical formal A-module defined over an A-algebra V_A which is explicity known and which is a BP_*-module.

Question Is there a spectrum S_A such that $BP_*(S_A) = V_A$?

If the answer is yes then the Novikov E_2-term for S_A has many interesting properties.

S_A is known to be an Eilenberg-MacLane spectrum when the field is algebraically closed.

The spectrum $S_A \wedge BP$ has been constructed in many cases by A. Pearlman.

25. (D.C. Ravenel) The spectra $T(n)$

$T(n)$ is a spectrum with $BP_* T(n) = BP_* [t_1, t_2 \cdots t_n]$. It figures in the proof of the nilpotence theorem.

It is not difficult to compute the Adams-Novikov E_2-term for $T(n)$ through dimension

$$2(p^2+1)(p^{n+1}-1).$$

It is in roughly this dimension that the first possible nontrivial differential occurs. In the case $n = 0$ this is the Toda differential which kills $\alpha_1 \beta^p$, for $p > 2$. For $n > 0$ the nontriviality of this differential is an open question. The usual extended power constructions do not settle the question.

26. (D.C. Ravenel) The order of β_1

For $p > 2$, $\beta_1 \in \pi^S_{2p^2-2p-2}$ is the first even dimensional stable homotopy element in positive dimensions.

Problem: Find the smallest k such that $\beta_1^k = 0$.

Toda showed (<1970?) that $k \le p^2 - p + 1$ for all $p > 2$.

For $p = 3$ $k = 6$ (*Toda*)

For $p = 5$ $k = 18$.

The relevant ANSS differentials

are $d_9(\alpha_1 \beta_4) = \beta_1^6$ $(p = 3)$

and $d_{33}(\gamma_{33}) = \beta_1^{18}$ $(p = 5)$.

Conjecture

For $p = 7$ $\beta_{139} = \gamma_3 \gamma_2$

$p = 11$ $\beta_1^{105} \in \langle \gamma_3, \gamma_2, \gamma_2, \gamma_2 \rangle$

etc

and these lead to $\beta\xi^{p^2-p} = 0$ for $p > 5$.

27. **(D.C. Ravenel) The chromatic filtration of the Burnside ring.**

For a finite group G, the Segal conjecture tells us that the stable cohomotopy group $[BG, S^0]$ is the completed Burnside ring of G, $A(G)^\wedge$.

Let $L_n S^0$ denote the Bousfield localization of S^0 with respect to $v_n^{-1} BP$. Let $F_n A(G)^\wedge$ denote the kernel of the map

$$A(G)^\wedge = [BG, S^0] \to [BG, L_n S^0].$$

The problem is to determine the ideals F_n.

Conjecture. A virtual finite G-set X is in F_n if $\#(X^H) = 0$ for all subgroups $H \subset G$ generated by $\leq n$ elements.

28. **(D.C. Ravenel) Morava K-theory of loop spaces**

Find a way to compute $K(n)_* \Omega^k \Sigma^k X$, either as a functor of $K(n)_* X$ or by showing that a suitable Eilenberg-Moore spectral sequence converges.

McClure has done this for $n = 1$ and $k = \infty$.

For $n = 1$ and $k < \infty$ one could then prove that the Smith map

$$\Omega_0^{2n+1} S^{2n+1} \to QRP^{2n}$$

is a $K(1)_*$-equivalence.

29. **(P. Shick)** Relate the v_n-torsion or v_n-periodic behaviour of $\alpha \in [X, S^0]_j$ to that of its root invariant

$$R(\alpha) \in [X, S^0]_{j+N}$$

in the sense of "Implications of Lin's theorem in stable and unstable homotopy theory", M.E. Mahowald and D.C. Ravenel, to appear.

(M.E. Mahowald and P. Shick have made some progress with this since the Arcata conference.)

H-spaces

1) N. Iwase

A) Determine the higher associativity of the pull-back of a sphere extension of a Lie group by the degree k mapping.

B) Find the example of a space which admits an A_n-structure but no A_n-primitive A_n-structure. This is unsolved even if $n = 2$.

C) Is there a three-connected homotopy associative H-space?

D) When does the Bar construction functor induce a weak equivalence from the space of homeomorphisms between compact Lie groups to the mapping space between their classifying spaces? This is not always true and not always false.

E) Justify the equivariant theory (homotopy theory, simple homotopy theory, algebraic K-theory, etc.) for non-compact Lie groups, or make clear the essential obstructions.

2) J. P. Lin

A) Prove a 14-connected finite H-space is acyclic.

B) Suppose $f: Y \longrightarrow Z$ factors as

$$Y \xrightarrow{(\overline{\Delta}\wedge 1)\overline{\Delta}} Y\wedge Y\wedge Y \longrightarrow Z ,$$

where Y is an H-space. If X is the fibre of f, does X split as $\Omega Y \times \Omega^1 Z$ as homotopy commutative H-spaces?

C) Are there any finite loop spaces whose mod 2 cohomology is not the mod 2 cohomology of a Lie group?

D) Suppose X is a 1-connected finite H-space and $A = H^*(X;\mathbb{Z}_2)$ is the corresponding cohomology Hopf algebra over $A(2)$. Are there "irreducible" Hopf algebras over $A(2)$ such that A splits as the tensor product of irreducible Hopf algebras over $A(2)$?

E) Can a finite loop space ΩB have $H^*(B;\mathbb{Q}) = Q[X_{n_1},\ldots,x_{n_r}]$ where $[n_1,\ldots,n_r] = [4,4,48,8,812,12,16,16,20,24,24,28]$? (This is an example of Adams-Wilkerson.)

F) Given X a 1-connected finite H-space, what can be said about the action of $A(2)$ on $PH^*(\Omega X;\mathbb{Z}_2)$?

G) Is a 6-connected finite H-space a product of seven spheres?

H) If X is a finite loop space is $H^*(X;\mathbb{Z}) = H^*(\text{Lie grp};\mathbb{Z})$?

K and L-theory

1. J. Milgram, L. Taylor, B. Williams

A) The Quinn-Ranicki assembly map has been attacked by a factorization which we can define on bordism as follows. There are pairings

$$(*) \qquad L_*(\mathbb{Z}[s]\pi) \otimes C \longrightarrow L_{*+1\mp 1}(\mathbb{Z}\pi)$$

where $C^{\pm}$ is the knot group of $S^{4k\mp 1} \longrightarrow S^{4k\pm 1}$ $(k \geq 2)$ and the involution on $\mathbb{Z}[s]$ is $s \longrightarrow 1-s$. Moreover, there is a "symmetric signature" $\Omega_*(B\pi) \longrightarrow L_*(\mathbb{Z}[s]\pi)$ such that the composite of the "symmetric signature" $\otimes$ {trefoil knot} is the obstruction to $M \times$(Kervaire problem). The E_8 knot $S^3 \longrightarrow S^5$ gives the obstruction to $M \times$(Milnor problem). Explain the presence of the knot group in the pairings (*).

B) In dimension 3 surgery can be done on homology though we don't have control of π_1 of the resulting manifold. For example, results of Madsen and Milgram on the spaceform problem for the groups $Q(8p,q,1)$, $(p,q) = 1$, p, q prime, show that for certain pairs such as (17,133), the surgery obstruction is trivial. Hence, there exists a homology 3-sphere M^3 with a free action of $Q(136,113,1)$. Find explicit examples of such actions. In particular what kinds of π_1 occur for M^3 ?

C) If π is finite, then the reduced surgery obstruction map

$$\bar{\sigma}: \Omega_n(B\pi \times G/TOP) \longrightarrow L'_n(\mathbb{Z}\pi)/L_n(\mathbb{Z})$$

factors through $\bigoplus_{i=1}^{4} H_i(\pi,\mathbb{Z}/2)$. (Here " ' " denotes $\ker(Wh_1(\mathbb{Z}\pi) \longrightarrow Wh_1(\mathbb{Q}\pi))$.

(a) Does $\bar{\sigma}$ factor through $\bigoplus_{i=1}^{3} H_i(\pi;\mathbb{Z}/2)$?

(b) Determine the corresponding L^s-results.

D) Understand the relationship between L-theory and Hermitian K-theory (in the sense of Quillen)

Conjecture: For any ring with involution (R,α) and $\varepsilon = \pm 1$, there exists a homotopy fibration

$$\Omega^{\infty}(S^{\infty}_{+} \hat{\wedge}_{\mathbb{Z}/2} K(R)) \xrightarrow{\tilde{H}} K\,\mathrm{Herm}(R,\alpha,\varepsilon) \longrightarrow \underline{L}^{*}(R,\alpha,\varepsilon),$$

where $\hat{\wedge}_{\mathbb{Z}/2}$ denotes $\mathbb{Z}/2\mathbb{Z}$-homotopy orbits and $\underline{L}$ denotes symmetric L-theory. Karoubi periodicity implies this conjecture is true when $\frac{1}{2} \in R$. The hyperbolic map $H: \underline{K(R)} \longrightarrow \underline{K\,\mathrm{Herm}}(R,\alpha,\varepsilon)$ factors through $\tilde{H}$.

2. V. Snaith

A finite dimensional representation of a finite Galois group, $G(L/K)$, where L/K is a local field extension is called a Galois representation The Deligne-Langlands local constants are homomorphisms $W_K: R(G(L/K)) \longrightarrow S^1$. (See Tate: Proceedings of the Durham conference (1977), editor. A. Fröhlich).

I have a general formula for $W_K(\rho)$ which is partially topological, partially number theoretic, and very complicated. However, if $\rho: G(L/K) \longrightarrow U_n(\mathbb{C})$ is the complexification of $\rho': G(L/K) \longrightarrow O_n(\mathbb{R})$, Deligne [Inventiones Math. (1976)] showed that $W_K(\rho) = (SW_2\ \rho')\cdot W_K(\det \rho) \in \{\pm 1, \pm\sqrt{-1}\}$ where $SW_2[\rho] \in H^2(K;\mathbb{Z}/2) \cong \{\pm 1\}$. Hence on $RSO(G(L/K))$, W_K is $ISO(G(L/K))$ - adically continuous, in fact $W_K(ISO(G(L/K))^3 = \{1\}$. On $RSp(G(L/K))$, the symplectic representation ring, W_K is $\{\pm 1\}$-valued. Is W_K trivial on some $IO(G(L/K))^N \cdot RSp(G(L/K))$?

Manifolds & Bordism

1) P. Gilkey

The eta invariant of Atiyah-Patodi-Singer defines maps

$$\eta: \tilde{K}(S^{2k-1}/G) \otimes \tilde{K}(S^{2n-1}/G) \longrightarrow Q/Z$$

$$\eta: \tilde{\Omega}^U_*(BG) \otimes R_o(G) \otimes R(U) \longrightarrow Q/Z$$

where G is a spherical space form group, $R_o(G)$ is the augmentation ideal of the representation ring of G, and $R(U)$ is the representation ring of the unitary group. The first is a perfect pairing and the second is non-singular in the first factor -- i.e. the eta invariant completely detects the K-theory of spherical space forms and equivariant unitary bordism of spherical space form groups.

Question one: What is the situation for constant curvature 0 or constant curvature -1 ? Why is the eta invariant so successful in this setting?

The second map can also be interpreted as giving a map

$$\eta: \tilde{\Omega}^U_*(BG) \otimes R(U) \longrightarrow [bu_*(BG)]$$

For $G = Z_p$ (p-prime) it is well known

$$\tilde{\Omega}^U_*(BZ_p) \simeq bu_*(BZ_p) \otimes Z[X_4, X_6, \ldots] .$$

We conjecture for spherical space form groups an additive splitting

Question two: $\tilde{\Omega}^u_*(BG) = bU_*(BG) \otimes Z[X_4, X_6, \ldots]$ and have proved it for $G = Z_4$, $G = \{\pm 1, \pm i, \pm j, \pm h\}$. The proof of this additive isomorphism is analytic; one wants a topological proof for spherical space form groups in general.

2) Problem (M. Kreck and S. Stolz):

Suppose that G and G' are compact simple Lie groups. Is it true that homeomorphic homogeneous spaces G/H and G'/H' are diffeomorphic? (If G and G' are not simple there are counterexamples (M. Kreck and S. Stolz: to appear in the Annals of Math.)) giving a negative answer to the corresponding general problem posed by W.C. and W.Y. Hsiang in 1966.)

3) M. Kreck, A. Libgober, and J. Wood

Is the diffeomorphism type of a complete intersection X_n in $\mathbb{C}P_{n+r}$ determined by its dimension, total degree, intersection pairing or Arf invariant (undefined or ±1), and Stiefel-Whitney and Pontrjagin classes?

4) A. Libgober and J. Wood

Is a compact Kähler manifold which is homotopy equivalent to $\mathbb{C}P_n$ necessarily analytically equivalent to $\mathbb{C}P_n$? [Yes $n = 2$ (Yau), $n = 4,6$ (Libgober-Wood)]

Transformation Groups

1. W. Browder

A) Under certain circumstances, I have shown that equivariant homotopy equivalence implies isovariant homotopy equivalence, in particular with a strong "gap" hypothesis that

(a) $\dim M^H < \frac{1}{2} \dim M^K$ and

(b) $(\dim M^H, \dim M^K) \neq (1,4)$

for every pair of isotropy groups $K \subset H$, where G is finite, acting PL. Weakening condition (a), tom Dieck and Löffler have shown that a linking invariant between fixed point sets can occur which is an obstruction to isovariant equivalence.

(i) Describe the first non-trivial obstructions for this problem, and can they be expressed as linking phenomena?

(ii) Remove the (1,4)-condition (b).

B) If $G = \prod^k \mathbb{Z}_p$ acts freely on $S^{n_1} \times \ldots \times S^{n_\ell}$, is $k \leq \ell$? (Classical)

If $n_1 = \ldots = n_\ell$, this is known by work of Carlsson (Inventiones, 1982) when G acts trivially on homology, and Adem-Browder (Inventiones, 1988) when $n_1 \neq 1,3,7$ or $p \neq 2$.

2. M. Morimoto

Let G be a compact Lie group (possibly a finite group). In the following we treat only smooth actions. If a G-manfiold has exactly one G-fixed point, then the action is called the one fixed point action. E. Stein [S], T. Petrie [P], E. Laitinen - P. Traczyk [L-T] and M. Morimoto [M1-2] studied one fixed point actions on spheres. We know that S^6 and S^7 and some higher dimensional spheres have one fixed point actions. Recently M. Furuta [F] showed that S^4 does not have one fixed point orientation preserving actions of finite groups, by observation of a moduli space of self-dual connections of some principal SO(3)-bundle over S^4, and his idea originates from Donaldson's work. Assuming Furuta's result, M. Morimoto showed that S^4 does not have one fixed point actions ([M2]).

Problem A. Does there exist a one fixed point action on S^3 ?

We note that some 3-dimensional homology spheres have one fixed point actions of A_5, the alternating group on five letters.

Problem B. Does there exist a one fixed point action on D^4 ?

Problem C. Does there exist a one fixed point action on S^5 ?

Problem D. Does there exist a one fixed point action on S^8 ?

From the results of R. Oliver [O] and W.-y. Hsiang - E. Straume [H-S], it holds that S^8 does not have one fixed point actions of compact connected Lie groups.

In the above S^n means the standard n-dimensional spheres and D^4 means the standard 4-dimensional disk.

References

[F] M. Furuta: A remark on fixed points of finite group actions on S^4, in preparation.

[H-S] W.-y. Hsiang-E. Straume: Actions of compact connected Lie groups on acyclic manifolds with low dimensional orbit spaces. J. reine angew. Math. 369, 21-39 (1986).

[L-T] E. Leitinen - P. Traczyk: pseudofree representations and 2-pseudofree actions on spheres. Proc. Amer. Math. Soc.97, 151-157, (1986).

[M1] M. Morimoto: On one fixed point actions on spheres, preprint.

[M2] M. Morimoto: S^4 does not have one fixed point actions, preprint.

[O] R. Oliver: Weight systems for SO(3)-actions. Ann. Math. 110, 227-241 (1979).

[P] T. Petrie: One fixed point actions on spheres, I and II. Adv. Math. 46, 3-14 and 15-70 (1982)

[S] E. Stein: Surgery on products with finite fundamental group. Topology 16, 473-493 (1977).

Note: Recently S. Demichelis has shown that a finite group acting locally linearly and preserving orientation on a closed $\mathbb{Z}$-homology 4-sphere has fixed point set a sphere.

3. R. Schultz

Problems on low-dimensional group actions

A) Let M^4 be a closed topological 4-manifold with a topological circle action. Is the Kirby-Siebenmann invariant of M trivial?

Comments: Work of Kwasik and Schultz shows that topological S^1-manifolds satisfy many of the same global restrictions as in the smooth category. Also, the answer is yes for free circle actions. Both of these suggest the answer is yes in general. Finally, the answer to this question will be yes in many cases if the same is true for the next question.

B) Is every topological circle action on a 4-manifold concordant to a smooth action?--This is related to Problem 6.9 in the list of problems in the proceedings of the Boulder Conference on Group Actions (A.M.S. Contemporary Mathematics Vol. 36, p. 544).

C) Let M^4 be the closed manifold homotopy equivalent but not homeomorphic to CP^2. Does M^4 admit a nontrivial involution?

Comments: Work of Kwasik and Vogel shows that there are no locally linear involutions, but for k odd there is a rich assortment of Z_4 actions (references are given below).

D) Classify all 4-dimensional h-cobordisms up to homeomorphism or diffeomorphism. In particular, if π is the fundamental group of a closed 3-manifold M^3, which elements of the Whitehead group of π can be realized as the Whitehead torsions of h-cobordisms with one end equal to M ?

Comments: The results of Freedman have led to new results in this direction when π is small, including the existence of exotic s-cobordisms (see Cappell-Shaneson, J. Diff. Geom. 22 (1986), 97-115, Kwasik, Comment. Math. Helv. 61 (1986), 415-428, and forthcoming work of Kwasik and Schultz). However, our overall understanding is far from complete. For π finite the realization question is connected to the existence of nonlinear free finite group actions on homotopy 3-spheres (Kwasik, Proc. Amer. Math. Soc. 97 (1986), 352-354). Here is a related question: Which elements of the projective class group can be realized as the finiteness obstructions for tame ends of topological 4-manifolds?

4. Tammo tom Dieck

A) A classical theorem of Jordan about finite subgroups G of $O(n)$ states: There exists an integer $j(n)$, independent of $G \subset O(n)$, such that G has an abelian normal subgroup A with $|G/A| < j(n)$. The following conjecture would be a homotopical generalization.

Conjecture. Given a natural number n. Let X be an n-dimensional homotopy representation of the finite group G with effective action. Then there exists an integer $J(n)$ such that G has an abelian normal subgroup A with $|G/A| < J(n)$.

T. tom Dieck and T. Petrie: Homotopy representations of finite groups, Publ. math. IHES. 56, 129-169 (1982).

B) The equivariant finiteness obstruction yields a homomorphism from the Picard group of the Burnside ring to projective class groups

$$s: \mathrm{Pic}\, A(G) \longrightarrow \prod_{(H)\subset G} \tilde{K}_0(\mathbb{Z}NH/H) .$$

The definition of s uses the geometric definition of Pic A(G) as a subgroup of the homotopy representation group.

Problem. Give an algebraic definition of s.

This should be a generalized Swan homomorphism (= boundary in a Mayer-Vietoris sequence) and would require a K_1-definition of Pic A(G).

T. tom Dieck: The Picard group of the Burnside ring. Journal für die reine u. angew. Math. 361, 174-200 (1985).

C) Problem. Give a classification of 3-dimensional homotopy representations.

5. K. Pawalowski

Let G be a compact Lie group whose identity connected component G_0 is abelian (i.e., G_0 is either a trivial gropu or a torus T^k with $k \geq 1$) and assume that the quotient group G/G_0 has a normal, possibly trivial, 2-Sylow subgroup. If G acts smoothly on $M = D^n$ on $\mathbb{R}^n$, there is the following restriction on the set F of points in M left fixed by G. Namely F is a stably complex manifold in the sense that F has a smooth embedding into some Eulcidean space such that the normal bundle of the embedding admits a complex structure; cf. Edmonds-Lee, Topology 14 (1975). In particular, F is orientable and all connected components of F are either even or odd dimensional.

Assume further that either (i) G/G_0 is of prime power order or (ii) G/G_0 has a cyclic subgroup not of prime power order. In case (i), it follows from Smith Theory that F is a $\mathbb{Z}$-acyclic when $G = G_0 = T^k$ with $k \geq 1$, and F is $\mathbb{Z}_p$-acyclic when $|G/G_0| = p^a$ with p prime, $a \geq 1$. In case (ii), if $M = D^n$, Oliver's work implies that $\chi(F) \equiv 1 \pmod{n_G}$, where n_G is the Oliver integer of G. It turns out that for G as above, these restrictions on F are both necessary and sufficient for a compact smooth manifold F (resp., a smooth manifold F without boundary) to occur as the fixed point set of a smooth action of G on a disk (resp., Euclidean space). This raises the question which smooth manifolds can occur as the fixed point sets of smooth actions of G on disks (resp., Euclidean spaces) for other compact Lie groups G. In particular, the following related problems are still unsolved.

Problem A. Let G be a compact Lie group such that G/G_0 is not of prime power order but each element of G/G_0 has prime power order. Is there a smooth action of G on a disk (resp., Euclidean spaces) such that the fixed point set is not a stably parallelizable manifold?

Problem B. Is there a compact Lie group G which can act smoothly on a disk (resp., Euclidean space) with fixed point set F consisting both of even and odd dimensional connected components? Can some of them be nonorientable manifolds? In particular, can F be the disjoint union of a point, a circle, and the closed (resp., open) Möbius band?

Problem C. Is there a compact Lie group G such that each component smooth manifold (resp., each smooth manifold without boundary) can occur as the fixed point set of a smooth action on G on a disk (resp., Euclidean space)?

Comments

Ad. 1. If such a G acts smoothly on D^n or $\mathbb{R}^n$, then at any two fixed points, the representations of G are equivalent. In particular, each fixed point set connected component has the same dimension; cf. Pawalowski, Math. Z. 187 (1984).

Ad. 2. According to the above discussion, if such a finite group G exists, G has a cyclic subgroup not of prime power order and G is an even order group whose 2-Sylow subgroup is not normal.

Ad. 3. Again, according to the above discussion, if such a finite group G exists, G is as in Ad. 2 and $n_G = 1$ in the case of smooth actions of G on disks.

Vol. 1201: Curvature and Topology of Riemannian Manifolds. Proceedings, 1985. Edited by K. Shiohama, T. Sakai and T. Sunada. VII, 336 pages. 1986.

Vol. 1202: A. Dür, Möbius Functions, Incidence Algebras and Power Series Representations. XI, 134 pages. 1986.

Vol. 1203: Stochastic Processes and Their Applications. Proceedings, 1985. Edited by K. Itô and T. Hida. VI, 222 pages. 1986.

Vol. 1204: Séminaire de Probabilités XX, 1984/85. Proceedings. Edité par J. Azéma et M. Yor. V, 639 pages. 1986.

Vol. 1205: B.Z. Moroz, Analytic Arithmetic in Algebraic Number Fields. VII, 177 pages. 1986.

Vol. 1206: Probability and Analysis, Varenna (Como) 1985. Seminar. Edited by G. Letta and M. Pratelli. VIII, 280 pages. 1986.

Vol. 1207: P.H. Bérard, Spectral Geometry: Direct and Inverse Problems. With an Appendix by G. Besson. XIII, 272 pages. 1986.

Vol. 1208: S. Kaijser, J.W. Pelletier, Interpolation Functors and Duality. IV, 167 pages. 1986.

Vol. 1209: Differential Geometry, Peñíscola 1985. Proceedings. Edited by A.M. Naveira, A. Ferrández and F. Mascaró. VIII, 306 pages. 1986.

Vol. 1210: Probability Measures on Groups VIII. Proceedings, 1985. Edited by H. Heyer. X, 386 pages. 1986.

Vol. 1211: M.B. Sevryuk, Reversible Systems. V, 319 pages. 1986.

Vol. 1212: Stochastic Spatial Processes. Proceedings, 1984. Edited by P. Tautu. VIII, 311 pages. 1986.

Vol. 1213: L.G. Lewis, Jr., J.P. May, M. Steinberger, Equivariant Stable Homotopy Theory. IX, 538 pages. 1986.

Vol. 1214: Global Analysis – Studies and Applications II. Edited by Yu.G. Borisovich and Yu.E. Gliklikh. V, 275 pages. 1986.

Vol. 1215: Lectures in Probability and Statistics. Edited by G. del Pino and R. Rebolledo. V, 491 pages. 1986.

Vol. 1216: J. Kogan, Bifurcation of Extremals in Optimal Control. VIII, 106 pages. 1986.

Vol. 1217: Transformation Groups. Proceedings, 1985. Edited by S. Jackowski and K. Pawalowski. X, 396 pages. 1986.

Vol. 1218: Schrödinger Operators, Aarhus 1985. Seminar. Edited by E. Balslev. V, 222 pages. 1986.

Vol. 1219: R. Weissauer, Stabile Modulformen und Eisensteinreihen. III, 147 Seiten. 1986.

Vol. 1220: Séminaire d'Algèbre Paul Dubreil et Marie-Paule Malliavin. Proceedings, 1985. Edité par M.-P. Malliavin. IV, 200 pages. 1986.

Vol. 1221: Probability and Banach Spaces. Proceedings, 1985. Edited by J. Bastero and M. San Miguel. XI, 222 pages. 1986.

Vol. 1222: A. Katok, J.-M. Strelcyn, with the collaboration of F. Ledrappier and F. Przytycki, Invariant Manifolds, Entropy and Billiards; Smooth Maps with Singularities. VIII, 283 pages. 1986.

Vol. 1223: Differential Equations in Banach Spaces. Proceedings, 1985. Edited by A. Favini and E. Obrecht. VIII, 299 pages. 1986.

Vol. 1224: Nonlinear Diffusion Problems, Montecatini Terme 1985. Seminar. Edited by A. Fasano and M. Primicerio. VIII, 188 pages. 1986.

Vol. 1225: Inverse Problems, Montecatini Terme 1986. Seminar. Edited by G. Talenti. VIII, 204 pages. 1986.

Vol. 1226: A. Buium, Differential Function Fields and Moduli of Algebraic Varieties. IX, 146 pages. 1986.

Vol. 1227: H. Helson, The Spectral Theorem. VI, 104 pages. 1986.

Vol. 1228: Multigrid Methods II. Proceedings, 1985. Edited by W. Hackbusch and U. Trottenberg. VI, 336 pages. 1986.

Vol. 1229: O. Bratteli, Derivations, Dissipations and Group Actions on C*-algebras. IV, 277 pages. 1986.

Vol. 1230: Numerical Analysis. Proceedings, 1984. Edited by J.-P. Hennart. X, 234 pages. 1986.

Vol. 1231: E.-U. Gekeler, Drinfeld Modular Curves. XIV, 107 pages. 1986.

Vol. 1232: P.C. Schuur, Asymptotic Analysis of Soliton Problems. VIII, 180 pages. 1986.

Vol. 1233: Stability Problems for Stochastic Models. Proceedings, 1985. Edited by V.V. Kalashnikov, B. Penkov and V.M. Zolotarev. VI, 223 pages. 1986.

Vol. 1234: Combinatoire énumérative. Proceedings, 1985. Edité par G. Labelle et P. Leroux. XIV, 387 pages. 1986.

Vol. 1235: Séminaire de Théorie du Potentiel, Paris, No. 8. Directeurs: M. Brelot, G. Choquet et J. Deny. Rédacteurs: F. Hirsch et G. Mokobodzki. III, 209 pages. 1987.

Vol. 1236: Stochastic Partial Differential Equations and Applications. Proceedings, 1985. Edited by G. Da Prato and L. Tubaro. V, 257 pages. 1987.

Vol. 1237: Rational Approximation and its Applications in Mathematics and Physics. Proceedings, 1985. Edited by J. Gilewicz, M. Pindor and W. Siemaszko. XII, 350 pages. 1987.

Vol. 1238: M. Holz, K.-P. Podewski and K. Steffens, Injective Choice Functions. VI, 183 pages. 1987.

Vol. 1239: P. Vojta, Diophantine Approximations and Value Distribution Theory. X, 132 pages. 1987.

Vol. 1240: Number Theory, New York 1984–85. Seminar. Edited by D.V. Chudnovsky, G.V. Chudnovsky, H. Cohn and M.B. Nathanson. V, 324 pages. 1987.

Vol. 1241: L. Gårding, Singularities in Linear Wave Propagation. III, 125 pages. 1987.

Vol. 1242: Functional Analysis II, with Contributions by J. Hoffmann-Jørgensen et al. Edited by S. Kurepa, H. Kraljević and D. Butković. VII, 432 pages. 1987.

Vol. 1243: Non Commutative Harmonic Analysis and Lie Groups. Proceedings, 1985. Edited by J. Carmona, P. Delorme and M. Vergne. V, 309 pages. 1987.

Vol. 1244: W. Müller, Manifolds with Cusps of Rank One. XI, 158 pages. 1987.

Vol. 1245: S. Rallis, L-Functions and the Oscillator Representation. XVI, 239 pages. 1987.

Vol. 1246: Hodge Theory. Proceedings, 1985. Edited by E. Cattani, F. Guillén, A. Kaplan and F. Puerta. VII, 175 pages. 1987.

Vol. 1247: Séminaire de Probabilités XXI. Proceedings. Edité par J. Azéma, P.A. Meyer et M. Yor. IV, 579 pages. 1987.

Vol. 1248: Nonlinear Semigroups, Partial Differential Equations and Attractors. Proceedings, 1985. Edited by T.L. Gill and W.W. Zachary. IX, 185 pages. 1987.

Vol. 1249: I. van den Berg, Nonstandard Asymptotic Analysis. IX, 187 pages. 1987.

Vol. 1250: Stochastic Processes – Mathematics and Physics II. Proceedings 1985. Edited by S. Albeverio, Ph. Blanchard and L. Streit. VI, 359 pages. 1987.

Vol. 1251: Differential Geometric Methods in Mathematical Physics. Proceedings, 1985. Edited by P.L. García and A. Pérez-Rendón. VII, 300 pages. 1987.

Vol. 1252: T. Kaise, Représentations de Weil et GL_2 Algèbres de division et GL_n. VII, 203 pages. 1987.

Vol. 1253: J. Fischer, An Approach to the Selberg Trace Formula via the Selberg Zeta-Function. III, 184 pages. 1987.

Vol. 1254: S. Gelbart, I. Piatetski-Shapiro, S. Rallis. Explicit Constructions of Automorphic L-Functions. VI, 152 pages. 1987.

Vol. 1255: Differential Geometry and Differential Equations. Proceedings, 1985. Edited by C. Gu, M. Berger and R.L. Bryant. XII, 243 pages. 1987.

Vol. 1256: Pseudo-Differential Operators. Proceedings, 1986. Edited by H.O. Cordes, B. Gramsch and H. Widom. X, 479 pages. 1987.

Vol. 1257: X. Wang, On the C*-Algebras of Foliations in the Plane. V, 165 pages. 1987.

Vol. 1258: J. Weidmann, Spectral Theory of Ordinary Differential Operators. VI, 303 pages. 1987.

Vol. 1259: F. Cano Torres, Desingularization Strategies for Three-Dimensional Vector Fields. IX, 189 pages. 1987.

Vol. 1260: N.H. Pavel, Nonlinear Evolution Operators and Semigroups. VI, 285 pages. 1987.

Vol. 1261: H. Abels, Finite Presentability of S-Arithmetic Groups. Compact Presentability of Solvable Groups. VI, 178 pages. 1987.

Vol. 1262: E. Hlawka (Hrsg.), Zahlentheoretische Analysis II. Seminar, 1984–86. V, 158 Seiten. 1987.

Vol. 1263: V.L. Hansen (Ed.), Differential Geometry. Proceedings, 1985. XI, 288 pages. 1987.

Vol. 1264: Wu Wen-tsün, Rational Homotopy Type. VIII, 219 pages. 1987.

Vol. 1265: W. Van Assche, Asymptotics for Orthogonal Polynomials. VI, 201 pages. 1987.

Vol. 1266: F. Ghione, C. Peskine, E. Sernesi (Eds.), Space Curves. Proceedings, 1985. VI, 272 pages. 1987.

Vol. 1267: J. Lindenstrauss, V.D. Milman (Eds.), Geometrical Aspects of Functional Analysis. Seminar. VII, 212 pages. 1987.

Vol. 1268: S.G. Krantz (Ed.), Complex Analysis. Seminar, 1986. VII, 195 pages. 1987.

Vol. 1269: M. Shiota, Nash Manifolds. VI, 223 pages. 1987.

Vol. 1270: C. Carasso, P.-A. Raviart, D. Serre (Eds.), Nonlinear Hyperbolic Problems. Proceedings, 1986. XV, 341 pages. 1987.

Vol. 1271: A.M. Cohen, W.H. Hesselink, W.L.J. van der Kallen, J.R. Strooker (Eds.), Algebraic Groups Utrecht 1986. Proceedings. XII, 284 pages. 1987.

Vol. 1272: M.S. Livšic, L.L. Waksman, Commuting Nonselfadjoint Operators in Hilbert Space. III, 115 pages. 1987.

Vol. 1273: G.-M. Greuel, G. Trautmann (Eds.), Singularities, Representation of Algebras, and Vector Bundles. Proceedings, 1985. XIV, 383 pages. 1987.

Vol. 1274: N. C. Phillips, Equivariant K-Theory and Freeness of Group Actions on C*-Algebras. VIII, 371 pages. 1987.

Vol. 1275: C.A. Berenstein (Ed.), Complex Analysis I. Proceedings, 1985–86. XV, 331 pages. 1987.

Vol. 1276: C.A. Berenstein (Ed.), Complex Analysis II. Proceedings, 1985–86. IX, 320 pages. 1987.

Vol. 1277: C.A. Berenstein (Ed.), Complex Analysis III. Proceedings, 1985–86. X, 350 pages. 1987.

Vol. 1278: S.S. Koh (Ed.), Invariant Theory. Proceedings, 1985. V, 102 pages. 1987.

Vol. 1279: D. Ieşan, Saint-Venant's Problem. VIII, 162 Seiten. 1987.

Vol. 1280: E. Neher, Jordan Triple Systems by the Grid Approach. XII, 193 pages. 1987.

Vol. 1281: O.H. Kegel, F. Menegazzo, G. Zacher (Eds.), Group Theory. Proceedings, 1986. VII, 179 pages. 1987.

Vol. 1282: D.E. Handelman, Positive Polynomials, Convex Integral Polytopes, and a Random Walk Problem. XI, 136 pages. 1987.

Vol. 1283: S. Mardešić, J. Segal (Eds.), Geometric Topology and Shape Theory. Proceedings, 1986. V, 261 pages. 1987.

Vol. 1284: B.H. Matzat, Konstruktive Galoistheorie. X, 286 pages. 1987.

Vol. 1285: I.W. Knowles, Y. Saitō (Eds.), Differential Equations and Mathematical Physics. Proceedings, 1986. XVI, 499 pages. 1987.

Vol. 1286: H.R. Miller, D.C. Ravenel (Eds.), Algebraic Topology. Proceedings, 1986. VII, 341 pages. 1987.

Vol. 1287: E.B. Saff (Ed.), Approximation Theory, Tampa. Proceedings, 1985–1986. V, 228 pages. 1987.

Vol. 1288: Yu. L. Rodin, Generalized Analytic Functions on Riemann Surfaces. V, 128 pages, 1987.

Vol. 1289: Yu. I. Manin (Ed.), K-Theory, Arithmetic and Geometry. Seminar, 1984–1986. V, 399 pages. 1987.

Vol. 1290: G. Wüstholz (Ed.), Diophantine Approximation and Transcendence Theory. Seminar, 1985. V, 243 pages. 1987.

Vol. 1291: C. Mœglin, M.-F. Vignéras, J.-L. Waldspurger, Correspondances de Howe sur un Corps p-adique. VII, 163 pages. 1987

Vol. 1292: J.T. Baldwin (Ed.), Classification Theory. Proceedings, 1985. VI, 500 pages. 1987.

Vol. 1293: W. Ebeling, The Monodromy Groups of Isolated Singularities of Complete Intersections. XIV, 153 pages. 1987.

Vol. 1294: M. Queffélec, Substitution Dynamical Systems – Spectral Analysis. XIII, 240 pages. 1987.

Vol. 1295: P. Lelong, P. Dolbeault, H. Skoda (Réd.), Séminaire d'Analyse P. Lelong – P. Dolbeault – H. Skoda. Seminar, 1985/1986. VII, 283 pages. 1987.

Vol. 1296: M.-P. Malliavin (Ed.), Séminaire d'Algèbre Paul Dubreil et Marie-Paule Malliavin. Proceedings, 1986. IV, 324 pages. 1987.

Vol. 1297: Zhu Y.-l., Guo B.-y. (Eds.), Numerical Methods for Partial Differential Equations. Proceedings. XI, 244 pages. 1987.

Vol. 1298: J. Aguadé, R. Kane (Eds.), Algebraic Topology, Barcelona 1986. Proceedings. X, 255 pages. 1987.

Vol. 1299: S. Watanabe, Yu.V. Prokhorov (Eds.), Probability Theory and Mathematical Statistics. Proceedings, 1986. VIII, 589 pages. 1988.

Vol. 1300: G.B. Seligman, Constructions of Lie Algebras and their Modules. VI, 190 pages. 1988.

Vol. 1301: N. Schappacher, Periods of Hecke Characters. XV, 160 pages. 1988.

Vol. 1302: M. Cwikel, J. Peetre, Y. Sagher, H. Wallin (Eds.), Function Spaces and Applications. Proceedings, 1986. VI, 445 pages. 1988.

Vol. 1303: L. Accardi, W. von Waldenfels (Eds.), Quantum Probability and Applications III. Proceedings, 1987. VI, 373 pages. 1988.

Vol. 1304: F.Q. Gouvêa, Arithmetic of p-adic Modular Forms. VIII, 121 pages. 1988.

Vol. 1305: D.S. Lubinsky, E.B. Saff, Strong Asymptotics for Extremal Polynomials Associated with Weights on ℝ. VII, 153 pages. 1988.

Vol. 1306: S.S. Chern (Ed.), Partial Differential Equations. Proceedings, 1986. VI, 294 pages. 1988.

Vol. 1307: T. Murai, A Real Variable Method for the Cauchy Transform, and Analytic Capacity. VIII, 133 pages. 1988.

Vol. 1308: P. Imkeller, Two-Parameter Martingales and Their Quadratic Variation. IV, 177 pages. 1988.

Vol. 1309: B. Fiedler, Global Bifurcation of Periodic Solutions with Symmetry. VIII, 144 pages. 1988.

Vol. 1310: O.A. Laudal, G. Pfister, Local Moduli and Singularities. V, 117 pages. 1988.

Vol. 1311: A. Holme, R. Speiser (Eds.), Algebraic Geometry, Sundance 1986. Proceedings. VI, 320 pages. 1988.

Vol. 1312: N.A. Shirokov, Analytic Functions Smooth up to the Boundary. III, 213 pages. 1988.

Vol. 1313: F. Colonius, Optimal Periodic Control. VI, 177 pages. 1988.

Vol. 1314: A. Futaki, Kähler-Einstein Metrics and Integral Invariants. IV, 140 pages. 1988.

Vol. 1315: R.A. McCoy, I. Ntantu, Topological Properties of Spaces of Continuous Functions. IV, 124 pages. 1988.

Vol. 1316: H. Korezlioglu, A.S. Ustunel (Eds.), Stochastic Analysis and Related Topics. Proceedings, 1986. V, 371 pages. 1988.

Vol. 1317: J. Lindenstrauss, V.D. Milman (Eds.), Geometric Aspects of Functional Analysis. Seminar, 1986–87. VII, 289 pages. 1988.

Vol. 1318: Y. Felix (Ed.), Algebraic Topology – Rational Homotopy. Proceedings, 1986. VIII, 245 pages. 1988

Vol. 1319: M. Vuorinen, Conformal Geometry and Quasiregular Mappings. XIX, 209 pages. 1988.